T0233722

Lecture Notes
in Business Information Processing   248

Series Editors

Wil van der Aalst
Eindhoven Technical University, Eindhoven, The Netherlands
John Mylopoulos
University of Trento, Povo, Italy
Michael Rosemann
Queensland University of Technology, Brisbane, QLD, Australia
Michael J. Shaw
University of Illinois, Urbana-Champaign, IL, USA
Clemens Szyperski
Microsoft Research, Redmond, WA, USA

More information about this series at http://www.springer.com/series/7911

Rainer Schmidt · Wided Guédria
Ilia Bider · Sérgio Guerreiro (Eds.)

Enterprise, Business-Process and Information Systems Modeling

17th International Conference, BPMDS 2016
21st International Conference, EMMSAD 2016, Held at CAiSE 2016
Ljubljana, Slovenia, June 13–14, 2016
Proceedings

 Springer

Editors
Rainer Schmidt
Munich University of Applied Sciences
Munich
Germany

Wided Guédria
Luxembourg Institute of Science and
 Technology
Esch-sur-Alzette
Luxembourg

Ilia Bider
Department of Computer and Systems
 Sciences
Stockholm University
Stockholm
Sweden

Sérgio Guerreiro
Universidade Lusófona de Humanidades e
 Tecnologias
Lisboa
Portugal

ISSN 1865-1348   ISSN 1865-1356 (electronic)
Lecture Notes in Business Information Processing
ISBN 978-3-319-39428-2   ISBN 978-3-319-39429-9 (eBook)
DOI 10.1007/978-3-319-39429-9

Library of Congress Control Number: 2016939916

This Springer imprint is published by Springer Nature
The registered company is Springer International Publishing AG Switzerland

Preface

This book contains the proceedings of two long-running events held along with the CAiSE conferences relating to the areas of enterprise, business-process, and information systems modeling: the 17th International Conference on Business Process Modeling, Development and Support (BPMDS 2016) and the 21st International Conference on Exploring Modeling Methods for Systems Analysis and Design (EMMSAD 2016). The two working conferences are introduced here.

BPMDS 2016

BPMDS has been held as a series of workshops devoted to business process modeling, development, and support since 1998. During this period, business process analysis and design have been recognized as a central issue in the area of information systems (IS) engineering. The continued interest in these topics on behalf of the IS community is reflected by the success of the last BPMDS events and the recent emergence of new conferences and workshops devoted to the theme. In 2011, BPMDS became a two-day working conference attached to CAiSE (Conference on Advanced Information Systems Engineering). The basic principles of the BPMDS series are:

1. BPMDS serves as a meeting place for researchers and practitioners in the areas of business development and business applications (software) development.
2. The aim of the event is mainly discussions rather than presentations.
3. Each event has a theme that is mandatory for idea papers.
4. Each event's results are usually published in a special issue of an international journal.

The goals, format, and history of BPMDS can be found on the website: http://www.bpmds.org/

The intention of BPMDS is to solicit papers related to business process modeling, development, and support (BPMDS) in general, using quality as the main selection criterion. As a working conference, we aim to attract papers describing mature research, but we still give place to industrial reports and visionary idea papers. To encourage new and emerging challenges and research directions in the area of business process modeling, development, and support, we have a unique focus theme every year. Papers submitted as idea papers are required to be of relevance to the focus theme, thus providing a mass of new ideas around a relatively narrow but emerging research area. Full research papers and experience reports do not necessarily need to be directly connected to this theme (they still needed to be explicitly relevant to BPMDS, however).

The focus theme for BPMDS 2016 idea papers was "Business Processes in a Connected World," in which we differentiate three subthemes: Business processes for connecting people that directly corresponds to the theme of CAiSE 2016, "Information

Systems for Connecting People," connecting intelligent objects to business processes, and connecting information/data/knowledge to business processes.

For the 17th edition of the BPMDS conference, we invited interested authors to engage, through their idea papers and the discussions during the two days of BPMDS 2016 in Ljubljana, in a deep discussion with all participants about how business processes and support system can enable connection between people, intelligent objects, information and knowledge.

BPMDS 2016 received 48 submissions from 30 countries (Algeria, Australia, Austria, Belgium, Brazil, Canada, China, France, Germany, Greece, Iran, Islamic Republic of, Israel, Italy, Latvia, The Netherlands, Norway, Portugal, Russian Federation, Saudi Arabia, Serbia, Slovenia, Spain, Sweden, Switzerland, Tunisia, Turkey, UK, Uruguay, Venezuela, and Vietnam). The management of the paper submission and reviews was supported by the EasyChair conference system. Each paper received at least three reviews. Eventually, 19 high-quality papers were selected. The accepted papers cover a wide spectrum of issues related to business process development, modeling, and support. They are organized under the following section headings:

- Process execution support
- Improving usability of process models
- Social and human perspective
- New directions in process modeling
- Consistency, correctness, and compliance
- Process and data mining
- Process variability

We wish to thank all the people who submitted papers to BPMDS 2016 for having shared their work with us, as well as the members of the BPMDS 2016 Program Committee, who made a remarkable effort in reviewing submissions. We also thank the organizers of CAiSE 2016 for their help with the organization of the event, and IFIP WG8.1 for the support.

April 2016   Ilia Bider
                                                                       Rainer Schmidt

EMMSAD 2016

Since 1995, the EMMSAD conference has focused on exploring, evaluating, and enhancing modeling methods and methodologies for the analysis and design of information systems, enterprises, and business processes.

Although the need for such studies is well recognized, there is a paucity of research in the literature.

The objective of the EMMSAD conference series is to provide a forum for researchers and practitioners interested in modeling methods for systems analysis and design to meet, and exchange research ideas and results. It also provides the participants with an opportunity to present their research papers and experience reports, and to take part in open discussions.

Whereas modeling techniques have traditionally been used to create intermediate artifacts in systems analysis and design, modern modeling methodologies take a more active approach. For instance in business process management (BPM), model-driven software engineering, domain-specific modeling (DSM), enterprise architecture (EA), enterprise modeling (EM), interactive models, and active knowledge modeling, the models are used directly as part of the information system of the organization. At the same time, similar modeling techniques are also used for sense-making and communication, model simulation, quality assurance, and requirements specification in connection with more traditional forms of information systems and enterprise development. Since modeling techniques are used in such a large variety of tasks with different goals, it is hard to assess whether a model is sufficiently good to achieve the goals. To provide guidance in this process, knowledge for understanding the quality of models and modeling languages is needed.

The basic principles of the EMMSAD series are:

1. EMMSAD serves as a meeting place for researchers and practitioners in the areas of information systems and business analysis.
2. The aim of the event is mainly discussions rather than presentations.
3. Each event's results are usually published in a special issue of an international journal.

The goals, format, and history of EMMSAD can be found on the website: http://www.emmsad.org/.

The intention of EMMSAD is to solicit papers related to the field of information systems analysis and design including numerous information modeling methods and notations (e.g., ER, ORM, UML, ArchiMate, EPC, BPMN, DEMO) that are typically evolving. Even with some attempts toward standardization (e.g., UML for object-oriented software design), new modeling methods are constantly being introduced, many of which differ only marginally from previous approaches. These ongoing changes significantly impact the way information systems, enterprises, and business processes are being analyzed and designed in practice.

EMMSAD 2016 received 19 submissions from 14 countries (Belgium, Bosnia and Herzegovina, Brazil, Canada, Colombia, France, Germany, Israel, Luxembourg, Norway, Portugal, Sweden, Tunisia, and the USA). The management of paper submissions and reviews was supported by the EasyChair conference system. Each paper received at least three reviews. Eventually, 11 high-quality papers were selected. Furthermore we take the opportunity to publish 1 paper, which has already been selected in 2015.

We wish to thank all the people who submitted papers to EMMSAD 2016 for having shared their work with us, as well as the members of the EMMSAD 2016 Program Committee, who made a remarkable effort in reviewing submissions. We also thank the organizers of CAiSE 2016 for their help with the organization of the event, and IFIP WG8.1 for the support.

April 2016   Wided Guedria
                                                              Sérgio Guerreiro

Organization

BPMDS 2016 Organization

Program Committee

Eric Andonoff	Irit/UT1
Judith Barrios Albornoz	University of Los Andes, Colombia
Ilia Bider	Stockholm University/IbisSoft, Sweden
Karsten Boehm	FH KufsteinTirol - University of Applies Science, Austria
Lars Brehm	Munich University of Applied Science, Germany
Dirk Fahland	Technische Universiteit Eindhoven, The Netherlands
Claude Godart	Loria, France
Paul Johannesson	Stockholm University, Sweden
Marite Kirikova	Riga Technical University, Latvia
Agnes Koschmider	Karlsruhe Institute of Technology, Germany
Marcello La Rosa	Queensland University of Technology, Australia
Jan Mendling	Wirtschaftsuniversität Wien, Austria
Michael Möhring	Aalen University, Germany
Pascal Negros	Université Paris I Sorbonne, France
Jens Nimis	University of Applied Sciences Karlsruhe, Germany
Selmin Nurcan	Université de Paris 1 Panthéon - Sorbonne, France
Elias Pimenidis	University of the West of England, UK
Gil Regev	Ecole Polytechnique Fédérale de Lausanne, Switzerland
Manfred Reichert	University of Ulm, Germany
Hajo A. Reijers	Eindhoven University of Technology, The Netherlands
Iris Reinhartz-Berger	University of Haifa, Israel
Stefanie Rinderle-Ma	University of Vienna, Austria
Colette Rolland	Université Paris 1 Panthéon Sorbonne, France
Shazia Sadiq	The University of Queensland, Australia
Rainer Schmidt	Munich University of Applied Sciences, Germany
Samira Si-Said Cherfi	CEDRIC - Conservatoire National des Arts et Métiers, France
Pnina Soffer	University of Haifa, Israel
Lars Taxén	Linköping University, Sweden
Roland Ukor	FirstLinq Ltd
Barbara Weber	University of Innsbruck, Austria

EMMSAD 2016 Organization

Co-chairs

Wided Guédria	Luxembourg Institute of Science and Technology, LIST, Luxembourg
Sérgio Guerreiro	Universidade Lusófona de Humanidades e Tecnologias, Lisbon, Portugal

Advisory Committee

John Krogstie	Norwegian University of Science and Technology (NTNU), Norway
Henderik A. Proper	Luxembourg Institute of Science and Technology, LIST, Luxembourg, and Radboud University Nijmegen, The Netherlands

Program Committee

David Aveiro	Madeira University, Portugal
Béatrix Barafort	Luxembourg Institute of Science and Technology, Luxembourg
João Barata	University of Coimbra, Portugal
Marija Bjekovic	Luxembourg Institute of Science and Technology, Luxembourg
Sybren De Kinderen	University of Luxembourg, Luxembourg
Dieter De Smet	Luxembourg Institute of Science and Technology, Luxembourg
Celine Decosse	Public Research Centre Henri Tudor, Luxembourg
Christophe Feltus	Luxembourg Institute of Science and Technology, Luxembourg
Khaled Gaaloul	Luxembourg Institute of Science and Technology, Luxembourg
Sepideh Ghanavati	Luxembourg Institute of Science and Technology, Luxembourg
Antonio Goncalves	Instituto Politecnico Setubal, Portugal
Wided Guédria	Luxembourg Institute of Science and Technology, Luxembourg
Sérgio Guerreiro	Universidade Lusófona de Humanidades e Tecnologias, Portugal
Janne J. Korhonen	Aalto University School of Science, Finland

Gabriel Leal	Luxembourg Institute of Science and Technology and University of Lorraine, France
Qin Ma	University of Luxembourg, Luxembourg
Pedro Malta	Universidade Lusófona de Humanidades e Tecnologias, Portugal
Diana Marosin	Luxembourg Institute of Science and Technology, Luxembourg
Rui Pedro Marques	Higher Institute of Accounting and Administration, University fo Aveiro; Algoritmi, University of Minho, Portugal
Carlos Mendes	Instituto Superior Técnico, Portugal
Owen Molloy	NUI Gallway, Ireland
Hervé Panetto	CRAN, University of Lorraine, CNRS, France
Robert Pergl	Czech Technical University, Czech Republic
Nuno Pombo	University of Beira Interior, Portugal
Carlos Páscoa	Portuguese Air Force, Portugal
Ivan Razo-Zapata	Luxembourg Institute of Science and Technology, Luxembourg
Alberto Silva	INESC-ID/Instituto Superior Técnico, Portugal
Dirk van der Linden	University of Haifa, Israel
Steven van Kervel	Formetis BV
Marielba Zacarias	Research Centre for Spatial and Organizational Dynamics, Universidade do Algarve, Portugal
Milan Zdravković	University of Nis, Serbia

Additional Reviewers

Gammaitoni, Loïc
Gouveia, Duarte
Yuan, Qixia

Contents

Process Variability

EMMSAD 2016

Fundamental Issues in Modeling

Requirements and Regulations

Enterprise and Software Ecosystem Modeling

Information and Process Model Quality

Meta-modeling and Domain Specific Modeling and Model Composition

Modeling of Architecture and Design

Process Execution Support

Enabling Self-adaptive Workflows
for Cyber-physical Systems

Ronny Seiger$^{(\boxtimes)}$, Steffen Huber, Peter Heisig, and Uwe Assmann

Institute of Software and Multimedia Technology,
Technische Universität Dresden, 01062 Dresden, Germany
{Ronny.Seiger,Steffen.Huber,Peter.Heisig,Uwe.Assmann}@tu-dresden.de

Abstract. The ongoing development of Internet of Things technologies leads to the interweaving of the virtual world of software with the physical world. However, applying workflow technologies for automating processes in these Cyber-physical Systems (CPS) poses new challenges as the real world effects of a process have to be verified to provide a consistent view of the cyber and physical world executions. In this work we present a synchronization and adaptation mechanism for processes based on the MAPE-K feedback loop for self-adaptive systems. By applying this loop, sensor and context information can be used to verify the real world effects of workflow execution and adapt the process in case of errors. The approach increases autonomy and resilience of process execution in CPS due to the self-adaptation capabilities. We present generic extensions to process meta-models and execution engines to implement the feedback loop and discuss our approach within a smart home scenario.

Keywords: Workflows for the Internet of Things · Cyber-physical Systems · Self-adaptive workflows · Real world processes

1 Introduction

With the advancement of Internet of Things (IoT) technologies, smart spaces and cloud computing, the Business Process Management (BPM) community faces new challenges resulting from the integration of business processes with real world objects, humans and digital services [18]. The class of *Cyber-physical Systems* (CPS) connects the real (*physical*) world of objects and things with the virtual (*cyber*) world of software and services by means of sensors, actuators and embedded computing devices. In contrast to traditional service-based BPM application fields, automating processes in process-aware CPS information systems requires consideration of processes' real world effects and context as well [30].

Solely relying on software feedback for evaluating the success of activity execution within a process is not sufficient enough to provide a consistent view of both the virtual and physical world due to unexpected failures in the real world (e.g., a broken light bulb, a malfunctioning actuator or the imprecise positioning of an object). In addition to feedback from services and execution engines, sensors

© Springer International Publishing Switzerland 2016
R. Schmidt et al. (Eds.): BPMDS/EMMSAD 2016, LNBIP 248, pp. 3–17, 2016.
DOI: 10.1007/978-3-319-39429-9_1

measuring real world properties can be incorporated to evaluate the results of physical world task executions. In case of errors and inconsistencies, the process has to be adapted with respect to the deviations and compensating actions have to be executed, which can be monitored again. This feedback loop enables self-awareness and self-adaptation for processes in cyber-physical systems–important features of emerging systems of systems [8]. However, current approaches mostly focus on extending *traditional* business processes into the real world of things and objects or on increasing the adaptiveness of purely virtual BPM systems. The issue of ensuring consistency between *cyber* and *physical* world using the generic approach of feedback loops for creating self-adaptive processes has not been discussed in detail yet.

In this work, we present an approach for enabling self-adaptive workflows based on the MAPE-K (Monitor, Analyze, Plan and Execute on a Knowledge base) control loop for self-adaptive systems [5]. In addition to feedback from the process management system and services, sensor data is used to monitor the real world effects of activities and to analyze the outcome of their execution. If an inconsistency between the *physical world* and the *assumed cyber world* can be detected, a compensation strategy is chosen and the adapted process is executed. That way, we achieve a certain level of self-healing for CPS workflows.

The paper is structured as follows: Sect. 2 introduces the concept of *Cyber-physical Consistency* for processes. Section 3 presents process meta-model extensions to enable self-adaptive workflows. Section 4 shows the application of the MAPE-K loop for process execution in detail. Section 5 demonstrates the concepts for self-adaptive workflows in a practical scenario from the smart home domain. Section 6 discusses the MAPE-K approach. Section 7 reviews related research. Section 8 concludes the paper.

2   Towards Cyber-physical Consistency for Processes

2.1   Cyber-physical Consistency

The ongoing trend towards *Cyber-physical Convergence* [6] in modern information systems leads to the emergence of the new class of *Cyber-physical Systems* (CPS), i.e., systems consisting of mutually-influencing digital and physical components [15]. Adding process-awareness to the properties of CPS promises a higher degree of flexibility and automation in various application domains for CPS (e.g., Smart Homes, Smart Factories or even Smart Cities) [24]. Traditional Business Process Management (BPM) systems rely on service responses and process related events triggered by the workflow engine to verify the execution and to check for conformance during process enactment. This software-based approach proved feasible for the monitoring and analysis of purely virtual and abstract business processes in accordance with the BPM lifecycle [1]. However, when extending the concept of business processes to also support the automation of workflow activities that influence the real world (*CPS workflows* [23]), monitoring and analysis capabilities of the BPM systems have to be extended to reflect the cyber-physical effects of these activities. Solely relying on direct software feedback for verifying the outcome of process execution in CPS is not feasible due to the lack of checking its actual real world effects. In case of unanticipated errors occurring in the real world (e.g., the malfunctioning of a software controlled actuator, real world obstacles preventing the process to proceed or unforeseen contexts), additional measures have to be taken to verify the actual physical effect of a process.

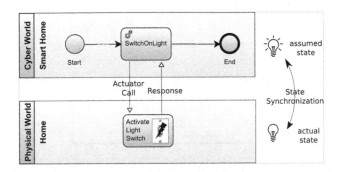

Fig. 1. Synchronization between cyber world and physical world states.

As an extension of the well-known ACID criteria for databases and distributed systems, we introduce the notion of *Cyber-physical Consistency* for a workflow in CPS as a matching of its assumed physical state (i.e., virtual state) and its actual physical state after execution. Cyber-physical consistency can be observed if the workflow's virtual state is in sync with the workflow's real world state. It is inconsistent if there is a mismatch between both states. Figure 1 illustrates the challenge of synchronizing the virtual and real world for a simple scenario in the smart home area: a process executed by the smart home control system is supposed to switch on the light in a certain room. A service call to the software, which controls the light switch actuator triggers the light to be switched on in the physical world. The software reports back the successful execution of the command and the process instance finishes assuming the light is switched on. However, a broken light bulb, hardware issues, real world obstacles or wrong parameters may lead to an undesired actual physical state of the lamp, which cannot be detected by the control software and therefore cyber-physical consistency is violated.

2.2 MAPE-K Control Loops for Ensuring CPS Consistency

In order to evaluate and ensure cyber-physical consistency for CPS workflows, we will apply the concept of the MAPE-K feedback loop known from the engineering of self-adaptive systems area [5] to process execution. This loop consists of the phases *Monitor* (M), *Analyze* (A), *Plan* (P) and *Execute* (E)–all repeatedly executed on the *Knowledge Base* (K). Figure 2 shows our adaptation of the MAPE-K concept for autonomous computing proposed by IBM [11]. We regard the process/process step in question as *Managed Element* (here: SwitchOnLight

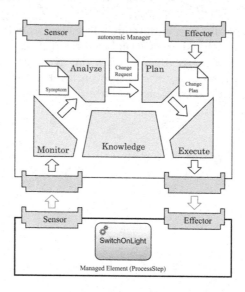

Fig. 2. Managing process steps with MAPE-K loops in accordance with [11].

task from Fig. 1). Besides software-controlled actuators/effectors influencing the real world, sensors are essential parts of CPS to measure physical properties and gather more complex real world context data. In the *Monitor* phase, we collect sensor data from the physical world. Relevant changes within this data (called *Symptoms*) are then analyzed to check for cyber-physical consistency after process execution to correlate changes in the real world to the effects of executing the process. In case an inconsistency is detected, a change request is triggered and a compensation strategy (change plan) is chosen and executed in order to try to restore CPS consistency and continue with process execution as planned. The MAPE-K loop may be executed repeatedly to verify the success of the compensating action(s). The following sections will explain model extensions and components necessary to implement the feedback loop for self-adaptive CPS workflows in detail.

3   Modelling Self-adaptive CPS Workflows

In order to define the physical effects of a workflow or an activity respectively, *traditional* process meta-models from the BPM domain (e.g., BPMN and WS-BPEL) have to be extended with additional information as they were not designed to be applied for real world processes. Figure 3 shows our generic proposal for extending process meta-models with specific attributes regarding their influence on the real world in UML notation. We propose the extension for the component-based process metal-model for CPS described in [24]. The *CpsStep* interface has to be implemented by the abstract class *ProcessStep*, which represents either an atomic process activity (cf. *Activity* class in BPMN) or a composite process step (i.e., a process or a subprocess containing multiple process steps) in accordance with the *Composite* design pattern.

The *CpsStep* interface adds the new attribute *cyberPhysical* to the process step stating if the step has an effect on the real world or not. In case this is true for the process step, a goal can be defined that will be used for analysis and adaptation in the MAPE-K loop. This goal has to be fulfilled to confirm cyber-physical consistency. We decided to use a more declarative goal-oriented approach for defining the domain-specific outcome of a process as proposed in [12], because modelling every possible outcome and error as well possible compensations for the process step in question is not feasible–especially not for complex systems of systems [13]. The *Goal* class may subsume several objectives that have to be met individually for the overall goal to be fulfilled. The *Objective* class contains the following attributes. The examples describe the objective that after triggering a light control switch via a process, the light intensity in the kitchen should be above 865 lux within 2 s. Otherwise, a compensation has to be searched in the *Plan* phase. The overall goal is to lighten up the room.

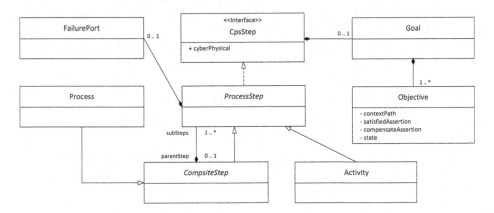

Fig. 3. Process meta-model extensions for self-adaptive CPS workflows.

- *contextPath*: Defines the context measuring points to be monitored (e.g., specific sensor and process data) in the *Monitor* phase. We base our concept on a graph-based representation of a context ontology [10], which acts as the *Knowledge Base K*. Therefore, this attribute defines one or more paths within this context graph. An exemplary context path representing the value of a light sensor in the kitchen could like this:

```
MATCH (kitchen)−[:instanceOf]−>(room)
MATCH (light)−[:instanceOf]−>(sensor)
MATCH (light)−[:isIn]−>(kitchen)
RETURN light.value AS lightIntensity
```

- *satisfiedAssertion*: Defines the required change within relevant context data as result of the execution of the process step. In the *Analyze* phase, this attribute is used to evaluate the success of the operation, i.e., to compare assumed reality with actual reality. The objective is satisfied if the assertion is evaluated to be *true*. An exemplary satisfied assertion requiring the light intensity to be above a certain threshold could look like this:

#lightIntensity > 865

- *compensationAssertion*: Defines an operation on context data and other data (e.g., a certain time frame), which is used to evaluate the need for finding a compensation action in the *Plan* phase (true/false). An exemplary compensation assertion testing if an objective was created more than 2 s ago could look like this:

#objective.created.isBefore(#now.minusSeconds(2))

- *state*: Is used as a runtime attribute to represent the state of an objective (e.g., satisfied, unsatisfied, compensation needed, or failed). In order for the goal to be fulfilled–and with it the related process step executed successfully–all its objectives have to be of state *satisfied*.

To cope with case that regular execution or MAPE-K execution are not able to fulfill all objectives for various reasons (state: failed), we extended the specific meta-model from [24] with the special port class *FailurePort*. This port activates an explicitly modelled process branch as a fallback solution to handle unresolvable errors manually (e.g., trigger a human task to ask for user interactions).

4   Executing Self-adaptive CPS Workflows

Using the MAPE-K feedback loop to add the capabilities of self-adaptivity and ensuring cyber-physical consistency to processes (or process steps), requires new components to be added to the basic process engines. Based on the model extensions described in the previous section, we will describe the necessary components and implementation of the MAPE-K control loop in accordance with the architectural blueprint for autonomic computing [11] in the following sections. Our implementation uses the process engine described in [23] as basic system.

4.1   MAPE-K Components

The feedback loop is regarded as one closed software component with interfaces to the physical world in the form of sensors and actuators. It receives goals and instance information about the cyber-physical process (step) to be executed from the process engine. Inside the *Feedback-Loop* component, there are components responsible for each of the MAPE phases (*Monitor, Analyzer, Planner* and *Executor*) as well as the *Knowledge Base*. Figure 4 presents the components and interfaces in UML notation. We implemented the *Feedback-Loop* as a dedicated web service that can be used by arbitrary process engines and applications.

Knowledge Base. The knowledge base holds model-based data about the physical and virtual context of the CPS based on sensor data, actuators, processes, goals and change plans used in the *Plan* phase in the form of an ontology. It is updated with every change within this data and accessed by all other components. In addition to represent information about the processes and their context as proposed in [22], we use the DogOnt ontology [4] to describe sensors, actuators and their relations for our applications in the smart home domain [10].

4.2 MAPE-K Phases in Process Execution

Upon execution of a process step, the engine evaluates the *cyberPhysical* flag of the process step in question. In case it is true, a request containing instance information and the goal is sent to the Feedback-Loop component to execute the MAPE-K loop for the process step instance. The process engine then executes the process step in the "regular" way and waits for the feedback service's reply. Meanwhile, the MAPE-K loop is initialized with the goal and objectives of the instance in parallel to test for cyber-physical consistency. In our implementation, the web service executes the MAPE-K loop internally also as a process based on the same process meta-model [24].

Monitor. The *Monitor* component constantly monitors context data from sensors and other entities. During the monitoring phase, the component performs a pre-analysis of the incoming data based on defined thresholds to evaluate the significance of changes within the data (filtering of jitter). In case there is a significant delta in relevant context data, the new values (*Symptoms*) are fed forward into the *Analyzer* component. In our prototype we support queries of the knowledge base as well as direct stream processing of event data. The Open-HAB middleware for IoT [26] serves as the main source of sensor data.

Analyze. The *Analyzer* component evaluates the symptoms received from the monitoring component with respect to the objectives contained in the process step. The satisfied assertion and compensation assertion define target values of relevant sensor data to be reached as effect of the execution, as well as additional conditions (e.g., time frames for changes) [11]. If the objectives and therefore the overall process goal cannot be satisfied, an inconsistency is assumed and a change request including a description of the mismatch is sent to the *Planner* component. In case the overall goal can be met (i.e., the patterns defined in the assertions can be detected within the symptoms), the process engine continues with the execution of the consistent process instance based on the process model.

Plan. In the *Plan* phase, compensations for the provided mismatch(es) are searched for. In accordance with the reference model for self-managed systems proposed by Kramer and Magee [14], the change management uses the plan request to find an alternative way of reaching the goal. We use the information about sensors and actuators contained in the smart home ontology to find a compensation action automatically by graph traversal. Sensors and actuators are directly linked to the physical values they measure and manipulate. This way, we can find actuators affecting the same physical values in the same context as replacement for misbehaving actuators.

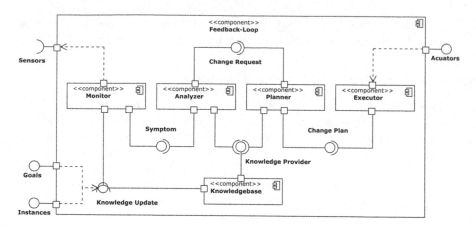

Fig. 4. The MAPE-K feedback loop for processes as a closed software component.

The mismatch information currently contains simplified descriptions of the deficits (e.g., value is *too high* or *too low*) that occurred. The *Planner* component is able to interpret these deficits semantically and to derive change plans containing parameters and commands/processes to be executed by the replacement component(s). For example, if the luminance values in a specific room have not reached a certain threshold after process step execution (*too low*), an alternative actuator (e.g., light switch or light dimmer) able to manipulate this physical value in the same room is selected. Based on the *too low*-mismatch and its capabilities, the actuator has to be switched on (light switch) or powered up (light dimmer) to increase the current luminance values–thus to restore cyber-physical consistency. To lift this approach on the process level, a repository of processes (process steps, activities, micro-processes) containing additional process information (e.g., goals and objectives the process is able to fulfill) could be used for finding compensating actions in the form of surrogate processes. A more advanced approach for deriving change plans and adaptations is, for example, *Case-based Reasoning* [27], which also considers historical change plans for the same or similar mismatches that have occurred.

Execute. Finally, the *Executor* component receives the change plan from the planner to execute the derived compensation/change actions. As we follow a process-based approach, this involves adapting the current process instance and requesting the basic process engine to execute this instance. During execution of the change plan, the MAPE-K loop is used again to check for objective and goal fulfillment with respect to the satisfied assertion of the replaced process step. This will also be done if the compensation process includes *cyber-physical* process steps with respect to their individual goals. If all goals can be satisfied by applying the MAPE-K loop, cyber-physical consistency is assumed and regular process execution continues. Implementing the MAPE-K loop as described, we are able to automatically react to unforeseen situations and errors to a certain degree–as long as a compensation for the occurred error can be found with the help of the underlying ontology.

In case not all objectives reach the *satisfied* state due to the MAPE-K loop not finding suitable compensations or the occurrence of other errors, the execution of the corresponding process step will reach a failure state and the *Failure Port* will be activated. This leads to the execution of the process step's failure branch, which has been modelled explicitly for the case of unresolvable errors. In our lighting example, a human task requesting the user to take care of the issue is triggered and also analyzed by the MAPE-K loop mechanism to see if the luminance level increases and process execution can be continued as planned. Also, the BPM systems internal exception handling mechanisms could be triggered to try dealing with errors the MAPE-K loop is not able to compensate.

5   Smart Lighting Case Study

The following case study in the Smart Home domain was conducted to provide a proof-of-concept evaluation and to illustrate the benefits and limitations of our approach. We used the extended PROtEUS editor and meta-model [24] to model and execute a continuous smart lighting process for a home office environment as depicted in Fig. 5. The process goal is to ensure minimal lighting conditions (i.e., at least 800 lux luminance) for the work environment in the case a resident is present in the office. The process steps are embedded into a loop to model a continuous automatic light control. The *TriggeredEvent* process step monitors sensor values from a light sensor and an occupation sensor located in the office. The event is activated according to the EPL pattern [23], when the luminance level is lower than 800 lux and at least one resident is present. This leads to the execution of the subsequent *RESTInvoke* process step calling a REST service that controls the light switch actuator. As this process step manipulates the real world state (i.e., the luminance level), the step is marked as *cyber-physical* to use the MAPE-K loop for verifying *cyber-physical consistency*. Depending on the success of process step execution (i.e., all objectives are satisfied) either the overall loop iteration is completed or a *HumanTask* process step is triggered as fallback solution. This explicitly modelled human task requests the user to manually restore cyber-physical consistency after the MAPE-K component failed.

Experiments. We ran the process in a controlled lab environment to monitor the execution and MAPE-K behaviour. The setup used one occupation sensor, one light sensor, one lamp controlled by a power switch and one lamp controlled by da dimmer. Basic REST service wrappers are used to integrate these sensors and actuators with the PROtEUS process engine [23]. To verify the intended process execution, we simulated three different scenarios (*Baseline*, *MAPE-K+* and *MAPE-K–*).

The *Baseline* scenario represents the basic process execution not using the MAPE-K loop (i.e., no *cyber-physical* process step). Consequently, the MAPE-K component and the failure port were never activated during process execution.

Whenever the luminance level was below 800 lux and the occupation sensor was triggered, the light was automatically switched on but this not always led to a sufficiently high luminance level. To illustrate another limitation of the *Baseline* process, we used a broken light bulb with the lamp. This cannot be detected by the light switch control software, which is why *cyber-physical consistency* is violated in this setup. Therefore, the outer loop process step was continuously executed as the lighting level was always below 800 lux.

In the *MAPE-K+* scenario, we marked the REST process step as *cyber-physical* and included a goal specification as described in Sect. 3. The goal and its objectives state that the controlled process step execution should lead to a luminance level of at least 870 lux within 2 s. Otherwise, the feedback loop will search for compensation actions to find and switch on other light sources in the same context. We made sure that another light source was available such that compensation actions can be found and executed. In case of a broken light bulb, the MAPE-K process detected the violation of *cyber-physical consistency* and was able to restore a consistent state by switching on the other light source.

We used the *MAPE-K-* scenario to illustrate that the MAPE-K enabled process has advantages over the baseline process even in case the MAPE-K loop fails and not all objectives can be satisfied. For this scenario we used the *MAPE-K+* process model and disconnected the alternative light source. When executed the MAPE-K loop fails to find a suitable compensation and reports an error. In contrast to the baseline process where the fail state information is unavailable, the MAPE-K component activates the failure port and the process engine continues to execute the failure process branch–in our case to request the user to fix the problem manually via a human task.

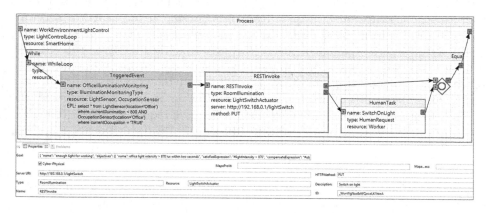

Fig. 5. Process model for the smart lighting scenario with MAPE-K.

6   Discussion

The previous sections show that with the introduction of the new *physical* dimension for business processes, there is a need to also verify the real world effects of automated processes in CPS. Feedback loops help to monitor, analyze, plan and execute real world processes to check for consistency between cyber and physical states at runtime. Sections 3 and 4 discussed necessary extensions for process meta-models and engines. Section 5 showed the applicability of our approach with the help of a proof-of-concept implementation for a smart home use case as one example for a cyber-physical system.

Process Modelling. Regarding the modelling of cyber-physical processes, we present a generic extension that can be applied to and reused in existing meta-models. With the introduction of goals and objectives for formalizing real world effects of a process, a strict separation of concerns is achieved as the "regular" (cyber) process can be modelled in a known way and the physical aspects are modelled as extensions to that. These extensions assume a composite process structure and can therefore be applied at the process, subprocess and activity level [24]. The more declarative approach of using goals and objectives reduces the overall modelling effort as not every possible error and unintended behaviour as well as corresponding failure handling processes have to be modelled explicitly. This is especially relevant for complex CPS with varying contexts and dynamic components resulting in a high level of unanticipated and emergent behaviour [13]. However, the modelling of the objectives requires detailed knowledge about contexts, sensors and actuators, and real world effects of the process. Using more sophisticated approaches for formalizing and deducing this knowledge may further reduce the modelling effort (e.g., as proposed in [17]).

Process Execution. The application of the MAPE-K loop for processes can be seen as a fine-grained realization of the BPM lifecycle at runtime [1]. The monitoring and analysis phases correspond to the diagnosis after process enactment, followed by process (re)design and system configuration in the plan phase, and finally the enactment (execute) of the adapted process. The domain, business and process knowledge is contained in the knowledge base. We propose to implement the MAPE-K loop using dedicated components for each phase of the loop. The component-based approach increases modularity and reusability for other processes and applications. Our implementation provides a web service component with standardized interfaces to be used by arbitrary services and applications to apply feedback loops to process-aware information systems. The execution of "regular" processes is extended indirectly with additional process steps according to the phases of the feedback loop. At runtime, process steps for each phase are generated and executed by the underlying process engine calling the specific MAPE-K components. This way, a coherent process-based view on the process execution can be achieved and the processes become self-aware of their execution and self-adaptive through the MAPE-K feedback [5].

Process Adaptation. The process-based MAPE-K loop presented in this paper can be used as a general framework for enabling self-adaptive workflows that influence physical world context properties. In the *Plan* phase of the MAPE-K subprocess, a suitable compensation for the occurred mismatch of cyber and physical world state is searched for and executed afterwards, i.e., the process adapts itself on the instance level–leading to a higher degree of autonomy for process execution in CPS. In case of a reoccurring error within multiple instances of the same process, the adaptation may also be done on the process model level (*process evolution*) [27]. However, the automated finding of replacement processes after a violation of cyber-physical consistency can be improved by more sophisticated reasoning and deduction mechanisms (e.g., as proposed in [17] or [27]). In general, various techniques can be plugged in to our component-based framework to realize the individual MAPE-K phases.

Process Consistency. The verification and matching of the assumed real world state with the actual physical world effects of an executed process is identified as one of the main challenges for real world processes [30]. To enable this verification, we introduce the notion of *cyber-physical consistency* as an extension to the well-known ACID criteria. The MAPE-K feedback loop provides means for automating the detection of inconsistent states based on sensor and context data and also for restoring consistency to a certain degree. Our approach increases resilience against failures and other unanticipated situations for processes that influence the real world. Eventually, this may lead to an increase of safety and flexibility, reduced resource consumption and an optimization of the overall work environment for smart homes and smart factories. Despite the goal of reaching a higher level of automation, the user can still be integrated into the workflows in case the MAPE-K loop fails and consistency has to be restored manually.

7   Related Work

As many existing "traditional" BPM systems and notations as well as adaptive process management systems are not designed to handle processes in the Internet of Things and CPS, real world processes have started to become a vibrant field of research within the BPM community [18]. From the self-adaptive and multi-agent systems communities, there have been several works proposed to combine Deming cycles with BPM systems to deal with failures in the *cyber* world and increase autonomy of process management systems [2,20]. However, these approaches work on a purely virtual and organizational level and do not consider the issue of automatic cyber-physical synchronization for operative processes [21]. In order to point out a need for considering real world effects of processes, Wombacher discusses ways to correlate business workflows and physical objects based on sensor data and process states [30]. Various approaches for integrating real world context data and objects into business processes were proposed with the emergence of smart spaces [3,7,9,19]. Weidlich et al. discuss an approach for optimizing event patterns for process monitoring by using process model knowledge and event processing [28], which can be applied to automatically adapt the goals and objectives in successive MAPE-K iterations (*meta-adaptation*). In [29] Wieland et al. present a workflow management system that uses situation recognition based on sensor data from production machines to adapt processes in case of errors and execute fault handling templates depending on the specific type of error situation. The SmartPM system by Marella et al. is able to adapt to errors during process execution based on models of expected reality and actual reality and recover from a potential gap between these two worlds using situation calculus and planning [17]. Especially the adaptation algorithms are worth to be investigated for integration into the MAPE-K loop.

In comparison with the aforementioned work on using BPM technologies to execute (partial) real world processes, our new approach of combining the worlds of BPM and self-adaptive system in CPS leads to a high level of autonomy and resilience to failures for cyber-physical workflows. We use the generic idea of the MAPE-K feedback loop for self-adaptive systems [5] as basis for extensions to existing process meta-models and engines, which is why our approach can be regarded as a more general, technology agnostic framework for implementing self-adaptive workflows for cyber-physical systems. Due to the meta-level extensions and component-based approach, arbitrary process notations can be used and various technologies for realizing the MAPE-K phases can be plugged in.

8   Conclusion

In this work, we presented a process-based framework for enabling self-adaptive workflows for cyber-physical systems (CPS). The application of BPM technologies to automate processes in CPS introduces the new requirement of providing a consistent view of processes' virtual world and real world effects (*Cyber-physical Consistency*). To achieve this goal, we propose to apply the MAPE-K feedback loop to monitor and analyze real world process execution using additional sensor and context data; and to find a compensation to be executed in case of inconsistencies. In comparison with related work, our approach reaches a high level of autonomy and resilience against failures for physical world process execution due to the capability of self-adaptiveness while still being able to keep the human in the loop. The model extensions and execution components for the MAPE-K loop can be applied to various process notations and engines, thus resulting in a generic framework for self-adaptive real world processes. Technologies used in the MAPE-K components can be exchanged easily.

With respect to future work, we will investigate the application of alternative algorithms in the *Analyse* and *Execute* phases to reduce the modelling effort and to increase autonomy via inference. Stream-based mining of processes combined with sensor data may lead to an increased accuracy of determining cyber-physical consistency [16]. We will also apply the generic MAPE-K framework to decentralized process execution environments as proposed in [25] to increase resilience of distributed execution of CPS workflows.

References

1. van der Aalst, W.M.P., ter Hofstede, A.H.M., Weske, M.: Business process management: a survey. In: van der Aalst, W.M.P., ter Hofstede, A.H.M., Weske, M. (eds.) BPM 2003. LNCS, vol. 2678, pp. 1–12. Springer, Heidelberg (2003)
2. Andonoff, E., Bouaziz, W., Hanachi, C., Bouzguenda, L.: An agent-based model for autonomous coordination of inter-organizational business processes. Informatica **20**(3), 323–342 (2009)
3. Baumgrass, A., Ciccio, C.D., Dijkman, R., Hewelt, M., Mendling, J., Meyer, A., Pourmirza, S., Weske, M., Wong, T.Y.: GET controller and UNICORN: event-driven process execution and monitoring in logistics (i) (2015)
4. Bonino, D., Corno, F.: DogOnt - ontology modeling for intelligent domotic environments. In: Sheth, A.P., Staab, S., Dean, M., Paolucci, M., Maynard, D., Finin, T., Thirunarayan, K. (eds.) ISWC 2008. LNCS, vol. 5318, pp. 790–803. Springer, Heidelberg (2008)
5. Brun, Y., Di Marzo Serugendo, G., Gacek, C., Giese, H., Kienle, H., Litoiu, M., Müller, H., Pezzè, M., Shaw, M.: Engineering self-adaptive systems through feedback loops. In: Cheng, B.H.C., Lemos, R., Giese, H., Inverardi, P., Magee, J. (eds.) Self-Adaptive Systems. LNCS, vol. 5525, pp. 48–70. Springer, Heidelberg (2009)
6. Conti, M., Das, S.K., Bisdikian, C., Kumar, M., Ni, L.M., Passarella, A., Roussos, G., Trster, G., Tsudik, G., Zambonelli, F.: Looking ahead in pervasive computing: challenges and opportunities in the era of cyberphysical convergence. Pervasive Mobile Comput. **8**(1), 2–21 (2012)
7. Dar, K., Taherkordi, A., Baraki, H., Eliassen, F., Geihs, K.: A resource oriented integration architecture for the Internet of Things: a business process perspective. Pervasive Mobile Comput. **20**, 145–159 (2015)
8. Gurgen, L., Gunalp, O., Benazzouz, Y., Gallissot, M.: Self-aware cyber-physical systems and applications in smart buildings and cities. In: Proceedings of the Conference on Design, Automation and Test in Europe, DATE 2013, EDA Consortium, San Jose, CA, USA, pp. 1149–1154 (2013)
9. Herzberg, N., Meyer, A., Weske, M.: An event processing platform for business process management. In: 17th IEEE International Enterprise Distributed Object Computing Conference, pp. 107–116 (2013)
10. Huber, S., Seiger, R., Schlegel, T.: Using semantic queries to enable dynamic service invocation for processes in the Internet of Things. In: 2016 IEEE International Conference on Semantic Computing (ICSC), pp. 214–221, February 2016
11. Kephart, J., Kephart, J., Chess, D., Boutilier, C., Das, R., Kephart, J.O., Walsh, W.E.: An architectural blueprint for autonomic computing. IBM (2003)
12. Koetter, F., Kochanowski, M.: Goal-Oriented model-driven business process monitoring using ProGoalML. In: Abramowicz, W., Kriksciuniene, D., Sakalauskas, V. (eds.) BIS 2012. LNBIP, vol. 117, pp. 72–83. Springer, Heidelberg (2012)
13. Kopetz, H.: System-of-systems complexity. arXiv preprint (2013). arxiv:1311.3629
14. Kramer, J., Magee, J.: Self-managed systems: an architectural challenge. In: Future of Software Engineering, FOSE 2007, pp. 259–268. IEEE (2007)
15. Lee, E.: Cyber physical systems: design challenges. In: 2008 11th IEEE International Symposium on Object Oriented Real-Time Distributed Computing (ISORC), pp. 363–369, May 2008
16. Leotta, F., Mecella, M., Mendling, J.: Applying process mining to smart spaces: perspectives and research challenges. In: Persson, A., Stirna, J. (eds.) CAiSE 2015 Workshops. LNBIP, vol. 215, pp. 298–304. Springer, Heidelberg (2015)

17. Marrella, A., Mecella, M., Sardina, S.: SmartPM: an adaptive process management system through situation calculus, indigolog, and classical planning. In: Principles of Knowledge Representation and Reasoning, pp. 1–10. AAAI Press (2014)
18. Meyer, S., Ruppen, A., Hilty, L.: The things of the Internet of Things in BPMN. In: Persson, A., Stirna, J. (eds.) CAiSE 2015 Workshops. LNBIP, vol. 215, pp. 285–297. Springer, Heidelberg (2015)
19. Meyer, S., Ruppen, A., Magerkurth, C.: Internet of Things-Aware process modeling: integrating IoT devices as business process resources. In: Salinesi, C., Norrie, M.C., Pastor, Ó. (eds.) CAiSE 2013. LNCS, vol. 7908, pp. 84–98. Springer, Heidelberg (2013)
20. Oliveira, K., Castro, J., España, S., Pastor, O.: Multi-level autonomic business process management. In: Nurcan, S., Proper, H.A., Soffer, P., Krogstie, J., Schmidt, R., Halpin, T., Bider, I. (eds.) BPMDS 2013 and EMMSAD 2013. LNBIP, vol. 147, pp. 184–198. Springer, Heidelberg (2013)
21. Perrin, O., Godart, C.: A model to support collaborative work in virtual enterprises. Data Knowl. Eng. **50**(1), 63–86 (2004). Advances in business process management
22. Saidani, O., Rolland, C., Nurcan, S.: Towards a generic context model for BPM. In: 2015 48th Hawaii International Conference on System Sciences (HICSS), pp. 4120–4129, January 2015
23. Seiger, R., Huber, S., Schlegel, T.: PROtEUS: an integrated system for process execution in cyber-physical systems. In: Gaaloul, K., Schmidt, R., Nurcan, S., Guerreiro, S., Ma, Q. (eds.) BPMDS 2015 and EMMSAD 2015. LNBIP, vol. 214, pp. 265–280. Springer, Heidelberg (2015)
24. Seiger, R., Keller, C., Niebling, F., Schlegel, T.: Modelling complex and flexible processes for smart cyber-physical environments. J. Comput. Sci. **10**, 137–148 (2015)
25. Seiger, R., Niebling, F., Schlegel, T.: A distributed execution environment enabling resilient processes for ubiquitous systems. In: 2014 IEEE International Conference on Pervasive Computing and Communications Workshops (PERCOM Workshops), pp. 220–223, March 2014
26. Smirek, L., Zimmermann, G., Ziegler, D.: Towards universally usable smart homes- how can myui, urc and openhab contribute to an adaptive user interface platform. In: IARIA, Nice, France, pp. 29–38 (2014)
27. Weber, B., Rinderle, S., Wild, W., Reichert, M.: CCBR–Driven business process evolution. In: Muñoz-Ávila, H., Ricci, F. (eds.) ICCBR 2005. LNCS (LNAI), vol. 3620, pp. 610–624. Springer, Heidelberg (2005). http://dx.doi.org/10.1007/11536406_46
28. Weidlich, M., Ziekow, H., Gal, A., Member, S., Mendling, J., Weske, M.: Optimising event pattern matching using business process models. IEEE Trans. Knowl. Data Eng. **26**(11), 1–14 (2013)
29. Wieland, M., Schwarz, H., Breitenbucher, U., Leymann, F.: Towards situation-aware adaptive workflows: Sitopta general purpose situation-aware workflow management system. In: 2015 IEEE International Conference on Pervasive Computing and Communication Workshops (PerCom Workshops), pp. 32–37. IEEE (2015)
30. Wombacher, A.: How physical objects and business workflows can be correlated. In: Proceedings - 2011 IEEE International Conference on Services Computing, SCC 2011, pp. 226–233 (2011)

Using the Guard-Stage-Milestone Notation for Monitoring BPMN-based Processes

Luciano Baresi, Giovanni Meroni$^{(\boxtimes)}$, and Pierluigi Plebani

Dipartimento di Elettronica, Informazione e Bioingegneria,
Politecnico di Milano, Piazza Leonardo da Vinci, 32, 20133 Milano, Italy
{luciano.baresi,giovanni.meroni,pierluigi.plebani}@polimi.it

Abstract. Business processes are usually designed by means of imperative languages to model the acceptable execution of the activities performed within a system or an organization. At the same time, declarative languages are better suited to check the conformance of the states and transitions of the modeled process with respect to its actual execution. To avoid defining models twice from scratch to cope with both the process enactment and its monitoring, this paper proposes an approach for translating BPMN process models to E-GSM ones: an extension of the Guard-Stage-Milestone artifact-centric notation. The paper also shows how a monitoring engine based on E-GSM specifications can detect anomalies during the execution of the process and classify them according to different levels of severity, that is, with respect to the impact on the outcome of the process.

Keywords: Guard-Stage-Milestone · Artifact-centric processes · BPMN · Process execution monitoring

1 Introduction

Process modeling represents one of the most crucial activities in Business Process Management and the goal of the resulting model is twofold. On the one hand, a business process model describes a portion of the world as it is (or as we want it to be) using a formalism easy to understand by all the relevant stakeholders (e.g., process owners and process users). On the other hand, a business process model — if properly defined in all of its parts — feeds the engine that will enact its execution. To this aim, imperative control-flow based languages are widely adopted, as their constructs and the underlying semantics are very intuitive. Among them, BPMN nowadays represents one of the most used notation adopted by both business and technical people.

However, when used for monitoring the execution of a process at run-time, imperative languages manifest a significant limitation: when a violation in the control flow occurs, an imperative process engine treats such a violation as an unhandled exception and stops monitoring the process until a user manually fixes the issue. This is not always desirable, especially when the engine has no control

R. Schmidt et al. (Eds.): BPMDS/EMMSAD 2016, LNBIP 248, pp. 18–33, 2016.
DOI: 10.1007/978-3-319-39429-9_2

on the monitored process, which would continue its execution even though the engine stopped. Declarative languages, on the other hand, do not have the notion of strict control flow. Therefore, declarative engines can both report deviations in the control flow and continue monitoring the process.

The goal of this paper is to mediate between these two perspectives by proposing a solution to monitor the execution of distributed control-flow processes modeled in BPMN, that relies on a monitoring system based on the artifact-centric Guard-Stage-Milestone (GSM) declarative language [5]. In particular, we start from a BPMN process, which is easy to conceive, and we transform it into a model defined using E-GSM, our extension of GSM. This transformation preserves control flow information, but such information, which is prescriptive in BPMN, becomes descriptive. Deviations from the "original" execution flow can easily be detected at run-time during the process enactment by analyzing the artifacts, that contain information about how the process is evolving, and represent the states through which the process should evolve during execution.

The adoption of E-GSM to drive the process monitoring introduces the following advantages. E-GSM allows one to define conditions both on the process and on external data to trigger the execution and termination of activities. Therefore, the monitoring platform can infer when activities are executed based on information coming from the environment, thus being not limited to explicit messages. Furthermore, E-GSM allows one to identify the results of the execution of the activities within the process model, and consequently it permits the identification of the activities that are incorrectly executed, if any.

The rest of the paper is structured as follows. Section 2 discusses how we extended GSM into E-GSM to enable a data-artifact driven process monitoring solution. Section 3 introduces the set of rules we defined to translate BPMN elements into equivalent E-GSM ones. Section 4 validates our work by showing how to apply the approach on a real business process in the domain of logistics. Section 5 surveys the state of the art, and Sect. 6 concludes the paper.

2 E-GSM

The GSM notation is a declarative language that allows one to model artifact-centric processes by defining conditions that determine the activation and termination of activities, called **Stages**. With respect to other declarative languages, like Declare [13], such conditions are not limited to dependencies among activities. Instead, they are based on events, which can be *external* (e.g., sent or received messages), or *internal* (e.g., termination of activities), to the process. Starting from the standard GSM notation and our preliminary work [2], we propose E-GSM, an extension to GSM where we distinguish between **Data Flow Guards** and **Process Flow Guards** and we add **Fault Loggers**.

The goal of this extension is to include information on the normal flow, that is, the expected behavior of the process, or happy path, in the artifact-centric process model. To this aim, the process model includes the dependencies among activities in terms of control flow. Being a declarative language, E-GSM does not

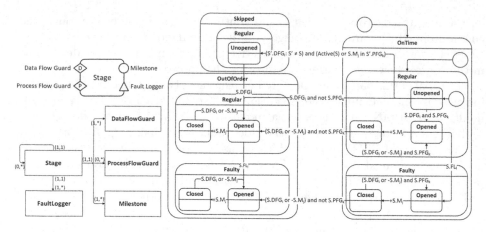

Fig. 1. E-GSM meta-model (bottom left), graphical representation (top left) and lifecycle of a Stage (right).

use control flow information to enforce a specific execution path among activities. Instead, it uses such information to let the process engine detect deviations between the happy path and how the process is actually executed.

The left portion of Fig. 1 shows a simplified version of the meta-model behind E-GSM, along with the graphical representation of its main elements. The original definition of GSM comprises **Stages**, **Guards**, and **Milestones**. A **Stage** represents the unit of work that can be executed in a process instance. A **Stage** can have one or more nested **Stages**, or it can be *atomic*, thus representing a single task. A **Stage** may be decorated with one or more **Guards** and **Milestones**.

A **Guard** (**Data Flow Guard** in E-GSM) is an Event-Condition-Action (ECA) rule[1]. If true, the associated **Stage** is declared opened. A **Milestone** is another ECA rule. If true, the **Stage** is declared closed. A **Milestone** may also have an *invalidator*: a boolean expression that can invalidate the **Milestone** and reopen the **Stage**.

In the proposed extension, a **Stage** can now also be decorated with **Process Flow Guards** and **Fault Loggers**. A **Process Flow Guard** is a boolean expression that predicates on the activation of the **Data Flow Guards** and **Milestones** used to map the expected control flow. The expression is evaluated once one of the **Data Flow Guards** of the associated **Stage** is triggered, and before the **Stage** becomes opened. If the expression is true, the **Stage** complies with the expected execution, otherwise the **Stage** has been activated without respecting the normal flow.

A **Fault Logger** is an ECA rule. If true, the associated **Stage** is declared faulty because something went wrong during the execution of the activity.

---

[1] An ECA rule is an [on e] [if c] expression, that is triggered when an event *e* occurs and the condition *c* is true. When [on e] is missing, the ECA is triggered once *c* becomes true, when [if c] is missing, the ECA is triggered once *e* occurs.

A faulty **Stage** does not imply its termination, as the termination is only determined by **Milestones**.

The right portion of Fig. 1 sketches the lifecycle of an E-GSM **Stage** organized around three main orthogonal execution perspectives: outcome, compliance, and status[2].

The *Execution outcome* captures the situation of a **Stage**, which can be either *regular* (none of its **Fault Loggers** has ever been triggered) or *faulty* (at least one of its **Fault Loggers** has been triggered, $A.FL_1$).

The *Execution compliance* captures the compliance of each **Stage** with the normal flow. A **Stage** is declared *onTime* by default. It can become *outOfOrder* (according to the normal flow) when one of its **Data Flow Guards** is triggered but none of its **Process Flow Guards** holds ($A.DFG$ and $not(A.PFG)$). If a **Stage** S' is declared *outOfOrder*, every other *onTime* **Stage** S that would trigger one of the **Process Flow Guards** of S' ($S.M_j$ or $Active(S) \in S'.PFG_k$) is declared *skipped*. If a **Stage** is *skipped*, once one of its **Data Flow Guards** is triggered ($S.DFG_i$), it becomes *outOfOrder*.

The *Execution status* captures the status of a **Stage**: *unopened*, *opened* or *closed*. A **Stage** is *unopened* if its **Data Flow Guards** have never been triggered. A **Stage** can become *opened* only if it is *unopened* or *closed* and the parent **Stage** is *opened*. In addition, at least one of its **Data Flow Guards** must be triggered ($S.DFG_i$). A **Stage** becomes *closed* if it is *opened* and a **Milestone** is achieved ($+S.M_j$), or if the parent **Stage** becomes *closed*.

The combination of these three perspectives says that the whole lifecycle assumes that a **Stage** is initially *onTime*, *regular*, and *unopened*. **Data Flow Guards** drive the change of execution status. **Fault Loggers** drive the outcome, while **Process Flow Guards** are in charge of the compliance. With respect to Standard GSM, E-GSM interprets reopening a *closed* **Stage** as a new iteration of that process portion. Therefore, once a parent **Stage** is reopened (i.e., it moves from *closed* to *opened*), the lifecycle of all its child **Stages** will restart from scratch.

Thank to these three perspectives, it is possible to detect at runtime when a deviation in the execution of a process occurs and which stages are involved. This enables a classification that predicates on the lifecycle of all stages to evaluate how severely variations during execution affect the outcome of the process. For example, Table 1 reports a possible classification of severity that can be modified according to any specific scenario: *None*, if all activities are executed at the right time and their execution was successful. *Low*, if the process terminated, the expected control flow was not respected, yet no activity was skipped and they were all successfully executed. *Medium-low*, if an activity was incorrectly executed, but the expected control flow was respected. *Medium*, if the process

---

[2] In this paper we use the notation introduced in [5], so we write $S.DFG_i$, $S.PFG_k$, $S.FL_1$ to indicate the activation of a Data Flow Guard, Process Flow Guard, or a Fault Logger associated with Stage S, $+S.M_j$ ($-S.M_j$) to indicate the achievement (invalidation) of a Milestone M_j, $S.M_j$ to indicate that Stage S is closed and a Milestone M_j is achieved, and $Active(S)$ to indicate that Stage S is opened.

Table 1. Severity levels. $S_x.o$, $S_x.c$ and $S_x.s$ indicate the state of **stage** S_x, along with the execution outcome, compliance and status respectively.

Severity	Execution outcome ($S_y.o$)	Execution compliance ($S_z.c$)	Execution status ($S_x.s$)
None	$\forall S_y : S_y.o = regular$	$\forall S_z : S_z.c = onTime$	$\forall S_x : S_x.s = unopened$ $\lor S_x.s = opened$ $\lor S_x.s = closed$
Low	$\forall S_y : S_y.o = regular$	$\exists S_z : S_z.c = outOfOrder$	$\forall S_x : S_x.s = unopened$ $\lor S_x.s = closed$
Medium-low	$\exists S_y : S_y.o = faulty$	$\forall S_z : S_z.c = onTime$	$\forall S_x : S_x.s = unopened$ $\lor S_x.s = opened$ $\lor S_x.s = closed$
Medium	$\forall S_y : S_y.o = regular$	$\exists S_z : S_z.c = outOfOrder$ $\lor S_z.c = skipped$	$\exists S_x : S_x.s = opened$
Medium-high	$\forall S_y : S_y.o = regular$	$\exists S_z : S_z.c = skipped$	$\forall S_x : S_x.s = unopened$ $\lor S_x.s = closed$
High	$\exists S_y : S_y.o = faulty$	$\exists S_z : S_z.c = outOfOrder$ $\lor S_z.c = skipped$	$\forall S_x : S_x.s = unopened$ $\lor S_x.s = opened$ $\lor S_x.s = closed$

is still in progress and, during execution, the expected control flow was not respected. *Medium-high*, if the process terminated and an activity was skipped. *High*, if an activity was incorrectly executed and no corrective action was taken (i.e., at least another activity was either skipped or incorrectly executed).

This classification assumes that all stages have the same importance. However, weights can be introduced to differentiate the influence of each specific stage on the process, or metrics taken from the conformance checking domain [16] can be adopted.

3   Transformation Rules

The aforementioned semantics of E-GSM is then used in 13 transformation rules [10] to translate a BPMN process model into an E-GSM one.

These transformation rules are applicable to every BPMN process model that complies with a workflow net [1], that is, the process has only one start event and only one end event, and it always terminates (soundness). Note that the control flow is always captured by Process Flow Guards, and as such it is never enforced. This allows the E-GSM model to continue monitoring a process even if violations in the control flow occur.

3.1   Basic Elements

The transformation rules defined for basic elements are presented in Fig. 2.

Rule 1. *A BPMN Activity* **A** *is translated into a Stage* **A** *with one or more Data Flow Guards (***A.DFG**$_i$*) and one or more Milestones (***A.M**$_j$*).*

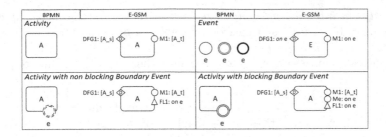

Fig. 2. BPMN to E-GSM transformation rules for basic elements.

Producing the conditions associated with those **Data Flow Guards** and **Milestones** is far from trivial [3]. They depend on the associated data objects and, if the activity is a task, on its type (i.e., *receive* or *user* task). In case of a generic task, placeholders A_s and A_t are associated with, respectively, A.DFG1 and A.M1 to represent the explicit start and termination of the activity. If the activity is a sub-process, A.DFG$_i$ and A.M$_j$ are then derived from the structure of the sub-process and from its elements, as explained in the following.

Rule 2. *A BPMN Start, End or Intermediate Event e is translated into a Stage E where E.DFG1 and E.M1 have the occurrence of the event as condition.*

Rule 3. *A BPMN Activity A with a non-interrupting Boundary Event e attached is translated into a Stage A according to Rule 1 with A.FL1 having the occurrence of the event as condition (i.e., on e).*

Rule 4. *A BPMN Activity A with an interrupting Boundary Event e attached is translated into a Stage A according to Rule 1 with an additional Milestone A.Me and A.FL1 having the occurrence of the event as condition.*

3.2 Normal Flow

The combination of the above rules for basic elements allows one to translate well-structured business process models [15]. In particular, we focus on five types of blocks, defined starting from the classical control flow patterns [17]:

– A *sequence block* is made of linked activities, events and other blocks without splits or merges. It corresponds to pattern *sequence*.
– A *parallel block* organizes activities, events, and other blocks in two or more parallel threads resulting from the combination of patterns *parallel split* and *synchronization*.
– A *conditional exclusive block* organizes activities, events, and other blocks in two or more branches resulting from a combination of patterns *exclusive choice* and *simple merge*.

- A *conditional inclusive block* organizes activities, events, and other blocks in two or more branches resulting from a combination of patterns *multi-choice* and *structured synchronized merge*.
- A *loop block* organizes activities, events, and other blocks according to pattern *structured loop*.

For each of these blocks, we delivered proper transformation rules in [10]. A graphical representation of them is reported in Fig. 3. Due to space constraints, in this paper we will only describe in detail how sequence, conditional exclusive, and loop blocks are translated.

Rule 5. *A sequence block corresponds to a Stage* Seq *that includes* S_x *inner Stages obtained by applying the transformation rules to all the elements (i.e., Activities, Events, inner blocks) that belong to the block.*

- *In addition to the existing Process Flow Guards, each inner stage has* S_x.PFG1 *to state that none of its Milestones is achieved, and at least one of the Milestones of the element that directly precedes it (if present) is achieved.*
- Seq *has a set* Seq.DFG *that includes all* S_x.DFG$_i$, *and a Milestone* Seq.M1 *that requires that, for all* S_x, *at least one* S_x.M$_j$ *be achieved.*

Rule 6. *A conditional exclusive block is translated into a Stage* Exc *that includes all the Stages obtained by applying Rule 5 to all its branches, which result in* S_x *inner Stages.*

- *For each* S_x, S_x.PFG1 *is added to check that no* S_x.M$_j$ *has already been achieved, that the condition on the branch from which* S_x *is produced (if present) is satisfied, and that none of the other inner Stages is opened (i.e.,* not Active(S_y) *where* $y \neq x$*).*
- Exc *has a set* Exc.DFG *that includes all* S_x.DFG$_i$, *and a Milestone* Exc.M1 *that requires that, for at least one* S_x, *one* S_x.M$_j$ *be achieved, and the condition on the branch from which* S_x *is produced (if present) be satisfied, as long as none of the other inner Stages is opened.*

Rule 7. *A loop block is translated into two Stages,* Ite *and* Loop. Ite *includes* S_x *inner Stages obtained by applying Rule 5 to all the branches within the loop block. One of these stages is a forward Stage, that is, its control flow goes in the same direction as the control flow that includes the loop block. The others are backward Stages.*

- *For all the inner Stages,* S_x.PFG1 *is added to check that no* S_x.M$_j$ *is already achieved. Moreover, if* S_x *is a backward stage,* S_x.PFG1 *also requires that the condition on the branch (if present) be satisfied, and that one of the Milestones of the forward stage be achieved.*
- Ite *has a set* Ite.DFG *that includes all* S_x.DFG$_i$, *and two Milestones, where:*
  - Ite.M1 *requires that one of the Milestones of the forward Stage be achieved and the exit condition of the loop (if present) be satisfied, as long as no backward Stage is opened.*

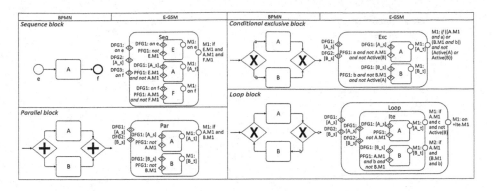

Fig. 3. BPMN to E-GSM transformation rules for normal flow blocks (due to space constraints, the conditional inclusive block is not presented).

- *Ite.M2* requires that one of the Milestones of the forward Stage be achieved and, for at least a backward Stage, one of its Milestones be achieved and the condition on that branch (if present) be satisfied, as long as none of the other backward Stages is opened.

Stage Loop includes Ite and has Loop.DFG = Ite.DFG and Loop.M = on Ite.M1 (i.e., the process can exit the loop).

The iteration **Stage Ite** has no **Process Flow Guards** since it is supposed to be executed multiple times and, every time it becomes opened, a new iteration of the loop is carried out. Thus, Ite is opened when at least one of its inner Stages can be opened too, and it is closed when either the process can exit the loop (Ite.M1 is achieved), or when an iteration is complete (Ite.M2 is achieved).

3.3 Exceptional Flow

BPMN supports the management of foreseen exceptions through boundary events, that is, events directly attached to activities. These events, like split gateways, determine a branching of the control flow into an *exceptional* flow, which leaves the boundary event, and a *normal* flow, to continue the execution from the activity. If the foreseen exception occurs while executing the activity, the attached boundary event activates the exceptional flow. A dedicated set of rules shown in Fig. 4 is thus required to preserve this behavior in E-GSM models. Again, we refer to [10] for the details.

Interrupting boundary events cause the normal and exceptional flows to be mutually exclusive, therefore we expect them to be merged by an exclusive merge gateway at the end. This requires that two additional blocks, called *forward exception handling* and *backward exception handling*, respectively, be defined. The forward exception handling block comprises an interrupting boundary event, and a *simple merge*, defined with a BPMN exclusive gateway, that merges the exceptional control flow and the portion of the normal control flow that follows the activity to

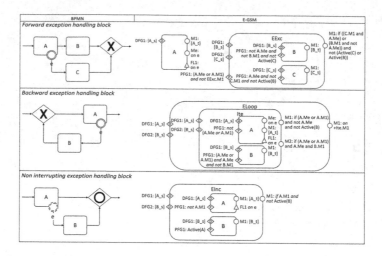

Fig. 4. BPMN to E-GSM transformation rules for handling exceptions.

which the boundary event is attached. Its behavior is similar to the one of the conditional exclusive block, with the exception of the branch condition, which predicates on the achievement of the milestone derived from the boundary event. The backward exception handling block comprises an interrupting boundary event and a *simple merge*, defined with a BPMN exclusive gateway, that merges the exceptional control flow and the portion of the normal control flow that precedes the activity to which the boundary event is attached. This block produces a loop that allows one to re-execute part of the normal control flow if the boundary event is triggered, and therefore it is translated similarly to a loop block.

In BPMN, boundary events could also be non interrupting, that is, they activate the exceptional control flow without terminating the associated activity. Therefore, the elements within the exceptional control flow can run in parallel with the normal flow that starts from the activity the boundary event is associated with. Since we expect these potentially simultaneous control flows be merged by an inclusive merge gateway, the transformation requires an additional block, called *non interrupting exception handling block*. This new block comprises a non interrupting boundary event to split the execution flow into an exceptional flow and the continuation of the normal one, and a *structured synchronized merge*, defined with a BPMN inclusive gateway, to merge the two flows in case the exception occurred.

4   Validation

The transformation rules introduced in the previous section allow any well-structured BPMN process model to be translated into E-GSM. To prove it, we developed a BPMN to E-GSM prototype translator[3], where the transformation rules

---

[3] The tool is publicly available at https://bitbucket.org/polimiisgroup/bpmn2egsm.

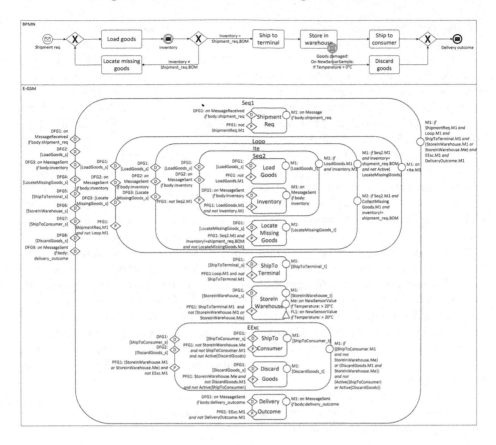

Fig. 5. BPMN and E-GSM models of the example shipping process.

are implemented in ATL (ATLAS Transformation Language [6]), and validated —and refined— the proposed rules against several BPMN business processes with different levels of complexity. A formal verification about the equivalence between BPMN processes and their correspondent E-GSM is under study and it aims to check if all the traces that a BPMN process can produce are also considered as satisfied in the E-GSM model.

Among these test processes, here we concentrate on an example taken from the logistics domain, which is shown at the top of Fig. 5, to better explain the advantages of adopting E-GSM to monitor the execution of complex (distributed) processes. A pharmaceutical company M has to ship drugs (that are highly susceptible to temperature variations) to one of its customers N. To do so, it relies on two shipping companies R and T for, respectively, rail and truck transportation, and on an inland terminal I for changing means. The shipping process starts when a shipment request by N is received, and comprises four main phases: (i) loading goods into a thermally-insulated shipping container; (ii) shipping such a container to I by rail; (iii) temporarily storing the container in a temperature-controlled warehouse;

(iv) delivering the goods to the customer's site by truck. Before starting phase (ii), an inventory report of the contents of the container must be produced, and it must be compared with the bill of materials included with the shipment request. If some products are missing, they must be located and loaded onto the container, and a new report must be produced. Furthermore, if the goods are exposed to a temperature higher than 20°C during phase (iii), they must be discarded and the whole process must be aborted. Our translator produces the E-GSM model shown at the bottom of Fig. 5.

Since all these activities interact with the shipping container, we can think of it as the process coordinator (i.e., the element that interacts with all the parties and has complete visibility on the whole process). To make the container process-aware, we can exploit the Internet of Things paradigm by equipping it with a single board computing device, sensors and a network interface, thus transforming it into a smart object (i.e., smart container). However, being the container completely passive, it cannot enforce the execution flow modeled in the process, and it needs information to identify when each activity is being executed. For this reason, a traditional process engine would be unsuited to monitor this process. On the other hand, an E-GSM engine[4] running onto the smart container would solve this problem: By predicating on on-board sensor values or explicit messages, **Data Flow Guards**, **Process Flow Guards**, **Milestones** and **Fault Loggers** can be triggered, and the execution of the process be monitored. This way, once a violation in the execution occurs, the E-GSM engine can report to stakeholders which activities are affected, and how severely the whole process is affected by such an incident.

To show how process monitoring can take advantage of the E-GSM model, we describe three possible scenarios.

4.1 An Error-Free Execution

Once the shipment request is received, `Seq1.DFG1` is triggered and, consequently, `Seq1` becomes opened (thus starting the process). This first triggers `Shipment-Req.PFG1`, then `ShipmentReq.DFG1`, which causes `ShipmentReq` to become *opened*, and finally `ShipmentReq.M1` be achieved, which moves `ShipmentReq` to the *closed* state, and triggers `Loop.PFG1`.

When R loads the goods onto the container, a notification is sent to the engine, which triggers `Loop.DFG1`, then `Ite.DFG1`, `Seq2.PFG1`, `Seq2.DFG1`, `Load-Goods.PFG1`, and finally `LoadGoods.DFG1`, which moves `Loop`, `Ite`, `Seq2` and `LoadGoods` to the *opened* state. After finishing loading the goods, the operator sends another notification, thus making `LoadGoods.M1` be achieved, which triggers `Inventory.PFG1` and moves `LoadGoods` to the *closed* state. It then produces the inventory of the loaded goods, which triggers `Inventory.DFG1`, and then makes `Inventory.M1` be achieved, which makes `Seq2.M1` achieved too, causing `Inventory` and `Seq2` to move to the *closed* state. Being the inventory consistent with the bill of materials included in the shipment request, also `Ite.M1` and,

---

[4] A prototype E-GSM engine is currently under development.

consequently, `Loop.M1`, are achieved, moving `Ite` and `Loop` to the *closed* state, and triggering `ShipToTerminal.PFG1`.

Once the rail shipping begins, R sends a notification, which triggers `ShipToTerminal.DFG1` and moves `ShipToTerminal` to the *opened* state. When the container is delivered to I, another notification is sent, which makes `ShipToTerminal.M1` become achieved, moving `ShipToTerminal` to the *closed* state and triggering `StoreInWarehouse.PFG1`. Similarly I sends a notification when the container is put in the warehouse and when T is ready to pick it up, thus triggering `StoreInWarehouse.DFG1`, achieving `StoreInWarehouse.M1`, triggering `EExc.PFG1`, and moving `StoreInWarehouse` to the *opened* state at first, and then to the *closed* state. After hooking the container to its truck, a notification is sent by T. That notification triggers `EExc.DFG1`, then `ShipToCustomer.PFG1`, and finally `ShipToCustomer.DFG1`, thus moving stages `EExc` and `ShipToCustomer` to the *opened* state. Once T delivers the goods to N, another notification is sent. That notification causes the achievement of `ShipToCustomer.M1`, which makes `EExc.M1` become achieved too, thus moving `ShipToCustomer` and `EExc` to the *closed* state and triggering `DeliveryOutcome.PFG1`.

Finally, once the goods are inspected by N, a report of the shipment is produced, which triggers `DeliveryOutcome.DFG1`, moving `DeliveryOutcome` to the *opened* state, and then causes the achievement of `DeliveryOutcome.M1`, which causes the achievement of `Seq1.M1` too, thus moving `DeliveryOutcome` and `Seq1` to the *closed* state and, since `Seq1` represents the whole process, terminating the monitoring activity. Once the process concludes, N queries the smart container and finds out that the severity level of the process is *None*, since all stages are in state either *unopened* or *closed*, their compliance is *onTime*, and their outcome is *regular*. Therefore, N accepts the goods.

4.2   A Catastrophic Execution

A second example shows how the system can monitor an incorrect execution of the process. During phase (iii), the warehouse cooling system breaks down, and the temperature of the goods goes beyond 20°C. Being the container equipped with a temperature sensor, the E-GSM engine is able to detect such an event and consequently triggers both `StoreInWarehouse.FL1` and `StoreInWarehouse.Me`, which move `StoreInWarehouse` to the *faulty* and *closed* states. This changes the severity level of the process from *none* to *medium-low*, since a *faulty* stage exists, but all stages are still *onTime*. Being `StoreInWarehouse` closed and `StoreInWarehouse.Me` achieved, `DiscardGoods.PFG1` is also triggered.

Instead of discarding the goods, I ignores that accident, and delivers the goods to N. This moves `ShipToConsumer` to state *outOfOrder*, since `ShipToConsumer.DFG1` is triggered before `ShipToConsumer.PFG1` becomes active. This causes the severity level of the process to become *high*, since there are both a *faulty* stage (`StoreInWarehouse`), and an *outOfOrder* one (`ShipToConsumer`).

Once N receives the goods, it queries the smart container and, since the severity level of the process is *high*, decides to immediately inspect its content, thus discovering that the goods have been spoiled. Therefore, it sends them back to M. In turn, M identifies that StoreInWarehouse is in the *faulty* state, and that ShipToConsumer is *outOfOrder*. Thank to this information, M is able to charge I a penalty for having spoiled the goods and not having reported that accident. Note that had T queried the smart container, it would have seen that the severity level was *medium-low*, since StoreInWarehouse was in *faulty* state, and could have avoided delivering the container to N.

4.3   A Troublesome yet Recoverable Execution

Let us now focus on a less critically incorrect execution of the process. In this case, the inventory of the container is not consistent with the bill of materials, which causes LocateMissingGoods.PFG1 to be triggered. However, R does not check the inventory and immediately begins shipping the container to I, which moves ShipToTerminal to the *outOfOrder* state, since ShipToTerminal.DFG1 is triggered before ShipToTerminal.PFG1 becomes active. This changes the severity level of the process from *none* to *medium*, as there are both *opened* stages (Seq1, Loop and Ite) and an *outOfOrder* one (ShipToTerminal).

Once N receives the goods, it queries the smart container and finds out that the severity level is still *medium* (since the missing goods were not collected and loaded onto the container, stages Seq1, Loop and Ite are still *opened*). So, it inspects the contents, discovers that some of them are missing, and asks M to ship the missing ones for free. By querying the smart container, M finds out that, even though the inventory did not match the shipping request, missing goods were never collected and loaded onto the smart container (i.e., LocateMissingGoods has not been executed even though LocateMissingGoods.PFG1 was satisfied), and the shipment continued anyway (i.e., ShipToTerminal is in state *outOfOrder*). Because of this information, M can blame R for having shipped the goods without checking the inventory first. Note that the severity level (*medium*) reflects the results of the process: being at least part of the goods successfully delivered, M did not experience a complete loss as in the previous case, where all the goods were spoiled, and the truck shipment was done pointlessly.

5   Related Work

Köpke et al. [7] propose transformation rules that transform a BPMN process model into a GSM equivalent. While we have borrowed from these rules the idea of transforming blocks into nested Stages, our transformation rules produce completely different expressions for Guards and Milestones. The reason behind such a discrepancy is that we are interested in identifying control flow violations, and not in forcing the process to rigidly follow a given execution flow, which is what is pursued in [7]. Eshuis et al. [4] define a semi-automated approach that starts from UML Activity Diagrams and produces a data-centric process model in GSM. They capture the lifecycle of the data artifacts referred to in the UML process model, and exploit control flow information to render it in GSM. Similarly, Kumaran et al. [8] and Meyer et al. [12] propose a language-agnostic algorithm to derive the lifecycle of artifacts based on an imperative process model. This is possible as long as each activity has input and output information entities explicitly defined in the model. Our work differentiates from [4,8,12], which use control flow information to model the interactions among data artifacts, by keeping such information in the target process model to assess compliance. Popova et al. [14] define a translator from Petri Nets to GSM. The main purpose of that translator is to transform the outcome of process mining algorithms, which is often represented as a Petri Net, to a GSM model. This way, process mining techniques can be used to identify business artifacts that the translator represents in a language that is easier to understand by domain experts than Petri Nets.

Concerning the integration of both activity and data-centric perspectives in business processes, Künzle et al. [9] propose a framework that maps portions of data structures to activities and use control flow information to define how such data objects should be manipulated. Similarly, Meyer et al. [11] propose a methodology to model both the control flow and data dependencies by extending BPMN data artifacts to define dependecines among all data items manipulated in a process. Both [9,11] use control flow information in a prescriptive way, while E-GSM uses it in a descriptive way to detect deviations from the original definitions during execution.

Conformance checking is the discipline that aims at identifying inconsistencies among a process model and its execution [16]. To do so, the process model is checked against high level execution logs, which report when and if activities have been executed. Our solution differs from this approach as it is able to autonomously identify when activities start or end, without relying on an execution log. Furthermore, it is able to detect deviations at runtime, whereas most process compliance techniques are applicable only when the process terminates.

6   Conclusions and Future Work

This paper extends the Guard-Stage-Milestone (GSM) notation to embed control flow information in the process model definition, presents a solution for transforming BPMN models into equivalent E-GSM ones, and shows how the derived E-GSM process model can be used to identify when activities are executed, to keep track of violations in the execution flow, and to evaluate the overall execution of a process along with different levels of severity.

As for our future work, we will investigate how to improve Rule 1 and Rule 2 by taking into account the nature of activities (i.e., receive tasks or user tasks) and events (i.e., timer, signal, etc.), and their associated data objects. We will also propose additional transformation rules to derive the *E-GSM Information Model*, which is not considered in this work, from data objects and implicit information defined in BPMN process models, which may also influence the definition of severity levels. In parallel, we will continue applying the proposed solution and assessing it on real industrial examples.

Acknowledgments. This work has been partially funded by the Italian Project ITS Italy 2020 under the Technological National Clusters program.

References

1. Van der Aalst, W.M.P.: Verification of workflow nets. In: Azéma, P., Balbo, G. (eds.) Application and Theory of Petri Nets 1997. LNCS, vol. 1248, pp. 407–426. Springer, Heidelberg (1997)
2. Baresi, L., Meroni, G., Plebani, P.: A gsm-based approach for monitoring cross-organization business processes using smart objects (2015). Accepted for publication
3. Cabanillas, C., Baumgrass, A., Mendling, J., Rogetzer, P., Bellovoda, B.: Towards the enhancement of business process monitoring for complex logistics chains. In: Lohmann, N., Song, M., Wohed, P. (eds.) BPM 2013 Workshops. LNBIP, vol. 171, pp. 305–317. Springer, Heidelberg (2014)
4. Eshuis, R., Van Gorp, P.: Synthesizing data-centric models from business process models. Computing **98**, 1–29 (2015)
5. Hull, R., Damaggio, E., Fournier, F., Gupta, M., Heath III, F.T., Hobson, S., Linehan, M., Maradugu, S., Nigam, A., Sukaviriya, P., Vaculin, R.: Introducing the guard-stage-milestone approach for specifying business entity lifecycles. In: Bravetti, M. (ed.) WS-FM 2010. LNCS, vol. 6551, pp. 1–24. Springer, Heidelberg (2011)
6. Jouault, F., Allilaire, F., Bézivin, J., Kurtev, I.: ATL: A model transformation tool. Sci. Comput. Program. **72**(1), 31–39 (2008)
7. Köpke, J., Su, J.: Towards ontology guided translation of activity-centric processes to GSM (2015). Accepted for publication
8. Kumaran, S., Liu, R., Wu, F.Y.: On the duality of information-centric and activity-centric models of business processes. In: Bellahsène, Z., Léonard, M. (eds.) CAiSE 2008. LNCS, vol. 5074, pp. 32–47. Springer, Heidelberg (2008)
9. Künzle, V., Reichert, M.: Philharmonicflows: towards a framework for object-aware process management. J. Softw. Maintenance Evol: Res. Pract. **23**(4), 205–244 (2011)
10. Meroni, G., Baresi, L., Plebani, P.: Translating BPMN to E-GSM: specifications and rules. Technical report, Politecnico di Milano (2016). http://hdl.handle.net/11311/976678
11. Meyer, A., Pufahl, L., Fahland, D., Weske, M.: Modeling and enacting complex data dependencies in business processes. In: Daniel, F., Wang, J., Weber, B. (eds.) BPM 2013. LNCS, vol. 8094, pp. 171–186. Springer, Heidelberg (2013)
12. Meyer, A., Weske, M.: Activity-centric and artifact-centric process model roundtrip. In: Lohmann, N., Song, M., Wohed, P. (eds.) Business Process Management Workshops. Lecture Notes in Business Information Processing, vol. 171, pp. 167–181. Springer, Switzerland (2013)
13. Pesic, M., Schonenberg, H., Van der Aalst, W.M.: Declare: full support for loosely-structured processes. In: Enterprise Distributed Object Computing Conference Proceedings. p. 287. IEEE (2007)
14. Popova, V., Dumas, M.: From Petri Nets to Guard-Stage-Milestone models. In: La Rosa, M., Soffer, P. (eds.) BPM Workshops 2012. LNBIP, vol. 132, pp. 340–351. Springer, Heidelberg (2013)

15. Reichert, M., Weber, B.: Enabling Flexibility in Process-Aware Information Systems: Challenges, Methods, Technologies. Springer Science & Business Media, Heidelberg (2012)
16. Rozinat, A., van der Aalst, W.M.: Conformance checking of processes based on monitoring real behavior. Inf. Syst. $33(1)$, 64–95 (2008)
17. Russell, N., Hofstede, A., Mulyar, N.: Workflow controlflow patterns: A revised view. Technical report BPM-06-22, BPM Center Report, BPMcenter.org (2006)

Modelling the Process of Process Execution: A Process Model-Driven Approach to Customising User Interfaces for Business Process Support Systems

Udo Kannengiesser[✉], Richard Heininger, Tobias Gründer, and Stefan Schedl

Metasonic GmbH, Münchner Straße 29 – Hettenshausen, 85276 Pfaffenhofen, Germany
`{udo.kannengiesser,richard.heininger,tobias.gruender,`
`stefan.schedl}@metasonic.de`

Abstract. This paper presents a process-driven approach for developing the user interfaces (UIs) of business process execution frontends. It allows customising the UIs to the needs of individual users and processes. The approach is based on viewing UI behaviour as a process that can be modelled and executed in the same way as the core process: as a sequence of steps, each of which is associated with a business object that describes the UI content in terms of the information displayed to the user. As both the UI process and the core process are run on the same business process engine, the two processes can interact smoothly using existing backend functionalities. The approach is demonstrated using a manufacturing scenario where shopfloor workers are provided with simple UIs on mobile devices to support the execution of a production process.

Keywords: Model-driven design · Customised user interfaces · Business process support systems · Subject-oriented Business Process Management (S-BPM)

1 Introduction

The design of user interfaces (UIs) is widely regarded as a critical factor for the acceptance of IT systems by users as well as for the acquisition of these systems by potential buyers. Although many system vendors today employ user experience (UX) designers to make UI design more effective, the fundamental problem remains that the space of possible UI designs can be too large to satisfy all users with a single solution. On the other hand, customising UIs is often tedious and costly, making many vendors opt for standardising their user interfaces, and limiting any customisations to broad user categories such as "simple users" and "advanced users".

This approach is also followed by most vendors of business process support systems. They usually have a single UI that has the same look and feel for most of its users. However, as customers become more demanding and the scope of these systems becomes broader to include more domains and applications [18], the ability to customise UIs to wider ranges of user preferences and skills becomes an important competitive advantage. One example includes workflow management applications in the industry 4.0 domain that seamlessly integrate business processes with shopfloor processes.

R. Schmidt et al. (Eds.): BPMDS/EMMSAD 2016, LNBIP 248, pp. 34–48, 2016.
DOI: 10.1007/978-3-319-39429-9_3

Here, workers with sometimes limited IT skills need to be provided with a very simple UI in order to adopt and master a given process support system.

This paper addresses the problem of customising the UIs of workflow systems by focussing on the "process of process execution" – i.e. the sequence of tasks to be performed, conjointly by a user and a user interface, necessary to execute the actual business process (called the "core process"). Specifically, the process of process execution (here called the "UI process") is modelled and executed in the same way as the core process. Each of the tasks in the UI process is associated with data (called business objects) composing the content and appearance of the UI. The UI process is run on the same execution engine as the core process, readily allowing for the dynamic interconnection between the two processes at runtime. The existing method of Subject-oriented Business Process Management (S-BPM) [6] provides a uniform modelling formalism for both the UI process and the core process. The approach can be seen as a process model-driven method for customising UIs to the needs of different users, devices and core processes to be executed.

The paper is structured as follows: Sect. 2 introduces the relevant foundations of S-BPM modelling. Section 3 describes how UIs can be modelled as the processes of (core) process execution. It shows how the S-BPM notation and a tool suite for S-BPM modelling and execution, the Metasonic Suite (www.metasonic.de/en), can be used for defining the UI workflow and UI content, and establishing the connection between UI and core processes. Section 4 demonstrates the use of our approach in a shopfloor scenario developed within an ongoing EU FP7 research project (www.so-pc-pro.eu). Section 5 concludes the paper with a summary of the approach, including a brief discussion of how it is the result of applying a "good theory" in practice, in reference to Lewin's quote that "nothing is more practical than a good theory" [10].

2 The S-BPM Approach to Business Process Modelling

Subject-oriented Business Process Management (S-BPM) is a method and notational approach to modelling and executing processes in a decentralised way. In S-BPM, processes are understood as interactions between process-centric roles (called "subjects"), where every subject encapsulates its own behaviour specification [6]. Subjects coordinate their individual behaviours by exchanging messages. The S-BPM approach is based on extensions of the Calculus of Communicating Systems by Milner [12] and Communicating Sequential Processes by Hoare [7]. Abstract State Machines (ASM) [1] are used as the underlying formalism to allow instant transformation of S-BPM models into executable software. S-BPM mostly targets those applications where a stakeholder-oriented, agile approach to business process management is preferred over more traditional methods based on global control flow. An increasing number of field studies demonstrate the benefits of S-BPM [5].

Based on the strong emphasis on the role concept and on the communication between roles, S-BPM shares some similarities with Role-Activity Diagrams (RAD) [13], UML communication diagrams, and the DEMO methodology [2]. However, there are also a number of significant differences with respect to these approaches. For instance, S-BPM has rigorously defined execution semantics, allows asynchronous communication, and supports end-user involvement in process modelling based on the simplicity of the S-BPM modelling constructs.

S-BPM models include two types of diagrams: a Subject Interaction Diagram (SID) specifying a set of subjects and the messages exchanged between them, and a Subject Behaviour Diagram (SBD) for every subject specifying the details of its behaviour. SBDs describe subject behaviour using state machines, where every state represents an action. There are three types of states in S-BPM: "receive states" for receiving messages from other subjects, "send states" for sending messages to other subjects, and "function states" for performing actions (typically operating on business objects) without involving other subjects. Examples of a SID and a SBD are shown in Figs. 1 and 2, respectively. They represent parts of a production process implemented in a Slovakian manufacturing company (in this paper referred to as "Company A") within the EU FP7 project SO-PC-Pro. Here, the SID in Fig. 1 includes subjects that coordinate (via messages directed to one another) to prepare the actual manufacturing subprocess. The SBD in Fig. 2 represents the internal behaviour specification of the subject "Work Task Preparation". The colours of the different states in the SBD indicate their types: green for receive states, yellow for function states, and red for send states. State transitions are represented as arrows, with labels indicating the outcome of the preceding state. For more details about the S-BPM notation readers may refer to Fleischmann et al. [6].

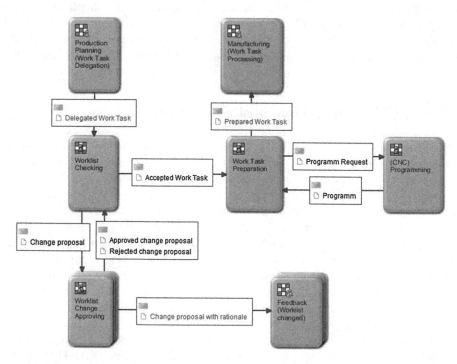

Fig. 1. Subject Interaction Diagram (SID) of a manufacturing preparation process on the shopfloor

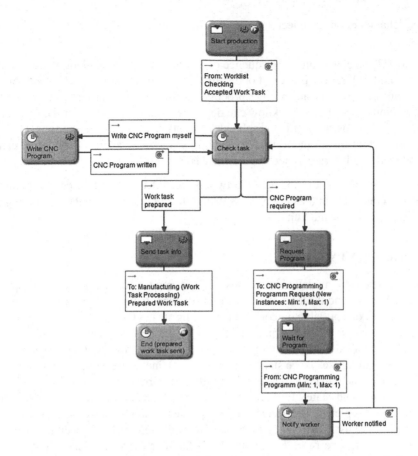

Fig. 2. Subject Behaviour Diagram (SBD) of the subject "Work Task Preparation" (Color figure online)

Subjects may be executed by human or computational agents [4]. When executed by a human worker, parts of the subject may also be automated by associating pieces of code (called "refinements") to individual states in the behaviour. These states are marked in Fig. 2 using a cogwheel icon in their top right corner. Refinements are always triggered from within the process in which they are defined, irrespective of whether that process is controlled by a traditional user interface or by another process.

3   Modelling Process Execution as a Process

The process of process execution is often a hybrid set of manual and automated tasks, the former commonly being guided by a UI. Using S-BPM, this process can be represented as a subject where some parts of its behaviour are executed by a human user and other parts are executed by a computational agent. We call this subject a "UI subject", and the behaviour of that subject accordingly "UI behaviour". The UI behaviour includes two aspects:

1. *UI workflow*: consisting of a sequence of generic steps independent of the specifics of the underlying core process. For example, a UI workflow may include a particular ordering of steps such as starting a process, displaying a list of user tasks, and editing a function state. The UI workflow can be modelled using an SBD for the UI subject.
2. *UI content*: consisting of the graphical elements (e.g. text fields, buttons, etc.) and appearance of the UI in terms of the layout, shapes and colours. UI content can be modelled as a business object within the SBD of the UI subject.

In this Section we describe how the two aspects of UI behaviour can be modelled and finally connected to the core process, using the commercial S-BPM modelling and execution tool Metasonic Suite.

3.1 Modelling UI Workflow

UI workflows can be modelled in various ways, depending on the needs of the specific users and devices, and the kind of core process they are to be connected with. The SBD in Fig. 3 shows one possible outcome of modelling such a UI workflow. The only types of states used in the SBD are function states, as the process of process execution is modelled using a single subject without any communication with other subjects.[1]

The state "Initialize" is the start state of the SBD, including a refinement to load the initial user interface. In case of a technical failure occurring in this state, a transition is followed to the end state "End (init failed)". In contrast, when the initialization is successfully completed, the states "Select process" and "Select task" need to be performed by the user. Depending on the nature of the selected task as either a function state, a send state or a receive state, the UI behaviour proceeds along separate paths ("Edit function state", "Edit send state" and "Edit receive state"), after which the UI automatically executes the state "Compute next step" to loop back to one of the three paths. During task execution, the user may also switch back to the task overview and select a different task (i.e. follow the transition back to "Select task"), and, while doing that, may also select a different process ("Start process"). Upon termination of the core process, the UI behaviour reaches its desired "End" state.

3.2 Modelling UI Content

The UI content is composed of two groups of data queried from the associated core process instance: the business objects handled in that process, and some of the meta-data needed for process instance management (e.g. subject instance ID and currently active states). Both groups of data are loaded at runtime from the core process and are represented in a business object handled by the UI process. The definition of this business

---

[1] This modelling decision is based on the fast response times required for the UI behaviour, which would not be reached with the current implementation of the Metasonic Suite if messaging was included.

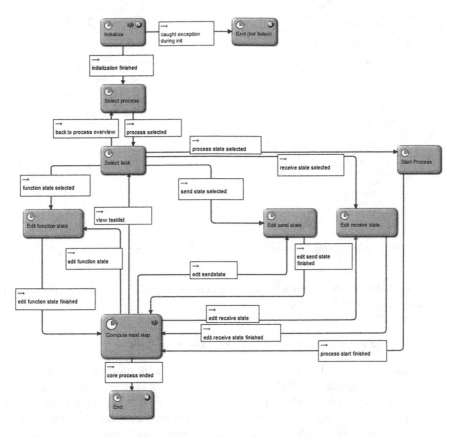

Fig. 3. Example of a SBD defining the UI workflow

object can use the existing data types provided by Metasonic's modelling editor (e.g. String, Number, Enumeration etc.), but also requires a new data type representing a placeholder for the business object of the core process. An example is provided in Fig. 4, showing the definition of a UI business object that contains data elements using standard data types and a placeholder for the core business object.

Out-of-the-box functionalities of the Metasonic Suite also allow defining custom views and layouts of the UI business object. Views [6] specify restrictions on the data elements, including whether an element is visible, hidden, or inactive for a particular state in a SBD. Views in the Metasonic Suite can be associated with client rules to define further attributes such as the colour to be used for displaying a data element. For every view a particular layout can be specified. In addition, the boot-strap framework (http://getbootstrap.com) is used for making the layout and shape of data elements responsive to different screen sizes, supporting conventional computer screens and mobile devices. All bootstrap functionalities such as CSS themes can be used to further customise the UI.

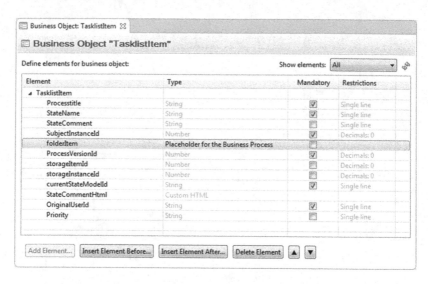

Fig. 4. Example for the definition of a business object in the UI process, containing a placeholder for the business object of the core process

3.3 Connecting the UI Process to the Core Process

UI processes can be modelled either generically for any core process, or for a specific core process. For example, the UI behaviour shown in Fig. 3 is very generic and may be used for all core processes. Other UI behaviours may be defined to tailor the UI to a specific core process and turn some of the "fixed" UI components such as generic menu items and navigation buttons into dynamically generated components that depend on where you currently are in the core process. For example, the "Next" button that is normally used in many workflow UIs to proceed from one user task to the next, may be turned into a set of buttons labelled according to the specific user options defined in the core process.

The concept of generic and specialised UI processes as well as their interplay is shown in Fig. 5. Instances of generic UI processes (designed for all core process models) may be used for providing the UI for instances of any core process. Instances of specific UI processes can be used for providing the UI only for instances of that core process they have been designed for. All process instances are run on the same runtime environment, the Metasonic execution engine.

Running core and UI processes on the same platform allows utilising a number of built-in mechanisms to establish the communication needed between the two processes. As shown in Fig. 6, the Metasonic frontend executing the UI process uses Java Remote Method Invocation (RMI) via API calls and connectors to access core process instance data from the Metasonic backend. That instance data is stored in a DBMS that is queried using Java Database Connectivity (JDBC). The frontend can be accessed by web browsers via HTTP.

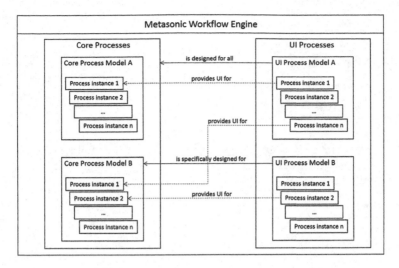

Fig. 5. Conceptual model of the interconnections between core process and UI process

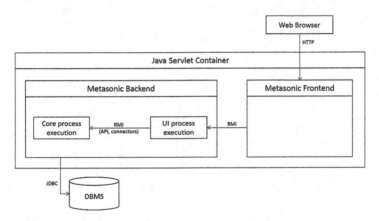

Fig. 6. Software architecture of the interconnected processes

4   Example: Customising UIs for a Shopfloor Process

This Section illustrates our approach based on a case scenario used in the SO-PC-Pro project. Parts of the core process in this scenario – a manufacturing preparation process at Company A – were already introduced in Sect. 2. We will focus on the subject "Work Task Preparation" whose behaviour (shown in Fig. 2) is to be executed by shopfloor workers, guided by a UI running on mobile devices. The standard UI provided in the Metasonic Suite for executing the state "Check task" in this subject is shown in Fig. 7. As it is not modelled as a UI process, it is always the same no matter who executes the process or what core process is executed.

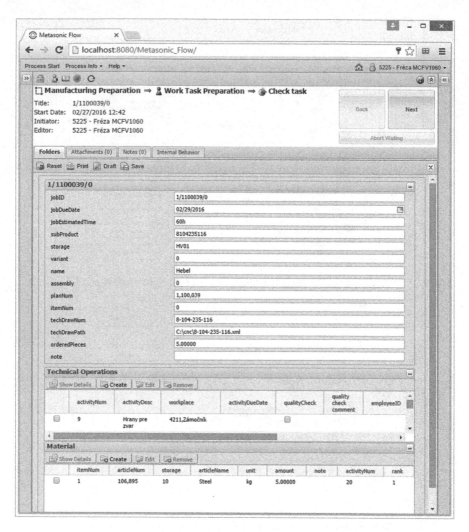

Fig. 7. Standard UI of metasonic flow, showing the data required to execute the function state "Check task"

In the specific case at Company A this UI was deemed too complex to be used by shopfloor workers, as many of them had only limited IT skills. Therefore, a UI process was modelled with the aim to simplify the UI for the workers. The model of this process is almost identical with the one in Fig. 3; it is a generic UI process that may be used for any core process. The appearance of the resulting UIs for the consecutive states "Select process", "Select task" and "Edit function state" in the UI process (cf. Fig. 3) is shown in Figs. 8, 9 and 10, respectively. The UI produced for the state "Edit function state" uses as a header the label of the state "Check task" imported from the core process.

Fig. 8. User interface for the "Select process" state in the generic UI process

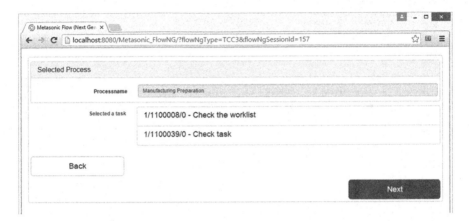

Fig. 9. User interface for the "Select task" state in the generic UI process

These UIs contain only those pieces of information and functionality that a worker would interact with, eliminating all those general menu items, buttons and tabs previously displayed in the standard UI that were regarded unnecessary for the workers. The examples also show that different UIs can be created for different core process steps.

The UI process was later specialised, as shown in Fig. 11, to increase comfort for workers when navigating in the core process from one state to another. For example, the resulting UI shown in Fig. 12 includes three new buttons – "Start production", "Request CNC Code" and "Write CNC Code myself" – to proceed from "Check task" along the corresponding transitions in the core process (cf. Fig. 2).

A final design of the workers' UIs was established after a few iterations in which the UIs were tested by workers in Company A using real process data. Desired UI adaptations were fairly easy to be implemented, simply by changing the UI process model without incurring major programming effort or changes to the core process.

A few technical limitations still exist related to the connection between core process components and the UI process. So far only (core) function states with one going transition can be controlled by the UI process, but not receive states, send states or functions states with multiple outgoing transitions. Development is already underway to address these limitations.

Fig. 10. UI for the "Check task" state (from the core process) used in the generic UI process

5   Related Work

Process modelling in the context of UI design has been proposed for a number of purposes, including usability analysis, requirements specification, and model-driven development [11, 21].

An early approach to modelling user tasks for UIs is the one by Parnas [14] based on state transition diagrams. These diagrams are fundamentally very similar to our simplified SBDs that have only function states but no send or receive states. However, the sole purpose of the models is to represent UI requirements that are then interpreted by human UI designers. The execution of state transition diagrams for automatically generating and controlling the behaviour of UI software is not within the scope of his work.

Dubé et al. [3] proposed hierarchically-linked statecharts (HIS) consisting of UML class diagrams and state machines for specifying the structure and behaviour of UIs, respectively. In particular, the state machine formalism has been chosen based on its

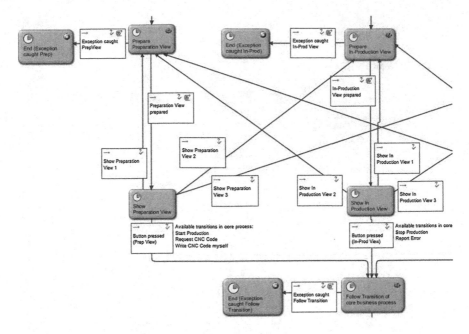

Fig. 11. Excerpt from the specialised UI process modelled for Company A

suitability for connecting UI responses with user events such as mouse clicks and key presses. Every visual element in the UI is defined with its own state machine. There are a number of similar approaches to model-driven design of UIs, using behavioural diagrams that are directly or indirectly drawn from UML [15, 16].

Work by Trætteberg and Krogstie [20] aims to realise individualised UIs by means of a model-based design approach that uses the core process model as a starting point. A task model representing the user's tasks is first extracted from the core process model, and then transformed into a dialog model representing the interaction logic of the UI. The core process and the task model are both modelled using BPMN, whereas the dialog model is modelled using the Diamodl notation [19] that is partially based on UML state charts. This approach requires significant manual work for the individual transformations. Transforming core process models into task models involves splitting lanes into pools to make explicit the data flow between them and annotating the task model with pre- and post-conditions. Transforming task models into dialog models then requires additional manual translation effort due to the separate notations used.

Kolb et al. [8] have proposed mappings between task-oriented process models, represented in BPMN, and the logic and contents of UIs, with the aim to generate UI components in a model-driven way. Based on that work, Schobel et al. [17] have implemented a system for designing the UIs of electronic questionnaires using process modelling, and for executing the UI logic on a workflow engine. Other work [9] proposes state-flow representations of data objects as micro-level processes that can be used for generating UIs. These approaches do not include modelling the core process separately from the UI process: There is only one process model that seems to represent both

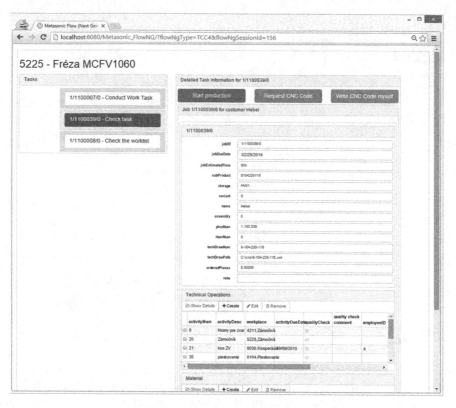

Fig. 12. User interface for "Check task", generated by a specialised UI process that has been designed only for the manufacturing preparation (core) process in Company A

processes at once. While such a tight coupling has advantages regarding UI maintenance (i.e. if the core process is modified, the UI process is automatically updated accordingly), it prevents customising UIs independently of the core process (i.e. the ability to model multiple UI processes for the same core process).

6   Conclusion

The UI design of business process support systems is a critical issue in the execution of human-centric processes. For a long time, standardised UIs have been favoured based on the high cost of UI customisation. However, as customer demands become more heterogeneous, partly driven by the increasing scope of business process management to cover new domains such as factory and supply chain processes, there is a clear need for more customised UI solutions. This paper has shown how process modelling can be leveraged to develop UIs that can be customised fairly easily using existing functionalities of a BPM suite and integrated frameworks such as bootstrap. The self-referential approach of using process modelling for specifying process execution enables using not only the same technical platform but also the same type of knowledge: the knowledge of process modelling. As a result, developers and integrators of process support systems can address many UI customisation needs without having to employ dedicated UI design specialists.

Our approach is an example for the practical application of a "good theory", using Lewin's [10] terms. Here the theory is the S-BPM methodology; it can be seen as "good" because it includes two concepts whose practical application has been demonstrated in this paper: the genericity and the formality of S-BPM process models. The concept of genericity results from the highly abstract modelling constructs in S-BPM: Processes are modelled independently of their embedding in particular organisations and IT infrastructures [4], using only five abstract symbols. This allows modelling any kind of process, including human-centric (business) processes, computational processes and manufacturing processes. In this paper we have shown how the process of process execution, which is often a mixture of human-centric and computational activities, can be modelled with S-BPM. The other concept, formality, is established by the well-defined execution semantics of S-BPM. It allows model-driven transformation of graphical process models into executable software. We have shown that this concept enables turning the modelled process of process execution directly, i.e. without manual intervention, into a running software – the UI of a process support system. We expect that this would be very difficult to be achieved using traditional BPM methodologies such as BPMN, due to their insufficient formal foundations.

Finally, the application of S-BPM for customising UI design can be seen as an example of the "eat your own dog food" principle: As a vendor of an S-BPM suite, we use our own methodology (S-BPM) and our own tool (Metasonic Suite) as a basis for generating customised UIs for our process execution frontend. The technical extensions needed for the realisation of this approach have now matured to product-level quality and will be available on the market with the next feature release (version 5.3) of the Metasonic Suite.

Acknowledgements. The research leading to these results has received funding from the EU Seventh Framework Programme FP7-2013-NMP-ICT-FOF(RTD) under grant agreement no 609190 (www.so-pc-pro.eu).

References

1. Börger, E., Stärk, R.: Abstract State Machines: A Method for High-Level System Design and Analysis. Springer, Berlin (2003)
2. Dietz, J.L.G.: DEMO: towards a discipline of organisation engineering. Eur. J. Oper. Res. **128**(2), 351–363 (2001)
3. Dubé, D., Beard, J., Vangheluwe, H.: Rapid development of scoped user interfaces. In: Jacko, J.A. (ed.) HCI International 2009, Part I. LNCS, vol. 5610, pp. 816–825. Springer, Berlin (2009)
4. Fleischmann, A., Kannengiesser, U., Schmidt, W., Stary, C.: Subject-oriented modeling and execution of multi-agent business processes. In: 2013 IEEE/WIC/ACM International Conferences on Web Intelligence (WI) and Intelligent Agent Technology (IAT), pp. 138–145, Atlanta, GA (2013)
5. Fleischmann, A., Schmidt, W., Stary, C.: S-BPM in the Wild: Practical Value Creation. Springer, Berlin (2015)

6. Fleischmann, A., Schmidt, W., Stary, C., Obermeier, S., Börger, E.: Subject-Oriented Business Process Management. Springer, Berlin (2012)
7. Hoare, C.A.R.: Communicating sequential processes. Commun. ACM **21**(8), 666–677 (1978)
8. Kolb, J., Hübner, P., Reichert, M.: Automatically generating and updating user interface components in process-aware information systems. In: Meersman, R., et al. (eds.) OTM 2012, Part I. LNCS, vol. 7565, pp. 444–454. Springer, Berlin (2012)
9. Künzle, V., Reichert, M.: A modeling paradigm for integrating processes and data at the micro level. In: Halpin, T., Nurcan, S., Krogstie, J., Soffer, P., Proper, E., Schmidt, R., Bider, I. (eds.) BPMDS 2011 and EMMSAD 2011. LNBIP, vol. 81, pp. 201–215. Springer, Berlin (2011)
10. Lewin, K.: Field Theory in Social Science: Selected Theoretical Papers. Harper & Brothers, New York (1951)
11. Limbourg, Q., Vanderdonckt, J.: Comparing task models for user interface design. In: The Handbook of Task Analysis for Human-Computer Interaction, pp. 135–154. Lawrence Erlbaum Associates, London (2004)
12. Milner, R.: Communicating and Mobile Systems: The Pi-Calculus. Cambridge University Press, Cambridge (1999)
13. Ould, M.A.: Business Processes: Modelling and Analysis for Re-Engineering and Improvement. Wiley, Chichester (1995)
14. Parnas, D.L.: On the use of transition diagrams in the design of a user interface for an interactive computer system. In: ACM/CSC-ER, pp. 379–385. ACM Press, New York (1969)
15. Paternó, F.: Towards a UML for interactive systems. In: Nigay, L., Little, M. (eds.) EHCI 2001. LNCS, vol. 2254, pp. 7–18. Springer, Berlin (2001)
16. Pinheiro da Silva, P., Paton, N.W.: User interface modelling with UML. Information Modelling and Knowledge Bases XII, pp. 203–217. IOS Press, Amsterdam (2001)
17. Schobel, J., Schickler, M., Pryss, R., Reichert, M.: Process-driven data collection with smart mobile devices. In: Monfort, V., Krempels, K.-H. (eds.) WEBIST 2014. LNBIP, vol. 226, pp. 347–362. Springer, Switzerland (2014)
18. Sinur, J., Odell, J., Fingar, P.: Business Process Management: The Next Wave. Meghan-Kiffer Press, Tampa (2013)
19. Trætteberg, H.: Dialog modelling with interactors and UML statecharts - a hybrid approach. In: Jorge, J.A., Jardim Nunes, N., Falcao e Cunha, J. (eds.) DSV-IS 2003. LNCS, vol. 2844, pp. 346–361. Springer, Heidelberg (2003)
20. Trætteberg, H., Krogstie, J.: Enhancing the usability of BPM-solutions by combining process and user-interface modelling. In: Stirna, J., Persson, A. (eds.) PoEM 2008. LNBIP, vol. 15, pp. 86–97. Springer, Heidelberg (2008)
21. van Welie, M., van der Veer, G.C., Eliëns, A.: An ontology for task world models. In: Markopoulos, P., Johnson, P. (eds.) Design, Specification and Verification of Interactive Systems 1998. Eurographics, pp. 57–70. Springer, Vienna (1998)

Improving Usability of Process Models

Integrating Textual and Model-Based Process Descriptions for Comprehensive Process Search

Henrik Leopold[1]([✉]), Han van der Aa[1], Fabian Pittke[2], Manuel Raffel[2],
Jan Mendling[2], and Hajo A. Reijers[1]

[1] Vrije Universiteit Amsterdam, De Boelelaan 1081,
1081 HV Amsterdam, The Netherlands
{h.leopold,j.h.vander.aa,h.a.reijers}@vu.nl
[2] WU Vienna, Welthandelsplatz 1, 1020 Vienna, Austria
{fabian.pittke,manuel.raffel,jan.mendling}@wu.ac.at

Abstract. Documenting business processes using process models is common practice in many organizations. However, not all process information is best captured in process models. Hence, many organizations complement these models with textual descriptions that specify additional details. The problem with this supplementary use of textual descriptions is that existing techniques for automatically searching process repositories are limited to process models. They are not capable of taking the information from textual descriptions into account and, therefore, provide incomplete search results. In this paper, we address this problem and propose a technique that is capable of searching textual as well as model-based process descriptions. It automatically extracts process information from both descriptions types and stores it in a unified data format. An evaluation with a large Austrian bank demonstrates that the additional consideration of textual descriptions allows us to identify more relevant processes from a repository.

1 Introduction

Business process models have proven to be an effective means for the visualization and improvement of complex organizational operations [7]. However, not all process-related information is available in the form of process models. On the one hand, because the creation of process models is a time-consuming endeavor that requires considerable resources [13]. On the other hand, because not all process information is best captured as a process model [1]. In particular, *work instructions* that describe tasks at a high level of detail are often documented in the form of textual descriptions, as this format is more suitable for specifying a high number of details [4]. As a result, process repositories in practice do not only consist of process models, but often also contain textual process descriptions. These are linked to individual activities of the process models in order to specify the detailed action items behind them.

The problem of this supplementary use of textual descriptions in process repositories is that automatic analysis techniques designed for process models, such as

© Springer International Publishing Switzerland 2016
R. Schmidt et al. (Eds.): BPMDS/EMMSAD 2016, LNBIP 248, pp. 51–65, 2016.
DOI: 10.1007/978-3-319-39429-9_4

weakness identification [5], service identification [20], or compliance queries [3], might provide incomplete results. Suppose a company aims to increase the share of digital communication, then it can query its process repository to find all processes that still include paper-based communication. The query results, however, will be limited to the process models that indicate the use of paper-based communication already in their activity text labels. Process models that describe the process at a higher level of abstraction, but link to textual descriptions revealing that this process is indeed associated with paper-based communication, will be ignored. Currently, there is no technique available that provides the possibility to search textual and model-based process descriptions in an integrated fashion. One explanation for the absence of such a technique might be the challenges that are associated with it. Among others, it requires the definition of an integrated data format that is able to represent both textual and model-based process descriptions.

Against this background, we use this paper to propose a technique that can search both text and model-based process descriptions. It combines natural language analysis techniques in a novel way and transforms textual as well as model-based process descriptions into a unified data format. By integrating technology from the semantic web domain, we facilitate the possibility of performing comprehensive search operations on this data format.

The remainder of this paper is organized as follows. Section 2 introduces the problem of searching textual and model-based process descriptions and discusses related work. Section 3 then introduces our proposed technique on a conceptual level. Section 4 presents the results of an evaluation with a large Austrian bank. Finally, Sect. 5 concludes the paper and provides an outlook on future research.

2 Background

This section introduces the background of our research. First, we illustrate the problem of searching textual and model-based process descriptions. Then, we reflect on related techniques that are currently available.

2.1 Motivating Example

In order to illustrate the importance of textual descriptions in the context of a process search, let us take a look at the implications of only taking process models into account. To this end, consider the example shown in Fig. 1. It shows a simple process model created using the Business Process Modeling and Notation (BPMN) and a small complementary text from a bank. We can see that the business process is triggered by the request to open a new bank account. Subsequently, the credit history of the customer is evaluated. The outcome of this evaluation can be either positive or negative. In case of a negative credit evaluation, the customer is rejected. If the credit history evaluated as positive, a new bank account is opened. Finally, the request is closed. In addition to the BPMN process model, there is complementary text. It further specifies the details of the activity *"Opening of new bank account for customer"*. Among others, it describes that the opening of a bank account is associated with a mail-based information exchange with the customer.

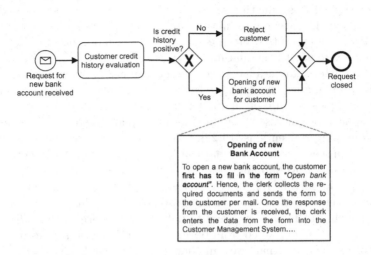

Fig. 1. Exemplary process model with complementary natural language description

Assume this process model is part of the process repository of an organization. If this organization was interested in all business processes that involve an interaction with a customer, automated search techniques would have no difficulties to identify the depicted process. That is, because the customer is explicitly mentioned in the activity labels, e.g. in *"Customer credit history evaluation"*. However, suppose the organization aims at improving its operations by replacing all mail-based correspondence with an electronic alternative. In this case, an automated search on the process model would not identify any potential for improvement. That is because the activities of the process model do not contain any words that might be associated with the activities of mailing or sending. Only the description attached to the activity *"Opening of new bank account"* explicitly refers to sending a form per mail.

This example illustrates the advantage of performing search operations that cover textual as well model-based process descriptions. As a respective technique is currently missing, it is our goal to define such a technique in this paper.

2.2   Related Work

The work from this paper relates to two major streams of research: process model search techniques and process model analysis techniques that employ Natural Language Processing (NLP) technology.

Techniques for *process model search* can be divided into two main groups. The first group consists of techniques focusing on *structure*. They compare query and process model with respect to behavioral properties, for instance, whether two activities occur in a particular order. Among others, such structural querying techniques have been defined based on temporal logical [2], the weak ordering formalism [16], and on indexing [14,28]. The limitation of these structural techniques is that they typically assume that semantically identical activities have identical or similar labels. The second group of search techniques focuses on the *content from text labels* and tries to overcome this problem. Among others, they employ NLP techniques to identify similar models based on their activity labels. Notable examples for such techniques have been defined in [3,27], where the authors use dictionaries and language modeling to retrieve semantically similar models.

NLP techniques are also often applied in the context of *process model analysis*. For instance, they are used to assure linguistic quality aspects of process models such as naming conventions [19,22]. Other application scenarios include the generation of process models from natural language texts and vice versa [11,17] and the detection of overlapping behavior of two process models [9,21].

Despite the important role of NLP technology for process model search and analysis, a conceptual solution for an integrated search technique is still missing. To develop such a technique, we need to define a data format that allows us to store the information extracted from both process description types in a unified way. Based on such a format, we can then perform search operations covering model-based as well as textual process descriptions.

3   Conceptual Approach

In this section, we introduce our approach for comprehensive process search by integrating textual and model-based process descriptions. We give an overview of the architecture of our approach. We then describe the unified format we use to integrate textual and model-based content. Afterwards, we show how to parse and transform textual and model-based descriptions into this unified format. Finally, we illustrate how the use of the unified data format supports comprehensive process search.

3.1   Overview

The main idea of our architecture is that the differing input sources of textual and model-based process descriptions must be stored in a unified way. Hence, two parsing components first extract the relevant information from the two input sources and then store it in the unified data store. Once the data store has been populated with all available process descriptions, it can be used to search processes. To this end, a user interface provides the possibility to specify queries in a user-friendly manner. Figure 2 illustrates our architecture graphically.

In the subsequent sections, we describe this architecture in detail. Because of the predominant role of the unified data store, we begin with the specification of the unified format.

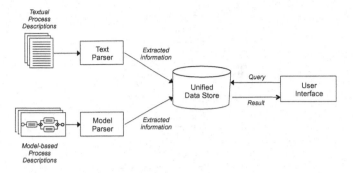

Fig. 2. Exemplary process model with complementary natural language description

3.2 A Unified Format for Integrating Textual and Model-Based Process Descriptions

To define a unified format for textual and model-based process descriptions, it is important to understand how each of these descriptions conveys the semantics of the business process it describes. In essence, textual process descriptions describe business processes by using sequences of proper natural language sentences structured into sections, subsections, and paragraphs. Process models, by contrast, also consist of graphical representations of modeling constructs such as activities, events, and gateways. An important share of the semantics of process models is, however, defined by the natural language labels that are attached to the activities [18]. These labels, however, do not necessarily represent proper sentences. As examples, consider the activity labels *"Opening of bank account for customer"* or *"Customer credit history evaluation"* from the BPMN model in Fig. 1. Therefore, a unified format must provide the possibility to store the essential information distilled from activity labels as well as proper natural language sentences.

According to [25], every activity label can be characterized by three components: an action, a business object on which the action is performed, and an optional additional information fragment that is providing further details. As an example, consider the activity label *"Opening of new bank account for customer"*. This activity consists of the action *"to open"*, the business object *"new bank account"*, and the additional information fragment *"for customer"*. In a proper natural language sentence, we can identify respective counterparts. Consider the sentence *"The clerk opens a new bank account for the customer"*. A grammatical analysis would reveal that this sentence contains the predicate *"opens"*, the object *"bank account"*, the subject *"clerk"*, and the adverbial *"for the customer"*. This example illustrates that the predicate corresponds to the action, the object to the business object, and the adverbial to the additional information fragment. A subject refers to the role executing the activity, which is typically specified outside the activity label.

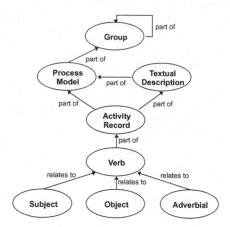

Fig. 3. Overview of unified data model

Based on these insights, we specify a format that stores the language content from sentences and activity labels in a unified way. Figure 3 illustrates this format. The core of this format is a so-called activity record, which might be part of a process model or a textual description. Each activity record may consist of one or more verbs, depending on the grammatical structure of the activity label or the sentence it refers to. Each verb relates to a subject, an object, and an adverbial. Note that each of these entities might be empty if the corresponding activity label or sentence does not contain this information. As indicated by the relations between the activity record, the process model, and the textual description, each process model and each textual description may contain several activity records. Moreover, a process model may consist of several textual descriptions. As process models in industry are typically organized in hierarchical process architectures [24], our format supports the organization of process models into groups and sub groups.

Practically, we implement this unified data format by building on the Resource Description Framework (RDF), an XML-based specification developed by the World Wide Web Consortium (W3C)[1]. RDF describes data in the form of triples that consist of two entities and a relation between them. As an example, consider the relation *"relates to"* between a object and an verb in our unified data format. A possible RDF triple for this relation would be (*"customer"*, *"relates to"*, *"reject"*). Similarly, all other relations from the unified data format can be represented as RDF triples. The advantage of storing data in the RDF format is that it can be easily and effectively accessed and queried [6]. Hence, it greatly contributes to our goal of providing a technique for integrated search.

In the subsequent sections, we describe how textual process descriptions as well as process models can be automatically transformed into this unified data format.

---

[1] http://www.w3.org/RDF.

3.3 Parsing Textual Process Descriptions

The text parser component takes a textual description as input and automatically extracts the process information required for the unified data format. This procedure consist of three subsequent steps:

1. *Linguistic analysis of sentences:* The first step concerns the identification of the grammatical entities such as subject, object, predicate, and adverbial for each sentence. What is more, we determine the relations between these entities, e.g. which verb relates to which object. To illustrate the required steps and the associated challenges, consider the sentence *"Hence, the clerk collects the required documents and sends the form to the customer per mail."* from Fig. 1. It contains one subject (*"clerk"*), two predicates (*"collects"* and *"sends"*), two objects (*"documents"* and *"form"*), and two adverbials (*"to the customer"* and *"per mail"*). Furthermore, it is important to note the relations between these entities. The predicate *"collects"* relates to the object *"documents"*, whereas the predicate *"sends"* relates to the object *"form"*. To automatically determine these grammatical entities and their relations, we again make use of the Stanford Parser. Besides the recognition of sentence borders, the Stanford Parser is also capable of producing word dependencies [8]. As an example, consider the following two object-related dependencies that the Stanford Parser generates for the considered example sentence:

   ```
   dobj(collects-5, documents-8)
   dobj(sends-10, form-12)
   ```

   These so-called direct-object dependencies (*dobj*) specify which words the Stanford Parser considers to be objects and which predicates relate to them. Thus, the first dependency tells us that *"documents"* (position 8 in the sentence) is an object that relates to the predicate *"collects"* (position 5 in the sentence). Analogously, *"form"* is an object that relates to the predicate *"sends"*. To make use of these generated dependencies, we developed an algorithm that automatically analyzes the Stanford Parser output and extracts the grammatical entities as well as their relations. Our component builds on the knowledge about existing dependencies and the consistent structure of these dependencies (name of the dependency followed by brackets that include two entities and their position). As a result, we are able to automatically obtain a set of grammatical entities and their relations from any given natural language sentence.

2. *Normalization of sentence components:* The words in sentences often do not occur in their base forms, i.e., verbs are not only used as infinitives; nouns are not always provided as singular nouns. This becomes a problem when entities are compared in the context of a search operation. For instance, *"send"*, *"sends"*, and *"sent"* all refer to the same base verb. However, an automated string comparison would indicate that these words differ from each other. To deal with such cases, we use the lexical database WordNet [26] to convert all words into their base form, i.e., predicates into infinitive verbs and subject as

well as objects into singular nouns. As a result, the predicates *"sends"* and *"sent"* are both transformed into *"send"*.

3. *Transformation of sentence components into RDF:* Once the entities have successfully been extracted and transformed into their base forms, the information is stored in the RDF format. To demonstrate this step, again consider the sentence *"Hence, the clerk collects the required documents and sends the form to the customer per mail."* from Fig. 1. For each predicate of the sentence, we create a *verb - activity record* RDF triple in order to capture the relation between the predicates and the sentence. The sentence is then represented by an activity record. Suppose this activity record has the identification number 2, then the respective RDF triples look as follows:

```
(collect, part of, ActivityRecord2)
(send, part of, ActivityRecord2)
```

In addition, we need to link the subjects, objects, and adverbials to the respective verbs:

```
(clerk, relates to, collect)
(clerk, relates to, send)
(document, relates to, collect)
(form, relates to, send)
(to customer, relates to, send)
(per mail, relates to, send)
```

3.4 Parsing Model-Based Process Descriptions

This component expects a set of process models as input and automatically extracts the information required for the unified data format. Similar to the parsing of textual process descriptions, it consists of three subsequent steps:

1. *Linguistic analysis of activity labels:* The linguistic analysis aims at properly deriving the activity components from the labels of the input model set. As discussed earlier in this section, the main challenge is to automatically detect the varying grammatical structures, even if the activity label does not contain a proper verb. As an example, consider the activity *"Customer credit history evaluation"* from Fig. 1. For this label it is necessary to automatically recognize that *"evaluation"* represents the action and *"customer credit history"* the business object. To properly derive these components from activity labels, we employ the label analysis technique introduced by [22]. It takes an activity label as input and respectively returns the comprised action(s), business object(s), and additional information fragment(s).

2. *Normalization of activity label components:* Similar to proper natural language sentences, activities often contain inflected words, i.e., verbs occurring in the third person form or nouns used in the plural form. What is more,

actions may even represent nouns (e.g., *"evaluation"* in *"Customer credit history evaluation"*). As pointed out for sentences, this has notable implications if two components are compared in the context of a search operation. Hence, we apply the lexical database WordNet [26] also on activity labels to convert all actions into infinitive verbs and all nouns into singular nouns. As a result, the action of the activity label *"Customer credit history evaluation"* is accordingly transformed into *"evaluate"*.

3. *Transformation of activity label components into RDF:* The storage of the extracted and normalized components as RDF works analogously to the storage of the sentence entities. We demonstrate this step using the activity *"Opening of new bank account for customer"*. As a result of applying the previous steps, we identified the action *"open"*, the object *"bank account"*, and the addition *"for customer"*. Suppose the resulting activity record has the identification number 3, then the RDF triples look as follows:

```
(open, part of, ActivityRecord3)
(bank account, relates to, open)
(for customer, relates to, open)
```

The example triples illustrate that the action is linked to the activity by using the *verb - activity record* triple. The business object and the addition are respectively associated with the verb by using the *object-verb* and the *adverbial-verb* relation.

In the next section, we show how we query the extracted data from the unified data format.

3.5 Querying the Unified Data Store

In order to query the extracted RDF triples, we use SPARQL (Simple Protocol and RDF Query Language). In essence, SPARQL is similar to SQL (Structured Query Language), the most popular language to query data from relational databases), but is specifically designed to query RDF data. As an example, consider the SPARQL query in Fig. 4, which retrieves all process models and textual process descriptions that contain an activity record relating to the verb *"send"* and the object *"form"*.

The example from Fig. 4 shows that a SPARQL query has the basic structure of an SQL query, i.e., it follows the *select - from - where* pattern. Before the actual query, however, it is required to define where the data model definition can be found (line 1). As SPARQL is designed for the semantic web, this is done via a Unified Resource Identifier (URI). In this example, for illustration purposes, we use the URI http://www.processsearch.com/Property/. After the definition of this prefix, the actual query starts. Line 2 specifies that we are interested in all process names of process models that fulfill the requirements stated as RDF triples in the block below. Using the variable *?verb*, we define that there must be a *Verb* according to our data model that carries the label *"send"* (lines 5 and 6). Moreover, we use the variable *?object* to define that *?verb* must be related to an *Object* that carries the label *"form"* (lines 8–10). Finally, we define that we are only interested in process models that contain an activity record that relates to an entity *Verb* as specified in *?verb*.

```
1    PREFIX ps: <http://www.processsearch.com/Property/>
2    SELECT ?processName
3    WHERE
4    {
5            ?verb ps:Type "Verb" .
6            ?verb ps:Label "send" .
7
8            ?object ps:RelatesTo ?verb .
9            ?object ps:Type "Object" .
10           ?object ps:Label "form" .
11
12           ?verb ps:PartOf ?activityRecord .
13           ?activityRecord ps:PartOf ?processModel .
14           ?processModel ps:Label ?processName .
15   }
```

Fig. 4. Exemplary SPARQL query to retrieve data from the RDF database

This exemplary query illustrates that an RDF-based unified data store can be easily queried for information we are interested in. To provide the users of our technique with an intuitive feature to search, we implemented a graphical interface in which users can specify the verbs, objects, subjects, and adverbials of the process descriptions they would like to retrieve. The input from the graphical user interface is then automatically inserted into a SPARQL query as provided above. As a result, the user can perform any search based on these four components and does not have to deal with any technical details.

4 Evaluation

In this section, we evaluate our technique with the process repository of an Austrian bank. Our goal is to demonstrate that the additional consideration of textual descriptions yields more comprehensive search results than the sole consideration of process models. We first discuss the setup of our evaluation experiment. Then, we introduce the process model collection we use. Afterwards, we explain the prototypical implementation of our technique. Finally, we present and discuss the results.

4.1 Setup

For the evaluation of our approach, we collaborated with a large Austrian bank. The Business Process Management department of this bank was struggling with two search scenarios that are of particular relevance to the work presented in this paper.

1. *Search for media disruptions:* Media disruptions occur when the information-carrying medium is changed, for example, when a clerk enters the data from a physical letter into an information system. Because media disruptions are often associated with errors, our evaluation partner had a considerable interest in identifying media disruptions in their process landscape. To design an exemplary query, we built on the insights from [5]. In a study on weakness patterns, they found that media disruptions are mainly indicated by the actions *"print"* and *"scan"* as well as the activity *"Enter data"*.
2. *Search for manual activities:* Manual activities are inevitable in most business processes. However, as automation is often associated with saving costs, the identification of automation candidates represents a key task in business process improvement [23]. Thus, our evaluation partner was also interested in identifying which automation potential their process repository exhibits. According to [5], manual activities are typically indicated by the actions *"document"*, *"record"*, and *"calculate"* as well by combinations of the actions *"verify"* and *"archive"* with the business objects *"document"* and *"information"*.

We use queries based on the weakness patterns discussed above to demonstrate the capabilities of our approach and to show the importance of taking text-based process descriptions into consideration.

4.2 Data

The process repository of our evaluation partner consists of 1,667 Event-driven Process Chains (EPCs). The process models cover various aspects of the banking business including the opening of accounts, the management and selling of financial products, as well as customer relationship management. On average, the process models contain 6.5 activities per model. The smallest model contains 1 activity, whereas the largest contains 181 activities. In addition to the process models, the repository contains 119 textual process descriptions in the PDF format. The textual descriptions complement the process models and mainly concern the area of credit management. Due to existing overlaps between the process models in the repository, a single textual description can be referred to by multiple process models. The size of these complementary process descriptions ranges from 119 to 60,558 words. Most of the description are rather long, resulting in an average size of 13,130 words. The language of both the process model elements and textual process descriptions is German.

4.3 Implementation

We implemented the approach defined in Sect. 3 as a Java prototype. To be able to deal with German process models and process descriptions, we integrated the German package of the Stanford Parser [15], the German component of the label analysis technique from [22], and a German implementation of WordNet called GermaNet [12]. In addition to these techniques, we use the Apache PDFBox to process PDF files and the import functionality from [10] to process different process model formats. Finally, to store the extracted RDF triples, we use the Apache Jena component TDB[2], which is a database optimized for RDF storage and querying.

4.4   Results

The results for both search scenarios are illustrated in Fig. 5. The figure shows the aggregated total for the search scenarios and the disaggregated number of retrieved processes for each weakness pattern (e.g., *Verb* = *"print"* or *Verb* = *"document"*). The light grey bars indicate the number of processes we retrieved from searching the model-based process descriptions. The dark grey bars indicate the number of processes we retrieved from searching the model-based as well as the textual process descriptions. The results illustrate that our proposition holds: we retrieve additional relevant process models if we also take the textual process descriptions accompanying the models into account. Interestingly, that does not only hold for the total of both search scenarios, but also for each of the weakness patterns as, for instance, for the verb *"print"* in the media disruption scenario or for the verb *"document"* in the manual activity search scenario. Altogether, the number of processes that are retrieved from the process repository increases from 83 to 151 for the media disruption scenario (an increase of 81.9 %) and from 213 to 359 (an increase of 68.5 %) for the manual activity scenario[3].

A detailed analysis of the results revealed that there is no overlap between the processes retrieved from the model-based and the textual descriptions in neither of the search scenarios. This shows that the details of some processes are fully described by process models, while the details of others are only captured in the accompanying textual descriptions. This again highlights the importance of considering both types of process descriptions.

The practical relevance of our technique is further demonstrated by the way it was perceived by our evaluation partner. The bank considered our technique to be highly useful for the analyses they are conducting and decided to integrate it with their ARIS platform. They set up a script that updates the database behind our technique on a daily basis. In this way, search operations can be conducted in an efficient way.

While these results are promising, they also have to be discussed in the light of some limitations. First, it is important to note that the investigated process repository is not statistically representative. That is, process repositories from other companies may consist of more or also fewer textual descriptions than the one we investigated. However, our technique does not rely on any specifics we encountered in the repository of our evaluation partner. Thus, we

---

[2] https://jena.apache.org/.

[3] Note that because a single textual description can be referred to by several process models, the identification of one relevant textual document may yield multiple relevant process models. This explains why the increase in the number of retrieved processes might be even higher than the total number of textual process descriptions.

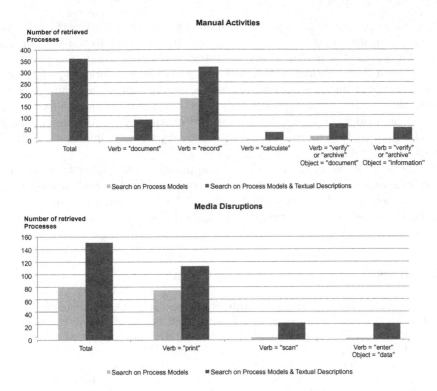

Fig. 5. Results for media disruptions search

are confident that our technique will perform comparably on other collections. Second, it should be noted that our technique cannot guarantee that all relevant information is identified. One reason is that the user has to define proper key words. Another reason is that our technique can only find information that is explicitly documented in one of the addressed description types.

5   Conclusion

In this paper, we introduced a comprehensive search technique that allows the user to identify information in textual as well as in model-based process descriptions. The technique combines natural language analysis tools in a novel way and builds on the transformation of textual and model-based process descriptions into a unified data format. We implemented the technique as a Java prototype that stores the extracted data in an RDF database and provides the user with a graphical interface to specify queries. An evaluation with a large bank showed that our solution can be successfully applied in industry and that the additional consideration of textual process descriptions indeed increases the number of identified processes.

From a research perspective, the proposed technique provides the foundations for integrating textual and model-based information. To the best of our knowledge, we are the first to define an integrated data format that allows to combine the process information from these two process description types. Hence, our technique can improve existing process search techniques and may help to increase their scope. From a practical perspective, our technique helps organization to perform more comprehensive search operations. As demonstrated in the evaluation, textual sources may contain equally relevant information about processes as model-based descriptions.

In future work, we plan to extend our approach with respect to structural process properties. To this end, we aim at integrating behavioral aspects from process models into the data format. In addition, we plan to define a technique that is capable of extracting such behavioral aspects from textual process descriptions.

References

1. van der Aa, H., Leopold, H., Mannhardt, F., Reijers, H.A.: On the fragmentation of process information: challenges, solutions, and outlook. In: Gaaloul, K., Schmidt, R., Nurcan, S., Guerreiro, S., Ma, Q. (eds.) BPMDS 2015 and EMMSAD 2015. LNBIP, vol. 214, pp. 3–18. Springer, Heidelberg (2015)
2. Awad, A., Decker, G., Weske, M.: Efficient compliance checking using BPMN-Q and temporal logic. In: Dumas, M., Reichert, M., Shan, M.-C. (eds.) BPM 2008. LNCS, vol. 5240, pp. 326–341. Springer, Heidelberg (2008)
3. Awad, A., Polyvyanyy, A., Weske, M.: Semantic querying of business process models. In: 12th International IEEE Enterprise Distributed Object Computing Conference, EDOC 2008, pp. 85–94. IEEE (2008)
4. Baier, T., Mendling, J.: Bridging abstraction layers in process mining by automated matching of events and activities. In: Daniel, F., Wang, J., Weber, B. (eds.) BPM 2013. LNCS, vol. 8094, pp. 17–32. Springer, Heidelberg (2013)
5. Becker, J., Bergener, P., Räckers, M., Weiß, B., Winkelmann, A.: Pattern-based semi-automatic analysis of weaknesses in semantic business process models in the banking sector (2010)
6. Candan, K.S., Liu, H., Suvarna, R.: Resource description framework: metadata and its applications. SIGKDD Explor. Newsl. 3(1), 6–19 (2001)
7. Davies, I., Green, P., Rosemann, M., Indulska, M., Gallo, S.: How do practitioners use conceptual modeling in practice? Data Knowl. Eng. 58(3), 358–380 (2006)
8. De Marneffe, M.C., Manning, C.D.: The stanford typed dependencies representation. In: Coling 2008: Proceedings of the Workshop on Cross-Framework and Cross-Domain Parser Evaluation, pp. 1–8. Association for Computational Linguistics (2008)
9. Dijkman, R., Dumas, M., García-Bañuelos, L.: Graph matching algorithms for business process model similarity search. In: Dayal, U., Eder, J., Koehler, J., Reijers, H.A. (eds.) BPM 2009. LNCS, vol. 5701, pp. 48–63. Springer, Heidelberg (2009)
10. Eid-Sabbagh, R.-H., Kunze, M., Meyer, A., Weske, M.: A platform for research on process model collections. In: Mendling, J., Weidlich, M. (eds.) BPMN 2012. LNBIP, vol. 125, pp. 8–22. Springer, Heidelberg (2012)

11. Friedrich, F., Mendling, J., Puhlmann, F.: Process model generation from natural language text. In: Mouratidis, H., Rolland, C. (eds.) CAiSE 2011. LNCS, vol. 6741, pp. 482–496. Springer, Heidelberg (2011)
12. Hamp, B., Feldweg, H.: Germanet - a lexical-semantic net for german. In: Proceedings of ACL workshop Automatic Information Extraction and Building of Lexical Semantic Resources for NLP Applications, pp. 9–15 (1997)
13. Indulska, M., Green, P., Recker, J., Rosemann, M.: Business process modeling: perceived benefits. In: Castano, S., Dayal, U., Casati, F., Oliveira, J.P.M., Laender, A.H.F. (eds.) ER 2009. LNCS, vol. 5829, pp. 458–471. Springer, Heidelberg (2009)
14. Jin, T., Wang, J., Wu, N., La Rosa, M., ter Hofstede, A.H.M.: Efficient and accurate retrieval of business process models through indexing. In: Meersman, R., Dillon, T.S., Herrero, P. (eds.) OTM 2010, Part I. LNCS, vol. 6426, pp. 402–409. Springer, Heidelberg (2010)
15. Klein, D., Manning, C.D.: Accurate unlexicalized parsing. In: 41st Meeting of the Association for Computational Linguistics, pp. 423–430 (2003)
16. Kunze, M., Weidlich, M., Weske, M.: Behavioral similarity – a proper metric. In: Rinderle-Ma, S., Toumani, F., Wolf, K. (eds.) BPM 2011. LNCS, vol. 6896, pp. 166–181. Springer, Heidelberg (2011)
17. Leopold, H., Mendling, J., Polyvyanyy, A.: Supporting process model validation through natural language generation. IEEE Trans. Softw. Eng. **40**(8), 818–840 (2014)
18. Leopold, H.: Natural Language in Business Process Models: Theoretical Foundations, Techniques, and Applications. LNBIP, vol. 168. Springer, Switzerland (2013)
19. Leopold, H., Eid-Sabbagh, R.H., Mendling, J., Azevedo, L.G., Baião, F.A.: Detection of naming convention violations in process models for different languages. Decis. Support Syst. **56**, 310–325 (2013)
20. Leopold, H., Mendling, J.: Automatic derivation of service candidates from business process model repositories. In: Abramowicz, W., Kriksciuniene, D., Sakalauskas, V. (eds.) BIS 2012. LNBIP, vol. 117, pp. 84–95. Springer, Heidelberg (2012)
21. Leopold, H., Niepert, M., Weidlich, M., Mendling, J., Dijkman, R., Stuckenschmidt, H.: Probabilistic optimization of semantic process model matching. In: Barros, A., Gal, A., Kindler, E. (eds.) BPM 2012. LNCS, vol. 7481, pp. 319–334. Springer, Heidelberg (2012)
22. Leopold, H., Smirnov, S., Mendling, J.: On the refactoring of activity labels in business process models. Inf. Syst. **37**(5), 443–459 (2012)
23. Limam Mansar, S., Reijers, H.A.: Best practices in business process redesign: use and impact. Bus. Process Manag. J. **13**(2), 193–213 (2007)
24. Malinova, M., Leopold, H., Mendling, J.: An empirical investigation on the design of process architectures. In: Wirtschaftsinformatik (2013)
25. Mendling, J., Reijers, H.A., Recker, J.: Activity labeling in process modeling: empirical insights and recommendations. Inf. Syst. **35**(4), 467–482 (2010)
26. Miller, G., Fellbaum, C.: WordNet: An Electronic Lexical Database. MIT Press, Cambridge (1998)
27. Qiao, M., Akkiraju, R., Rembert, A.J.: Towards efficient business process clustering and retrieval: combining language modeling and structure matching. In: Rinderle-Ma, S., Toumani, F., Wolf, K. (eds.) BPM 2011. LNCS, vol. 6896, pp. 199–214. Springer, Heidelberg (2011)
28. Yan, Z., Dijkman, R., Grefen, P.: Fast business process similarity search. Distrib. Parallel Databases **30**(2), 105–144 (2012)

Application of Business Process Diagrams' Complexity Management Technique Based on Highlights

Gregor Jošt[(✉)] and Gregor Polančič

Faculty of Electrical Engineering and Computer Science, University of Maribor,
Smetanova 17, 2000 Maribor, Slovenia
{gregor.jost, gregor.polancic}@um.si

Abstract. The purpose of business process diagrams is mainly to make the communication between process-related stakeholders more effective. Currently, BPMN (Business Process Model and Notation) is the leader and de-facto standard for business process modeling. However, the notation and their corresponding diagrams have been perceived as complex by many different researchers. Much work was already done in order to manage such complexity by changing or extending the existing BPMN notation. In this paper, we will propose a solution that aims to decrease the complexity of business process diagrams without changing the notation or existing approaches by introducing opacity-driven graphical highlights.

1 Introduction

A business process diagram (hereinafter referred to as BPD) is a visual process model, which is typically expressed in a graph-like process modeling notation. The main purpose of a BPD is to provide a mean for a standardized and more effective communication between process analysts, where the effectiveness of 'diagrammatic communication' is measured by the level of common understanding of the intended message (i.e. how the sender understands a BPD) and received message (i.e. how the receiver understands the same BPD) (Fig. 1) [1].

To perform effective 'diagrammatic communication' it has to be ensured that BPDs remain simple to read, understood and properly maintained [2], which is often challenging, because business processes and the corresponding workflows commonly represent complex systems [3]. The level of complexity of a BPD affects the time and effort one needs for an effective understanding, maintenance and modification of a diagram. The complexity of a BPD depends on the modeled business process (i.e. business process complexity). However, since the same business process can be modeled in different ways, the BPD complexity also depends on the level of abstraction of a process model, modeling notation, modeling approach, modelers' experiences, etc. As such, it is reasonable to identify the least complex visual representation (BPD) of a business process, since such representation will most likely be the simplest to interpret and understand. Furthermore, less complex diagrams are less likely to contain syntactical errors.

© Springer International Publishing Switzerland 2016
R. Schmidt et al. (Eds.): BPMDS/EMMSAD 2016, LNBIP 248, pp. 66–79, 2016.
DOI: 10.1007/978-3-319-39429-9_5

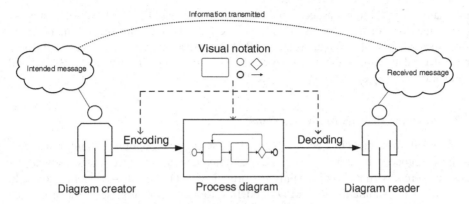

Fig. 1. Diagrammatic communication, as defined by Moody [1]

Since the complexity of a BPD can be (at least partially) addressed with proper tools and techniques, several of them have been proposed, as well as implemented in standardized process modeling languages (i.e. notations), for example: modularity, abstraction and hierarchy. However, since there are several evidences that current process modeling notations and the resulting BPD are still difficult to learn, understand and interpret [4–6], approaches for managing complexity or improving understand-ability of BPD have been implemented in supportive modeling tools (e.g. dictionary support, birds-eye view, automatic layout, syntax validation, etc.). One of the main benefits of tool-based complexity management techniques is that they do not affect the specified notation and remain simpler to improve or innovate. In this paper, we propose a novel tool-based complexity management technique based on opacity-driven graphical highlights and a prototype IT solution which implements the proposed technique for managing the complexity of BPMN diagrams.

2   Research Background

2.1   Business Process Diagrams and Notations

A BPD can be defined as a structured description of a real ("AS IS") or proposed ("TO BE") business process that defines and explains the sequence of activities and events within the business process, as well as all the relevant relations that occur between them. A BPD is the most common representation of a business process model. However, a modeler can also define a business process model by using plain or structured text, computer simulation, computer readable file etc. A visual notation, which is necessary for constructing a BPD, consists of graphical symbols and corre-sponding composition rules. Nowadays many different process modeling notations are in use, such as Petri nets, Workflow Process Description Language (WPDL), Unified Modeling Language 2.0 Activity Diagram (UML AD), Business Process Model and Notation (BPMN), Event Driven Process Chain (EPC), Yet Another Workflow Lan-guage (YAWL) and Integrated DEFinition Method 3 (IDEF3). Thus a business process model can be represented with different notations and different diagrams, representing the same or different aspects of the process, such as its flow of activities (sequence-flow) or data-flow. Currently, the de-facto standard for process modeling is Business Process Model and Notation (BPMN). According to the survey, BPMN is mainly used for documenting purposes (52 %), following by execution (37 %) and simulation (11 %) of business processes [7].

2.2 Business Process Modeling Tools

BPDs result from modeling activities that strive to improve, optimize and/or renovate a business process in order to improve its efficiency and reduce costs. Nowadays, business process modeling is mainly performed with IT support – modeling tools. In order to create standard-based BPD, modeling tools need to precisely implement the specified notation. For example, process modeling tools that support BPMN are clearly defined in BPMN 2.0 specification [8], where it states that a software can claim compliance or conformance with BPMN if, and only if, the software fully matches the applicable compliance points as stated in the specification. As an alternative to full process modeling conformance, the BPMN 2.0 specification defines three conformance sub-classes: descriptive, analytic and common executable. Descriptive and analytic sub-classes focus on different subsets of visible BPMN elements, whereas common executable sub-class focuses on what is required for executable process models.

In addition to a wide variety of compliant BPMN modeling tools, modelers have also the choice of creating BPD using generic diagramming tools - i.e. tools that support several modeling techniques or notations. Such tools (e.g. Microsoft Visio or Dia) are commonly extensible with templates (i.e. stencils), used to draw or paint symbols, shapes or patterns. In contrast to BPMN tools, they commonly do not implement a standardized meta-model, so they have limited capabilities in light of syntactical and semantic verification as well as exchange of BPDs. In general, dedicated modeling tools can reduce the time and effort needed for developing solutions in regard to particular problem types, whilst generic diagramming tools can be applied to different notations and a wider variety of problems, yet have difficulty finding results that are relevant to a specific problem [9].

2.3 Business Process Diagrams' Complexity

Complexity, as a measurable property, often appears in scientific literature. Sometimes it is precisely defined, but often also vaguely and in different ways. Cardoso derived his definition for process complexity from IEEE Standard Glossary of Software Engineering Terminology, defining it as: *'the degree to which a process is difficult to analyze, understand or explain. It may be characterized by the number and intricacy of activity interfaces, transitions, conditional and parallel branches, the existence of loops, roles, activity categories, the types of data structures and other process characteristics.'* [10]. Edmonds [11] strictly separates complexity of a real process from complexity of the corresponding BPD. He connects complexity of a BPD with complexity or difficulty to understand the notation in which the BPD is described in. Furthermore, he elaborates that complexity also depends on the type of difficulty, which depends on the modeling goals. In this manner, Edmonds defines the complexity of a BPD as: *'That property of a language expression which makes it difficult to formulate its overall behavior, even when given almost complete information about its atomic components and their inter-relations.'* [11]. Thus the complexity of a real (business) process is measured indirectly by measuring the complexity of its BPD, which is represented with a specific language or diagram technique [12].

Since BPD formally presents a directed attributed graph [13], the complexity of a BPD is fundamentally expressed with the number of diagram elements and relations between them. Their quantity affects the understandability of the diagram [14].

When we are modeling diagrams of existing (i.e. AS IS) business processes, we have no influence on the process complexity itself, but we can still improve the understandability of the corresponding diagram (e.g. we use a layout algorithm or modularization of more complex sections). On the contrary, when we plan and model future (i.e. TO BE) business processes, we do not only define the complexity of their diagrams, but also directly influence the complexity of the real process.

2.4   Business Process Diagrams Complexity Management

There have been several approaches at decreasing the complexity of BPMN diagrams. In this section we will review several suggested approaches.

A pattern based approach for reducing BPMN diagram complexity was done by (La Rosa et al. [15]). In the paper they identify patterns to reduce the perceived model complexity and to consequently simplify the representation of the process model. They defined patterns as a capture of features to manage process model complexity and distinguished them between abstract syntax (various types of process elements and the structural relationships between them) and concrete syntax (representational aspects such as symbols, colors and position). The paper focused only on concrete syntax. By analyzing existing BPM literature, standards, features of tools and consulting with BPM experts and practitioners they identified eight patterns (Table 1).

Table 1. Patterns for concrete syntax modifications

Pattern name	Hypernym	Purpose
Layout guidance	/	Modify the diagram layout
Enclosure	Highlight	Outline visual mechanisms to emphasize certain aspects or parts of the diagram
Graphical		
Pictorial	Annotation highlight	
Textual		
Explicit	Representation	Explicit and alternative visual representations for modeling constructs
Alternative		
Naming guidance	/	Naming conventions to be used in process model

The purposed patterns were evaluated using the technology acceptance model (TAM), where 15 process modeling experts participated. A seven-point scale was used to measure the participant's perception on usefulness and ease of use for each of the patterns. Industry experts found the patterns useful and in general easy to use, yet have noted, that the beginners would need additional training to utilize them in an efficient and effective way (La Rosa et al. [15]).

3   Proposed Solution

Based on the research background and current struggles with managing the complexity of both, BPMN notation and BPMN-based BPDs, we propose a solution that introduces several opacity-driven graphical highlights (hereinafter referred to as opacity-driven highlights) into BPDs. Each such opacity-driven highlight emphasizes a specific part of a BPD by changing the opacity levels of its constituting elements. In this section we will define the theoretical foundations of the proposed solution, provide the rationale for each specific opacity-driven highlight and perform analysis of an example diagram.

3.1   Theoretical Foundations

The general idea for our proposed solution is based on the aforementioned work of La Rosa et al. [15], where authors identified eight patterns for managing process model complexity. The focus of our research is on 'graphical highlights pattern', which was identified in order to enrich the representation and consequentially reduce the cognitive overhead when reading the diagrams. This can be achieved by changing the visual appearance of model elements, e.g. shape, line thickness or background color. Moreover, introducing colors increases both readability and understandability of process diagrams [16].

Taking this into the consideration, we propose a novel solution that exploits the strengths of colors by manipulating the opacity of BPMN elements. This is reasonable, since applying a level of opacity results in a different color, which is commonly a lighter tone of its original color. We applied this concept to BPDs by introducing opacity-driven highlights, where each such highlight represents a specific part (i.e. a sub-graph) of the BPD. Based on the type of the opacity-driven highlight, a BPD is divided into a set of relevant elements that obtain 100 % opacity, and a set of irrelevant elements, which have their opacity reduced by a certain value x. This creates an impression that the irrelevant elements are transparent while the relevant ones are highlighted. As such, our proposed solution preserves the strengths of colors, while reducing their shortcomings.

3.2   Proposed Highlights for Business Process Diagrams

Based on the existing work in the fields of measurements of BPDs, complexity coping mechanisms and process improvement methodologies, we propose nine opacity-driven highlights. Furthermore, we classified each such highlight into two categories, namely structural and behavioral graphical highlights. This is in accordance with process model quality assurance, more specifically, syntactic quality and verification [17]. The structural opacity-driven highlights are represented in Table 2.

Table 2. Structural opacity-driven highlights

Structural opacity-driven highlight	Description
Input documents	Highlights all Data Objects that are inputs to Tasks
Output documents	Highlights all Data Objects that are outputs from Tasks
Message flows	Highlights Message Flows between the Pools and its corresponding Flow Objects or collapsed Pools
Errors	Highlights all Boundary Error Events in the BPD and their corresponding Tasks

The behavioral opacity-driven highlights are represented in Table 3.

Table 3. Behavioral opacity-driven highlights

Behavioral opacity-driven highlight	Description
Start event	A path that begins at the selected Start Event and ends at all possible End Events
End event	A path that begins at the selected End Event and ends at all possible Start Events
XOR Gateway	A path that begins at the selected Sequence Flow, which directly follows a selected diverging XOR Gateway, and ends at all possible End Events
Dynamic error	A path that begins at the selected Boundary Error Event and ends at all possible End Events
Dynamic document	All Flow Objects that either produce or consume a selected Data Object

3.3   Analysis of the Case Diagram

In this subchapter we will provide an example of opacity-driven highlights. To this end, we chose a sample BPMN diagram that represented a help-desk process (Fig. 2).

The help-desk process was chosen because it included all five basic categories of BPMN elements. For the purpose of the case study, we will apply the Message flows structural opacity-driven highlight, which focuses on Message Flows between the Pools and its corresponding Flow Objects. These relevant elements obtain their default 100 % opacity, whereas all the other elements have their opacity reduced to 20 % (this value was arbitrary chosen and is a subject of further experiments). In order to achieve

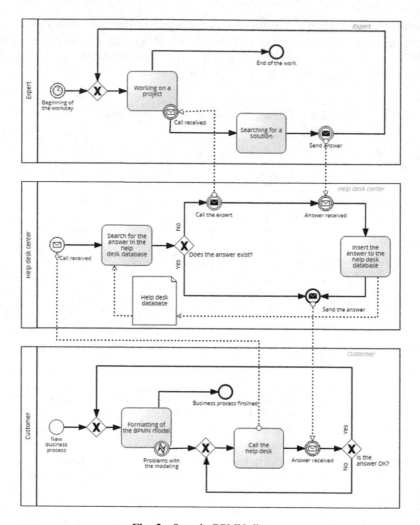

Fig. 2. Sample BPMN diagram

this, we needed to calculate and apply the new color for elements that are not part of the graphical highlight. Figure 3 represents the result of applying Message flows structural opacity-driven highlight.

As can be seen from the figure above, by reducing opacity of the elements that are deemed irrelevant, we consequentially highlight the relevant elements, which form a specific structural or behavior graphical highlight. In this way, our solution is independent of any specific color, while still preserving one of its main advantages, namely highlighting important information within a BPDs.

3.3.1   Complexity Analysis of the Case Diagram

Since the goal of our proposed solution is to reduce the complexity of BPDs, we further analyzed the structural complexity of both approaches (standard BPMN diagrams and

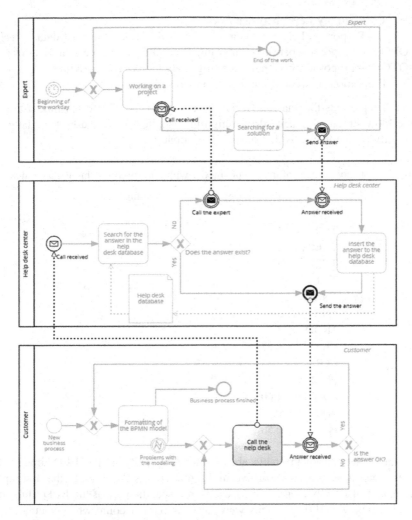

Fig. 3. Message flow structural opacity-driven highlight

BPMN diagrams, which have opacity-driven highlights enabled). To this end, we used
structural metrics as proposed by Rolón et al. [18], which are as follows:

- TNSE – Total Number of Start Events
- TNIE – Total Number of Intermediate Events
- TNEE – Total Number of End Events
- TNE – Total Number of Events
- TNT – Total Number of Task
- TNCS – Total Number of Collapsed Sub-Process
- TNG – Total Number of Gateways
- TNDO – Total Number of Data Objects
- CLA – Connectivity Level between Activities

- CLP – Connectivity Level between Pools
- PDOPIn – Proportion between Incoming Data Object and the total data objects
- PDOPOut – Proportion between Outgoing Data Object and the total data objects
- PDOTOut – Proportion between Outgoing Data Object and activities
- PLT – Proportion between Pools/Lanes and activities

Table 4 represents the analysis of the both original model and the model that has Message Flow structural opacity-driven highlight applied. In the case of the latter, we focused only on the elements that have 100 % opacity.

Table 4. Complexity analysis of the sample model with opacity-driven highlights and without

Metric	Base model	Opacity-driven highlights
TNSE	3	1
TNIE	6	5
TNEE	3	1
TNE	12	7
TNT	6	1
TNCS	0	0
TNG	5	0
TNDO	1	0
CLA	0.27	0
CLP	1.33	1.33
PDOPIn	1.00	0
PDOPOut	1.00	0
PDOTOut	0.17	0
PLT	0.50	3

As can be seen from the table above, the complexity of the BPD dramatically decreases when applying opacity-drive highlight. This is done under the assumption that the reader of BPMN diagram is focused on a specific part of the BPD (thus, only BPMN elements with 100 % opacity were taken into the account when calculating the metrics). Our solution therefore focuses on the elements in question and consequentially decreases the overall complexity of the BPD.

4 Proof and Analysis of a Prototype IT Solution

In order to achieve the desired behavior of our proposed approach, we implemented a prototype IT solution (hereinafter referred to as prototype) that renders BPMN diagrams and supports the opacity-driven highlights in order to reduce the complexity of BPDs. The prototype is able to parse any BPMN diagram, compliant with the BPMN metamodel. The desired diagram is then displayed to the user, along with all the necessary functionalities to support the opacity-driven highlights. In this way, users are able to upload their existing BPMN diagrams to our prototype as long as the tool, in which the diagrams are modelled, supports the BPMN 2.0 compliant export of the diagrams.

4.1   Technical Aspects

The prototype consists of server and client side. The server side is implemented in the PHP: Hypertext Preprocessor (PHP) programming language and has three main functionalities. First, it supports the upload of Extensible Markup Language (XML) files that store the information about BPMN diagrams. Second, the server side takes care of parsing these XML files in order to retrieve the necessary data about a diagram. Third, the server side exposes representational state transfer (REST) services that accept HTTP GET requests. Each such service corresponds to a specific opacity-driven highlight and calls the necessary algorithms that search for elements, relevant to a specific highlight.

On the client side, the model is drawn by using Scalable Vector Graphics (SVG), which is an XML-based vector image format for two-dimensional graphics. Each element is grouped together with its corresponding label (if defined) by using the "g" SVG element. This is reasonable, since the transformations that apply to the "g" element are performed on all of its child elements. So, by reducing the opacity of "g" element, all the corresponding child elements will have a reduced opacity as well.

In this phase, the behavioral opacity-based graphical highlights are rendered as well. They visually resemble standard HyperText Markup Language (HTML) checkboxes, yet they behave like radio buttons. This was done in order to let the users intuitively know that each graphical highlight can be "turned off" by unchecking the checkbox. While such behavior could still be applied to radio buttons, it would not feel intuitive to users, since radio buttons generally cannot be "unchecked". A prototype example of the SVG code for a Task is as follows:

```
<svg xmlns="http://www.w3.org/2000/svg" style="width:
300px; height: 100px">
  <g id="DE276A138492">
    <rect x="20.0" y="10" rx="10" ry="10" class="normal
non-highlighted"></rect>
    <text x="53" y="53" class="normal-text non-
highlighted-text"> Task A </text>
  </g>
</svg>
```

For the sake of simplifying the representation, we left out all the necessary Cascading Style Sheets (CSS) information, which is defined within the class attribute of both "rect" and "text" elements.

After the modeled is rendered, the tool provides also a graphical interface for accessing both the structural and behavioral opacity-driven highlights by calling the aforementioned REST services. Since we wanted to achieve the execution of each highlight to be as seamless as possible, we used AngularJS, an open-source JavaScript framework, maintained by Google. Using a JavaScript-based framework enabled us to prevent the page from reloading when the user activated a specific highlight, by dynamically calling REST services within AngularJS application on the client's side. When an array of elements is retrieved, the client takes care of setting the correct opacity for each element.

The user interface that enables users to manipulate with graphical highlights was built by using Bootstrap, a free front-end framework for web development. Bootstrap allowed us to create a responsive and consistent user interface, which is easy to use. Besides, we introduced the structural opacity-driven highlights in form of buttons, since they always highlight a predictable set of elements. Each structural opacity-driven highlight has a corresponding button at the top of the page and has a Bootstrap-specific appearance. Figure 4 represents the prototype with structural (top of the page) and behavioral (checkboxes within the model) opacity-driven highlights.

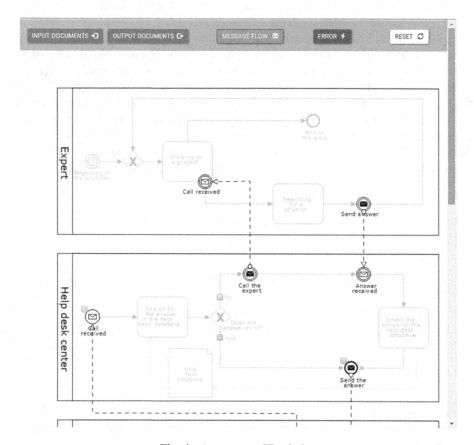

Fig. 4. A prototype IT solution

As can be seen from the image above, the Message flows behavioral opacity-driven highlight is activated. By using the aforementioned web-technologies we achieved almost identical visual appearance of the diagram to the one in Fig. 3.

4.2   Functional Aspect

Our prototype implementation of the proposed solution supports the following functionalities. Besides the upload and parse of XML documents and rendering the corresponding model, the tool supports both structural and behavioral opacity-driven highlights, defined in the "Proposed solution" chapter.

As already stated, the behavioral opacity-driven highlights are realized in form of checkboxes that are part of the following elements: Start Events, End Events, Sequence Flows, which follow diverging Exclusive Gateways, Data Objects and Boundary Error Events. In case of Start and End Events, the checkbox is placed in the upper right corner, while the Boundary Error Event has a checkbox placed in the lower left corner. In case of Sequence Flows that follow diverging Exclusive Gateways, the checkbox is placed on the Sequence Flow whereas Data Objects have checkbox placed on the border at the upper left corner.

User can choose any of the defined Structural opacity-driven highlights by clicking a dedicated button as the top section of the prototype. If a model does not include the elements, which are part of a specific highlight (e.g. there are no Message Flows in the diagram and the user selects Message flow opacity-driven highlight), the user is notified that the desired elements are not in the diagram. Besides the buttons for structural opacity-driven highlights, additional reset button is provided in order reset the model to its initial state.

Finally, it is also important to stress that the selection of each highlight is mutually exclusive, meaning that only one such highlight can be activated at a given moment. Each selection of a highlight therefore overrides the previous highlight, if it is already activated.

4.3   Analysis of a Use Case Scenario

Combining technical and functional aspect, a typical use case scenario is described step-by-step in the Table 5. The scenario is divided into user and system columns and it represents user's inputs, system's processing and its outputs.

Table 5. A use case scenario for the Message flow structural opacity-driven highlight, which represents the communication between the user and the system.

Sequence	User	System
1	The user opens the prototype application and selects one of the existing BPMN diagram	
2		The system parses the corresponding XML document and renders the BPMN diagram
3	The user wants to see how messages are passed between the Pools. To this end, the user clicks the button that activates the Message flow opacity-driven highlights	

(Continued)

Table 5. (*Continued*)

Sequence	User	System
4		The system runs the algorithm that searches for the relevant elements within the diagram and sets opacity to BPMN elements according to the results
5	The user examines the model, which has Message Flows highlighted and gathers the desired information. Afterwards, user wants to see the complete model again, so the user clicks Reset button	
6		The system sets the opacity back to 1 for every element in the BPMN diagram, thus resetting the BPMN diagram back to its initial state
7	The users obtains the complete overview of the BPMN diagram	

The scenario in the table above describes the communication between the user's input and the system's outputs for the case of Message flow opacity-driven highlights. As can be seen from the description, the user has complete control on when a specific opacity-driven highlight will be activated and can reverse the state of the diagram back to its initial state when he desires so.

5   Discussion

In this paper we presented a novel approach at managing complexity of BPDs. In order to achieve this, we developed a theoretical background that is based on a specific property of the color, namely the opacity. Opacity enabled us to enhance the BPMN diagrams in light of structural and behavioral opacity-driven highlights, while not interfering with BPMN specification or existing complexity-coping mechanisms. The analysis of the case diagram demonstrated that the usage of opacity-driven highlights decreases the complexity of the BPD by reducing the size of the model to only a set of relevant elements. After the theoretical aspect of our proposed solution, we implemented a prototype tool, which is used to render the BPMN diagrams and supports the afore-mentioned opacity-driven highlights. The tool was implemented in PHP programming language, along with AngularJS and Bootstrap.

We plan to further investigate both structural and behavioral opacity-driven highlights. With additional literature overview in the field of complexity coping mechanisms and complexity of BPDs in general, we plan to obtain additional candidates for opacity-driven highlights. Additionally, we plan to conduct an empirical investigation of proposed solution as well.

References

1. Moody, D.: The 'physics' of notations: toward a scientific basis for constructing visual notations in software engineering. IEEE Trans. Softw. Eng. **35**(6), 756–779 (2009)
2. Gruhn, V., Laue, R.: Complexity metrics for business process models. In: 9th International Conference on Business Information Systems, BIS 2006, pp. 1–12 (2006)
3. Cardoso, J.: Approaches to compute workflow complexity. Business **06291**, 16–21 (2006)
4. Decker, G., Puhlmann, F.: Extending BPMN for modeling complex choreographies. In: Meersman, R., Tari, Z. (eds.) OTM 2007, Part I. LNCS, vol. 4803, pp. 24–40. Springer, Heidelberg (2007)
5. Fernández, H.F., Palacios-González, E., García-Díaz, V., Pelayo G-Bustelo, B.C., Sanjuán Martínez, O., Cueva Lovelle, J.M.: SBPMN — an easier business process modeling notation for business users. Comput. Stand. Interfaces **32**(1–2), 18–28 (2010)
6. Muehlen, Mz, Recker, J.: How much language is enough? Theoretical and practical use of the business process modeling notation. In: Bellahsène, Z., Léonard, M. (eds.) CAiSE 2008. LNCS, vol. 5074, pp. 465–479. Springer, Heidelberg (2008)
7. Kocbek, M., Jost, G., Hericko, M., Polancic, G.: Business process model and notation: the current state of affairs. Comput. Sci. Inf. Syst. **12**(2), 509–539 (2015)
8. Object Management Group (OMG): Business process model and notation (BPMN) version 2.0. Business, vol. 50, p. 170 (2011)
9. Terry, M., Mynatt, E.D.: Enhancing general-purpose tools with multi-state previewing capabilities. Knowl.-Based Syst. **18**(8), 415–425 (2005)
10. Cardoso, J.: How to measure the control-flow complexity of web process and workflows. In: Workflow Handbook, vol. **2005**, pp. 199–212 (2005)
11. Edmonds, B.: What is Complexity - The philosophy of complexity per se with application to some examples in evolution, p. 13. Manchester Metropolitan University (1999)
12. Latva-Koivisto, A.M.: Finding a complexity measure for business process models. Complexity, pp. 1–26 (2001)
13. Dijkman, R., Dumas, M., García-Bañuelos, L.: Graph matching algorithms for business process model similarity search. In: Dayal, U., Eder, J., Koehler, J., Reijers, H.A. (eds.) BPM 2009. LNCS, vol. 5701, pp. 48–63. Springer, Heidelberg (2009)
14. Mendling, J., Reijers, H.A., Cardoso, J.: What makes process Smodels understandable? In: Alonso, G., Dadam, P., Rosemann, M. (eds.) BPM 2007. LNCS, vol. 4714, pp. 48–63. Springer, Heidelberg (2007)
15. La Rosa, M., ter Hofstede, A.H.M., Wohed, P., Reijers, H.A., Mendling, J., van der Aalst, W.M.P.: Managing process model complexity via concrete syntax modifications. IEEE Trans. Ind. Inform. **7**(2), 255–265 (2011)
16. Erol, S.: Coloring support for process diagrams: a review of color theory and a prototypical implementation. Vienna University of Economics and Business (2015)
17. Dumas, M., La Rosa, M., Mendling, J., Reijers, H.A.: Fundamentals of Business Process Management. Springer, Heidelberg (2013)
18. Rolón, E., Ruiz, F., García, F., Piattini, M.: Applying Software metrics to evaluate business process models. CLEI Electron. J. **9**, 15 (2006)

A Methodology and Implementing Tool for Semantic Business Process Annotation

Beniamino Di Martino, Antonio Esposito$^{(\boxtimes)}$, Salvatore Augusto Maisto, and Stefania Nacchia

Department of Industrial and Information Engineering,
Seconda Università degli Studi di Napoli, via Roma 4, 81031 Aversa, Italy
`beniamino.dimartino@unina.it, antonio.esposito@unina2.it,`
`{salvatoreaugusto.maisto,stefania.nacchia}@studenti.unina2.it`

Abstract. The task of accurately model Business Process is steadily growing in complicatedness, partially due to the ever changing and dynamic contexts such processes are defined into, and the complexity of domain-specific concepts characterizing today's global economic environment. Even though the modern IT provides several tools to help the Business Process modellers, they often do not offer sufficient support to the definition and interpretation of domain concepts or relationships, due to a general lack of precise domain knowledge and ambiguities in the terms used to define such concepts. Such semantic ambiguity negatively affects the efficiency and quality of Business Process modelling. To address these issues, an ontology based approach is proposed to mitigate semantic ambiguity, and a means to capture rich, semantic information on complex Business Processes through domain specific ontologies is presented. Also, a prototype tool which allows users to annotate existing BPMN models is described.

Keywords: BPMN · Business process · Ontology · OWL · Semantic · Semantic annotator

1 Introduction

Business Process models and diagram, once mostly used in many companies simply as documentation, are now at the base of automatic model transformation and code generation, becoming every day more and more integrated in the information technology environment. The need for automation and integration has fuelled the development of several machine readable standards, such as the Business Process Execution Language (BPEL), and the Business Process Model Notation (BPMN), which aim to formalize business processes in order to make their definition uniform and easily shareable among domain experts, but also to create a shared notation for everyone to use. Such standards are based on XML and are free to use and customize, so many tools have also been developed to support the creation, execution and simulation of Business Processes: BPMN-based tools, in particular, have invaded the market, with each tool proposing

R. Schmidt et al. (Eds.): BPMDS/EMMSAD 2016, LNBIP 248, pp. 80–94, 2016.
DOI: 10.1007/978-3-319-39429-9_6

some particular customization. Among these, remarkable examples are represented by Camunda, Activiti, Bonita or jBPMN. Despite all these efforts to integrate IT and Business have been made, the gap between the two is still remarkable: every standard formalization has its own particular way to describe and define the flows and tasks of processes, and moreover the level of abstraction changes from language to language. BPEL is an orchestration and execution language, which aims to show the execution level of a process that involves message exchanges, interactions between systems and so on. Business Process Model and Notation (BPMN), on the other hand, is a standard for Business Process modelling that provides a graphical notation for specifying Business Processes in a Business Process Diagram (BPD), based on a flowcharting technique very similar to Activity Diagrams from Unified Modelling Language (UML). The objective of BPMN is to support business process management, for both technical and business users, by providing a notation that is intuitive to business users, yet able to represent complex processes. These standards, yet so different, share the total lack of semantic information: the same process, if different terms and visions are applied during its definition, can appear as two different processes when described using both of the standards and seen separately. Conversely two different processes can look as being the same to the eyes of a developer, even when described using the same standard, if ambiguous semantics is applied without any additional domain information. Thus, the need to semantically annotate Business Processes with meaningful information which can give a clearer vision of the concepts involved in each of tasks, roles and rules defining them. In this paper we focus on the description of the methodology and the subsequent tools developed in order to enrich the process description with semantic annotations.

The structure of this paper is as follows: in Sect. 2 we will describe the goals we want to achieve with a semantic annotation of BPMN and the motivations for using semantic-based technologies; in Sect. 3 we will report a brief overview of the main projects related to the application of semantics to processes; then in Sect. 4 the core methodology of the whole search will be explained and in Sect. 5 the annotation tool will be presented with all its features; a practical example where the whole semantic annotation procedure is outlined is presented in Sect. 6, and at last, some considerations and an outlook on future research will be presented in Sect. 7.

2 Goals and Motivations

The standards currently available to describe Processes are surely machine-readable, but definitely not "machine-understandable". All the information regarding the real meaning of a Process and its domain have no place in the standard languages: however they are fundamental if the designed processes are to be executed automatically or used as a base for automatic code generation. Moreover as they are not built on expressive, logic-based representation techniques, they also fail at making the whole business process space accessible to intelligent queries and machine reasoning that would facilitate analysis, search and validation of Business Processes.

This insufficient degree of machine-accessible representations of the processes and data about processes inside organizations hinders the enablement of the following features, mentioned in [3], that are always valid and unrelated to the specific domain for which the process is being defined:

- **Roles validation:** once the process has been modelled, and the various actors who play a part in its execution are defined, it is necessary to make sure that a person or an information system alike, is really entitled to perform the task.
- **Inspection:** as the process might be a starting point for a code generation or an automatic execution, the requirement compliance is a crucial aspect that must be verified.
- **Optimization:** since the processes are nowadays more and more integrated in the IT environment, optimization is an aspect that must not be forgotten, but facilitated as much as possible. This particular point may require the need to replace partially or wholly a process, according to some rules of thumb, but it also may involve the integration of many processes.
- **Re-use:** Often the same process or sub-process can be used in different contexts, if properly adapted to the domain it is deployed into. The possibility to re-use, even partially, the definition of a business process, can reduce the errors and bugs in new models, especially if the re-used components have been exhaustively tested.
- **Enhancement:** Once a process has been defined and a context specific implementation of it has been selected, it is still possible to enhance its performances, security and scalability features by exploiting services which offer the needed characteristics. This is even truer if we consider automatic tasks. Once a process has been defined and a context specific implementation of it has been selected, it is still possible to enhance its performances, security and scalability features by exploiting services which offer the needed characteristics.

All these problems features can be addressed and easily implemented by a semantic enrichment of the process models. However, the introduction of semantics must be done at two levels: the **Structural level** and the **Domain level**. In this way each element of a process has a semantic representation and, thanks to the domain level, it can be placed in an real context. The main goal of the presented research effort is to provide an easy and user-friendly way to annotate the Process model, no matter what standard is used to define it, in order to enable structural and domain analysis of the annotated model.

3   State of the Art

Several research efforts have been carried out to analyse the benefits of applying semantics to business processes definitions and, in some cases, models and tools have been provided. The goals sometimes vary from research to research. The practical benefits deriving from the application of semantics to Business Process Modelling are presented in [13], in which the authors stress the important role played by semantic annotations in easing the design and development of new business processes. In [3] the main motivations behind the use of semantics applied to business process modelling are presented, via a discussion on the features enabled by semantics which are not natively supported by Business Process standards.

The work presented in [11] provides the definition of an ontology for the semantic description of a BPMN's structure: all the graphical elements of the standard are represented by OWL classes, with object properties marking their exact relationship. Such an ontology is then used to validate the process model and ensure that all of the constraints imposed by the notation are respected. The approach proposed in our previous work [3] exploits such a structural representation, but it also extends it in order to support the recognition of specific process patterns and to provide a context-aware analysis of the process. In [2] the authors propose a model and an implementing tool to enrich the BPMN syntax with tags, which enable the connection to existing ontologies. However, such an approach can cause incompatibilities with other BPMN based tools which, in the best case, will simply ignore the tags. The approach and practical implementation we are providing here completely avoid such issues by providing external annotations of the BPMN, which do not add new syntactical elements to the existing documentation.

Models and tools for the semantic annotation of BPMN have also been developed within the European funded project **SUPER - Semantics Utilized for Process management within and between EnteRprises** [1]. Such a project has provided a complex and exhaustive framework, composed by a set of tools for the annotation of BPMN documents and a multi-layered ontology structure for the description of both structural aspects of the BPMN and domain specific concepts. The framework is completely based on the **Web Service Modeling Ontology** (WSMO) [10] to provide the semantic support: the approach presented here relies instead on the **Web Ontology Language** OWL [9] for semantics, which enables the possibility to leverage OWL-S [8] for the description of BPMN tasks as semantic web services and also makes it possible to accurately describe their orchestration. OWL-S is more mature than WSMO [5] and, being completely based on OWL, it can be managed with the same tools used to create, manage and share ontologies.

The tool SeMFIS [4] is bases on a meta-model guided approach, for the semantic management and annotation of semi-formal conceptual models. The presented approach consists in enriching the conceptual models with a mapping to reference ontologies, expressed in OWL. However, such models are intended to be used by humans for communication and understanding and not machines and, even if BPMN models could be defined via a specialization of supported conceptual models, there is no hint to the possibility of exporting them to be used on a different tool or platform. On the opposite side, the meta-model presented in this paper mainly addresses context specific machine readable formalisms, BPMN in particular, and will offer the possibility to export semantically annotated models which could be interpret by other tool seamlessly.

The tool Pro-Seat [6] uses an approach which seems very similar to ours, as regards the definition of a meta-model for generic process models definitions and semantic enrichment via OWL and OWL-S. Support of BPMN is also provided explicitly, but it was not possible to test the tool in order to assess the actual differences with our approach, since it was not available any-more and the authors did not respond to our e-mails.

4 Methodology

The methodology we are going to describe in this section is based on an ontology-based semantic annotation approach, aiming to enrich and reconcile the semantics of Process models for managing the relative knowledge. After much consideration we came to the conclusions that there are two fundamentals dimensions in a process model that need to be considered:

- The **Structural** dimension, which defines: all the activities involved in the process and their connections; how the tasks and activities are connected and to whom they are assigned; the orchestration and choreography of the entire process, which represent the pillars of the Structural dimension. These aspects are well covered by various standards and formalisms.
- The **Domain** dimension, which contains information on the process' goals, conditions and requirements, and it is scarcely covered (if not completely neglected) by the current used standards.

In [3], and in a more deep way in Sect. 2, we have extensively justified the reasons why the representation of a business process proposed by the current standards lacks in expressiveness and has many flaws when it comes to aspects like requirements compliance, roles compatibility, optimization and reuse. Our methodology takes an all-encompassing approach to perform an analysis and to model a process in the best way possible. For this reason it has been decided to

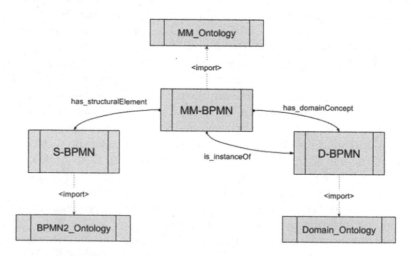

Fig. 1. The methodology abstract model

apply a hybrid methodology, which provides for the use of both structural and domain information. The core of the methodology is to produce a knowledge base in which the process model, no matter which standard is used to describe it, is transformed and translated into two separate ontologies, **"D-BPMN"** and **"S-BPMN"**. Such ontologies represent, respectively, the Domain and the Structural aspects. Figure 1 shows how such ontologies are interrelated.

4.1   Meta-Model

Before describing any further the ontologies production process, there is a crucial aspect that must be addressed in the methodology, that is the heterogeneity of various models:

- Since models are created in different modelling languages, the same business phenomenon is represented differently in different models.
- The same applies to Terminology, which varies from model to model.
- Conceptualization mismatches include different classifications, aggregations, attribute assignments and value types.

In order to cope with semantic interoperability problems, common and shared semantics should be referenced to annotate the heterogeneous representations in process modelling languages and business process model contents, in a human and machine understandable manner. An ontology is considered a kind of agreement on a domain representation and can help reconciling the representations provided by heterogeneous models.

To make all the process models reconciled and comparable we have extended a *Meta Model*, referred below as *MM*, in which the general concepts used in a process model are expressed, as explained in [7] and shown in Fig. 2. This meta-model is itself an ontology and is supposed to provide common and core semantics of process modelling constructs.

The main concepts of the Meta-Model Ontology are:

- **Activity** is a synonym of a Process, and can be atomic or composed by other activities.
- **Artifact** is something involved in an Activity such as a tool or a software.
- **Actor-role** represents the entity which interacts with or performs an Activity.
- **Input** and **Output** define the simple information needed by an Activity or produced by it.
- **Preconditions** and **Postconditions** describe general representations of constraints.
- **Exception** define not-ordinary situations which need to be addressed by ad-hoc Activities.
- **Workflow Patterns** represent orderings of different Activities.

All these concepts describe the general aspects of a Process: once the models become instances of the meta-model, than our analysis can focus on describing the domain associated to them.

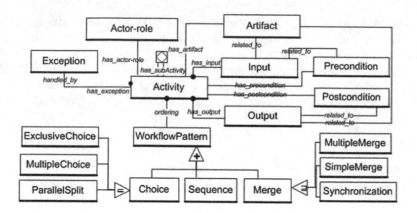

Fig. 2. General process ontology as described in [7]

Different mapping strategies can be used to connect concepts defined in the meta-model, in the structural ontology and in the domain specific ontology. They can be simple rules, defined as annotations in the meta-model, which refer specific model contents in modelling constructs to corresponding domain concepts. More complicated mappings can be defined through refined semantic relationships between concepts used in models and concepts defined in a domain ontology. We opted for a simple reference to map the individuals instantiated in the meta-model ontology and the ones in the structural ontology, assuming that almost all the concepts in the model have equal or approximately equal concepts in the ontology.

The semantic relationship is described by the object property ***has_structural Link***, in this way, we keep trace of the connection between the individuals in the MM ontology and their equivalent in the **"S-BPMN"** ontology.

Equally a simple reference is implemented to obtain the semantic connection between the MM ontology and the **"D-BPMN"**: the object properties created are ***has_domainConcept*** and ***is_instanceOf***.

The decision to define and outline two object properties, that at first sight seem redundant, stems from the need to represent two domain concepts clearly separated: the object property "has_domainConcept" provides a semantic relationship that allows to contextualize an abstract element in a particular and actual domain so that, e.g., a task has a real meaning in the context of a process. The other object property, "is_istanceOf", provides a more specific information on who is really entitled to perform a particular task.

4.2 Structural and Domain Ontologies

To make the methodology work at its best we need to analyse the modelling standard, in this case BPMN, and to elicit as much information as possible regarding the structure of its composite processes: thus it has been necessary to start from a valuable knowledge base that included all the aspects of the BPMN standard. So we based our representation on an existing full scale semantic depiction of the Business Process notation, presented in [11]. Once all the informations regarding the process have been annotated, the produced **"S-BPMN"** ontology contains every and each element of the process: all the graphic components in the BPMN have their corresponding individual, with all its data and object properties asserted according to its relations with the other elements in the process. What our methodology aims to offer, compared to other approaches, it is to use the information related to the domain the process is placed in, in order to enable a context-aware analysis. To do so, a domain ontology **"D-BPMN"** can be added, which contains within it all the information concerning the application context. Such information would be simply lost if only the structure of BPMN was analysed. With our approach each element present in the model of the process is associated to domain specific information. Several domain ontologies already exist in literature, dealing with different topics and contexts. Using such ontologies to enrich the original BPMN model would reduce the time needed to produce a new ad-hoc one. Of course, existing ontologies can be easily imported and/or extended to better fit the requirements. Using a domain ontology would enable the research for information on the roles of people involved in the process. In this way it is possible to associate to the BPMN element, corresponding to the person in interest, additional information such as the exact role in the company, personal information, possible links with other roles, and any other information that is considered necessary to a more in-depth analysis. Furthermore it is possible to associate additional information with each activity present in BPMN.

At the moment, as also described in Sect. 6, it is up to the user to correctly annotate the analysed BPMN with information deriving from the chosen domain ontology. Of course, she would be supported by the tool in making such annotations.

5   Implementation

The prototypical tool as far developed, shown in Fig. 3, is web based: in this way it is practically accessible to everyone from everywhere and it is completely platform-independent to increase its interoperability; one of its main features is the possibility to store all the uploaded files, both bpmn and owl, in a shared remote data-store.

Once the files are uploaded into the system, the core feature is to visualize the structure of the ontology and the Business Process and the user, in a very intuitive and simple way, can define the semantic annotations.

The first step in implementing the software has been achieved developing a "converter", using the Java programming language: to put it simply, the converter has a .bpmn file as input and, in return, it produces two .owl files as outputs, one for the structural representation the other for the meta-model representation.

Clearly that's a quite reducing definition but it is a first approach to understand that the component perfectly manages to match the alleged classes in the ontology and the various components of the input process and, furthermore, it ensures that no valuable information is lost in the conversion procedure. The structure of the converter comprehends three major components:

– BPMN model
– OWL model
– BPMN2OWL converter

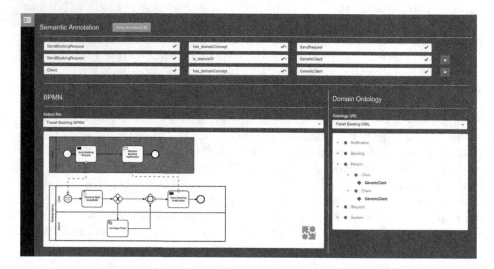

Fig. 3. Business process semantic annotator developed tool

5.1 BPMN Model

The BPMN model has been realized according to the classes found in "The BPMN 2.0 ontology" [11], trying and succeeding to create a one-to-one correlation between the components. Thanks to this approach it has been rather easy locating the essential individuals and all the interconnected information used to populate the newly created ontology. The entire model depends on a single class, "BPMNAbstractElement", that is inherited by all other classes of the model, creating different levels of abstraction Fig. 4.

Every level of abstraction corresponds to a distinct set of BPMN elements: in this way, in each consecutive level the information about a particular set of element is worked out. This procedure eventually leads to the leave classes, which represent the actual BPMN graphical elements, illustrated in the previous chapter.

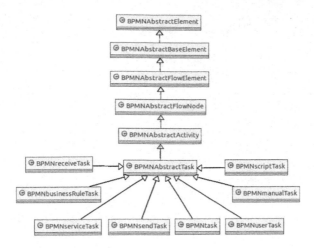

Fig. 4. Levels of abstraction for BPMN task elements

5.2 OWL Model

The OWL model developed for the software has its roots in the "OWLAPI" model, which is extended and enhanced by the software model.

The OWL API is a Java API and reference implementation for creating, manipulating and serialising OWL Ontologies. The latest version of the API is focused towards OWL2.

New features have been added to better support the fast creation of individuals in the ontology.

5.3 BPMN2OWL Converter

The core of the analysis software converter lies in the BPMN2OWL class, which allows to implement the actual conversion from the initial BPMN model to the final OWL model.

Via the Camunda Public API, it has become possible to analyse a .bpmn file, input of the converter, and produce the objects inside the software structure, briefly illustrated in the section about the BPMN Model.

Initially the BPMN2OWL class loads into the system "The BPMN 2.0 ontology" [11], which sets up the first knowledge base the software operates on, afterwards the .bpmn file is brought into the system and analysed to determine every element and related features of the process to be instanced.

To make the instance of the bpmn elements easier and more efficient, without an unnecessary code duplication, a Design Pattern has been used: to be precise a "Factory Method Design Pattern" Fig. 5 has been implemented, and its mixture with the Java Reflection grants the BPMN2OWL an effortless way to instance the concrete element without the need to specify the corresponding class of the element.

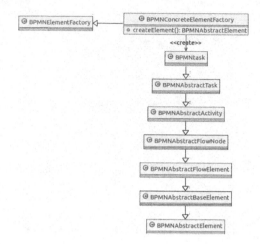

Fig. 5. Implementation of factory method design pattern

During the BPMN structural analysis, individuals in the meta-model are instantiated too, so that for each element in the BPMN diagram there is a corresponding individual in the meta-model ontology.

There is a one to one relationship between the individual in the meta-model and the one instantiated in the ontology representing the structural composition of the BPMN, S-BPMN, and this relationship is represented by the object property "has_structuralElement", described earlier in Sect. 4.

Once all the informations regarding the process have been annotated, the converter produces two .owl files: one for the structural representation, while the other allows the mapping between the BPMN elements and the generic concepts forming the meta-model ontology. But in both knowledge bases every and each element of the process is present, and it corresponds to an individual; all its data properties have been asserted, and according to its relations with the other elements in the process, the object properties have been asserted too.

The second step in implementing the software has been achieved developing a GUI.

The tool provides a graphic and intuitive annotator, which allows the user to create the relationships between, on the one hand, the elements of the BPMN diagram loaded into the system and on the other, the individuals in the domain ontology also stored in the system. Through the utilization of the tool, users can connect every element of the process with domain concepts, using the expressly created object property "has_domainConcept"; in this way, also actors defined as a simple lane in a diagram can be connected and related to actual actors, using the object property "is_istanceOf".

6   Example

To better explain the Process annotation procedure, we apply it to a process modelled using BPMN: the example represents a simple travel booking process, shown in Fig. 6, where a client interacts with a clerk and an automated system to book a travel. The workflow described by the BPMN diagram is the following: the client sends a request for booking and then waits a notification on whether it will be accepted or not. The clerk interacts with a system to check the availability and, eventually, she informs the client on the outcome of the booking. The first step of the annotation procedure is to produce the S-BPMN ontology, i.e., the structural semantic representation of the process. Listing 1.1 reports an example of representation of the task **SendBookinRequest** within the structural ontology. Connections with other elements of the BPMN diagram, sequence flows in particular, are represented by ad-hoc properties *has_SequenceFlow*.

```
<owl:NamedIndividual rdf:about=#SendBookingRequest">
    <rdf:type rdf:resource="#sendTask"/>
    <BPMN2:implementation rdf:datatype="#DTtImplementation">##
        WebService</BPMN2:implementation>
    <BPMN2:completionQuantity rdf:datatype="#integer">1</
        BPMN2:completionQuantity>
    <BPMN2:startQuantity rdf:datatype="#integer">1</
        BPMN2:startQuantity>
    <BPMN2:name rdf:datatype="#string">Send Booking Request</
        BPMN2:name>
    <BPMN2:id rdf:datatype="#ID">SendBookingRequest</BPMN2:id>
    <BPMN2:isForCompensation rdf:datatype="#boolean">false</
        BPMN2:isForCompensation>
    <BPMN2:has_sequenceFlow rdf:resource="#SequenceFlow_0upbp9p"/>
    <BPMN2:has_sequenceFlow rdf:resource="#SequenceFlow_1bnnn10"/>
</owl:NamedIndividual>
```

Listing 1.1. Representation of the **SendBookingRequest** Element of type **sendTask**

At the same time all these constructs used to represent the process are also annotated in the meta-model ontology. The case of the *SendBookingRequest* task is reported in Listing 1.2, where it is defined as an Activity in the meta-model.

```
<owl:NamedIndividual rdf:about="#SendBookingRequest">
    <rdf:type rdf:resource="MM-BPMN#Activity"/>
    <MM_Ontology:has_structuralElement rdf:resource="S-BPMN#
        SendBookingRequest"/>
</owl:NamedIndividual>
```

Listing 1.2. Annotation of the task represented in listing 1.1 in the meta-model ontology

On the other hand the **Client**, which is the label of a lane in BPMN, is annotated as an Actor in the meta-model, as also reported in Listing 1.3.

```
<owl:NamedIndividual rdf:about="#Client">
    <rdf:type rdf:resource="MM-BPMN#Acotr"/>
    <MM_Ontology:has_structuralElement rdf:resource="S-BPMN#
        Client"/>
</owl:NamedIndividual>
```

Listing 1.3. Annotation of the Client as an Actor in the meta-model ontology

It is easy to see that the task has become an individual belonging to the class that represents a generic activity, and in the same way, the client has become a generic actor.

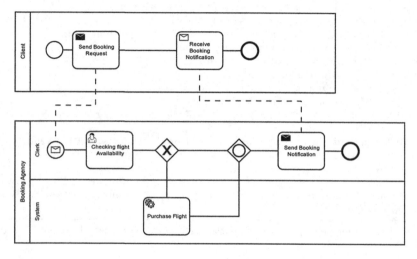

Fig. 6. Travel booking BPMN diagram

Next, a domain ontology is employed to annotate the model contents which are described in the process annotation model. In this phase it is the user himself who, exploiting the features provided by the tool, can decide the correspondence between the abstract element and the domain concepts. Once annotated, in the final meta-model ontology the annotated domain concepts can be found, as shown in Listing 1.4.

```
<owl:NamedIndividual rdf:about="#SendBookingRequest">
    <rdf:type rdf:resource="MM-BPMN#Activity"/>
    <MM_Ontology:has_structuralElement rdf:resource="S-BPMN#
        SendBookingRequest"/>
    <MM_Ontology:is_istanceOf rdf:resource="D-BPMN#
        GenericClient"/>
    <MM_Ontology:has_domainConcept rdf:resource="D-BPMN#
        SendRequest"/>
</owl:NamedIndividual>
```

Listing 1.4. Connections between the structural and domain ontology

7   Conclusions and Future Works

Semantic ambiguity in Business Process modelling and execution is a very prominent issue which often mines the understanding and sharing of the Business Processes applied to many enterprise applications. In this paper, we addressed the main issues and problems that originate from this semantic ambiguity, and highlighted the features that a shared semantic representation would enable.

Since the business process models are reusable knowledge and resources, they are required to be understandable and adaptable to other enterprise modelling users. We discern two main levels of semantic interoperability of process models. To discover the desired process models, we propose a semantic annotation method for annotating process model fragments and modelling languages.

Moreover, the paper presented the semantic annotation approach by a simple example illustration. It disclosed the technical possibility of the approach. With the semantic annotation of process models, process model designers can search their desired models and, e.g., reuse them in their specific projects without bother of semantic mismatch of various models.

The next step would be to extend the methodology and improve both it and the tool features: more precisely we would like to extend the research and add the possibility to define, to annotate the process with semantic informations regarding the presence of possible work-flow patterns [12]; in this way it would be possible to implement matching algorithm to discover new patterns into processes, and a more complete analysis can be performed.

Moreover another dimension of the business processes will be investigated, the **goal dimension**: goal models will be integrated in the current knowledge base, and they will be used to link process models with the goal hierarchy, which helps connect the users' query desires and the potential process models.

Acknowledgements. This research has been supported by the European Community's Seventh Framework Programme (FP7/2007-2013) under grant agreement n 256910 (mOSAIC Project), by PRIST 2009, "Fruizione assistita e context aware di siti archeologici complessi mediante dispositivi mobili" and CoSSMic (Collaborating Smart Solar-powered Micro-grids - FP7-SMARTCITIES-2013).

References

1. SUPER - Semantics Utilized for Process management within and between EnteRprises. http://projects.kmi.open.ac.uk/super/
2. Di Francescomarino, C., Tonella, P.: Supporting ontology-based semantic annotation of business processes with automated suggestions. Int. J. Inf. Syst. Model. Des. (IJISMD) **1**(2), 59–84 (2010)
3. Di Martino, B., Esposito, A., Nacchia, S., Maisto, S.A.: Semantic annotation of bpmn: current approaches and new methodologies. In: Proceedings of the 17th International Conference on Information Integration and Web-Based Applications & Services (iiWAS2015), pp. 95–99. ACM (2015)
4. Fill, H.-G.: Semfis: a tool for managing semantic conceptual models (2012)

5. Lara, R., Roman, D., Polleres, A., Fensel, D.: A conceptual comparison of WSMO and OWL-S. In: Zhang, L.-J., Jeckle, M. (eds.) ECOWS 2004. LNCS, vol. 3250, pp. 254–269. Springer, Heidelberg (2004)

6. Lin, Y.: Semantic annotation for process models: facilitating process knowledge management via semantic interoperability (2008)

7. Lin, Y., Ding, H.: Ontology-based semantic annotation for semantic interoperability of process models. In: International Conference on Computational Intelligence for Modelling, Control and Automation, 2005 and International Conference on Intelligent Agents, Web Technologies and Internet Commerce, vol. 1, pp. 162–167. IEEE (2005)

8. Martin, D., Burstein, M., Hobbs, J., Lassila, O., McDermott, D., McIlraith, S., Narayanan, S., Paolucci, M., Parsia, B., Payne, T., et al.: Owl-s: semantic markup for web services, W3C Member submission 22, 2004–2007 (2004)

9. McGuinness, D.L., Van Harmelen, F., et al.: Owl web ontology language overview, W3C Member submission 10 (2004)

10. Roman, D., Keller, U., Lausen, H., de Bruijn, J., Lara, R., Stollberg, M., Polleres, A., Feier, C., Bussler, C., Fensel, D., et al.: Web service modeling ontology. Appl. Ontology $1(1)$, 77–106 (2005)

11. Rospocher, M., Ghidini, C., Serafini, L.: An ontology for the business process modelling notation. In: Formal Ontology in Information Systems: Proceedings of the Eighth International Conference (FOIS 2014), vol. 267, p. 133. IOS Press (2014)

12. van Der Aalst, W.M., Ter Hofstede, A.H., Kiepuszewski, B., Barros, A.P.: Workflow patterns. Distrib. Parallel Databases $14(1)$, 5–51 (2003)

13. Wetzstein, B., Ma, Z., Filipowska, A., Kaczmarek, M., Bhiri, S., Losada, S., Lopez-Cob, J.-M., Cicurel, L.: Semantic business process management: a lifecycle based requirements analysis. In: SBPM (2007)

Social and Human Perspective

Considering Social Distance as an Influence Factor in the Process of Process Modeling

Michael Zimoch$^{(\boxtimes)}$, Jens Kolb, and Manfred Reichert

Institute of Databases and Information Systems, Ulm University, Ulm, Germany
{michael.zimoch,jens.kolb,manfred.reichert}@uni-ulm.de

Abstract. Enterprise repositories comprise numerous business process models either created by in-house domain experts or external business analysts. To enable a widespread use of these process models, high model quality (e.g., soundness) as well as a sufficient level of granularity are crucial. Moreover, they shall reflect the actual business processes properly. Existing modeling guidelines target at creating correct and sound process models, whereas there is only little work dealing with cognitive issues influencing model creation by process designers. This paper addresses this gap and presents a controlled experiment investigating the construal level theory in the context of process modeling. In particular, we investigate the influence the *social distance* of a process designer to the modeled domain has on the creation of process models. For this purpose, we adopt and apply a gamification approach, which enables us to show significant differences between low and high social distance with respect to the quality, granularity, and structure of the created process models. The results obtained give insights into how enterprises shall compose teams for creating and evolving process models.

1 Introduction

Due to the increasing adoption of process-aware information systems (PAIS), contemporary enterprise repositories comprise large collections of process models [1]. Usually, process models vary in respect to their quality and level of granularity. Further, they face a wide range of problems affecting model understandability and error probability [2]. However, high quality of process models is crucial for enterprises to guarantee proper process implementation and execution in a PAIS [3]. As a prerequisite, process models should reflect the actual business processes properly and at the right level of granularity [4]. To address this issue, considerable work on criteria related to process model quality and comprehensibility has been conducted [5,6]. In addition, modeling guidelines exist that support process designers in creating process models of high quality [7,8].

There is only little work evaluating the influence of cognitive aspects on the *process of process modeling* [9] as well as their effects on the resulting process models [10,11]. If we do not understand these cognitive aspects, however, process modeling projects might not deliver proper artifacts or even fail. This paper

© Springer International Publishing Switzerland 2016
R. Schmidt et al. (Eds.): BPMDS/EMMSAD 2016, LNBIP 248, pp. 97–112, 2016.
DOI: 10.1007/978-3-319-39429-9_7

investigates a fundamental factor presumably influencing the *process of process modeling*, i.e., *social distance* [12]. The latter is a well-established notion in the *Construal Level Theory (CLT)*, constituting an important part of *psychological distance* [13]. In this context, studies have shown that human thinking and acting are both strongly influenced by psychological distance [14]. According to CLT, we experience only the *here and now*, and form an abstract mental *construal* of distant objects or events [12,14]. For example, when attending a music festival, one is able to undergo the whole festival atmosphere. In turn, watching the festival on television, the focus is more on the line-up and, hence, the performances of the bands, i.e., experience is more superordinate.

Section 2 introduces CLT. Gamification and the considered process scenario are described in Sect. 3. Section 4 introduces the research question addressed and defines the experiment setting. Section 5 deals with experiment preparation and its execution. Results are presented and analyzed in Sect. 6. Finally, Sect. 7 discusses related work and Sect. 8 summarizes the paper.

2   Background on Construal Level Theory

Construal Level Theory (CLT) describes the effects *psychological distance* has on objects or events [12,14]. Generally, CLT states that increasing psychological distance affects our mental representation of these objects or events. In turn, this influence on human perception has a strong impact on our actions and thoughts [13]. The reason behind this phenomenon is the so-called *level of construal (LOC)*, which describes how individuals interpret and perceive objects and events. Increasing psychological distance affects the cognitive abilities and leads to a change in the perception of an object or event.

CLT describes two levels of thinking: *low-* and *high-level construal. High-level construals* are abstract, decontextualized, coherent, and superordinate representations compared to *low-level construals*. If an object or event is further away, we think about it in terms of high-level construals. However, the smaller the distance to objects or events is, the more we think in low-level construals. Moreover, these two levels of construals are influenced by *psychological distance*. While *objective distance* describes the quantitative spatial distance in the real world, *psychological distance* describes our feelings, thoughts, and emotions in relation to an object or event. In turn, the latter is considered as *psychologically distant*, if it is not experienced physically. For this case, a mental representation must be constructed.

Psychological distance can be further subdivided into *social, spatial, temporal,* and *hypothetical distance* [13,15–17]. *Social distance*, on which we focus in this paper, describes our relation to other individuals or accrues for events not being self-experienced (cf. Fig. 1); e.g., whether or not choosing a seat in a bus being more distant from a particular individual is directly reflected by the latter [18].

In previous research we already addressed the first characteristic, i.e., the relation to other individuals [19]. More precisely, results showed a significant influence of social distance on the quality and level of granularity of created process models. In accordance with CLT, process models created by process designers with a low social distance revealed a higher quality as well as granularity compared to process models created by process designers with a higher social distance. Furthermore, process designers were more self-confident about the process models they had create. Hence, the latter characteristic, i.e., event which is not self-experienced, is evaluated in this paper.

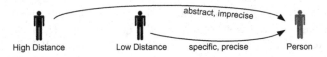

Fig. 1. Construal Level Theory - Social Distance

3 Gamification, Virtual World, and 3D Scenario

In order to simulate variability with respect to social distance, a *gamification approach* is applied, i.e., the benefits of gamification in a virtual world are used in the context of process modeling. First, this allows for an adequate reflection of the real world problem. Second, the motivation of subjects (i.e., participants of an experiment) may increase. Third, an occurrence of the effects of social distance may be ensured.

Gamification is the technique of using game elements, designs, and thinkings in a non-game context to engage and motivate employees [20]; e.g., achievements known from computer games are interpreted in enterprise software. As a consequence, work becomes more enjoyable, thus resulting in higher efficiency [21]. Moreover, a *virtual world* constitutes a computer-simulated environment, using the metaphor of the real world, but without its physical limitations [22]. In a virtual world, individuals act as textual, 2D, or 3D *avatars*, i.e., as a controllable proxy in the virtual world. Thus, they experience a degree of telepresence, i.e., an experience of presence in a remote location [23].

In the context of our experiment, relative to a real-world process from a manufacturer of gardening tools, a process scenario related to the *processing of an order in a warehouse* is contrived, which may be either experienced actively or passively (cf. Sect. 4). The entire process takes place in a full 3D virtual environment taking elements of gamification into account; e.g., *exploring* (i.e., learning more about the virtual construct) and *puzzle elements* (i.e., motivating subjects to solve a problem). The 3D warehouse scenario is implemented with *Unity*, a game development platform. In the realized scenario, subjects interact with a 3D avatar using *point and click* game mechanics.

Following this, a description of the *processing of an order in the warehouse* is provided. Figure 2 shows the layout as well as the chronological progress through the warehouse. The scenario starts in the office of the warehouse ①. First, an order is taken providing information on the items to be processed. Generally, several items need to be processed by subjects in this context. At the storage racks (cf. Fig. 3), subjects have the choice to get the items either with the forklift or the picking system ②. Since the forklift can carry only one pallet at a time, the items must be collected sequentially. The picking system comprises several grapplers that allow collecting all items either separately or at once. Then, items are disclosed at the collection point and checked for completeness ③. Following this, the items need to be packed in appropriate boxes, which are then palletized ④. After placing each box on a pallet, subjects may decide on how to transport the pallets to the shipping area, i.e., either by using the forklift or the automatic loading system ⑤. While the forklift can transport the pallets only sequentially, the automatic loading system takes care of everything automatically. As advantage of the automatic loading system, the subjects can print the required delivery documents (i.e., bill of delivery and pallet receipts) in parallel ⑥. Thereafter, pallets are labeled with the printed pallet receipts and are loaded on the trailer with the forklift ⑦. Finally, the bill of delivery is placed in the trailer and doors are closed.

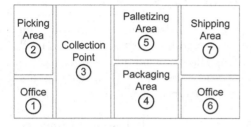

Fig. 2. Layout of the Warehouse Scenario

Fig. 3. Storage Racks

4   Research Question and Experiment Definition

This section introduces the definition and planning of the experiment for measuring the influence of the social distance on the *process of process modeling* and the resulting *artifacts*. Section 4.1 explains the context of the experiment and defines its goal. Section 4.2 introduces the hypothesis considered for testing, and Sect. 4.3 presents the experimental setup. Section 4.4 explains the design of the experiment. Finally, Sect. 4.5 discusses factors threatening the validity of results.

4.1   Context Selection and Goal Definition

Business processes are either modeled by in-house process designers or external ones. In this context, process designers are responsible for interviewing process stakeholders and participants as well as for capturing the gathered knowledge in process models. Usually, the process designers are not directly involved in the processes to be modeled; e.g., they may be member of the quality assurance department. In other cases, due to limited resources, enterprises assign such modeling and analysis tasks to external resources; e.g., business analysts.

So far, it has not been well understood how an increased social distance affects the quality, granularity, and structure of the resulting process models. To close this gap, this paper investigates the following research question:

Is the process of process modeling, i.e., the quality, granularity, and structure of the process models resulting from it, affected by the social distance the process designers have on the respective business processes?

Despite existing work on the quality [2,8,24,25], granularity [26], and structure [27] of process models there is only little research addressing cognitive aspects of process modeling [10,11,28]. In particular, it is not well understood whether certain cognitive aspects lead to minor process quality, i.e., deficiencies regarding the pragmatic, semantic, perceived, and syntactic model quality.

Based on previous research (cf. [19]), this paper continues investigating the influence social distance has on the *process of process modeling* and its *outcomes*. As opposed to the previous experiment, where social distance was experienced by the relation to other individuals, the presented experiment varies social distance with a scenario (i.e., processing of an order in a warehouse) that may either be experienced actively (i.e., low) or passively (i.e., high social distance) using gamification. The goal can be formulated as:

Analyze	*process models*
for the purpose of	*evaluating*
with respect to their	*level of construal*
from the point of view of	*the researchers*
in the context of	*students and research staff.*

4.2   Hypothesis Formulation

Based on the goal definition and taking CLT into account, six hypotheses are derived. In detail, they investigate whether social distance influences the level of construal during the *process of process modeling* or, more precisely, the quality, granularity, and structure of the resulting process models:

4.3   Experimental Setup

This section describes *subjects*, *object*, and *response variables* of the experiment as well as its *instrumentation* and *data collection procedure*.

Subjects. Ideally, process designers are modeling experts. However, they usually obtain only basic training and have limited process modeling skills [29]. From subjects (i.e., students and staff members) we require that they are familiar with process modeling although they were not experts in this area. A replication of the experiment with modeling experts might lead to different results [30]. Hence, results might not be generalizable for the entire population of process designers.

Does the social distance influence the **pragmatic quality** when creating process models?
$H_{0,1}$: There are no significant differences in the pragmatic quality when modeling processes with low social distance compared to high social distance.
$H_{1,1}$: There are significant differences in the pragmatic quality when modeling processes with low social distance compared to high social distance.

Does the social distance influence the **semantic quality** when creating process models?
$H_{0,2}$: There are no significant differences in the semantic quality when modeling processes with low social distance compared to high social distance.
$H_{1,2}$: There are significant differences in the semantic quality when modeling processes with low social distance compared to high social distance.

Does the social distance influence the **perceived quality** when creating process models?
$H_{0,3}$: There are no significant differences in the perceived quality when modeling processes with low social distance compared to high social distance.
$H_{1,3}$: There are significant differences in the perceived quality when modeling processes with low social distance compared to high social distance.

Does the social distance influence the **syntactic quality** when creating process models?
$H_{0,4}$: There are no significant differences in the syntactic quality when modeling processes with low social distance compared to high social distance.
$H_{1,4}$: There are significant differences in the syntactic quality when modeling processes with low social distance compared to high social distance.

Does the social distance influence the **level of granularity** when creating process models?
$H_{0,5}$: There are no significant differences in the level of granularity when modeling processes with low social distance compared to high social distance.
$H_{1,5}$: There are significant differences in the level of granularity when modeling processes with low social distance compared to high social distance.

Does the social distance influence the **process model structure** when creating process models?
$H_{0,6}$: There are no significant differences in the process model structure when modeling processes with low social distance compared to high social distance.
$H_{1,6}$: There are significant differences in the process model structure when modeling processes with low social distance compared to high social distance.

Object. The object is the outcome resulting from a stated modeling task, i.e., a *process model* expressed in terms of the *Business Process Model and Notation (BPMN)*. To ensure familiarity of subjects with BPMN and to guarantee that differences in response variables are not caused due to a lack of familiarity with BPMN, but rather due to differences in social distance, we choose an easy and understandable scenario. More precisely, the modeling task deals with the *processing of an order in a warehouse* (cf. Sect. 3). Task descriptions are created reflecting low and high social distance. One group is directly involved (i.e., low) in the process, while the other is only indirectly involved (i.e., high social distance). For low social distance, subjects are actively playing the warehouse scenario. In turn, regarding high social distance, subjects are watching the warehouse scenario in a video. To ensure that there exist no interferences and there is sufficient clearance between the two social distances, two pilot studies for each social distance are performed. Respective task descriptions are kept rather abstract to give subjects the possibility to model as detailed as they like.

Factor and Factor Levels. The factor considered in the experiment is *social distance* with levels *low* and *high social distance*. Accordingly, the task description is adjusted to vary social distance, i.e., to model the order process either after playing (i.e., low) or watching (i.e., high social distance) the scenario.

Response Variable. As response variable, we consider the *level of construal* that cannot be directly measured. Everything being distant from us is expressed more abstractly (cf. Sect. 2). We assume that the level of construal impacts the *quality*, *granularity*, and *structure* of the resulting process model. For this purpose, *process model quality* is characterized by four dimensions, i.e., *pragmatic, semantic, perceived*, and *syntactic quality* making use of semiotic theory, i.e., SEQUAL framework [31,32]. *Pragmatic quality* describes process model comprehension. It is measured by the *level of understanding*. In turn, *semantic quality* covers *correctness, relevance, completeness*, and *authenticity* of a process model. *Correctness* expresses that all elements of a process model are correct. *Relevance* signifies that all elements in the process model are relevant for the process. Moreover, *completeness* implies that relevant aspects about the domain are not missing, i.e., superfluous elements are considered as well. Finally, *authenticity* expresses that the chosen representation gives a true impression of the domain. *Pragmatic quality* and *semantic quality* are rated by two modeling experts in a consensus-building process based on a 7-point Likert scale [33], i.e., from 0 (strongly disagree) to 6 (strongly agree). In turn, *perceived quality* depends on the degree to which a subject agrees with the resulting process model. It can be subdivided into *agreement, missing aspects, accurate description, mistakes*, and *satisfaction* [34]. Perceived quality is rated by each subject on a 5-point Likert scale, ranging from 0 (strongly disagree) to 4 (strongly agree), after finishing the modeling task. *Agreement* expresses to which degree the process model matches with the actual business process. *Missing aspects* rates whether significant aspects are missing in the resulting process model. In turn, *accurate description* expresses how accurately the process model matches the real world process. *Mistakes* corresponds to the subject rate indicating whether there are serious mistakes in the resulting process model. Finally, *satisfaction* expresses the degree subjects are satisfied with the process models created by them. *Syntactic quality* of a process model is measured by counting syntactical rule violations of the applied modeling language, i.e., BPMN. *Process granularity* is measured through the complexity of the resulting process models, i.e., simple metrics like *number of activities, gateways, nodes, edges, elements*, and *execution paths*. *Process model structure* is analyzed with the following process metrics: *separability, sequentiality, cyclicity*, and *diameter* [3,35]. *Separability* is defined as the ratio of the number of cut-vertices to the total number of nodes in the process model. *Sequentiality*, in turn, is the degree to which the process model is constructed of pure sequences of tasks. Moreover, *cyclicity* relates to the number of nodes on cycles to all nodes in the process model. *Diameter* gives the length of the longest path from a start node to an end node in the process model. Figure 4 summarizes the response variables we consider in a research model.

4.4   Experimental Design

We apply guidelines for designing experiments as described in [36], and conduct a *randomized, balanced*, and *blocked single factor experiment*. The experiment is randomized since subjects are assigned to groups randomly and it is ensured that both groups have same size (i.e., balanced). Moreover, subjects are grouped (i.e., blocked) to not mix social distance. Finally, only a *single factor* varies, i.e., the level of construal. Figure 5 illustrates this setup.

Instrumentation and Data Collection Procedure. To precisely measure response variables in a non-intrusive manner, we use the *Cheetah Experimental Platform (CEP)* [9]. CEP provides a BPMN modeling environment that records modeling steps and their attributes; e.g., timestamps and type of modeling action. Resulting process models are then stored. Finally, demographic data and qualitative feedback is gathered from subjects based on questionnaires.

4.5 Risk Analysis

Generally, any experiment bears risks that might affect its results. Thus, its validity or, more precisely, its levels of validity need to be checked, i.e., *internal validity* ("Are effects caused by independent response variables?") and *external validity* ("May results be generalized?").

Risks to *Internal Validity*. Risks that might influence the modeling outcome include process modeling experience of involved subjects and uneven distributions of subjects over two groups. Furthermore, post data validation ensures that in both groups subjects are at least moderately familiar with process modeling (cf. Sect. 5.3). It is assured that both groups show the same or similar familiarity level, i.e., median is 3 for both groups on a 5-point Likert scale. Further, the chosen modeling task constitutes a risk to internal validity. To ensure familiarity of subjects and to guarantee that differences in quality, granularity, and structure are due to social distance, we choose an easy and comprehensible scenario (cf. Sect. 3). To further ensure that subjects are not negatively influenced by tiredness, boredom, or hunger, the experiment is conducted at a time of the day for which the mentioned frame of mind can be excluded.

Risks to *External Validity*. On one hand, the subjects have academic background (i.e., students and research staff), which might limit generalizability of results. On the other, they rather have profound knowledge in process modeling (cf. Sect. 5.3). We may consider them as proxies for professionals who have obtained basic training so far. Further, process model quality may depend on the appropriateness of the chosen modeling languages and tools. To mitigate this risk, both groups use an intuitive process modeling tool as well as an established modeling language (cf. Sect. 4.3). Finally, a potential risk for external validity is that we measure social distance with one modeling task. To mitigate this and to allow for generalizability, varying experiments need to be conducted.

5 Experiment Operation

Based on the provided experiment definition, Sect. 5.1 summarizes the experiment preparation. Section 5.2 describes the execution of the experiment, and Sect. 5.3 deals with the validation of the data collected during the experiment.

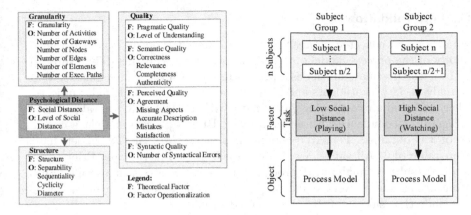

Fig. 4. Research Model   **Fig. 5.** Experiment Design

5.1 Experiment Preparation

Students and research staff familiar with process modeling are invited to join the experiment. Subjects are not informed about the aspects we want to investigate. However, they are aware that the experiment takes place in the context of a thesis. For all subjects, anonymity is guaranteed. Before conducting the experiment, for each level of social distance two pilot studies are performed to eliminate ambiguities and misunderstandings as well as to improve modeling tasks. Further, it is checked whether the social distance between the tasks is sufficiently large. Finally, an evaluation sheet is created to assess the level of construal by analyzing quality, granularity, and structure of resulting process models.

5.2 Experiment Execution

The experiment is executed in a computer lab at Ulm University. All in all, 95 students and staff members participate. Due to spatial constraints, up to 10 subjects conduct the experiment at the same time and several sessions within a period of two weeks are offered. Each session lasts about 60 min and runs as follows: The procedure of the experiment is explained and worksheets with task descriptions are handed out. Thereby, subjects are randomly assigned to one of the subject groups (cf. Sect. 4.4). Then, subjects start playing or watching the warehouse scenario. Subsequently, they fill out an initial questionnaire capturing their modeling experience. This information is used to test whether subjects are familiar with process modeling. Then, subjects are asked to model the warehouse scenario based on their own experience and in a way they think it is appropriate. Finally, subjects provide their rating for perceived quality and may give feedback.

5.3   Data Validation

In total, data is collected from 95 subjects. One of them is excluded due to invalidity of the process model obtained, i.e., the process model differs substantially from the postulated task description. Hence, 94 subjects are considered for data analysis, i.e., 84 students and 10 staff members (with 33 female subjects). Further, the median concerning *familiarity* with BPMN is 3, i.e., above average. Regarding confidence with *understanding* BPMN process models, a median value of 3 is obtained. Perceived competence in *creating* BPMN models has a median value of 3. All values are based on a 5-point Likert scale. Prior to the experiment, subjects analyzed 19 process models and created 7 in average.[1] Since all values range above average and subjects are familiar with process modeling, we conclude that subjects fit to the targeted profile.

5.4   Threats to Validation

Apparently, the experiment conducted faces the limitation that we did not involve and compare professional process modelers and IT experts from industry, but prospective ones (i.e. students). Although various investigations have shown that students are proper substitutes for professionals in empirical studies (e.g. [37,38]) the results for professionals may differ.

6   Data Analysis & Interpretation

Section 6.1 presents descriptive statistics of the data gathered during the experiment. Section 6.2 discusses whether a data set reduction is needed. Section 6.3 tests the hypotheses. Finally, Sect. 6.4 discusses results.

6.1   Data Analysis and Descriptive Statistics

Figure 6 displays box plots (i.e., median, min, and max values as well as 1st and 3rd quartiles) of measurements for the *pragmatic, semantic, perceived*, and *syntactic quality*. Further, the items of semantic and perceived quality are combined into an aggregated variable [39], i.e., *validity &completeness* and *agreement of subjects*. As a prerequisite, all response variables must show high reliability. For this purpose, *Cronbach's* α is calculated.[2] For semantic quality, a Cronbach with $\alpha = 0.84$ and for perceived quality a Cronbach with $\alpha = 0.77$ results.

As shown in Fig. 6, process models created by subjects with low social distance present a better level of understanding and contain less syntactical errors. Regarding high social distance, in turn, process models seem to give a better account of the domain. Moreover, perceived quality does not differ between the subject groups. Further, Fig. 7 presents calculated values for the *process model structure*. There are only minimal differences in process model structure between

---

[1] The full data set can be found in http://bit.ly/1VB2aS3.
[2] According to [39], $\alpha > 0.6$ acceptable reliability; $0.7 < \alpha < 0.9$ good reliability.

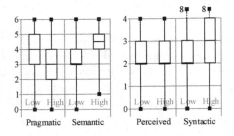 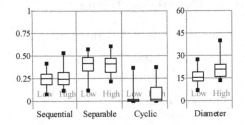

Fig. 6. Measurements for Quality   **Fig. 7.** Measurements for Structure

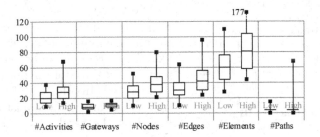

Fig. 8. Measurements for Granularity

the process models. However, the diameter shows a clear difference depending on the level of social distance. Process models whose subjects show a high social distance contain notable longer paths (median of 23 for low, 30.5 for high).

Figure 8 shows results related to the *granularity* of process models, i.e., *number of activities, gateways, nodes, edges, total process elements*, and *possible execution paths*. As a result, process model granularity is higher if subjects have a high social distance. Especially, differences in the numbers of total process elements are large. Note that low social distance results have a median of 60, whereas high social distance leads to a median of 82 process elements.

6.2   Data Set Reduction

In general, the results of statistical analyses depend on the quality of the input data, i.e., faulty data might lead to incorrect conclusions. Therefore, it is important to identify outliers and to evaluate whether these shall be excluded. Note that the latter might be critical due to potential loss of information. In the experiment, several outliers can be identified, but we decide to not remove them since we consider them as correct, not being the result of wrong modeling. Hence, removing them would bias results.

6.3 Hypothesis Testing

Table 1. Results of Hypotheses Testing

Response Variable	p-value
Pragmatic Quality $H_{1,1}$	
Level of Understanding	< 0.01 (< 0.05)
Semantic Quality $H_{1,2}$	
Validity &Completeness	< 0.01 (< 0.05)
Perceived Quality $H_{1,3}$	
Agreement of Subjects	0.410 (> 0.05)
Syntactic Quality $H_{1,4}$	
Number of Syn. Errors	0.046 (< 0.05)
Level of Granularity $H_{1,5}$	
Number of Activities	< 0.01 (< 0.05)
Number of Gateways	0.039 (< 0.05)
Number of Nodes	< 0.01 (< 0.05)
Number of Edges	< 0.01 (< 0.05)
Number of Elements	< 0.01 (< 0.05)
Number of Paths	0.148 (> 0.05)
Process Model Structure $H_{1,6}$	
Sequentiality	0.326 (> 0.05)
Separability	0.092 (> 0.05)
Cyclicity	0.258 (> 0.05)
Diameter	< 0.01 (< 0.05)

Section 6.1 indicates differences regarding low and high social distance. In the following, we test whether observed differences are statistically significant. We test the response variables with the *Mann-Whitney-U-test* [40]. A successful u-test (with $p < p_0$ at risk level $\alpha = 0,05$) will reject a null hypothesis. Table 1 shows the results of hypothesis testing (cf. Sect. 4.2). In summary, hypotheses $H_{1,1}$, $H_{1,2}$, and $H_{1,4}$ can be accepted. Despite the number of significant results, like $H_{1,6}$, $H_{1,5}$ is only partially supported, and thus both hypotheses cannot be accepted. In addition, $H_{1,3}$ shows no significance and, hence, must be rejected. Based on the results, we may conclude that social distance (i.e., event which is not self-experienced) leads to a change in the *quality*, *granularity*, and *structure* of resulting process models.

6.4 Discussion

The results indicate that process designers showing a high social distance (i.e. passive participation) to a particular business process tend to create a more fine-grained, detailed, and complete process model, i.e., reflecting a high *semantic quality* and *granularity*. In turn, process designers showing a low social distance (i.e. active participation) create a more course-grained and abstract, but easy to understand process model with less syntactical errors, i.e., reflecting a high *pragmatic* and *syntactic quality*. Regarding *perceived quality* and *process model structure*, final results do not show any or only small differences.

Interestingly, the results only partially comply with CLT (cf. Sect. 2) and our previous experiment [19]. It appears that the investigated factor of the social distance (cf. Sect. 4) has a different impact on the *process of process modeling* and, hence, resulting outcomes differ in several aspects (cf. Sect. 6.1). As possible explanation an active participation results in major attention devoted to actions performed by oneself, while a passive participation results to equal attention paid to all details [41]. BPMN knowledge might be a critical moderator reversing the relationship between construal level and distance (i.e., social distance) leading to circumstances where the abstract seems near and the concrete seems far [42].

However, combining previous results, in general, one can assume that social distance leads to a change in the *quality, granularity*, and *structure* of resulting process models. It is noteworthy that results differ depending on how a process designer experiences social distance, i.e., relation to other individuals or events which are not self-experienced. For enterprises, it is thus recommended to evaluate the modeling domain and, hence, to involve specific process designers to ensure desired outcomes; e.g., to achieve a high process model quality, it is thus recommended to involve process designers being more confident with corresponding business processes.

7   Related Work

This paper investigates the impact of social distance on the quality, granularity, and structure of process models. The work is related to frameworks and guidelines dealing with process model quality. SEQUAL uses semiotic theory for identifying various aspects of process model quality [25], whereas GoM describes quality considerations for process models [7]; 7PMG, in turn, characterizes desirable properties of a process model [8]. Moreover, research on comprehensibility and maintainability exists. The influence of model complexity on process model comprehensibility is investigated in [5]. [35] discusses factors for errors in process models; [43] discusses the impact of different quality metrics on error probability.

[44] provides prediction models for true usability and maintainability of process models. How and at which level of granularity a designer models a particular process is described in [26]. In the context of process modeling only little work exists that takes cognitive aspects into account. [28] presents the effects of reducing cognitive load on end user understanding of conceptual models, whereas [11] describes the cognitive difficulty of understanding different relations between process model elements.

Common to all these approaches is the focus on the created process model (i.e., the *product of process modeling*), while little attention has been paid on the *process of the process* modeling itself. Nautilus complements related work by investigating the process of process modeling for tracing model quality back to modeling strategies resulting in process models of different quality [45].

The effectiveness of gamification based on a quality service model analyzing the social and psychological motivations of participants is discussed in [46]. Agile and efficient responds to changing requirements and consequential amendments to corresponding business processes are provided in [47], based on a gamification and BPM approach incorporated into a social network. Finally, [48] provides preliminary evidence that blending process management to gamification concepts may be beneficial.

Considerable work involving conceptual modeling of processes in a 3D virtual world can be found in [49]. In addition, [50] provides an approach for collaborative process modeling using a 3D environment. A similar use case in a 3D scenario to visualize storyboards for business process models is proposed in [51].

8   Conclusion

This paper investigated whether social distance affects the *process of process modeling* and its *outcomes*, i.e., the *quality, granularity,* and resulting *process model structure.* In particular, an experiment using gamification in a virtual world was conducted showing that there are significant differences depending on whether a process designer has a low or high social distance to the modeled domain. While first results look promising, further investigations are desirable. More precisely, their generalization needs to be confirmed by additional empirical experiments to obtain more accurate results allowing for such a generalization.

As a next step, we will focus on psychological distance (i.e., social, spatial, temporal, hypothetical) as well as the use of gamification and virtual worlds to learn more about the particular effects on the *process of process modeling.* Combining experiment results enables us to extract guidelines on how modeling teams in enterprises should be composed and optimal process models can be obtained. Finally, experiments with practitioners are planned to validate results in real-world scenarios.

References

1. Weber, B., Reichert, M., Mendling, J., Reijers, H.A.: Refactoring large process model repositories. Comput. Ind. **62**(5), 467–486 (2011)
2. Mendling, J.: Metrics for Process Models: Empirical Foundations of Verifiation, Error Prediction, and Guidelines for Correctness. Springer, Heidelberg (2008)
3. Reijers, H.A., Mendling, J.: A study into the factors that influence the understandability of business process models. In: Systems, Man and Cybernetics, Part A: Systems and Humans. pp. 1–14 (2011)
4. Moody, D.L.: Theoretical and practical issues in evaluating the quality of conceptual models: current state and future directions. Data Knowl. Eng. **55**(3), 243–276 (2005)
5. Mendling, J., Reijers, H.A., Cardoso, J.: What makes process models understandable? In: Alonso, G., Dadam, P., Rosemann, M. (eds.) BPM 2007. LNCS, vol. 4714, pp. 48–63. Springer, Heidelberg (2007)
6. Mendling, J., Strembeck, M.: Influence factors of understanding business process models. In: Abramowicz, W., Fensel, D. (eds.) BIS 2008. LNBIP, vol. 7, pp. 142–153. Springer, Heidelberg (2008)
7. Becker, J., Rosemann, M., von Uthmann, C.: Guidelines of business process modeling. In: Aalst, W.M.P., Desel, J., Oberweis, A. (eds.) Business Process Management. LNCS, vol. 1806, pp. 30–49. Springer, Heidelberg (2000)
8. Mendling, J., Reijers, H.A., van der Aalst, W.M.: Seven process modeling guidelines (7PMG). Inf. Softw. Technol. **52**(2), 127–136 (2010)
9. BPM Research Cluster: Cheetah Experimental Platform (2016). http://bpm.q-e.at/?page_id=56
10. Figl, K., Weber, B.: Individual creativity in designing business processes. In: Bajec, M., Eder, J. (eds.) CAiSE Workshops 2012. LNBIP, vol. 112, pp. 294–306. Springer, Heidelberg (2012)
11. Figl, K., Laue, R.: Cognitive complexity in business process modeling. In: Mouratidis, H., Rolland, C. (eds.) CAiSE 2011. LNCS, vol. 6741, pp. 452–466. Springer, Heidelberg (2011)
12. Trope, Y., Liberman, N., Wakslak, C.: Construal levels and psychological distance: effects on representation, prediction, evaluation, and behavior. J. Consum. Psychol. **17**, 83–95 (2007)

13. Todorov, A., Goren, A., Trope, Y.: Probability as a psychological distance: construal and preferences. J. Exp. Soc. Psychol. **43**, 473–482 (2007)
14. Trope, Y., Liberman, N.: Construal-level theory of psychological distance. Psychol. Rev. **117**, 440–463 (2010)
15. Day, S., Bartels, D.: Representation over time: the effects of temporal distance on similarity. Cognition **106**, 1504–1513 (2008)
16. Liberman, N., Sagristano, M.D., Trope, Y.: The effect of temporal distance on level of mental construal. J. Exp. Soc. Psychol. **38**, 523–534 (2002)
17. Fujita, K., Henderson, M.D., Eng, J., Trope, Y., Liberman, N.: Spatial distance and mental construal of social events. Psychol. Sci. **17**, 278–282 (2006)
18. Pronin, E., Olivola, C.Y., Kennedy, K.A.: Doing unto future selves as you would do unto others: psychological distance and decision making. Pers. Soc. Psychol. Bull. **34**, 224–236 (2008)
19. Kolb, J., Zimoch, M., Weber, B., Reichert, M.: How social distance of process designers affects the process of process modeling: insights from a controlled experiment. In: Proceedings of the SAC 2014. pp. 1364–1370 (2014)
20. Deterding, S., Dixon, D., Khaled, R., Nacke, L.: From game design elements to gamefulness: defining "gamification". In: MindTrek 2015. pp. 9–15 (2011)
21. Zichermann, G., Cunningham, C.: Gamification by Design: Implementing Game Mechanics in Web and Mobile Apps. O'Reilly, Sebastopol (2011)
22. Bartle, R.A.: Designing Virtual Worlds. New Riders, Berkeley (2004)
23. Davis, A., Khazanchi, D., Murphy, J., Zigurs, I., Owens, D.: Avatars, people, and virtual worlds: foundations for research in metaverses. J. Assoc. Inf. Syst. **10**, 90–117 (2009)
24. Moody, D.L.: The "physics" of notations: toward a scientific basis for constructing visual notations in software engineering. Softw. Eng. **35**(6), 756–779 (2008)
25. Krogstie, J.: Model-Based Development and Evolution of Information Systems. Springer, London (2012)
26. Holschke, O., Rake, J., Levina, O.: Granularity as a cognitive factor in the effectiveness of business process model reuse. In: Dayal, U., Eder, J., Koehler, J., Reijers, H.A. (eds.) BPM 2009. LNCS, vol. 5701, pp. 245–260. Springer, Heidelberg (2009)
27. Polyvyanyy, A., Smirnov, S., Weske, M.: On application of structural decomposition for process model abstraction. In: Proceedings of the BPSC 2009. pp. 110–122 (2009)
28. Moody, D.L.: Cognitive load effects on end user understanding of conceptual models: an experimental analysis. In: Benczúr, A.A., Demetrovics, J., Gottlob, G. (eds.) ADBIS 2004. LNCS, vol. 3255, pp. 129–143. Springer, Heidelberg (2004)
29. Wolf, C., Harmon, P.: The state of business process management 2012. In: BPTrends Report (2012)
30. Petre, M.: Why looking isn't always seeing: readership skills and graphical programming. Commun. ACM **38**, 33–44 (1995)
31. Lindland, O.I., Sindre, G., Solvberg, A.: Understanding quality in conceptual modeling. IEEE Softw. **11**(2), 42–49 (1994)
32. Krogstie, J., Sindre, G., Jørgensen, H.: Process models representing knowledge for action: a revised quality framework. J. Inf. Syst. **15**, 91–102 (2006)
33. Recker, J., Safrudin, N., Rosemann, M.: How novices model business processes. In: Hull, R., Mendling, J., Tai, S. (eds.) BPM 2010. LNCS, vol. 6336, pp. 29–44. Springer, Heidelberg (2010)
34. Rittgen, P.: Quality and Perceived Usefulness of Process Models. In: Proceedings of the 24th Symposium on Applied Computing (SAC 2010). pp. 65–72 (2010)

35. Mendling, J., Neumann, G.: Error Metrics for Business Process Models. In: Proceedings of the CAISE 2007. pp. 53–56 (2007)
36. Wohlin, C., Runeson, P., Höst, M., Ohlsson, M.C., Regnell, B., Wesslen, A.: Experimentation in Software Engineering - An Introduction. Kluwer, Norwell (2000)
37. Höst, M., Regnell, B., Wohlin, C.: Using students as subjects a comparative study of students and professionals in lead-time impact assessment. Empirical Softw. Eng. **5**, 201–214 (2000)
38. Svahnberg, M., Aurum, A., Wohlin, C.: Using students as subjects - an empirical evaluation. In: ESEM 2008, ACM. pp. 288–290 (2008)
39. Kline, P.: Handbook of Psychological Testing, vol. 2. Routledge, New York (1999)
40. Sirkin, M.: Statistics for the Social Sciences, vol. 3. Sage, Thousand Oaks (2005)
41. Tversky, A., Kahnemann, D.: Availability: a heuristic for judging frequency and probability. Cogn. Psychol. **5**, 207–232 (1973)
42. Kyung, E.J., Menon, G., Trope, Y.: Construal level and temporal judgments of the past: the moderating role of knowledge. Psychon. Bull. Rev. **21**, 734–739 (2014)
43. Mendling, J., Verbeek, H.M.W., van Dongen, B.F., van der Aalst, W.M.P., Neumann, G.: Detection and prediction of errors in EPCs of the SAP reference model. Data Knowl. Eng. **64**(1), 312–329 (2008)
44. Aguilar, E.R., Sanchez, L., Carballeira, F.G., Ruiz, F., Piattini, M., Caivano, D., Visaggio, G.: Prediction models for BPMN usability and maintainability. In: Proceedings of the CEC 2009. pp. 383–390 (2009)
45. Pinggera, J., Soffer, P., Fahland, D., Weidlich, M., Zugal, S., Weber, B., Reijers, H., Mendling, J.: Styles in business process modeling: an exploration and a model. Soft Syst. Model. **14**, 1055–1080 (2013)
46. Aparicio, A.F., Vela, F.L.G., Sánchez, J.L.G., Montes, J.L.I.: Analysis and application of gamification. In: New Trends in Interaction, Virtual Reality and Modeling, Human-Computer Interaction Series. pp. 113–126 (2013)
47. Santorum, M., Front, A., Rieu, D.: ISEAsy: a social business process management platform. In: Lohmann, N., Song, M., Wohed, P. (eds.) BPM 2013 Workshops. LNBIP, vol. 171, pp. 125–137. Springer, Switzerland (2014)
48. Brito, T.P., Paes, J., Moura, A.B.: Game-based learning in IT service transition. In: Proceedings of the CSEDU 2014. pp. 110–116 (2014)
49. Brown, R.A.: Conceptual modelling in 3D virtual worlds for process communication. In: Proceedings of the APCCM 2010. pp. 25–32 (2010)
50. West, S., Brown, R.A., Recker, J.C.: Collaborative Business Process Modeling Using 3D Virtual Environments. In: Proceedings of the APCCM. pp. 51–60 (2010)
51. Kathleen, N., Brown, R.A., Kriglstein, S.: Storyboard augmentation of process model grammars for stakeholder communication. In: Proceedings of the IVAPP 2014. pp. 114–121 (2014)

What Have We Unlearned Since the Early Days of the Process Movement?

Gil Regev[1,2(✉)], Olivier Hayard[2], and Alain Wegmann[1]

[1] School of Computer and Communication Sciences,
Ecole Polytechnique Fédérale de Lausanne (EPFL), 1015 Lausanne, Switzerland
{gil.regev,alain.wegmann}@epfl.ch
[2] Itecor, Av. Paul Cérésole 24, cp 568, 1800 Vevey 1, Switzerland
{g.regev,o.hayard}@itecor.com

Abstract. The vision brought forth by Michael Hammer in the late 1980s was to save struggling American companies by getting them to focus on the creation of value for clients by reorganizing their operations and structure around the use of IT systems. This was the Business Process Reengineering (BPR) movement. It spawned most if not all the business process work since then, including BPMDS. The main principle behind BPR was to design the envisioned process around outcomes (value), not tasks. In this paper, we show that this principle was not heeded by the followers of BPR. This is plainly visible when you look into any example of a process model done with modern process modeling notations, such as BPMN. What one sees is mostly a set of interconnected tasks, with mostly an implicit outcome. It is about time we went back to the early principles of BPR and connected people by explicitly showing the collaboration between the actors of the process, the outcome of the process, and only then designing the activities and their sequence.

1 Introduction

The BPR (Business Process Reengineering or Redesign) movement erupted on the business scene in the early 1990s promising to save organizations, especially large corporations, from certain decline by radically transforming their work practices with the help of IT systems. Business Process Management (BPM) as a business and research discipline followed a few years later.

The founding fathers of the BPR movement, Davenport, Hammer, Short and Champy [3, 4, 7–9] saw the transition to a business process view as a way of parting from the industrial age way of organizing work for predictable markets.

Whereas in the industrial age the main focus was on scaling production to meet an ever growing demand for generic products, businesses now needed a way of guaranteeing affordable quality, on-time delivery of products and service that meet (often individual) customer needs.

Existing ways of working, whether in manufacturing or in the service industries, were seen as too complicated and intricate. A radical simplification was seen as necessary. This simplification was powered by the possibilities of sharing information across time and space afforded by IT systems. It became possible to design work with the

© Springer International Publishing Switzerland 2016
R. Schmidt et al. (Eds.): BPMDS/EMMSAD 2016, LNBIP 248, pp. 113–121, 2016.
DOI: 10.1007/978-3-319-39429-9_8

outcome in mind, the product or service matching expectations delivered on-time and within a budget. The intricacies of delivering these products and services could be greatly simplified with an emphasis on the collaboration between individuals across departments.

In this paper we show that present day most common BPM tools focus more on the intricacies of delivering a product and service rather than on the outcome and collaboration envisioned by the founding fathers. We argue that it is both desirable and possible to add a modeling phase where the focus is on the collaboration and outcome before modeling the details of the process. Our purpose is not to propose yet another BP modeling notation, nor to trivialize BP modeling. Our aim is to help BP modelers to realize that BP modeling is not solely about describing tasks and their sequencing, that more high-level descriptions enables them to design better processes.

2   Outcome vs. Tasks

From the early writings of Davenport and Short [4], and Hammer [7] to their subsequent books [3, 8, 9], the vision for BPR remained remarkably steady. The organization of work as a fragmentation of the assembly of a complete product into a series of routine tasks, while necessary and useful during the industrial age, was preventing companies from offering the products and services customers were expecting in the 1980s.

This is beautifully said by Hammer in the following quote [7]:

> "Conventional process structures are fragmented and piecemeal, and they lack the integration necessary to maintain quality and service. They are breeding grounds for tunnel vision, as people tend to substitute the narrow goals of their particular department for the larger goals of the process as a whole. When work is handed off from person to person and unit to unit, delays and errors are inevitable. Accountability blurs, and critical issues fall between the cracks. Moreover, no one sees enough of the big picture to be able to respond quickly to new situations. Managers desperately try, like all the king's horses and all the king's men, to piece together the fragmented pieces of business processes."

The managers and supporting staff are called by Hammer and Champy [9]: "the glue that holds together the people who do the real work". They state that in many organizations the cost of the glue has surpassed the cost of direct labor. This, they argued, leads to "Inflexibility, unresponsiveness, the absence of customer focus, an obsession with activity rather than result." Likewise Davenport and Short [4] say that "difficult inter-functional" (inter-departmental) issues were hampering many quality improvement efforts in manufacturing companies. They report on an example where different departments within a company each optimized its performance but the overall process "was quite lengthy and unwieldy" [4]. Hammer and Champy [9] therefore advocate that processes "must be kept simple." Designing processes with many handovers from one person to another and what's more crossing department boundaries is unlikely to yield good results.

The remedy envisioned in BPR was to see the whole process, from end to end, and focus more on the results created by the process than the tasks and their coordination. The definition of the concept of business process, as set forth by Hammer and Champy [9] embodies both views:

"a collection of activities that takes one or more kinds of input and creates an output that is of value to the customer."

It is easy to get carried away by the focus on the activities comprising the process at the expense of the input, output and value. There is probably a natural tendency for modelers to think of a process in terms of a sequence of activities. There was also the need to implement an IT system that supports the process, hence the need to give the details of the activities.

This was, however, not at all what was meant by Hammer and Champy. Addressing this problem at a later stage, Hammer provided an updated definition of a business process [8]: "a group of tasks that together create a result of value to a customer" and specified that "The key words in this definition are 'group,' 'together,' 'result,' and 'customer'." Note that tasks is not in the list of key words. This is because for Hammer the tasks are less important. He defines the process perspective as seeing the [8] "collection of tasks that contribute to a desired outcome" rather than isolated tasks. He goes as far as saying that [8]: "the essence of a process is its inputs and its outputs [..] Everything else is detail." therefore recommends that a process be seen as a black box that creates value by transforming inputs into outputs.

This black box view can be understood as the direct result of the clean slate approach advocated by Hammer [7, 8] and Hammer and Champy [9]. In this approach there is no reason to analyze the existing process because it will only tie the analysts' minds into the old ways of working. The process should be created anew by focusing only on the desired results. This is a marked difference from the other BPR current by Davenport [3] and Davenport and Short [4]. They emphasize the need for a process vision, outcome and objectives, but still recommend the analysis of the existing process citing four reasons: (1) Developing a common understanding among the people involved in the process redesign; (2) Easing the migration to the new process; (3) Recognizing problems in the existing process so as not to repeat them; (4) Providing a baseline for measuring the new process.

Business Process Reengineering, more than Business Process Redesign, sought to make radical transformation from a clean sheet. If only minor improvements were needed, e.g. 10 % increase in profit or 10 % decrease in cost, no reengineering was called for according to Hammer and Champy, rather, a quality program or some such was entirely sufficient [9]. Our point here is not to argue for or against radical change. We do argue in favor of an initial step in business process modeling where the process activities are not modeled at all. Process modeling should begin by describing the process as a black box, as advocated by Hammer and Champy, focusing solely on the outcomes of the process for its stakeholders and not activities, their sequencing or their attribution to roles.

In essence, we seek to model how the business process brings suppliers, customers and regulators together to create value for them all. It is useful to remember that Davenport and Short envisioned early on the use of groupware to improve interpersonal processes [4]. Likewise, Hammer and Champy saw companies, not as asset portfolios but as "people working together to invent, make, sell, and provide service." [9].

3   Scope of the Process or Who is the Customer

The focus on customer satisfaction as the focal point of process design is common to both currents of BPR. Davenport and Short [4] identify two main process characteristics. Processes have customers (internal or external) and they cross organizational boundaries. For Hammer [8] the first principle of process design is to be customer-driven.

The customer-driven focus ties back to the outcome of the process. The first accounts of these outcomes were quite simple. Hammer and Champy [9], for example, define that the order fulfillment process ends when the goods are delivered. Hammer [8] moves this end point to when the customer has paid the bill, defining three outputs for the process: the delivered goods, the satisfied customer and the paid bill. The paid bill, according to Hammer [8], is the best indication that the customer is satisfied. Moving the end point of the process and including more outputs in the outcome is synonymous with the expansion of the scope of the process. Davenport and Short [4] consider this scope expansion to be a key issue of process analysis, for example adding order entry into the sales process.

Hammer [8] advocates to analyze the process from the outside-in so that customer requirements drive the process design. This means beginning by defining what quality, flexibility, delay, price and such that customers expect from the process. There is an inherent limitation in this reasoning in that it considers the customer in the singular form. With respect to his earlier work, Hammer [8] stretches this notion of single customer to multiple customers. He lists all of the following as customers of a pharmaceutical company selling a medicine [8]:

A. "The patient who takes a medicine.
B. The physician who prescribes it.
C. The pharmacist who dispenses it.
D. The wholesaler who distributes it.
E. The Food and Drug Administration scientists and officials who approve its use.
F. The insurance company that pays for it"

The question is then, what makes these people customers of the pharmaceutical company? What is the determinant of a customer? Hammer definition of a customer is [8]: "Customers are people whose behavior the company wishes to influence by providing them with value." It so happens that the people whose behavior the company wishes to influence include those we call suppliers. Indeed, Suppliers who do not receive what they expect from a client can also refuse to be part of its process and thus avoid being influenced and given value by the process. Hammer [8] acknowledges this by stating that the process should include measures that ensure that the company doesn't go broke trying to satisfy customers. This means that despite earlier claims by the BPR proponents, the company cannot simply design a process that caters to all the wishes of its customers. Most business processes include steps that cater to the different needs of the stakeholders involved [15]. All the stakeholders participating in the process have to somehow receive some value from their participation. Notice that this may be negative value for so called disfavored users [16].

4   BPM Missing the Big Picture

In BPM the accepted definition of a business process is quite similar to the one proposed by Hammer and Champy (see above), namely [11]: "a set of partially ordered activities intended to reach a goal."

Unfortunately, most of the attention of practitioners and researchers alike has been on the set of activities rather than on the goal (or outcome). As a result, the vast majority of notations used for business process modeling e.g., BPMN, YAWL, UML, IDEF, Petri nets, EPC, have elaborate constructs for modeling activities, their sequence, exceptions, messages, roles etc.

Figure 1 shows a typical business process model designed with BPMN. This is the level of detail at which most business processes are modeled. We see the sequence of activities assigned to the patient and to the doctor's office. What cannot be seen in this model, at least not explicitly, is why these two stakeholders (the patient and the doctor's office) engage in this process. What is the outcome for them? With enough scrutiny, we can see that an "illness occurs" for the patient who then begins on a long track of dialoging with the doctor's office. By the end of the process, the doctor's office sends a medicine that is received by the patient. What is the effect of the medicine on the patient and what the doctor's office receives in return for its involvement in this process, is not part of the model.

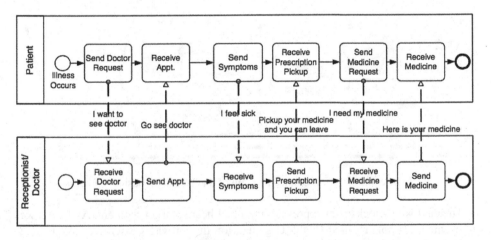

Fig. 1. Example of a BPMN business process model (source: [12])

As we can see very little or no provision is given to modeling the outcome, goal or value for the stakeholders who participate in the process. This is essentially the same as putting together the industrial era tasks, dividing them among those who perform the process, and hoping for the best. Remember Hammer's definition that a process must be first viewed as something that transforms an input into an output. Where are the collection of activities, the togetherness of the stakeholders? Where is the value of the process and for whom?

Also not part of the process in Fig. 1 are the other stakeholders mentioned by Hammer, e.g. the pharmacist, the wholesaler, the FDA. Adding them into the process model, at the level it is described in Fig. 1, will render an already complicated model very cumbersome. By modeling the process as a black box, it is possible to show its effect on multiple stakeholders while keeping the model quite simple. This is the subject of the following section.

5   Collaboration Over Task Orchestration

In Fig. 2, we show one way of modeling processes as black boxes and their outcome for the stakeholders involved. The modeling notation used is SEAM [19] but others can be used too (See Related Work).

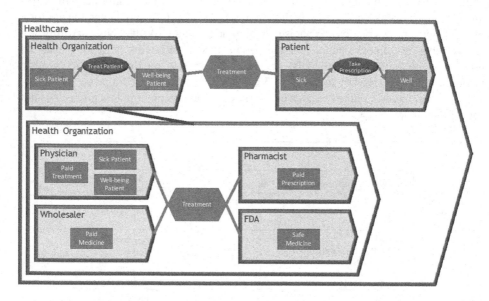

Fig. 2. A collaboration and outcome view of the Treatment business process

The enclosing block arrow represents the Healthcare market segment. At the top we represented a supplier called Health Organization that provides a service called Treat Patient to a Stakeholder called Patient. The two stakeholders are bound by a relationship that we call Treatment. It is this relationship that represents the business process as a black box. The business process creates changes on both the Health Organization and the Patient at the same time. In this example the Patient goes from the initial state of being Sick to the final state of being Well. Simultaneously, the Patient's record in the Health Organization is transformed from Sick Patient to Well-being Patient. The Well-being Patient state and the Well state are the final states of the process and together define its outcome. How these transformations are done, through what sequence of actions is not modeled here. Only the inputs and outputs as described by Hammer and Champy [9] and Hammer [8].

The upper Health Organization and the Patient are shown as black boxes. Only their activities and states are shown. The bottom block arrow names Health Organization is a white box representation. It shows the stakeholders participating in the provision of the Treat Patient service. Here again, we see a process called Treatment. This process is internal to the Health Organization. It connects the internal stakeholders that the Patient may or may not see. We described those found in Hammer's description quoted above, namely: Physician, Pharmacist, Wholesaler and Food and Drug Administration (FDA). Each one or more states corresponding to the outcome of the process. The Physician, for example, has three states, she records the Sick Patient, the Well-Being patient and the Paid Treatment. Notice that we don't show the activities with an input and output as in the upper level stakeholders. This is to show that the notation can be used with even more parsimony.

With the global view afforded by the model of the process as a black box and its outcomes it is possible to expand or shrink the scope of the project as envisioned by the founding fathers of BPR.

We'd like to emphasize that our point is not to "trivialize" process modeling by reducing it to the black box view. Our proposal is to augment BPMN style process models with with a more abstract view that affords to reflect on the process as managing a set of relationships between stakeholders. It has been shown in [20] that with SEAM it is possible to drill down from the black box view of the process to a BPMN like process diagram embedded in its context.

6 Related Work

The representation of the process as a collaboration between several stakeholders was imported into SEAM from Catalysis [2] where it is called a joint action: an abstraction of "multiple interactions" that shows "the net effect on all participants." Note that there is a concept called black-box pool in BPMN. It is mostly used to represent external stakeholders or when there is no need to represent the set of activities in the pool. This however does not show the collaboration and its effects. In BPMN, the closest concept to our description of a collaboration (joint action) is the Transaction a Sub-Process that [12], "leads to an agreed, consistent, and verifiable outcome across all participants."

As we have said in Sect. 5, SEAM is not the only method that can be used in order to more clearly view the net effect of a business process on its stakeholders. Other methods or frameworks include, the state-oriented business process modeling [10], BMM [13], SIPOC [18], e3Value [6] and BMG [5, 14].

A simple example is Samarin's [17] proposal to begin by modeling a business process as a black box, showing only its input, output, guidance and resources with an IDEF0 diagram. This goes a long way toward modeling the essence of the process, but stops short of explicitly showing the collaboration between stakeholders and the outcome for each one.

7   The Loss of the Big Picture in BPMDS

A direct parallel can be drawn between the road taken by the BP modeling tools and the Working Conference on Business Process Modeling Development and Support (BPMDS). BPMDS was created in 1998 by Ilia Bider and Maxim Khomyakov as workshops whose goal was to [1], "facilitate discussions of the topics relevant to the practice of modeling and building computerized support [to business processes]." Unlike most academic conferences, the organizers of BPMDS made serious efforts to bring together practitioners and researchers in the field of business process modeling and to maintain open discussions about business process issues. Access to practitioners was greatly facilitated with help available for transforming their papers into academic articles, and sessions were devoted to open discussions and brainstorming about business process issues. However, as BPMDS enjoyed increasing success with academics, which ensured its survival, it became harder to invite practitioners and to discuss general purpose business process issues. BPMDS became a regular working conference devoted almost exclusively to technical academic paper presentation. This evolution made a powerhouse of academic publishing, but at the loss of its identity as a meeting place for practice and research devoted to discussing business process topics.

8   Conclusions

In this short paper we discussed the vision of the pioneers of the BPR movement, who viewed business processes as connecting and providing value to their stakeholders. We showed that this vision seems to have been overlooked by business process modelers with the result that business model modeling notations offer scarcely any modeling construct for that purpose. We then described, ever so briefly, a modeling notation that has explicit elements for modeling a business process as a collaboration among its stakeholders, including the value it provides them. We hope that this work will be a first step toward a rebirth of the BPR vision for BPM practitioners and researchers.

One interesting research that can be spawned from this was of thinking is to trace every task in a business process to some outcome for a stakeholder. This would establish traceability links between tasks and outcomes and will go toward one of the wishes of BPR, namely to remove all tasks that do not bring value. Without knowing which value is created by which task it is a guessing game to remove tasks from a process.

References

1. Bider, I.: Short Bio of Dr. Maxim Khomyakov. In: Proceedings of the HCI*02 Workshop on Goal-Oriented Business Process Modeling (2002)
2. D'Souza, D.F., Cameron, W.A.: Objects, Components and Frameworks with UML: The Catalysis Approach. Addison-Wesley, Boston (1999)
3. Davenport, T.H.: Process Innovation: Reengineering Work Through Information Technology. Harvard Business School Press, Boston (1993)

4. Davenport, T.H., Short J.E.: The new industrial engineering: information technology and business process redesign. MIT Sloan Management Review, July 15 1990
5. Fritscher, B., Pigneur, Y.: Business IT alignment from business model to enterprise architecture. In: Salinesi, C., Pastor, O. (eds.) CAiSE Workshops 2011. LNBIP, vol. 83, pp. 4–15. Springer, Heidelberg (2011)
6. Gordijn, J., Akkermans, J.M.: Value-based requirements engineering: exploring innovative e-commerce ideas. Requirements Eng. **8**(2), 114–134 (2003)
7. Hammer, M.: Reengineering Work: Don't Automate Obliterate. Harvard Business Review, Boston (1990)
8. Hammer, M.: Beyond Reengineering. HarperCollins, New York (1996)
9. Hammer, M., Champy, J.: Reengineering the Corporation: A Manifesto for Business Revolution. Nicholas Brealey, Boston (1993)
10. Khomyakov, M., Bider, I.: Achieving workflow flexibility through taming the chaos. In: Patel, D., Choudhury, I., Patel, S., de Cesare, S. (eds.) OOIS 2000, pp. 85–92. Springer, Heidelberg (2000)
11. Kueng, P., Kawalek, P.: Goal-based business process models: creation and evaluation. Bus. Process Manag. **3**(1), 17–38 (1997)
12. OMG: Business Process Model and Notation (BPMN) (2011)
13. OMG: Business Motivation Model (BMM) (2015)
14. Osterwalder, A., Pigneur, Y.: Business Model Generation (2010)
15. Regev, G., Alexander, I.F., Wegmann, A.: Modelling the regulative role of business processes with use and misuse cases. Bus. Process Manag. **11**(6), 695–708 (2005)
16. Regev, G., Hayard, O., Gause, D.C., Wegmann, A.: Toward a service management quality model. In: Glinz, M., Heymans, P. (eds.) REFSQ 2009 Amsterdam. LNCS, vol. 5512, pp. 16–21. Springer, Heidelberg (2009)
17. Samarin, A.: Improving Enterprise Business Process Management Systems. Trafford, Bloomington (2009)
18. Thomas, D.: Identifying High-Level Requirements Using SIPOC Diagram. http://www.isixsigma.com/tools-templates/sipoc-copis/identifying-high-level-requirements-using-sipoc-diagram/ (undated), Accessed 3 April 2016
19. Wegmann, A.: On the systemic enterprise architecture methodology (SEAM). In: 5th International Conference on Enterprise Information Systems, ICEIS 2003, pp. 483–490 (2003)
20. Wegmann, A., Regev, G., Rychkova, I., Lê, L.-S., de la Cruz, J.D., Julia, P.: Business and IT alignment with SEAM for enterprise architecture. In: 11th IEEE International EDOC Conference on EDOC 2007 (2007)

New Directions in Process Modeling

Towards Impact Analysis of Data in Business Processes

Arava Tsoury[✉], Pnina Soffer, and Iris Reinhartz-Berger

University of Haifa, Mount Carmel, Haifa 3498838, Israel
{atsoury, spnina, iris}@is.haifa.ac.il

Abstract. Business processes heavily rely on data. Data is used as input for activities; it is manipulated during process execution and it serves for decisions made during the process. Thus, changes (in values or structure) of data may influence large portions of the business process. We introduce in this paper the concept of 'data impact analysis' which analyzes the effects of data elements on other business process elements, including activities, routing constraints, and other data elements. This type of analysis is important in scenarios such as process or database redesign and unexpected changes in data values. The paper further proposes a set of primitives depicting impacts of data within business processes, and demonstrates the use of these primitives to query the overall impact of a data element within a business process.

Keywords: Business processes · Impact analysis · Data inaccuracy · Process redesign · Database

1 Introduction

Business processes highly influence the business success of organizations and their interaction with the environment [15]. While business processes are manifested through activities and decisions, these essentially rely on data, typically stored in information systems and manipulated during process execution. Data elements are not stable, i.e., their values essentially change over time. Moreover, their structure may also be changed occasionally. Due to the fundamental role of data throughout the process, such changes cannot be considered as local changes.

In this paper, we introduce the concept of *data impact analysis* in business processes. We borrow this term from a wider concept of business impact analysis, which deals with identification and examination of possible changes in business conditions (both natural and human-caused) and their effects on critical business functions [18]. Business impact analysis is particularly used for risk management and disaster recovery. Data impact analysis aims to play a similar role, focusing on data elements involved in business processes. In other words, data impact analysis identifies for each data element its effects on other business process elements within a single process or across processes.

Data impact analysis can be useful in a variety of situations. At redesign, modifications are made to the database schema or the process model due to, e.g., changes in the requirements or the organizational regulations. Examples of modification in the

© Springer International Publishing Switzerland 2016
R. Schmidt et al. (Eds.): BPMDS/EMMSAD 2016, LNBIP 248, pp. 125–140, 2016.
DOI: 10.1007/978-3-319-39429-9_9

database schema include changes in the types of elements, adding or omitting attributes or constraints, or changing relations; modification in the process model may involve changes in the control flow. In these cases, it is required to understand how the "local" changes globally affect the business process. In particular, we relate to effects on and of data elements. At runtime, data impact analysis can be useful when unexpected changes in the values of data elements occur due to errors, exceptions, or last minute changes. Data impact analysis can be useful in such cases for analyzing the effect of the changes within the process, enabling risk analysis, and management. In this sense, if an inaccurate or mistaken value is used through the process, the goal of the process might not be achieved. Furthermore, the value might change after many actions have been performed based on the previous value, and all these need to be recognized, corrected, and sometimes compensated for. For this, it is essential to encompass and fully address all the impacts of the data item and other elements that depend on it.

Change management in the area of business processes has mainly been studied in the context of control flow changes [2, 5, 9, 13]. Process adaptation and flexibility approaches support and enable changes [10, 21]. However, a detailed examination of data elements and data values in this context is missing. Dependencies among data and other process elements have also been studied, e.g., [4, 7, 17], but the granularity in which these elements are addressed is coarse and does not support a full analysis of the impact of their values.

To overcome the above gaps, this paper discusses the importance of data impact analysis at the attribute level, introducing a set of primitives depicting impacts of atomic data elements, termed data items, in business processes. These primitives, which are derived from analyzing the relationships between different elements involved in the business process, are the basis for querying the impact of a data element within a process and across processes.

The remainder of the paper is organized as follows: Sect. 2 presents a running example to illustrate and motivate the need for our approach. Sections 3 and 4 present basic definitions and the suggested primitives, respectively, while Sect. 5 elaborates on the data impact analysis approach along with preliminary results. Section 6 discusses the related work and, finally, in Sect. 7 we discuss the contribution and the limitations of the work, paving a road to future research directions.

2 Running Example

To illustrate the need for data impact analysis, consider a make-to-order process, in which customers order products from a catalog, and the products are produced upon demand. Figure 1 depicts a process model of this (ordering) process[1]. The process operates on top of a relational database, which mainly includes tables related to customers and orders (including their life-cycle transitions, e.g., approvals, payments,

---

[1] Due to length limitations, the process model does not specify the data elements that participate in the process. A full model of the process along with its data flow can be found at http://hevra.haifa.ac.il/is-web/staff/data_impact_analysis/ordering_process_example.pdf.

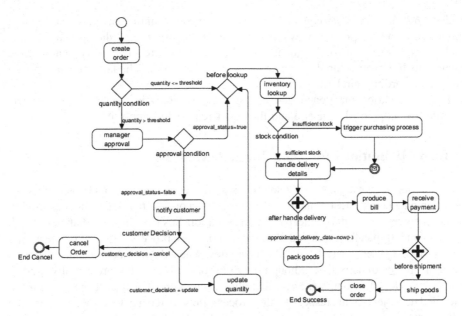

Fig. 1. A process model of the ordering process

shipments). The database also includes tables related to products, their bill of materials, suppliers and material-supplier relations.[2]

Concentrating on redesign and value change scenarios, we next illustrate these scenarios through examples from the make-to-order process.

Redesign scenarios: Assume that the organization has to add an attribute called *unit of measure* that is specified when creating the order and indicates quantities in various units. This is a change to the database schema. However, this change may also cause modifications in calculating the total price when creating the order, handling the delivery details, and packing the goods. Actually, this change may impact all the activities and decisions that involve *quantity* and hence analyzing the impact of this data element is required to understand the potential consequences of the change.

Value change scenarios: Consider that the customer asks to change the shipping address to one that is different than appearing in the order. The value of this attribute is used in various activities, such as *handle delivery details*, and may influence additional data elements along the process, such as *approximate delivery date* and the *packing type* (these attributes may be influenced by the distance to the destination, directly derived from the shipping address).

As another example, consider an error occurred when typing the ordered quantity. As noted, this data element is used throughout the process, e.g., for determining whether manager approval is required or not, as an input to the *inventory lookup* activity, as an input to the purchasing process (in case of insufficient material), and as an input to the

---

[2] The database structure diagram is available online at http://hevra.haifa.ac.il/is-web/staff/data_impact_analysis/ordering_process_data_structure.pdf.

produce bill activity. Consequently, any error in the quantity can have substantial consequences on the process as well as on related processes (e.g., the purchasing process). An incorrect value of the quantity might lead to incorrect decisions regarding the quantity to be manufactured, wrong planning of the work schedule, and misguided decisions regarding purchasing.

In the scenarios mentioned above, data impact analysis would support understanding the scope of the change and the effort required to apply it.

3   Basic Definitions for Data Impact Analysis

Various studies investigated the relationships between processes and data, concentrating on data flows. Most of them use the term 'data object' to refer to the data used in business processes. A data object is an information business artifact that is created, evolved, and (typically) archived through a business process [3]. Data objects are characterized by sets of (related) data attributes. Such data objects usually serve as inputs or outputs of activities, although only part of their attributes are actually used as inputs or produced as outputs. Several studies such as [6, 8, 12] further represent the state of data objects with respect to the process flow, referring to a specific attribute (called "state" or "status"), which changes its value during process execution to reflect the progress in the process performed on the object. Some studies refer to the structural relations between data objects, as specified in the database schema [7].

In practice, process model elements, such as activities, gateways, and routing constraints, use values of specific attributes of the data objects rather than complete data objects. Moreover, Sun et al. [16] claim that current workflow models do not provide details on how different data attributes are used in the process. Consequently, it is difficult to understand the exact impact of each attribute on the whole process. In other words, we claim that the notion of data objects is too coarse-grained for the aim of data impact representation and analysis. Thus, in this paper, we primarily refer to the attributes of data objects. We call them data items and represent them as follows.

Definition 1 (Data item). A data item is a variable or an attribute. It is represented by a pair (n, t), where n is a string identifying the data item name[3] and t represents its type[4].

Data objects can be considered as structures of related data items. Generally, data items can be related to each other irrespectively of the data objects to which they belong. For example, in the make-to-order process, approximate delivery time of orders may be related to the lead-time of the relevant materials delivered by the different suppliers. Therefore, we assume a set of relations between data items.

Definition 2 (Data items relation). Let d_i and d_j be two data items. A data items relation is a pair (d_i, d_j), where d_i relates to d_j[5].

---

[3] For simplicity, we assume that the data item name is unique within the process.

[4] A data item type denotes the possible range of values the data item can assume. It can be considered as a (finite or infinite) set of values. During process execution, a data item has a specific value from this range at a certain time.

[5] Ternary relations and relations of higher degrees are relaxed to binary relations.

These pairs can be directly derived from the database schema.

As noted, data items form a basis for data impact analysis, which reveals their impacts on other elements in a business process, as reflected in a process model. Different definitions of a process model exist in the literature such as in [6, 7]. We refine those definitions by explicitly referring to data items rather than data objects. The following definition is used as the basis for our data impact analysis approach.

Definition 3 (Process model). A process model is a tuple PM = (A, G, F, DI, R, AO, AI), where:

- A is a finite (non-empty) set of activities.
- G is a set of gateways.
- F ⊆ (A ∪ G) × (A ∪ G) is a non-empty set of control flows.
- DI is a finite set of data items.
- R is a set of routing constraints represented as pairs (p, f) where p is a predicate over the power set of DI p: $\wp(DI) \rightarrow$ {true, false} and f∈F.
- AO ⊆ (A × DI) is a set of data flows representing activity outputs.
- AI ⊆ (DI × A) is a set of data flows representing activity inputs.

Figure 1 is an example of a process model, in which DI, AO and AI were omitted (see footnote 1). Even in process models that do represent those elements, the relations between inputs and outputs in a specific activity are not explicitly specified. For example, in the *create order* activity, *total payment* is calculated based on *quantity* and *product price*, ignoring other attributes of the product and the order which may be relevant to the order creation in general. We thus assume the existence of triplets of inputs-activities-outputs, specified by process designers.

Definition 4 (Input-activity-output relation). An input-activity-output relation is a triplet, *(d_i, a, d_j)*, where d_i and d_j are data items, *a* is an activity, and the value of d_i (directly) influences the value of d_j in activity *a*.

d_i or d_j may be Null, allowing cases where the output does not use any specific data input or the input is used without affecting any specific outcome, respectively.

Based on the above definitions, Fig. 2 depicts the elements of a business process and their relations in the form of an Entity Relationship Diagram (ERD). Note that gateways and activities are abstracted as behavioral nodes, to enable jointly associating their connecting edges (flows) with routing constraints. Four of the relations are derived from the process model (*source, destination, for, used in*), three are derived from the input-activity-output relations (*input for, output of, affected through*), and one is derived from the data items relations (*related to*).

Based on the process model definition and the derived ERD (Fig. 2), two kinds of flow relations exist among model elements: control flow (CF) and data flow (DF). CFs allow following the order of the activities in the process, to trace temporal relations between data manipulations. Yet, they do not directly affect data values. For example, gateways are pure control flow elements, passing or receiving control to or from edges, whose routing constraints are evaluated as true. Hence, they can be affected by data items through routing constraints on their incoming flows, but have no data flow effect of their own.

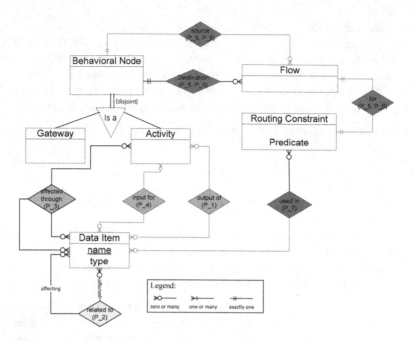

Fig. 2. An ERD depicting the constituents of a business process and their relations

Considering the DFs, which are the focus of this paper, we define seven primitives for conducting data impact analysis. These primitives are derived from the model in Fig. 2. For traceability, the primitives' identities (P_1-P_7) appear in that model close to the relations from which they are derived. Note that primitives P_5 and P_6 are derived from three relations: *source, destination,* and *for.*

4 Primitives for Data Impact Analysis

We elaborate here on each one of the seven primitives by describing its essence, illustrating it through the running example, and suggesting a notation. The uses of these primitives for conducting data impact analysis are discussed in the next section.

Primitive 1 (P_1): An activity affects a data item

Description: a data item d_j is manipulated (created, deleted, or value-updated) by an activity a_x. Activity a_x does not use any data item d_i for this manipulation.
Example: The *manager approval* activity writes a data item *approving employee number* that documents the manager who handled the approval of the order, independently of the inputs to this activity (i.e., the order details).
Notation: $O_{a_x}(d_j)$, d_j is the output of a_x.

Primitive 2 (P_2): A data item affects another data item

Description: when a data item d_i changes its value, another data item d_j is manipulated due to structural constraints imposed in the database, irrespectively of the business process activities.

Example: A Boolean data item *stock below reorder level* is reevaluated whenever the data item *amount in stock* is changed: if the amount in stock is decreased below a certain threshold, *stock below reorder level* becomes true.

Notation: $S_{d_i}(d_j)$, change in the value of data item d_i causes a change in the value of data item d_j, irrespectively of specific activities.

Primitive 3 (P_3): A data item affects another data item through an activity

Description: Data item d_j is an output of activity a_x. Activity a_x uses the value of data item d_i for manipulating the value of d_j.

Example: The data items *quantity* and *product price* affect the data item *total payment* through the activity *Produce Bill*.

Notation: $I_{a_x}O(d_i, d_j)$, data item d_i is used by a_x as an input for setting the value of data item d_j as an output.

Primitive 4 (P_4): A data item affects an activity

Description: activity a_x uses the value of data item d_i. The value can affect the internal execution of the activity a_x, but does not affect the value of its outputs.

Example: The activity *pack goods* is performed differently for single units and for multiple units. The packing changes according to the quantity that is ordered. Therefore, the activity *pack goods* is affected by the data item *quantity,* but none of its data outputs is affected by this input.

Notation: $V_{d_i}(a_x)$, the value of data item d_i affects the execution of activity a_x.

Primitive 5 (P_5): A routing constraint affects an activity

Description: A specific activity a_x with an incoming edge f_i will be executed only if the routing constraint $r_i = (p_i, f_i)$ satisfies $p_i = $ True.

Example: The activity *trigger purchasing process* will be executed if and only if *quantity* is greater than *amount in stock.*

Notation: $r_i(a_x)$, r_i imposes constraints on the execution of a_x.

Primitive 6 (P_6): A routing constraint affects a gateway[6]

Description: A specific gateway g_x with an incoming edge f_i will be executed only if the routing constraint $r_i = (p_i, f_i)$ satisfies $p_i = $ True.

Example: The gateway *before lookup* will be executed if *quantity* is less than or equal to *threshold* or if the *approval status* is *'true'.*

Notation: $r_i(g_x)$, r_i imposes constraints on the execution of g_x.

---

[6] Note that the impact of a data item on a gateway is indirect, through a routing constraint.

Primitive 7 (P_7): A data item directly affects a routing constraint

Description: Data item d_i is used in the routing constraint $r_x = (p_x, f_x)$ for evaluating the predicate p_x. Therefore, d_i has an impact on r_x.

Example: The routing constraint *sufficient stock* is affected by *quantity and amount in stock*, as *sufficient stock* is true if and only if *quantity is greater than amount in stock*.

Notation: $V_{d_i}(r_x)$ the value or the existence of data item d_i is used by the routing constraint r_x.

Based on the primitives above, we can answer a variety of questions related to data impacts. Examples include: (1) what are all the impacts of a specific data item across the process? (2) What are the impacts of a data item starting from a specific place in the process? We elaborate next how to answer these questions using our approach.

5   The Data Impact Analysis Approach

In this section, we first describe the infrastructure of the approach, namely the queries used for analyzing data impact (Sect. 5.1). Then we present variants of algorithms that use the queries and can serve for impact analysis during redesign (Sect. 5.2) and upon unexpected value changes (Sect. 5.3). Finally, we refer to implementation and preliminary results (Sect. 5.4).

5.1   The Infrastructure of the Approach

The approach assumes a relational database derived from the model depicted in Fig. 2. On top of this database, we defined eight generic queries (Fig. 3) whose aim is to extract the impacts based on the suggested primitives. Each of the queries receives a process model element in the form of (@key) and returns the elements affected by it following a certain primitive. Besides the element keys, the query returns the element type, which can be data item (d), activity (a), routing constraint (r), or gateway (g) and the primitive used (P_1-P_7).

5.2   Analyzing Data Impact for Redesign Scenarios

In redesign scenarios, we are interested in all the impacts of a specific data item within the process. The input in these cases is a data item key (name). The effect of this data item is analyzed using all primitives. To detect indirect effects, the analysis is repeated for every element that was identified, until **new** effects cannot be found.

Listing 1 provides the pseudo code of this algorithm. Note that if a data item affects an activity, but does not affect other data items through this activity, the algorithm returns the activity but does not continue seeking its impacts further. If, on the other hand, an activity is affected by a routing constraint (primitive P_5), this effect relates to

Q1 for P_1 – returns data items affected by an activity	Q2 for P_2 - returns data items affected by a data item	Q3.1 for P_3 - returns data items affected by a data item through an activity (outputs)
Select name, 'd', 'P_1' From output_of Where activity_Id=@key	Select affected_name, 'd', 'P_2' From related_to Where effecting_name= @key	Select affected_name, 'd', 'P_3' From affected_through Where effecting_name= @key

Q3.2 for P_3 - returns activities affected by a data item to create an output	Q4 for P_4 - returns activities affected by a data item	Q5 for P_5 - returns activities affected by a routing constraint
Select activity_Id, 'a', 'P_3' From affected_through Where effecting_name=@key	Select activity_Id, 'a', 'P_4' From input_for Where name=@key	Select activity_Id, 'a', P_5' From Flow Where routingC_Id=@key

Q6 for P_6 – returns gateways affected by a routing constraint	Q7 for P_7 - returns routing constraints affected by a data item	
Select gateway_Id, 'g', 'P_6' From Flow Where routingC_Id=@key	Select routingC_Id, 'r', P_7 from used_in where name=@key	

Fig. 3. Generic SQL queries for extracting impacts of process elements

triggering its execution, and hence the algorithm will continue analyzing its impacts. As previously noted, since gateways do not percolate data impacts, they can be retrieved (as affected), but without additional effects sought.

Algorithm 1: the overall impact of a given data item *di* (relevant for redesign scenarios)
Input: *di* – a data item
Used structures:

-    S is a list of triples (element, type, primitive), where element is the process element,
      type ∈ {d (data), a (activity), g (gateway), r (routing constraint)}, and primitive is one of
      P_1-P_7 used for retrieving the process element

-    The database is built based on the model presented in Fig. 2 and the specified process
      model, database schema, data item relations, and input-activity-output relations. The dif-
      ferent queries are executed on this database.

S = new List()
S.add (di, d, null)
e ← S.firstNotTraversed()   *// returns the first element from the list that is not traversed and*
                                   // marks it as traversed
Do until e=null
         switch *e. type*
                     case *(d)*   *// For data items, primitives P_2, P_3, P_4, P_7 are relevant*
                             S.insert(Q2)   *// inserts only elements not existing in S*
                             S.insert(Q3.1)
                             S.insert(Q3.2)
                             S.insert(Q4)
                             S.insert(Q7)
                     case *(r)*   *// For routing constraints, primitives P_5, P_6 are relevant*
                             S.insert(Q5)
                             S.insert(Q6)
                     case *(a)*   *// Primitive P_1 is relevant only if the activity is extracted*
                             //using primitive P_5
                             if e.primitive='P_5'
                                       S.insert(Q1)
                     Case *(g)*   *// ignore, data items affect gateways only though their*
                             //incoming routing constraints
     e ← S.firstNotTraversed()
End Do

Listing 1. Algorithm 1 for analyzing data impact in redesign scenarios

5.3   Analyzing Data Impact for Value Change Scenarios

In scenarios of unexpected value changes, the impacts of a specific data item, given a particular (partial) process trace, are of interest. The relevant algorithm is based on the following definitions.

Definition 5 (Process trace). A process trace is a sequence $t = <n_1, ..., n_m>$, where n_i is a behavioral node (an activity or a gateway) and the order of n_i reflects the temporal order of a single process instance.

At run time, it is possible to look at a partial process trace, which represents a specific path traversed by the process up to a given moment. We refer to the closure of a trace, $\bar{t}$, as all process elements directly connected to trace t. Formally expressed:

Definition 6 (Closure of trace). The closure of a trace $t = <n_1, ..., n_m>$ is defined as: $\bar{t} = \{n_i\}_{i=1..m} \cup \{d \mid d$ is an input or an output of $n_i\} \cup \{r = (p, f) \mid f$ connects $n_i, n_{i+1}\}$.

Our approach analyzes the impact of a certain data item until the end of the partial trace, and returns possible future impacts, analyzing the elements reachable from the partial trace (n_m).

Definition 7 (Elements reachable from a trace). Let t be a (partial) process trace $<n_1, ..., n_m>$. The elements reachable from trace t are all elements existing in the closure of any trace that includes t, namely, $t' = <n_1, ..., n_m, n_{m+1} ..., n_k>$ such that n_k directly leads to a final state of the process.

Listing 2 provides the pseudo code of the algorithm that returns the impacts of a data item starting at a given point in the process (given as a partial trace). We assume that the input data item might have already affected other elements in the partial trace, and can still have impacts (direct and indirect) in the "future" parts of the process. We use a function Reachable ($\{e\}$, t), where $\{e\}$ is a set of process elements and t is a trace, returning the largest sub-set of $\{e\}$ whose elements are all reachable from t.

5.4   Implementation and Preliminary Results

For getting insights into the strengths and weaknesses of the approach, we implemented the algorithms presented in Sects. 5.2 and 5.3 in MS-SQL environment and applied them to three cases described below. To populate the database, we used process models specified in BPMN (using its XML format). In addition, we used two tables: one that holds the pairs of related data items (see Definition 2) and the other that holds the data item-activity-data item triplets (see Definition 4).

The cases we considered were based on the ordering process depicted in Fig. 1: addition of an attribute highly related to an existing commonly used attribute (case 1, exemplifying a redesign scenario), and unexpected value changes of attributes used in different places in the process (cases 2 and 3).

Case 1 (attribute addition, a redesign scenario): Assume the addition of the attribute *unit of measure*, to enable specifying quantities (e.g., ordered, delivered) in various units. This decision entails changes in the database as well as changes in the activities and possibly in the process. To assess the extent of changes that are required, a time consuming and error prone analysis is needed. Since *unit of measure* is relevant in connection to *quantity*, the impacts of *quantity* in the process model need to be examined. Using Algorithm 1, we retrieve nine activities (*create order, manager approval, update quantity, inventory lookup, trigger purchasing process, handle delivery details, produce bill, pack goods, ship goods*), nine data items (*manager code, approval status, max approved quantity, total payment, approximate delivery date, shipping price, packing type, material id, required quantity*), one gateway (*before lookup*) and five routing constraints (*quantity <= threshold, quantity > threshold, approximate delivery date = now()-3, insufficient stock, sufficient stock*).

Some of the retrieved elements are straightforward and could easily be identified manually. Examples of such elements are the routing constraints on which *quantity* explicitly appears (*quantity <= threshold* and *quantity > threshold*) and the data item *total payment* which obviously depends on the ordered quantity. Yet, some retrieved elements are not immediate and require a careful analysis and a profound under-standing of the process. The data items *approximate delivery date* and *shipping price*, for example, which are outputs of the activity *handle delivery details*, are both affected by the ordered quantity in a non-trivial manner. Following the approximate delivery date, another indirect effect is on the routing constraint whose condition is *approximate delivery date = now()-3*. Other examples of routing constraints are *insufficient stock* and *sufficient stock*, comparing quantity in stock with the ordered quantity.

Case 2 (an unexpected value change scenario): Assume that the value of the data item *lead time* is changed after delivery details have been handled. This data item indicates the approximate time for delivery of a specific material, required for pro-ducing the ordered product. The value of *lead time* is mainly used in the process to calculate the approximate delivery date of the order (the activity *handle delivery details*). However, it also affects other data items and activities, e.g., *shipping price*, which depends on *delivery date* (with an agreed upon discount if the delivery date is over a month from the order date).

To assess the impact of this unexpected change we use Algorithm 2 with the data item *lead time* and the trace <*create order, quantity condition, before lookup, inventory lookup, stock condition, handle delivery details*, after *handle delivery*>. We receive two types of impact for this case: the current impact within the closure of the trace and the future impact within elements reachable from the trace. The current impact includes one activity (*handle delivery details*) and two data items (*shipping price, approximate delivery date*), whereas the future impact includes one activity (*pack goods*), one routing constraint (*approximate_delivery_date = now()-3*), and one data item (*packing type*). The implication of these outcomes is that, when such change takes place, the organization should be concerned with the *handle delivery details* activity (which has been performed using the previous value of *lead time*) and examine the values of *shipping price* and *approximate delivery date*. In the remaining yet unperformed parts of the process, the condition *approximate_delivery_date = now()-3* should rely on the updated value. Furthermore, packing should be replanned and rescheduled, considering the expected delivery date and the *packing type*.

Algorithm 2: the impacts of a given data item *di* given a process trace *t*

Inputs: *di* – a data item
         t – a trace

Used structures:

- S_{before}, S_{after} are lists of triples (element, type, primitive), where element is the process element, type $\in$ {d (data), a (activity), g (gateway), r (routing constraint)}, and primitive is one of P_1-P_7 used to retrieve the process element. S_{before} and S_{after} holds elements directly related to trace t or reachable from that trace, respectively.

- The database used by the different queries is as in Algorithm 1.

Used function:

- Reachable ({e}, t) gets a set of process elements and a trace and returns a subset of elements which are reachable from the trace.

S_{before} = new List(); S_{after} = new List()

S_{before}.*add (di, d, null)*

$S = S_{before} \cup S_{after}$

$e \leftarrow$ *S.firstNotTraversed()*   *// returns the first element from the list that is not traversed and*
                                  // marks it as traversed

Do until e=null

      switch *e. type*

           case (d)   *// For data items, primitives P_2, P_3, P_4, P_7 are relevant*

               S_{before}.insert(Q2 $\cap$ $\bar{t}$); S_{after}.insert(reachable(Q2$-\bar{t}$, t))

               S_{before}.insert(Q3.1 $\cap$ $\bar{t}$); S_{after}.insert(reachable(Q3.1$-\bar{t}$, t))

               S_{before}.insert(Q3.2 $\cap$ $\bar{t}$); S_{after}.insert(reachable(Q3,2$-\bar{t}$, t))

               S_{before}.insert(Q4 $\cap$ $\bar{t}$); S_{after}.insert(reachable(Q4$-\bar{t}$, t))

               S_{before}.insert(Q7 $\cap$ $\bar{t}$); S_{after}.insert(reachable(Q7$-\bar{t}$, t))

           case (r)   *// For routing constraints, primitives P_5, P_6 are relevant*

               S_{before}.insert(Q5 $\cap$ $\bar{t}$); S_{after}.insert(reachable(Q5$-\bar{t}$, t))

               S_{before}.insert(Q6 $\cap$ $\bar{t}$); S_{after}.insert(reachable(Q6$-\bar{t}$, t))

           case (a)   *// Primitive P_1 is relevant only if the activity is extracted*

                          // using primitive P_5

               if e.primitive='P_5'

               S_{before}.insert(Q1$\cap$ $\bar{t}$); S_{after}.insert(reachable(Q1$-\bar{t}$, t))

           case (g) *//ignore, no affected elements derived from gateways*

      $S = S_{before} \cup S_{after}$

      $e \leftarrow$ *S.firstNotTraversed()*

End DO

Case 3 (an unexpected value change scenario): Assume that an error has been made in typing the *shipping address* when the order is created. This error is accidently discovered after delivery details have been handled.

To assess the impact of this error we once again use Algorithm 2, but this time with the data item *shipping address* and the trace <*create order, quantity condition, before lookup, inventory lookup, stock condition, handle delivery details, after handle delivery*>. The current impact this time includes two activities (*create order, handle delivery details*) and three data items (*total payment, shipping price, approximate delivery date*). The future impact includes three activities (*pack goods, ship goods, produce bill*), one routing constraint (*approximate_delivery_date = now()-3*), and one data item (*packing type*).

With the given unexpected value change, the company should be concerned with actions made in the *handle delivery details* activity using the wrong value. In particular, examination of the values of the data items *shipping price* and *approximate delivery date* is required, since they might depend on the shipping distance. Furthermore, the value of the data item *total payment* that includes the shipping price, needs to be examined. The value of this data item is further used in the remaining yet unperformed *produce bill* activity. Similarly to case 2, the condition *approximate_delivery_-date = now()-3* should rely on the updated value. Here again, packing should be replanned and rescheduled, considering the expected delivery date and the *packing type*.

In summary, all three cases entail changes whose full impacts can only be discovered through a detailed analysis. We have shown that using both algorithms is helpful in analyzing the effects of these changes. The examined cases are typical of redesign that concerns data in the process and of unexpected data value changes that occur at runtime. The infrastructure we propose is generic, so additional algorithms can be developed, using the same set of generic queries for answering additional questions related to impact analysis. Relevant questions can relate to the impacts of data items across processes and across different instances of the same process.

6 Related Work

Related work relevant to this paper resides mainly in three areas: data aware process modeling, process modifications, and business impact analysis.

In the context of data aware process modeling, many efforts have attempted to address the relations between processes and data. Several studies, such as [1, 14], propose workflow design approaches that help design a process model with respect to the data perspective, in order to avoid data flow problems. The approaches are fundamentally centered on data rather than on activity flows. However, these approaches support the design of workflow models with respect to data and do not aim to analyze the impact of data. The study in [12] addresses the control-flow of the process and the related object life cycles together. When the state of an object changes, this becomes explicit in the process model. Based on the state changes of the object, the process model specifies the life cycle of each object type. The consistency between business process models and life cycles of business objects is checked. However, the impacts of the data object on the process are not handled. The study in [7] proposes an approach to model processes that include complex data dependencies. The goal of this research is to enable automated process enactment from process models. The approach extends BPMN and combines concepts of relational data modeling.

The studies in [19, 20] propose a set of "anti-patterns" supporting the detection of data flow errors in a process model with respect to workflow-nets with data. These patterns take dependencies among process elements into account. The usage of these patterns allows verification of the control flow and the data flow at design time.

Over the years, several studies proposed techniques and heuristics for changing processes, e.g., [10, 11, 21]. Particularly, in [21] 17 change patterns for processes are proposed. However, these patterns do not address data aspects. In [10], the authors refer to related data problems, e.g., missing data, unnecessary data, and lost data, when changing the process. However, their approach does not consider indirect impacts of the data, and considers only local changes.

Business impact analysis and business processes change management have been studied mostly in the context of workflows [2, 4, 9, 13]. The most relevant study for our research is [4], which defines several dependency types between process elements for analyzing the impact of change considering elements in the same process. While addressing data dependencies, this study does not refer to the impact of data items on routing constraints or other data items.

In summary, the literature related to data-aware process analysis has a major focus on data flow correctness and on the consistency of control and data flows. This is also true for data considerations while introducing changes to processes. The literature related to business impact analysis does not focus directly on the impacts of data. To fill this gap, our focus is on impact analysis of data, attempting to provide a systematic approach for analyzing both direct and indirect data impacts along the process.

7   Conclusion and Future Work

Data in business processes mainly refers to persistent entities, whose values are manipulated over their life time by the process. The persistent existence of data is one of the sources of dependencies among different parts of the process, so changes cannot be considered locally in most cases. This non-local nature requires consideration at process redesign. Moreover, at runtime, understanding the dependencies and impacts of data items in a specific point of time (after a certain trace in the process was followed) is essential when their values changes unexpectedly. In such cases, manual examination of the impacts is error-prone and time consuming.

In this paper, we introduce the concept of data impact analysis. In particular, we proposed a set of primitives that represent dependencies among process model elements. The primitives were used for designing a relational database, which enables extracting impacts of model elements by querying the database. The database and the defined queries are generic. Two algorithms that use different combinations of these queries are introduced to enable answering different data impact-related questions at design time as well as at runtime. The proposed approach contributes to data aware business process management as well as to business impact analysis.

The current design is basic and does not consider additional model elements such as events and possible effects across concurrently active process instances and cross process effects. We plan in future to address these aspects. We further plan to suggest algorithms for additional scenarios and to evaluate the results of the data impact analysis empirically.

Acknowledgment. This research is supported by the Israel Science Foundation under grant 856/13.

References

1. Bhattacharya, K., Hull, R., Su, J.: A data-centric design methodology for business processes. Handbook of Research on Business Process Modeling, pp. 503–531 (2009)
2. Casati, F., Ceri, S., Pernici, B., Pozzi, G.: Workflow evolution. Data Knowl. Eng. **24**(3), 211–238 (1998)
3. Cohn, D., Hull, R.: Business artifacts: a data-centric approach to modeling business operations and processes. Bull. IEEE Comput. Soc. Tech. Comm. Data Eng. **32**(3), 3–9 (2009)
4. Dai, W., Covvey, D., Alencar, P., Cowan, D.: Lightweight query-based analysis of workflow process dependencies. J. Syst. Softw. **82**(6), 915–931 (2009)
5. Künzle, V., Reichert, M.: PHILharmonicFlows: towards a framework for object-aware process management. J. Softw. Maint. Evol. Res. Pract. **23**(4), 205–244 (2011)
6. Meyer, A., Weske, M.: Extracting data objects and their states from process models. In: 17th IEEE International Enterprise Distributed Object Computing Conference (EDOC), pp. 27–36 (2013)
7. Meyer, A., Pufahl, L., Fahland, D., Weske, M.: Modeling and enacting complex data dependencies in business processes. In: Daniel, F., Wang, J., Weber, B. (eds.) BPM 2013. LNCS, vol. 8094, pp. 171–186. Springer, Heidelberg (2013)
8. Reichert, M.: Process and data: two sides of the same coin? In: Meersman, R., et al. (eds.) OTM 2012, Part I. LNCS, vol. 7565, pp. 2–19. Springer, Heidelberg (2012)
9. Reichert, M., Dadam, P.: ADEPTflex—supporting dynamic changes of workflows without losing control. J. Intell. Inform. Syst. **10**(2), 93–129 (1998)
10. Reichert, M., Weber, B.: Enabling Flexibility in Process-Aware Information Systems: Challenges, Methods, Technologies. Springer Science & Business Media, Heidelberg (2012)
11. Reijers, H.A., Mansar, S.L.: Best practices in business process redesign: an overview and qualitative evaluation of successful redesign heuristics. Omega **33**(4), 283–306 (2005)
12. Ryndina, K., Küster, J.M., Gall, H.C.: Consistency of business process models and object life cycles. In: Kühne, T. (ed.) MoDELS 2006. LNCS, vol. 4364, pp. 80–90. Springer, Heidelberg (2007)
13. Sadiq, S.W., Orlowska, M.E., Sadiq, W.: Specification and validation of process constraints for flexible workflows. Inf. Syst. **30**(5), 349–378 (2005)
14. Sadiq, S., Orlowska, M., Sadiq, W., Foulger, C.: Data flow and validation in workflow modelling. In: Proceedings of the 15th Australasian Database Conference, vol. 27, pp. 207–214. Australian Computer Society, Inc. (2004)
15. Soffer, P.: Mirror, mirror on the wall, can I count on You at all? Exploring data inaccuracy in business processes. In: Bider, I., Halpin, T., Krogstie, J., Nurcan, S., Proper, E., Schmidt, R., Ukor, R. (eds.) BPMDS 2010 and EMMSAD 2010. LNBIP, vol. 50, pp. 14–25. Springer, Heidelberg (2010)
16. Sun, S.X., Zhao, J.L., Nunamaker, J.F., Sheng, O.R.L.: Formulating the data-flow perspective for business process management. Inf. Syst. Res. **17**(4), 374–391 (2006)
17. Sun, S.X., Zhao, J.L.: Formal workflow design analytics using data flow modeling. Decis. Support Syst. **55**(1), 270–283 (2013)
18. Tjoa, S., Jakoubi, S., Quirchmayr, G.: Enhancing business impact analysis and risk assessment applying a risk-aware business process modeling and simulation methodology. In: Third IEEE International Conference on Availability, Reliability and Security (ARES), pp. 179–186 (2008)

19. Trčka, N., van der Aalst, W.M., Sidorova, N.: Data-flow anti-patterns: discovering data-flow errors in workflows. In: van Eck, P., Gordijn, J., Wieringa, R. (eds.) CAiSE 2009. LNCS, vol. 5565, pp. 425–439. Springer, Heidelberg (2009)
20. von Stackelberg, S., Putze, S., Mülle, J., Böhm, K.: Detecting data-flow errors in BPMN 2.0. Open J. Inf. Syst. (OJIS) **1**(2), 1–19 (2014)
21. Weber, B., Reichert, M., Rinderle-Ma, S.: Change patterns and change support features–enhancing flexibility in process-aware information systems. Data Knowl. Eng. **66**(3), 438–466 (2008)

Semi-automatic Derivation of RESTful Interactions from Choreography Diagrams

Adriatik Nikaj[1]([✉]), Fabian Pittke[2], Mathias Weske[1], and Jan Mendling[2]

[1] Hasso Plattner Institute at the University of Potsdam, Potsdam, Germany
{adriatik.nikaj,mathias.weske}@hpi.de
[2] Institute for Information Business, WU Vienna, Vienna, Austria
{fabian.pittke,jan.mendling}@wu.ac.at

Abstract. Enterprises reach out for collaborations with other organizations in order to offer complex products and services to the market. Such collaboration and coordination between different organizations, for a good share, is facilitated by information technology. The BPMN choreography diagram is a modeling language for specifying the exchange of information and services between different organizations at the business level. Recently, there is a surging use of the REST architectural style for the provisioning of services on the Web, but no systematic engineering approach to design their collaboration. In this paper, we address this gap by defining a semi-automatic method for the derivation of RESTful interactions from choreography diagrams. The method is based on natural language analysis techniques to derive interactions from the textual information in choreography diagrams. The proposed method is evaluated in terms of effectiveness and considered to be useful by REST developers.

Keywords: Choreography diagram · RESTful interactions · Natural language analysis

1 Introduction

Traditionally, research in BPM has focused on internal processes of organizations. The trend towards more complex services extends BPM towards a view of interactions between multiple processes. Such interactions, enabled by information technology, require standard models, like BPMN [1], which can be understood by all the participants. In particular, BPMN business process choreography specifies the interactions between two or more participants and the order in which these interactions take place at the business level. On the technical level, REST [2] is increasingly becoming the architectural style of choice for providing services on the Web leading to the mainstream development of RESTful APIs.

However, taking business interactions, modeled by choreography diagrams, down to the level of RESTful interactions is challenging. The designers of choreography diagrams are usually business-process domain experts and do not have knowledge of software development. The same holds for IT developers with

© Springer International Publishing Switzerland 2016
R. Schmidt et al. (Eds.): BPMDS/EMMSAD 2016, LNBIP 248, pp. 141–156, 2016.
DOI: 10.1007/978-3-319-39429-9_10

respect to the business process choreographies. While there is first research into bridging this gap [3], a methodical approach for deriving REST interactions from business process choreographies is missing.

In this paper, we address this research gap and present an approach that takes as input a standard BPMN choreography diagram and it generates as output a RESTful choreography. Our approach is based on natural language processing techniques, which use textual descriptions of the choreography task to map to the most suitable REST verb with a corresponding REST URI. We implemented our approach in a research prototype and applied it on a set of choreography diagrams from different domains. The derived REST requests have also been evaluated by REST experts confirming the usefulness of our approach. The created REST choreography is used to derive code skeletons which facilitate the development of REST APIs.

This paper is organized as follows. Section 2 introduces choreography diagrams and the RESTful choreography diagram. These concepts are illustrated by a running example. Section 3 presents our semi-automatic approach of deriving RESTful choreography diagrams. Section 4 discusses the setup and the results of our user evaluation. Section 5 provides the related work before Sect. 6 concludes the paper and describes future work.

2   Preliminaries

This section briefly describes choreography diagrams by the help of an example. Additionally, REST architectural style is explained before the concept of RESTful choreography diagram is introduced.

2.1   Choreography Diagram

The business process choreography diagram introduced in BMPN 2.0 [1] is a modeling language that focuses on the specification of the interactions between two or more participants, who, in general, are business actors, e.g., enterprises, customers, or organizations. Compared to business process models, the choreography diagram abstracts from the participants' internal processes and specifies the order in which the messages are exchanged between the participants. Figure 1 depicts an example of a choreography diagram that is also a RESTful choreography (see next subsection). This diagram describes the interaction between different participants involved in the submission, review, and organization processes with the goal of arranging a scientific conference. Some of the main stakeholders in a conference include the organizers, authors, and reviewers. The diagram depicts the interactions between these three participants starting from issuing a call-for-papers (CFP) and ending, in the best case, with the confirmation of the paper publication.

To facilitate these interactions, the participants make use of a Review Management System (RMS) which, in our case, is inspired by http://easychair.org. The assumption here is that all the participants are subscribed to the RMS

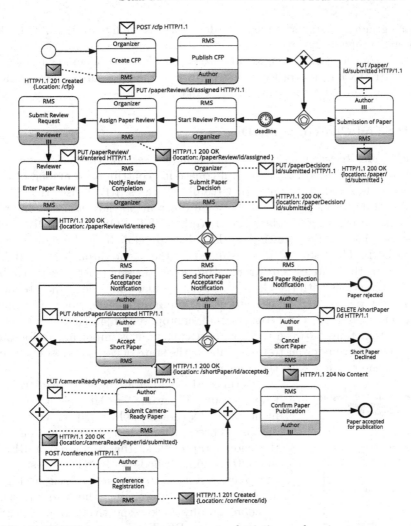

POST /cfp HTTP/1.1

HTTP/1.1 201 Created
{Location: /cfp}

PUT /paperReview/id/assigned HTTP/1.1

PUT /paper/
id/submitted HTTP/1.1

PUT /paperReview/
id/entered HTTP/1.1

HTTP/1.1 200 OK
{location: /paperReview/id/assigned }

PUT /paperDecision/
id/submitted HTTP/1.1

HTTP/1.1 200 OK
{location: /paper/
id/submitted }

HTTP/1.1 200 OK
{location: /paperDecision/
id/submitted}

HTTP/1.1 200 OK
{location: /paperReview/id/entered}

PUT /shortPaper/id/accepted HTTP/1.1

DELETE /shortPaper
/id HTTP/1.1

HTTP/1.1 200 OK
{location: /shortPaper/id/accepted}

PUT /cameraReadyPaper/id/submitted HTTP/1.1

HTTP/1.1 204 No Content

HTTP/1.1 200 OK
{location:/cameraReadyPaper/id/submitted}

POST /conference HTTP/1.1

HTTP/1.1 201 Created
{Location: /conference/id}

Fig. 1. RESTful choreography for paper submission and review management

and notified via email for any relevant information. The RMS is responsible for coordinating these three participants throughout the entire collaboration.

As it can be seen in Fig. 1, the main element of a choreography diagram is the choreography task (graphically depicted as a rounded rectangle). It shows message exchanges between two participants. The participant initiating the message exchange is called the initiator, while the other participant is called the recipient. The return message is optional and can be sent from the recipient to the initiator. To graphically distinguish the initiator from the recipient, the latter is always highlighted in grey. The same applies for the initiating and return messages (although the messages are not required to be graphically depicted). For example, the choreography task *Create CFP* has only the initiating message, while the choreography task *Submit Paper* has also a return message as a confirmation.

Choreography diagrams define the order in which the interactions are carried out. Choreography tasks have an order dependency that is modeled via sequence flows. The sequence flows, events and gateways are used in a similar fashion as in regular BPMN business process models. However, only a subset of events and gateways can be used in the choreography diagram.

2.2 RESTful Choreography Diagram

The REST architectural style [2] is increasingly used for the development of RESTful web services. Its architectural constrains contribute to among others, better scalability and portability. In virtually all cases, REST uses the HTTP protocol as a means of interactions between different participants. The interaction is achieved by using standard HTTP verbs (GET, POST, PUT, DELETE) on resources. The resources resting on the server are globally and uniquely identified via URL. Their state can be changed by the client through these REST verbs. Due to the stateless constraint, the server does not need to remember previous interactions with the client in order to understand the client's request.

Business processes can be used to model the internal behaviour of the participants involved in RESTful conversation as proposed in [4]. However, when it comes to interactions between multiple participants, it is important to focus on a global perspective in order to reason about the state of common resources and the allowed interactions with these resources.

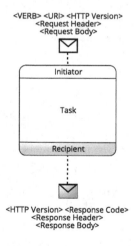

Fig. 2. The annotation of the choreograph task in RESTful choreography [3]

To this end, Nikaj et al. [3] introduce RESTful Choreography Diagrams - a lightweight enhancement of BPMN choreography diagram with REST details. These details include annotations for the choreography tasks that represent a RESTful interaction, called a RESTful task, like the *Submit review* choreography task in Fig. 1. This is realized by refining the two messages of choreography task respectively into a REST request and a REST response like depicted in Fig. 2. Figure 1 depicts the RESTful choreography that can be manually derived by the same business process choreography (the choreography without the REST notation) as an input. However, the person responsible for the enhancement of the choreograph task with REST notations has to understand both the business aspect of the choreography and the implementation aspect of the RESTful interaction. This problem is addressed in our paper by proposing a semi-automatic approach for deriving RESTful choreographies from business process choreographies.

3 Semi-Automatic Generation of RESTful Choreography

This section presents our semi-automatic approach to generate RESTful choreographies. Section 3.1 discusses the relevant concepts of the approach. Section 3.2 then explains how the concepts are employed to identify the type of REST request that is expressed in a choreography task label. Section 3.3 shows how choreography tasks are finally enriched with RESTful information.

3.1 Foundations

This subsection starts with a formal specification of a choreography diagram as our core artifact of our approach. We consider a choreography diagram to be a tuple $C = (N, S, P, L, label)$, such that:

- $N = T \cup E \cup G$ is a set of nodes;
- T is a set of choreography tasks;
- E is a set of events;
- G is a set of gateways;
- S is a set of sequence flows;
- P is a set of participants;
- L is a set of natural language text labels;
- $label : T \mapsto L$ is a function which assigns a text label to a choreography task.

In order to process the textual information of the labels, it is necessary to access this information in a structured way. As a starting point we observe that choreography tasks are similarly labeled as activities, often referring to the corresponding send task in a business process model [1]. Thus, we assume that each label of a choreography task contains the following components: an action and a business object on which the action is applied [5]. As an example, consider the label *Submit paper review* from Fig. 1. It contains the action *to submit* and the business object *paper review*. It is important to note that these components can be communicated in different grammatical variations. For instance, the labels *Paper submission* or *Conference registration* communicate the action in a different grammatical structure by using nouns, which express the actions *to submit* or *to register*, respectively.

In order to be independent of grammatical labeling structures, we rely on the label annotation approach of Leopold et al. [6] which identifies actions and business objects with a decent degree of accuracy (Avg. precision: 91 %, avg. recall: 90.6 %). Considering $l = label(t) \in L$ to be the label of an arbitrary choreography task and considering W_V and W_N to describe the set of all verbs and nouns respectively, we refer to the action and the business object of l as follows:

- $\alpha : L \mapsto W_V$ is a function that assigns an action to a choreography task label;
- $\beta : L \mapsto W_N$ is a function that assigns a business object to a choreography task label.

As an example, consider the choreography task labeled with *Submit paper review* from Fig. 1. According to the prior conceptualization, the action is given by α(Submit paper review) = *to submit* and the business object is given by β(Submit paper review) = *paper review*. We will use these label components in the following to derive the respective REST requests and to generate the RESTful annotations from the text labels of choreography tasks.

3.2   REST Verb Derivation via Natural Language Analysis

The general idea of deriving REST requests via natural language analysis is based on the assumption that the choreography task label provides all relevant information. Specifically, we focus on the actions of the labels since they describe which specific activities have to be carried out and how these activities affect the system. The REST verb derivation applies two steps. The first step compares the action of the respective choreography task label with synonym words that reflect the meaning of the different REST verbs. The second step involves a linguistic similarity analysis of the action of the choreography task label and the synonym words, in case the action of the label does not exactly match with any of the synonym words. In the following, we discuss these two steps in further detail.

First, we require a set of synonym words before we can conduct the derivation. A challenge is that the REST verbs are associated with a specific technical meaning that does not necessarily correspond with the linguistic meaning of a word. For example, the REST verb *POST* instructs the server to create a new distinguishable resource, while the verb *to post* typically describes the act of publicizing news on bulletin boards. Therefore, it is necessary to define a set of synonym words that reflect the meaning of *POST* in a technical sense. For this purpose, we asked REST experts for natural language verbs that best resemble the meaning of the REST verbs. The result of this process is shown in Table 1. For example, the experts agreed that the meaning of *POST* is best reflected by the verbs *to create* or *to request*. As the identified verbs might not capture all the variation in language, we further consider additional synonyms that may be extracted from computational lexicons, such as WordNet [7]. For example, a *POST* verb might also be related to the verbs *to produce* or *to make*. Other examples may be retrieved from the previous table in edged brackets.

The *synonym analysis step* investigates whether or not the action of a choreography task label equals one of the synonym words of the REST verbs. If this condition evaluates to true, we map the respective REST verb to the choreography task. Otherwise, no REST verb is mapped to this task. As an example, consider the choreography tasks *Create CFP* and *Accept Short Paper*. The first task would map to *POST* because its action *to create* is a member of the synonym words of the set Syn_{POST}. The second task would map to *PUT* since its action *to accept* is a member of the set Syn_{PUT}. This logic is expressed by the following function, assuming *REST* to be the set of REST verbs:

$$syn(l) = \begin{cases} POST & \text{, if } \alpha(l) \in Syn_{POST} \\ PUT & \text{, if } \alpha(l) \in Syn_{PUT} \\ GET & \text{, if } \alpha(l) \in Syn_{GET} \\ DELETE & \text{, if } \alpha(l) \in Syn_{DELETE} \\ \emptyset & \text{, otherwise} \end{cases} \tag{1}$$

The *similarity analysis step* serves as a fallback strategy in case the synonym analysis step fails to assign a REST verb to a choreography task. In this case, it is necessary to find a REST verb that is most closely related to the action. Therefore, it is necessary to determine the relatedness of an action with the synonym words. In our approach, we use the notion of semantic similarity (see e.g. [8–10]) to quantify this relatedness. We utilize the distributional similarity of the DISCO word similarity tool [11], denoted with sim_{DISCO}, because it outperforms existing similarity measures [12]. Given a choreography task label l, its action $\alpha(l)$, and the set of synonym words of an arbitrary REST verb Syn_{REST}, the relatedness of an action of a choreography task label and a synonym REST set is given as follows:

$$rel(\alpha(l), Syn_{REST}) = \max_{w \in Syn_{REST}} sim_{DISCO}(\alpha(l), w) \tag{2}$$

As an example, we consider the choreography task *Enter paper review* from Fig. 1. Since the action *to enter* is not member of the synonym sets of the REST verbs, we determine its relatedness to each synonym set. Using the 2nd order distributional similarity, we receive the following relatedness values: rel(enter, Syn_{POST}) = 0.48, rel(enter, Syn_{PUT}) = 0.92, rel(enter, Syn_{GET}) = 0.92, rel(enter, Syn_{DELETE}) = 0.55.

Finally, we consider all of the relatedness scores to derive the most suitable REST verb for a given choreography task label. In this case, we assume that the highest relatedness score reflects the most suitable REST verb for a given choreography task. Accordingly, we assign this REST verb to the highest relatedness score. However, it might be the case that several relatedness scores are equal which consequently leads to more than assignment of a REST verb emphasizing the necessity of a user to choose the correct REST verb. Formally, we describe the similarity analysis step as follows:

Table 1. Synonym word sets of the REST verbs

REST verb	Description	Synonym word sets
POST	creation of a new resource on the server	Syn_{POST} = {create, request, [produce, make, ...] }
PUT	editing an existing resource	Syn_{PUT} = {confirm, edit, accept, [support, redact, ...] }
GET	retrieving an existing resource from the server	Syn_{GET} = {retrieve, read, [get, find, recover, ...] }
DELETE	deleting an existing resource	Syn_{DELETE} = {cancel, delete, [erase, postpone, ...] }

$$sim(l) = \{r \in REST | \max rel(\alpha(l), Syn_r)\} \qquad (3)$$

As an example, consider again the choreography task *Enter paper review* and its relatedness scores. Since the scores of PUT and GET are equal, the similarity analysis strategy assigns both REST verbs PUT and GET to the choreography task. The following section will explain how RESTful requests are generated for the choreography tasks using the respective REST verb.

3.3 REST Request Generation

The task of generating REST requests involves the generation of a URI explaining how the resource is addressed via the HTTP. In order to generate a URI, we consider its generation as a language generation problem that uses the available information of the choreography task and the REST verb derivation from the previous step. Many language generation systems take a three-step pipeline approach that first determines the required information of a sentence, second plans the expression of this information, and third transforms them into correct sentences [13]. In contrast to these systems, we do not require a fully flexible approach, since the final links follow regular structures [3]. Therefore, we use a template-based approach [14–16] to generate REST URIs. In particular, we use the choreography task together with the REST verb from the previous step and select the respective link template. Afterwards, we fill the template with the necessary information, i.e. action and business object of a choreography task label. It has to be noted that there are cases in which a choreography is not associated with another REST link when the request derivation reveals more than one or no REST verbs requiring the user to correct the links.

Table 2 shows link templates for the different REST verbs and gives examples created from the choreography tasks of Fig. 1. The templates emphasize that the business object of a choreography task label ($\beta(l)$) plays an important role for the REST links since it resembles the server resource that needs to be addressed by a REST verb. We therefore associate the business object together with a unique identifier. In case the state of a specific resource has to be changed, the link also explains how its state changes with the REST verb. This change is expressed by the past participle of the action of a choreography task label.

Table 2. Link templates for REST requests based on [3]

Link template	Example
POST /<$\beta(l)$> <HTTP Version>	POST /CFP HTTP/1.1
PUT /<$\beta(l)$>/id/<Past Participle of α> <HTTP Version>	PUT /paperReview/id/entered HTTP/1.1
GET /<$\beta(l)$>/id <HTTP Version>	GET /paperReview/id HTTP/1.1
DELETE /<$\beta(l)$>/id <HTTP Version>s	DELETE /shortPaper/id HTTP/1.1

4   Evaluation

This section describes our evaluation. First, we explain the architecture of our prototypical implementation. Then, we present results on the accuracy of the derivation steps for a set of 172 choreography diagrams form practice.

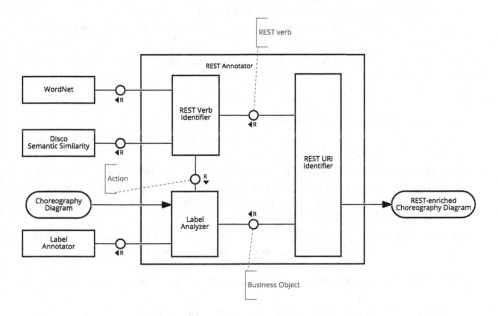

Fig. 3. REST Annotator software architecture

4.1   Evaluation Setup

For evaluating our approach, we developed a tool, called REST Annotator. The architecture of the REST Annotator is depicted in Fig. 3 as an FMC diagram [17]. The REST Annotator takes a set of choreography diagrams as an input and it outputs a set of REST-enriched choreography diagrams. The tool makes use of three external components: the Label Annotator by Leopold et al. [6], WordNet [7], and the distributional similarity component of the DISCO tool [11]. The main component constituting the tool is composed of three sub-components: Label Analyzer, REST Verb Identifier and REST URI identifier.

The Label Analyzer is responsible for extracting all the labels from the model and analysing them with the help of Label Annotator. The latter is used to notate the action and the business object of a choreography task label. The Label Analyzer maps the action and the business object for each label to the REST Verb Identifier and the REST URI Identifier components. The REST Verb Identifier component requires the action provided by the Label Extractor and the synonyms of WordNet resembling the respective REST verb. If no synonym is found, the component requires the semantic similarity score between the action and the synonym sets of the REST verbs from the Disco Semantic Similarity component. Once the semantic relation of the action with each of the REST verbs is identified, the REST verb and its respective score is passed to the REST URI Identifier component. This component generates as outputs the set of choreography diagrams enriched with REST annotation.

As evaluation data, we use choreography diagrams from the BPM Academic Initiative. The initiative offers a rich set of process models from different domains. Overall, we retrieve 424 BPMN choreography diagrams. Since these diagrams have not been used for REST purposes, it is necessary to *clean the data*. In particular, we apply the following criteria:

1. *English-only Diagrams.* We include only diagrams with English text labels. This criteria is necessary because most of the natural language analysis components only support English.
2. *Non-trivial Diagrams.* We select those diagrams that describe a meaningful interaction between actors. In particular, we exclude diagrams with only one or two choreography tasks since they do not give sufficient context to judge their relevance for REST.
3. *Syntactically correct Diagrams.* Diagrams which have syntax errors with respect to the BPMN 2.0 choreography diagram specification are excluded.

After the cleaning, we ended up with a set of 172 choreography diagrams that satisfied all the criteria.

With regard to the *evaluation procedure*, we chose three human evaluators, who have extensive knowledge regarding REST APIs. The evaluators had to perform a three-step evaluation for each choreography task: (a) the REST relevance of a choreography task, (b) the correctness of the REST verb, and (c) the suitability of the generated REST URI. Their judgment on (a) is based on the contextual information they can obtain from the choreography diagram containing the task under assessment, e.g., the involved participants, the exchanged messages, the description of events, the entirety of all choreography tasks. This evaluation step is necessary because our test data contains task labels that do not describe an interaction at all. In case (a) holds true, the evaluator further has to rate if the identified REST verb is correct (b) and if the generated URI is suitable (c).

4.2   Evaluation Results

This section discusses the results which are summarized in Table 3. The 172 models contain 1213 choreography task labels in total. From these labels, 698 labels (57.54 %) actually describe a RESTful interaction and have been considered for the human assessment. In the following discussion, we only focus on those labels that are relevant for the REST context and discuss how the verb identification and the link generation performs in these cases.

Table 3. Quantitative results of the user evaluation

Total No. of Labels	1213
No. of REST-relevant Labels	698 (57.54%)
No. of REST-irrelevant Labels	515
Total No. of Correctly Identified REST Verbs	523 (74.93%)
.. with the Synonym Identification Strategy	265
.. with the Similarity Identification Strategy	258
Total No. of Incorrectly Identified REST Verbs	175 (25.07%)
.. requiring a human decision among alternatives	55
.. requiring full human correction	120
Total No. of Correct Links	424 (60.74%)
.. POST Links	56
.. PUT Links	182
.. GET Links	170
.. DELETE Links	16
Total No. of Incorrect Links	274 (39.26%)
.. POST Links	26
.. PUT Links	76
.. GET Links	143
.. DELETE Links	29

The *verb identification strategies* have identified the correct REST verb in 523 labels which amounts to almost 75% of all REST-relevant labels. Among these labels, we further distinguish between the verbs that have been identified with the synonym strategy and the similarity strategy. The synonym strategy is capable to derive the correct REST verb in 265 labels, while the similarity strategy derives the correct REST verb for 258 choreography labels. The results emphasize the need for the similarity identification strategy of the REST verb. In total, 175 choreography labels (25.07%) have been annotated with the wrong REST verb. We identify two classes of errors that lead to the wrong annotation. The first class subsumes choreography labels for which the similarity strategy revealed two or more equal similarity scores. This has been the case for 55 choreography labels. Here, our approach does not make a decision for one particular alternative, but it presents all alternatives to the user for selection. The second class covers such cases in which our approach identified the wrong REST verb. The evaluation has revealed 120 choreography labels for which our approach did not find the correct REST verb. These cases have also to be corrected by the user.

The approach to *generate RESTful links* has created 424 correct and 274 incorrect links. The results for the POST and PUT links are satisfactory because the amount of correct cases is clearly outnumbering the amount of incorrect one. For example, 56 POST links have been generated correctly, while 26 links are incorrect. In case of the GET links, the ratio of correct and incorrect links is balanced. However, the evaluation showed that the DELETE links have issues in terms of correctness. We identified the labeling quality as a main cause for the incorrect links. For example, we found choreography tasks that have not been specified correctly by referring to a particular state, e.g. *payment confirmed* or *invoice sent*, or by not recognizing the business object correctly, e.g. *payment report* or *customer invoice*. Nevertheless, we conclude that the link generation works satisfactory and produces a large number of correct REST links.

We also exemplify the results by applying our approach to the example brought in Sect. 2. Figure 1 shows the derived RESTful choreography after the human correction is applied to the generated choreography from our tool. The REST Annotator generates correct REST URIs for 7 out of 9 REST-relevant choreography tasks.

4.3   Discussion

Three main observations emerge from the quantitative evaluation results. The first observation relates to the correct annotation of choreography tasks with REST URIs. For example, it identifies PUT to be the correct REST verb for the task *accept short paper* and generates the URI PUT */shortPaper/id/accepted*. However, we also encounter problems for cases, in which the approach retrieves several possibilities for REST verbs and fails to make a decision for one particular REST verb. In consequence, we require human interaction to choose among the possibilities. In the example, the choreography task *enter paper review* falls into this group. The approach identifies the REST verbs PUT and GET because the action *to enter* is not a member of any REST verb synonym list and the semantic similarity score is equal for both REST verbs. Based on this result, the link generator component creates two possible links, among which the user has to choose. Nevertheless, the links themselves have been created correctly. The third observation covers such links that are incorrect and that need to be manually corrected by the user. As an example, consider the choreography task *conference registration*, for which the our approach creates a GET link. However, we would expect a $POST$ or a PUT request. Incorrect links of this type may have several error sources. On the one hand, the label annotation component might have misclassified the choreography task and erroneously changed action and business object. On the other hand, the verb identification component might have caused the error because the action is either a direct member of the synonym word lists or its similarity score with the synonym words is highest for one of the other REST verbs. In our example, the former applies. The REST verb GET has been identified, since the action *to register* is a WordNet synonym of *to read* and thus a member of the synonym word set Syn_{GET}. Hence, the other alternatives are not considered so far, which finally requires the user to correct this REST URI.

At last, Fig. 4 depicts a concrete instance of the RMS RESTful interaction. The part in bold and the order of REST interactions are generated by the REST Annotator tool and provided to the developer as a skeleton to follow for developing the RESTful API. In the RSM context, the two rectangles represent respectively

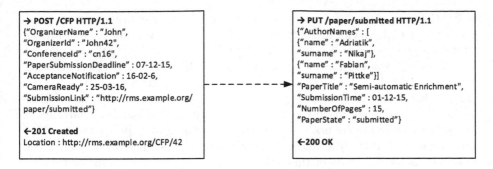

```
→ POST /CFP HTTP/1.1
{"OrganizerName" : "John",
"OrganizerId" : "John42",
"ConferenceId" : "cn16",
"PaperSubmissionDeadline" : 07-12-15,
"AcceptanceNotification" : 16-02-6,
"CameraReady" : 25-03-16,
"SubmissionLink" : "http://rms.example.org/
paper/submitted"}

←201 Created
Location : http://rms.example.org/CFP/42
```

```
→ PUT /paper/submitted HTTP/1.1
{"AuthorNames" : [
{"name" : "Adriatik",
"surname" : "Nikaj"},
{"name" : "Fabian",
"surname" : "Pittke"}]
"PaperTitle" : "Semi-automatic Enrichment",
"SubmissionTime" : 01-12-15,
"NumberOfPages" : 15,
"PaperState" : "submitted"}

←200 OK
```

Fig. 4. A concrete skeleton instance of RMS implementation

the concrete instances of the *Create CFP* and *Paper Submission* choreography tasks from Fig. 1. The dashed arrow expresses that the second instance can only be executed only after the first one is executed. For a given RESTful choreography, a skeleton diagram can be derived for each participant who offer a RESTful API. Hence, we jump from a global choreography view, to at least one orchestration view that focuses only on the REST behavioral interface i.e. the order in which the REST requests and responses are performed within a single participant application. The benefit of applying our approach is in that the same URI generation logic is used across all participants contributing to a better understandability, maintenance and evolution of REST APIs [18]. The automation of deriving skeletons from a RESTful choreography is left as a future work.

5   Related Work

We identify two major groups of research related to our approach. First, our approach is related to model-driven approaches that focus on the process of designing and engineering REST APIs or RESTful services. Examples include the work from Valverde and Pastor [19] or Schreier [20], who support this process by providing metamodels. While the former metamodel focuses on the specification of REST services and the generation of machine-readable specifications, the latter approach addresses formal aspects of a REST application, such as application structure and behavior. Laikorpi et al. [21] consider the design of a RESTful API as a model transformation problem and describe necessary transformations and intermediate models for developing RESTful services. Our approach contributes to model-driven approaches by deriving REST information from choreography diagrams in a semi-automatic way. In contrast to these approaches, our approach is based on the BPMN choreography standard, which specifies business interactions from a global perspective to derive REST skeletons with implementation details.

Second, our approach relates to the idea of bridging the gap between the business process model with its underlying orchestration system. With this regard, Decker et al. [22] propose an extension of BPEL web service composition standard [23] for closing the gap between composition and choreographies. The aim of the BPEL4Chor extension is to orchestrate process choreographies by integrating existing BPEL service orchestrations. BPEL4Chor is a bottom-up approach and it is based on web services standards like SOAP and WSDL [24]. Opposite to that, we take a top-down approach for deriving RESTful interactions. Another approach establishes the relation between BPMN and REST [4]. The author suggests that a part of a business process, per se, can be published as a REST resource. While this approach focuses on the internal behavior of the participant involved in a RESTful interaction, we focus on the global perspective, which allows reasoning about the allowed interactions at the implementation level. Moreover, the added value of this work consists in providing a semi-automatic methodical approach to derive RESTful choreographies from original business process choreographies.

6   Conclusions and Future Work

The paper defines a semi-automatic approach for deriving RESTful interactions from BPMN Choreography Diagram. The proposed approach is based on natural language analysis techniques to derive the most suitable REST verb for the interaction and to generate a REST URI for the derived REST verb. Our approach was evaluated by developing the REST Annotator tool and applying it to choreography diagrams from different domains. The output of the tool was assessed by three REST experts. They agreed that the verb identification is correct in 74.93 % of cases, while the URI is correct in 60.74 % of cases. This work contributes an additional step towards the research gap between business process choreographies and their implementation.

Our approach also has limitations in terms of the imprecise nature of natural language and the capabilities of the employed language processing tools. These issues mainly lead to wrongly identified REST verbs and incorrectly generated REST URIs that have to be corrected by users. In future work, we plan to make use of word sense disambiguation technology and of behavioral aspects of the choreography diagram. The former explicitly considers the semantics of words in a given context and improves the quality of the synonyms and the accuracy of semantic similarity. The latter relates to the sequential order of choreography tasks and might be helpful to resolve conflicts with several alternatives. For example, if a POST and a GET request have been identified and the respective choreography task is at the beginning of the interaction, then it is more likely to be a POST request. In this way, we aim to improve the accuracy of the proposed method.

References

1. OMG: Business Process Model and Notation (BPMN), Version 2.0. http://www.omg.org/spec/BPMN/2.0/
2. Fielding, R.T.: Architectural Styles and the Design of Network-based Software Architectures. PhD thesis (2000)

3. Nikaj, A., Mandal, S., Pautasso, C., Weske, M.: From choreography diagrams to RESTful interactions. In: Service Oriented Applications, WESOA 2015, co-located with ICSOC 2015. Springer (2015)
4. Pautasso, C.: BPMN for REST. In: Dijkman, R., Hofstetter, J., Koehler, J. (eds.) BPMN 2011. LNBIP, vol. 95, pp. 74–87. Springer, Heidelberg (2011)
5. Mendling, J., Reijers, H.A., Recker, J.: Activity labeling in process modeling: empirical insights and recommendations. Inf. Syst. **35**(4), 467–482 (2010)
6. Leopold, H., Eid-Sabbagh, R., Mendling, J., Azevedo, L.G., Baião, F.A.: Detection of naming convention violations in process models for different languages. Decis. Support Syst. **56**, 310–325 (2013)
7. Miller, G.A.: WordNet: a lexical database for English. Commun. ACM **38**(11), 39–41 (1995)
8. Wu, Z., Palmer, M.: Verbs semantics and lexical selection. In: Proceedings of the 32nd Annual Meeting on Association for Computational Linguistics, pp. 133–138 (1994)
9. Resnik, P.: Using information content to evaluate semantic similarity in a taxonomy. In: Proceedings of the 14th International Joint Conference on Artificial Intelligence, pp. 448–453 (1995)
10. Lin, D.: An information-theoretic definition of similarity. In: ICML, vol. 98, pp. 296–304 (1998)
11. Kolb, P.: Disco: a multilingual database of distributionally similar words. In: Proceedings of KONVENS 2008, Berlin (2008)
12. Kolb, P.: Experiments on the difference between semantic similarity and relatedness. In: Proceedings of the 17th Nordic Conference on Computational Linguistics (2009)
13. Reiter, E., Dale, R.: Building applied natural language generation systems. Nat. Lang. Eng. **3**(1), 57–87 (1997)
14. Denger, C., Berry, D.M., Kamsties, E.: Higher quality requirements specifications through natural language patterns. In: 2003 IEEE International Conference on Software - Science, Technology and Engineering, pp. 80–90 (2003)
15. Leopold, H., Mendling, J., Polyvyanyy, A.: Generating natural language texts from business process models. In: Ralyté, J., Franch, X., Brinkkemper, S., Wrycza, S. (eds.) CAiSE 2012. LNCS, vol. 7328, pp. 64–79. Springer, Heidelberg (2012)
16. Leopold, H., Mendling, J., Polyvyanyy, A.: Supporting process model validation through natural language generation. IEEE Trans. Softw. Eng. **40**(8), 818–840 (2014)
17. Knöpfel, A., Gröne, B., Tabeling, P.: Fundamental modeling concepts. Effective Communication of IT Systems, England (2005)
18. Palma, F., Gonzalez-Huerta, J., Moha, N., Guéhéneuc, Y.-G., Tremblay, G.: Are RESTful APIs well-designed? detection of their linguistic (Anti)patterns. In: Barros, A., Grigori, D., Narendra, N.C., Dam, H.K. (eds.) ICSOC 2015. LNCS, vol. 9435, pp. 171–187. Springer, Heidelberg (2015). doi:10.1007/978-3-662-48616-0_11
19. Valverde, F., Pastor, O.: Dealing with rest services in model-driven web engineering methods. In: V Jornadas Científico-Técnicas en Servicios Web y SOA, JSWEB (2009)
20. Schreier, S.: Modeling restful applications. In: Proceedings of the Second International Workshop on RESTful Design, pp. 15–21. ACM (2011)
21. Laitkorpi, M., Selonen, P.: Towards a model-driven process for designing restful web services. In: IEEE International Conference on Web Services, pp. 173–180. IEEE (2009)

22. Decker, G., Kopp, O., Leymann, F., Weske, M.: Bpel4chor: extending BPEL for modeling choreographies. In: IEEE International Conference on Web Services 2007, pp. 296–303 (2007)
23. Jordan, D., Evdemon, J., Alves, A., Arkin, A., Askary, S., Barreto, C., Bloch, B., Curbera, F., Ford, M., Goland, Y., et al.: Web services business process execution language version 2.0. OASIS Standard **11**, 1–10 (2007)
24. Alonso, G., Casati, F., Kuno, H., Machiraju, V.: Web Services. Data-Centric Systems and Applications. Springer, Heidelberg (2004)

Controlling Time-Awareness
in Modularized Processes

Andreas Lanz[1], Roberto Posenato[2(✉)], Carlo Combi[2], and Manfred Reichert[1]

[1] Institute of Databases and Information Systems, Ulm University, Ulm, Germany
{andreas.lanz,manfred.reichert}@uni-ulm.de
[2] Department of Computer Science, University of Verona, Verona, Italy
{roberto.posenato,carlo.combi}@univr.it

Abstract. The proper handling of temporal process constraints is crucial in many application domains. A sophisticated support of time-aware processes, however, is still missing in contemporary information systems. As a particular challenge, temporal constraints must be also handled for modularized processes (i.e., processes comprising subprocesses), enabling the reuse of process knowledge as well as the modular design of complex processes. This paper focuses on the representation and support of such time-aware modularized processes.

Keywords: Process-aware information system · Temporal constraints · Subprocess · Process modularity · Controllability

1   Introduction

The proper support of temporal process constraints is indispensable in many application domains. Although it has received increasing attention in the research community [1,6,8], a sophisticated support of time-aware processes is still missing in contemporary process-aware information systems (PAIS). It is further widely acknowledged that the capability to modularly design process schemata constitutes a fundamental requirement for obtaining comprehensible and re-usable process schemas [14].

At first glance, temporal process constraints and process modularity seem to be orthogonal features that may be managed in an independent way. When taking a closer view on them, however, it turns out that modularity in combination with the reuse of time-aware processes requires the ability to represent the overall temporal behavior of a process. This way, temporal constraints of a process containing time-aware subprocesses can be evaluated in a *true modular* way, i.e., without replacing the subprocess tasks with their (temporal) components.

To the best of our knowledge, the issue of representing the overall temporal properties of a process has not been considered in literature so far. This paper, therefore, focuses on the representation and support of time-aware modularized processes. In particular, we introduce a sound and complete method to derive the duration restrictions of a time-aware process in such a way that

© Springer International Publishing Switzerland 2016
R. Schmidt et al. (Eds.): BPMDS/EMMSAD 2016, LNBIP 248, pp. 157–172, 2016.
DOI: 10.1007/978-3-319-39429-9_11

(a) Main process.

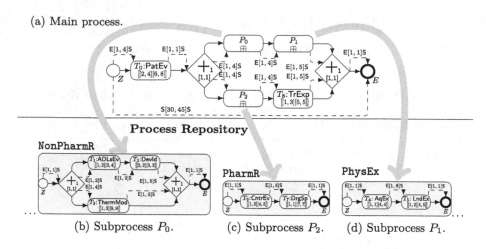

(b) Subprocess P_0.   (c) Subprocess P_2.   (d) Subprocess P_1.

Fig. 1. Motivating example: The process for managing osteoarthritis.

its temporal properties are *completely* described. Then, we show how this characterization of a process can be merged with other temporal constraints when re-using it as a subprocess of a modularized process. In accordance with recent research contributions, we focus on the *dynamic controllability* (DC) of time-aware processes [10]. In general, DC corresponds to the capability of a process engine to execute a process schema for *all* allowed durations of *all* tasks, while still satisfying *all* temporal constraints; i.e., DC ensures that it is possible to execute a process schema without any need to restrict the allowed durations of a task for satisfying all temporal constraints. In this context, task durations are called *contingent* as they are not under the control of the process engine.

As a motivating scenario, consider a high-level specification of an excerpt of a clinical guideline related to the management of osteoarthritis of the hand, hip and knee [7]. A possible schema for this process is depicted in Fig. 1. After completing the initial *Patient Evaluation* (task T_0: PatEv) two parallel branches become activated. The first one is composed of process *Non-Pharmacologic Recommendation* (P_0: NonPharmR) followed by process *Specification of Physical Exercises* (P_1: PhysEx). The second one consists of process *Pharmacologic Recommendation* (P_2: PharmR) followed by a *Treatment Explanation* to the patient (task T_8: TrExp). As depicted in Fig. 1, P_0, P_1, and P_2 constitute subprocesses from a process repository which, in turn, are composed of other tasks and are reusable in other clinical processes (e.g., related to other pathologies). In detail, *Non-Pharmacologic Recommendation* P_0 consists of two parallel branches: The first one evaluates the patient's ability to perform activities of daily live (task T_1: ADLsEv) followed by the identification of needed assistive devices (task T_2: DevId). The second branch consists of giving instructions to the patient related to the use of thermal modalities (task T_3: ThermMod). In turn, the *Specification of Physical Exercises* (i.e., P_1) consists of the specification of aquatic exercises (task T_4: AqEx) followed by the specification of land exercises (task T_5: LndEx).

Finally, *Pharmacologic Recommendation* (i.e., P_2) consists of the evaluation of contraindications (task T_6: CntrEval) followed by a drug specification (task T_7: DrgSp).

We enrich these process schemas with temporal constraints that need to be obeyed to guarantee the successful completion of each step of the therapy. They allow for the temporal characterization of tasks, edges and gateways, according to the concepts introduced in [10]. Note that the durations of tasks are not completely under the control of the process engines as these tasks are carried out by human users (e.g., doctors, nurses). Therefore, task durations are represented as *guarded ranges*. Such a duration range may be partially restricted by the system during process execution to ensure successful completion of the processes. For example, task T_6 has temporal constraint $[\![1,2]\![4,5]\!]$ meaning that prior to the execution of the task its duration may be restricted, but in any case the minimum required duration must not exceed 2 time units and the maximum duration cannot be constrained below 4 (e.g., a duration of $[3,5]$ or $[1,2]$ would not be allowed). As another example consider task T_7 with temporal constraint $[\![1,1]\![7,7]\!]$. The latter means that this task may last 1 to 7 time units and all possible durations shall be allowed during process execution. This ensures that the user executing the task has enough flexibility to successfully complete the task. Constraints on gateways and edges are standard temporal constraints, specifying the possible durations (within a range), which are under the control of the process engine. The two main research questions addressed in this paper are:

1. How can the overall temporal behavior of a process be represented (cf. Sect. 3)? Addressing this question is a fundamental prerequisite for being able to provide some kind of modularity from the temporal perspective as well. *Note that without such characterization, it would be necessary to re-compute the temporal features of a subprocess each time it is used in a modularized process.* As will be shown, a subprocess can be represented as a kind of extended guarded range. On one hand the duration of the subprocess can be controlled to some extent due to the nature of the contained temporal constraints; on the other, it cannot be completely controlled since the contingent durations of the contained tasks must be guaranteed.
2. How to apply such knowledge when using a process as a subprocess inside a modularized process, in order to avoid having to re-analyze the internal constraints of the subprocess (cf. Sect. 4)? This will, for example, enable us to store time-aware processes including their overall temporal properties inside a process repository and to reuse them in a truly modular fashion.

2   Background and Related Work

In literature, there exists considerable work on managing temporal constraints for business processes [1,3,8]. These approaches focus on issues like the modeling and verification of time-aware processes. In [5], an extended version of the *Critical Path Method* known from project planning is used. *Simple Temporal Networks with Uncertainty* (STNU) [13] are used as basic formalism in [3], whereas authors in [2,8] use *Conditional Simple Temporal Networks with Uncertainty* for checking the DC of process schemas. This paper relies on *Simple Temporal Network with Partially Shrinkable Uncertainty* (STNPSU), an extension of STNU where contingent links are extended for a more flexible management of temporal constraints [10].

An STNPSU [10] is a directed weighted graph (cf. Fig. 2) where nodes represent time-point variables (timepoints), usually corresponding to the start or end of activities, and edges $A \xrightarrow{[x,y]} B$, called *requirement links*, represent a lower and an upper bound constraint on the distance between the two timepoints it connects; e.g., $A \xrightarrow{[x,y]} B$ represents the constraint that timepoint B has to occur between x and y time units after the occurrence of A (i.e., $x \leq B - A \leq y$). In an STNPSU, it is possible to characterize certain timepoints as *contingent timepoints*, meaning that their value cannot be decided by the system executing the STNPSU, but is decided by the environment at run time. Each contingent timepoint has one incoming edge, called *guarded link*, drawn with a double line, e.g., $A \xRightarrow{[[x,x'][y',y]]} C$. A guarded link $A \xRightarrow{[[x,x'][y',y]]} C$ consists of a pseudo-contingent duration range $[x,y]$ augmented with two *guards*, the *lower guard* x' and the *upper guard* y' [10]. A is called the *activation timepoint*. Before executing a guarded link, its duration range $[x,y]$ can be modified. However, any modification must be done in a way respecting the corresponding guards, i.e., $x \leq x'$ and $y \geq y'$. When activating a guarded link $A \xRightarrow{[[x^*,x'][y',y^*]]} C$ (i.e., when executing timepoint A), the current value $[x^*,y^*]$ of the duration range becomes a fully contingent range, which is then made available to the environment for executing timepoint C. That is, once A is executed, C is guaranteed to be executed such that $C - A \in [x^*,y^*]$ holds. However, the particular time at which C is executed is *uncontrollable* since it is decided by the environment; i.e., it can be only observed when it happens.

More formally, an STNPSU is a triple $(\mathcal{T},\mathcal{C},\mathcal{G})$, where $\mathcal{T}$ is a set of *timepoints*, $\mathcal{C}$ is a set of *requirement links* $X \xrightarrow{[u,v]} Y$, and $\mathcal{G}$ is a set of *guarded links* each having the form $A \xRightarrow{[[x,x'][y',y]]} C$ with A and C being timepoints and $0 < x \leq y < \infty$, $x \leq x'$, and $0 < y' \leq y$. It is noteworthy that guarded links may be used to represent two different types of constraints: If $x' < y'$ holds, a guarded link represents a temporal constraint with a *partially contingent range*. Particularly, the guarded link represents a constraint with a *contingent* (i.e., *unshrinkable*) *core* $[x',y'] \subseteq [x,y]$. In turn, if $x' \geq y'$ holds, a guarded link represents a temporal constraint with a *partially shrinkable range* with a *guarded core* $[y',x']$.

Furthermore, each STNPSU is associated with a *distance graph* $\mathcal{D} = (\mathcal{T},\mathcal{E})$, derived from the upper and lower bound constraints [10,13]. In the distance graph, each link between a pair of timepoints A and B is represented as two *ordinary edges* in $\mathcal{E}$: $A \xrightarrow{y} B$, representing the constraint $B \leq A + y$, and $A \xleftarrow{-x} B$, for the constraint $B \geq A + x$, $x,y \in \mathbb{R}$. Moreover, for each guarded link between a pair of timepoints A and C, $\mathcal{E}$ contains two other labeled edges, called *lower* and *upper case labeled values*. A lower case labeled value, $A \xrightarrow{c:x'} C$, represents the fact that C cannot be forced to be executed at a time greater than x' after

---

Algorithm 1. STNPSU-DC-Check(G)

Input: $G = (\mathcal{T}, \mathcal{C}, \mathcal{G})$: STNPSU graph instance to analyze.
Output: the dynamic controllability of G.

1 $D :=$ distance graph of G;
2 **for** 1 to *CutOffBound* **do** // `CutOffBound`=$O(|\mathcal{T}|)$
3 | $D' := D$ without lower case labels and with upper ones as normal labels ;
4 | **if** *(D' has a negative cycle)* **then return** *false*;
5 | Generate new edges in D using edge-generation rules
6 |_ **if** *(no edges generated)* **then return** *true*;
7 **return** *false;*

A, i.e., it is not possible to add a constraint $A \xleftarrow{-x''} C, x' < x''$ to the network. In turn, an upper case labeled value, $A \xleftarrow{C:-y'} C$, represents the fact that C cannot be forced to be executed at a time less than y' after A, i.e., it is not possible to add a constraint $A \xrightarrow{y''} C, y'' < y'$ to the network.

These two kinds of labels are fundamental for determining the dynamic controllability of the network as explained in the following. Note that these two representations of an STNPSU can be used interchangeably.

An STNPSU is denoted as *dynamically controllable* (DC), if there exists a strategy for executing its timepoints in such a way that: (i) all constraints in the network can be satisfied, no matter how the execution of any guarded link turns out, and (ii) for any other guarded link $A \xrightarrow{[[x,x'][y',y]]} C$ the lower bound x never must be increased beyond its lower guard x' and the upper bound y never must be decreased below its upper guard y' [10]. Note that in [10], we showed that it is possible to adapt Morris et al.'s edge-generation rules and algorithm MM5 for STNU [13] to check the DC of a STNPSU in polynomial time. Due to lack of space, we do not report the adapted edge-generation rules (cf. [10] for details), but only the new version of the algorithm (cf. Algorithm 1).

For each process exhibiting temporal constraints, a *time-aware process schema* needs to be defined [8]. In the context of this work, a process schema corresponds to a directed graph that comprises a set of *nodes*—representing *tasks* and *gateways* (e.g., AND-Split/Join)—as well as a set of *control edges* linking these nodes and specifying precedence relations between them. Each process schema contains a unique start and end node, and may be composed of control flow patterns like sequence, parallel split, and synchronization. Moreover, [12] elaborated the need for a proper run-time support of time-aware processes. In this work, we focus on the most fundamental category of time patterns, i.e., *durations* and *time lags*.

3   Characterization of Time-Aware Processes

This section shows how to determine a proper representation for the duration of a process. For this purpose, we consider a process schema P with a single start

Table 1. STNPSU transformation rules.

Process Schema	STNPSU	Process Schema	STNPSU
Start/End node $\bigcirc_Z \quad \bullet_E$	$Z \xrightarrow{[0,\infty]} \quad \xrightarrow{[0,\infty]} E$	Time Lag end-start $\rightarrow A \rightarrow B \rightarrow$ $E[t,u]S$	$\rightarrow A_S \Rightarrow A_E \xrightarrow{[t,u]} B_S \Rightarrow B_E \rightarrow$
Task $\rightarrow \boxed{\begin{array}{c} A \\ [\![x,x']\!][y',y]\!] \end{array}} \rightarrow$	$\xrightarrow{[0,\infty]} A_S \xrightarrow{[\![x,x']\!][y',y]\!]} A_E \xrightarrow{[0,\infty]}$	start-start $\rightarrow A \rightarrow B \rightarrow$ $S[t,u]S$	$\rightarrow A_S \Rightarrow A_E \xrightarrow[{[t,u]}]{[0,\infty]} B_S \Rightarrow B_E \rightarrow$
ANDsplit $\rightarrow \begin{array}{c} + \\ [1,1] \end{array}$	$\xrightarrow{[0,\infty]} +_S \xrightarrow{[1,1]} +_E \begin{array}{c} \nearrow^{[0,\infty]} \\ \searrow_{[0,\infty]} \end{array}$	end-end $\rightarrow A \quad B \rightarrow$ $E[t,u]E$	$\rightarrow A_S \Rightarrow A_E \xrightarrow[{[t,u]}]{[0,\infty]} B_S \Rightarrow B_E \rightarrow$
ANDjoin $\begin{array}{c} + \\ [1,1] \end{array} \rightarrow$	$\begin{array}{c} \searrow^{[0,\infty]} \\ \nearrow_{[0,\infty]} \end{array} +_S \xrightarrow{[1,1]} +_E \xrightarrow{[0,\infty]}$	start-end $\rightarrow A \quad B \rightarrow$ $S[t,u]E$	$\rightarrow A_S \Rightarrow A_E \xrightarrow[{[t,u]}]{[0,\infty]} B_S \Rightarrow B_E \rightarrow$
Control Edge $A \rightarrow B$	$A_S \Rightarrow A_E \xrightarrow{[0,\infty]} B_S \Rightarrow B_E$		

and a single end node. Note that in this paper we do not consider the choices pattern, but we are currently extending STNPSU to support choices as well. Moreover, preliminary analysis shows that the results presented in this paper will be applicable to this extended kind of STNPSU. First, we show how to verify the *dynamic controllability* (DC) of process schema P and, if P is DC, how to derive its minimal constraints. This can be done by transforming P into an STNPSU S using the transformation rules depicted in Table 1. The resulting STNPSU is characterized by having a single *initial timepoint* that occurs before any other one—called Z—and a single *ending timepoint*—called E—that occurs after any other timepoint. This STNPSU is then checked for DC by applying the standard algorithm for DC checking [10]. In particular, using a constructive proof analogous to the one presented in [8], one can easily show that the process will be DC if and only if the corresponding STNPSU is DC.

Note that the DC checking algorithm also derives the minimum and maximum duration between timepoints Z and E, i.e., the minimum and maximum durations of the process. However, these bounds are not sufficient for characterizing the temporal behavior of the process as they do not represent its possible non-restrictable duration ranges. As an example consider the STNPSU depicted in Fig. 2c, which corresponds to process P_2 of Fig. 1. One can easily show that the duration range between Z and E corresponds to $[5, 19]$. However, this range cannot be reduced to $[5, 10]$, for example, since the internal task T_7 has a contingent duration of 1 to 7, which cannot be controlled (i.e., restricted) by the process engine. In particular, if T_7 lasts exactly 7, process P_2 lasts at least 11 time units. On the other hand, representing a subprocess by considering the duration range between Z and E to be a contingent one would make the overall process over-constrained, and thus limit the overall temporal flexibility of the modularized process.

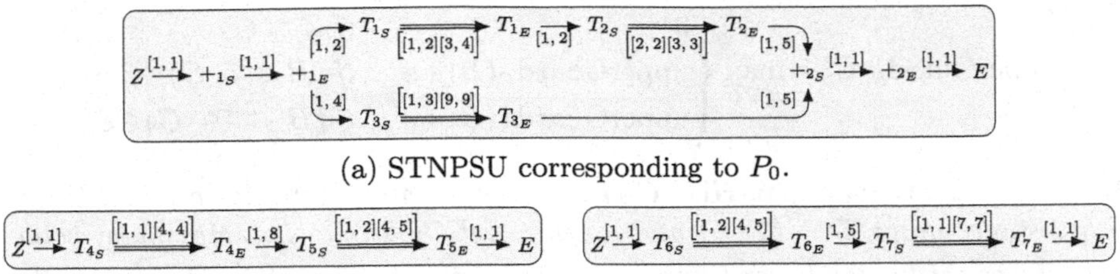

(a) STNPSU corresponding to P_0.

(b) STNPSU corresponding to P_1.

(c) STNPSU corresponding to P_2.

Fig. 2. STNPSUs corresponding to subprocesses P_0, P_1 and P_2 depicted in Fig. 1.

We, therefore, suggest representing the duration of a process by a guarded range with proper guards in order to prevent unacceptable restrictions of the duration range of the process. In the following, we propose a method to determine the lower and upper guard of such guarded range based on the STNPSU representation of the process schema. In this context, the *upper guard* for the duration range of a process P represents the lowest value the maximum duration of the process may be decreased to. In other words, considering the corresponding STNPSU S of P, the upper guard corresponds to the lowest value the upper bound of the requirement link, which is derived between Z and E by the DC checking algorithm, may be decreased to. It can be determined considering the maximum guards of any guarded link and the lower bounds of any requirement link in S as outlined in Example 1.

Example 1 (Upper Guard). Consider the STNPSU depicted in Fig. 2c. While the upper bounds of the internal requirement links may be restricted to their lower bounds (i.e., 1) by the process engine, the upper bounds of the two guarded links cannot be restricted below their upper guards (i.e., 4 and 7, respectively). Therefore, the value we obtain when summing the lower bound values of the requirement links and the upper guards of the guarded links, i.e., $1+4+1+7+1 = 14$, represents the minimal value the upper bound of the link between Z and E may be restricted to.

In turn, the *lower guard* for the duration range of a process P represents the greatest value the minimum duration of the process may be increased to. In the STNPSU S, therefore, the lower guard corresponds to the greatest value the lower bound of the requirement link between Z and E may be increased to.

If there are several paths leading from Z to E, it is necessary to consider the maximum/minimum such value considering all paths. Therefore, Definitions 1 and 2 specify the concept of lower/upper guard for any timepoint of an STNPSU.

Definition 1 (Upper Guard). *Given a dynamically controllable STNPSU S with distance graph $\mathcal{D} = (\mathcal{T}, \mathcal{E})$ and a timepoint C. Then: The minimum value that may be set for the upper bound v of a requirement link $Z \xrightarrow{[u,v]} C$ is called the upper guard of C:*

$$\text{upperGuard}_S(C) = \max_{B \in \mathcal{T}} \begin{cases} 0 & \text{if } Z \equiv C \\ \text{upperGuard}_S(B) + x & \text{if } (B \xleftarrow{-x} C) \in \mathcal{E} \\ \text{upperGuard}_S(B) + y' & \text{if } (B \xleftarrow{D:-y'} C) \in \mathcal{E} \end{cases}$$

Definition 2 (Lower Guard). *Given a dynamically controllable STNPSU S with distance graph $\mathcal{D} = (\mathcal{T}, \mathcal{E})$ and a timepoint C. Then: The maximum value that may be set for the lower bound u of a requirement link $Z \xrightarrow{[u,v]} C$ is called the lower guard of C:*

$$\text{lowerGuard}_S(C) = \min_{B \in \mathcal{T}} \begin{cases} 0 & \text{if } Z \equiv C \\ \text{lowerGuard}_S(B) + y & \text{if } (B \xrightarrow{y} C) \in \mathcal{E} \\ \text{lowerGuard}_S(B) + x' & \text{if } (B \xrightarrow{d:x'} C) \in \mathcal{E} \end{cases}$$

Definitions 1 and 2 allow determining to which extent the upper/lower bound of the derived requirement link between Z and a timepoint C in an STNPSU S may be reduced/increased, without affecting the DC of S (cf. Lemmas 1 and 2).

Lemma 1 (Upper Guard). *Let S be a dynamically controllable STNPSU, Z be the initial timepoint and C be a timepoint in S. Then: The upper bound v of the distance $Z \xrightarrow{[u,v]} C$ between Z and C may be reduced to at most $\text{upperGuard}_S(C)$, preserving the DC of S.*

Lemma 2 (Lower Guard). *Let S be a dynamically controllable STNPSU, Z be the initial timepoint and C be a timepoint in S. Then: The lower bound u of distance $Z \xrightarrow{[u,v]} C$ between Z and C may be increased to at most $\text{lowerGuard}_S(S)$, preserving the DC of S.*

Sketch of Proof (see a technical report [9] for the full proof). By considering the *AllMax-Projection* D' used by Algorithm 1, one can show that if y is restricted beyond its guard $\text{upperGuard}_S(C)$, S can no longer be DC. On the other hand, assuming that y is restricted to $\text{upperGuard}_S(C)$ and the network is not DC, one can show that in this case the value of $\text{upperGuard}_S(C)$ must have been greater than assumed, which contradicts the assumption. The proof of Lemma 2 is analogous considering the AllMin-Projection. The AllMin-Projection is similar to the AllMax-Projection D', but considers only ordinary and lower-case edges. □

Using Definitions 1 and 2, it now becomes possible to determine to which extent the lower/upper bound of the duration range of a process can be restricted, while preserving its DC as illustrated by Example 2.

Example 2. The minimum and maximum durations of the processes depicted in Fig. 1 are determined by the DC checking algorithm as P_0: $[11, 20]$, P_1: $[5, 19]$, and P_2: $[5, 19]$. Using Definitions 1 and 2, it now becomes possible to determine to which extent these duration ranges may be restricted: the minimum duration of P_0 may be restricted to $\text{lowerGuard}_{P_0}(E) = 15$ at most, while its maximum duration may be restricted to $\text{upperGuard}_{P_0}(E) = 15$; the duration of P_1 may be restricted to $\text{lowerGuard}_{P_1}(E) = 13$ and $\text{upperGuard}_{P_1}(E) = 11$, respectively; and the duration of P_2 to $\text{lowerGuard}_{P_2}(E) = 10$ and $\text{upperGuard}_{P_2}(E) = 14$.

Based on the definitions of lowerGuard and upperGuard, one can easily verify that their value is always non-negative. Moreover, it is easy to verify that the upperGuard(C) value is given by value u of edge $Z \xleftarrow{-u} C$ in the AllMax-Projection graph of the network, while lowerGuard(C) value is given by value v of edge $Z \xrightarrow{v} C$ in the AllMin-Projection graph. Using standard STN algorithms [4], therefore, the computational cost of determining lowerGuard(C) and upperGuard(C) is at most $O(n^3)$, with n being the number of timepoints in the considered STNPSU.

Given a range $[u, v]$ that represents the overall duration of a DC process, Definitions 1 and 2 assure that it is always possible to reduce one of the two bounds of the respective duration range to the corresponding guard (i.e., upperGuard(E) or lowerGuard(E)) without affecting the DC of the process. However, it is not possible to restrict both bounds simultaneously since the restriction of one bound may change the guard of the other bound as shown by Example 3.

Example 3. Let us consider the STNPSU from Fig. 2c that corresponds to subprocess P_2. One can easily determine that lowerGuard$_{P_2}(E) = 10$ and upperGuard$_{P_2}(E) = 14$ hold. Moreover, the duration range of the process is $[5, 19]$ as determined by the DC checking algorithm. Considering Lemmas 1 and 2, it then can be easily shown that the minimum duration of the process may be increased to 10 or its maximum duration may be restricted to 14. However, for process P_2 it is not possible to increase the minimum duration to 10, while at the same time restricting the maximum duration to 14. In particular, if the minimum duration is increased to 10, due to the *partially contingent guarded link* between timepoints T_{7_S} and T_{7_E} (representing task T_7), the maximum duration must not be decreased below 16 to further guarantee the DC of the process. On the other hand, the maximum duration may be decreased to 14, but then the minimum duration must not be increased beyond 8. In detail, a span of at least 6 must be ensured for the final duration range of the process.

To fully represent the overall temporal properties of a process we suggest considering an additional value that represents the minimal span to be guaranteed for the duration range. We denote this value as the *contingency span* of the process. It can be defined using the *link contingency span* and *path contingency span* of the corresponding STNPSU.

Definition 3 (Link Contingency Span). *A positive link contingency span Δ corresponds to the span that needs to be guaranteed for a link in order to ensure the DC of an STNPSU. In turn, a negative link contingency span corresponds to the maximum span provided by a link that can be used to reduce the contingency span of previous guarded link.*

(a) *For a guarded link $A \xrightarrow{[[a, a'][b', b]]} B$, the link contingency span Δ_{AB} is defined as $\Delta_{AB} = b' - a'$.*

(b) *For a requirement link $A \xrightarrow{[a, b]} B$, the link contingency span Δ_{AB} is defined as $\Delta_{AB} = a - b$.*

Next, we need to find a way to determine the contingency span of a path based on the link contingency span of its links. First, let us consider a guarded link $A\xrightarrow{[\![a,a']\!][b',b]\!]}$ followed by a requirement link $B\xrightarrow{[c,d]}C$. In this case, the contingency span required by the guarded link can be partially or fully compensated by the subsequent requirement link, as the duration of the latter can be decided based on the actual duration of the former. Thus, the contingency of the path from A to C is given by $\varDelta_{AB}+\varDelta_{BC}$. In turn, for a requirement link $A\xrightarrow{[a,b]}B$ followed by a guarded link $B\xrightarrow{[\![c,c']\!][d',d]\!]}C$ we must differentiate two subcases: If the guarded link is *partially contingent* (i.e., $c' < d'$) the previous requirement link cannot be used to compensate its contingency span as the duration of the requirement link must be decided before executing the guarded link. Therefore, the contingency span of the path from A to C is given by $\varDelta_{BC}$. However, if the guarded link is *partially shrinkable* (i.e., $d' \leq c'$), its link contingency $\varDelta_{BC}$ is negative. In this case, the contingency span of the path from A to C is again given by $\varDelta_{AB}+\varDelta_{BC}$ as both links could be used to reduce the contingency of a previous guarded link. Finally, the combination of two requirement links (guarded links) is similar to the above cases. When considering a path that consists of more than two links, the link contingency spans need to be combined in an incremental way starting from the inital timepoint Z. When considering two or more parallel paths, in turn, it becomes necessary to consider the most demanding case, i.e., the path with the largest contingency span. This leads to the following recursive approach for calculating the contingency span of a path.

Definition 4 (Path Contingency Span). *Let S be a dynamically controllable STNPSU and Z be its initial timepoint. By definition the path contingency span of Z is* $\mathrm{cont}_S(Z) = 0$. *Then: The path contingency span $\mathrm{cont}_S(C)$ of any other timepoint C is given by*

$$\mathrm{cont}_S(C) = \max\left\{0, \max_{B\in\mathcal{T}}\{\mathrm{cont}_S(B) + \varDelta_{BC}\}\right\}$$

It is noteworthy that the path contingency span of any timepoint is always greater or equal to zero, i.e., $\mathrm{cont}_S(C) \geq 0$. Moreover, the problem of determining the value of $\mathrm{cont}_S(C)$ can be reduced to the problem of finding the minimal distance between Z and C in a weighted graph considering the negative link contingency spans as edge values [9]. Using the Bellman–Ford algorithm, the computational cost of determining $\mathrm{cont}_S(C)$ is at most $O(n^3)$, with n being the number of timepoints in the STNPSU.

Example 4. Regarding the STNPSUs from Fig. 2, the path contingency span of timepoints E are as follows: $\mathrm{cont}_{P_0}(E) = 2$, $\mathrm{cont}_{P_1}(E) = 2$, and $\mathrm{cont}_{P_2}(E) = 6$.

Based on Definition 4, it becomes possible to describe the admissible duration ranges between two timepoints in an STNPSU.

Lemma 3. *Let S be a dynamically controllable STNPSU, Z be its initial timepoint, and C be any other timepoint. Then: In order to preserve the DC of S, any* *restriction* $Z \xrightarrow{[u^*, v^*]} C$ *(u $\leq u^* \leq$ lowerGuard$_S(C)$, upperGuard$_S(C) \leq v^* \leq v$) of the distance between Z and C must be done in such a way that $v^* - u^* \geq$ cont$_S(C)$ holds.*

Sketch of Proof (see [9] *for the full proof).* By induction it can be shown that when restricting $[u, v]$ to $[u^*, v^*]$ (with $v^* - u^* <$ cont$_S(C)$), S is no longer DC. □

From the previous observations, we can derive important relationships between lowerGuard(C), upperGuard(C) and cont(C) values:

Lemma 4. *Let S be a dynamically controllable STNPSU, Z be its initial timepoint and C be any other timepoint. If T is the network derived from S by restricting upper bound v of the distance $Z \xrightarrow{[u, v]} C$ between Z and C to v^*, with upperGuard$_S(C) \leq v^* \leq v$, in T it holds*

$$\text{lowerGuard}_T(C) = \min\{\text{lowerGuard}_S(C); v^* - \text{cont}_S(C)\}$$

Lemma 5. *Let S be a dynamically controllable STNPSU, Z be its initial timepoint and C be any other timepoint. If T is the network derived from S by restricting the lower bound u of the distance $Z \xrightarrow{[u, v]} C$ between Z and C to u^*, with $u \leq u^* \leq$ lowerGuard$_S(C)$, in T it holds*

$$\text{upperGuard}_T(C) = \max\{\text{upperGuard}_S(C); u^* + \text{cont}_S(C)\}$$

Sketch of Proof (see [9] *for the full proof).* The proofs of Lemmas 4 and 5 are similar. In particular, assuming that u/v is restricted to lowerGuard$_T(C)$/upperGuard$_T(C)$ and the resulting network is not DC, one can show that in this case cont$_S(C) < 0$ holds. □

The previous results give rise to the following theorem that enables a complete description of the overall temporal properties of a process.

Theorem 1 (Overall Temporal Properties of a Process). *Considering a process P and the corresponding STNPSU S, let Z and E be the single start and single end timepoints of S. Then: The overall temporal properties of P can be described by a guarded range with contingency $[\![x, x'][y', y]\!] \updownarrow c$, where*

- *x and y are the bounds of the requirement link $Z \xrightarrow{[x, y]} E$ between initial timepoint Z and ending timepoint E in S, as derived by the DC checking algorithm,*
- *$x' = $ lowerGuard$_S(E)$ and $y' = $ upperGuard$_S(E)$, and*
- *$c = $ cont$_S(E)$.*

Proof. Definitions 1 and 2 show how to use the values of lowerGuard$_S(E) = x'$ and upperGuard$_S(E) = y'$ to specify the possible restrictions regarding the lower and upper bounds of the duration range $[x, y]$ of a process (i.e., its minimum and maximum duration). This way, we can fully represent the possible duration ranges of the process as a guarded range $[\![x, x'][y', y]\!]$. Moreover, Lemmas 3–5 show how to use the path contingency span cont$_S(E) = c$ in order to ensure that any possible restriction of the duration range $[\![x, x'][y', y]\!] \updownarrow c$ of the process preserves its DC. □

Based on Theorem 1, it becomes possible to represent the overall temporal properties of a process using a single guarded range with contingency, as illustrated by Example 5.

Example 5. First, consider process P_1 as depicted in Fig. 1 together with the corresponding STNPSU shown in Fig. 2. The overall temporal properties of this process may be described by guarded range with contingency $[\![5, 13]\![11, 19]\!]\updownarrow 2$. Since the contingency span of this process corresponds to 2, it is possible to restrict the overall duration range of the process to $[13, 15]$ or $[9, 11]$, while still preserving its DC. In turn, the overall temporal properties of process P_2 (cf. Figs. 1 and 2) can be described by a guarded range with contingency $[\![5, 10]\![14, 19]\!]\updownarrow 6$. For example, the duration range of the process, therefore, can be restricted to $[6, 14]$, $[10, 17]$, or $[8, 14]$. However, due to the required contingency span of 6, for example, it must not be restricted to $[10, 14]$, or $[10, 15]$.

Such kind of compact representation of the overall temporal properties of a process schema is crucial for being able to reuse it as part of a modularized process. In particular, when adding a subprocess task to a process schema, a duration range must be specified. Based on the guarded range with contingency determined for the subprocess it is now possible to determine a proper duration range for it when it is insert in the main process. This duration range ensures that, without having to reanalyze the subprocess schema, any restriction of the duration of the subprocess task in the main process will be made in such a way that the subprocess remains dynamically controllable.

4   DC-Checking of Modularized Time-Aware Processes

As shown in the previous section, for each time-aware process, it is possible to derive a *guarded range with contingency* that fully describes the overall temporal properties of the process. In this section we show how this knowledge may be utilized for enabling a sophisticated support of modularized time-aware processes in a PAIS.

In a PAIS, the available process schemas are generally stored in a central process model repository. Based on the results presented in Sect. 3, it now becomes possible to enhance the information about the process schemas in such a repository with the overall temporal properties of the process schema represented as a guarded range with contingency. Such information can then be utilized when re-using a process schema as part of a modularized time-aware process. In particular, during design time a process designer may select a process schema from the repository to be used as a subprocess task. Similar to an atomic task, the designer then has to configure the subprocess task within the process schema; i.e., he must specify the duration range of the particular subprocess task. In order to ensure the executability of the modularized process the designer must guarantee that the duration range set for the subprocess task is compliant with the overall temporal properties of the (sub-)process schema. In this context, the repository information about the overall temporal properties of the (sub-)process schema may be used to support the process designer in choosing a proper duration range for the respective subprocess task. In other words, the designer must select a guarded range as duration range of the subprocess task, which satisfies the guards as well as the contingency of the guarded range with contingency representing the overall temporal properties of the (sub-)process schema as stored in the repository.

In general, the duration range $[\![x, x'][y', y]\!]$ of a subprocess task needs to be selected with respect to the overall temporal properties of the respective (sub-) process schema $[\![u, u'][v', v]\!] \updownarrow c$ such that $u \leq x \leq x' \leq u'$ and $v \geq y \geq y' \geq v'$ hold. Moreover, if $c > 0$ holds, $y' - x' \geq c$ must hold as well. When observing these constraints, it is guaranteed that, during the execution of a subprocess task of a modularized process, the respective subprocess instance may be completed without violating any of its temporal constraints (i.e., the subprocess is DC).

Example 6. Figure 3 depicts the modularized process from Fig. 1 where proper duration ranges have been selected for the three subprocess tasks P_0, P_1 and P_2, which are related to (sub-)process schemas NonPharmR, PhysEx and PharmR. For example, for subprocess task P_0, duration range $[\![10, 14][16, 20]\!]$ is used. This range has the same outer bounds as the overall temporal properties of the respective process schema, i.e., $[\![10, 15][15, 20]\!] \updownarrow 2$. Moreover, the lower and upper guard of the duration range ensure that the guards as well as contingency value determined for the process schema are observed. In turn, for subprocess task P_1 the designer decides to further restrict the upper bound of the duration range to 9 (thus also decreasing the lower guard to 9). Note that this still guarantees the DC of subprocess schema PhysEx as it complies with the respective guards and contingency. Finally, for subprocess P_2, the designer increased the lower bound to 8 and the upper guard to 17, thus providing a possible contingency of 7 instead of the required contingency of 6.

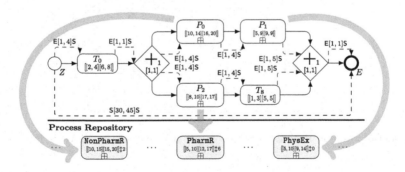

Fig. 3. Modularized process.

After completing the design of the modularized process schema, the dynamic controllability of the parent process schema itself needs to be verified. Then, the overall temporal properties of the modularized process schema may be determined based of the approach presented in Sect. 3.

Finally, the modularized process itself may be added to the process repository. It may then be reused as a subprocess in the context of another modularized process. This enables the definition of hierarchically structured modularized time-aware process schemas comprising multiple levels.

5 Proof of Concept

The presented approach was implemented as a proof-of-concept prototype in the ATAPIS Toolset [11]. This prototype enables users to create time-aware process schemas and to automatically transform them to a corresponding STNPSU. The STNPSU can then be checked for dynamic controllability. Moreover, the overall temporal properties of the process can be determined.

The screenshot from Fig. 4 shows the ATAPIS Toolset[1]: at the top, the process schema from Fig. 1b is shown. At the bottom, the automatically generated STNPSU and its minimal network are depicted. Finally, the dialog in the middle shows the overall temporal properties of the process schema which have been determined based on the STNPSU.

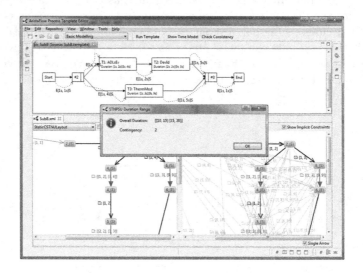

Fig. 4. Determining Process Overall Temporal Properties in ATAPIS Toolset.

Moreover, using the ATAPIS prototype it becomes possible to create modularized time-aware processes and to assign a proper duration range to each subprocess task based on the overall temporal properties of the respective (sub-) process schema. The resulting modularized time-aware process schema can then

---

[1] A screencast demonstrating the toolset is available at http://dbis.info/atapis.

be checked for dynamic controllability and its overall temporal properties be determined. It is then possible to reuse this modularized time-aware process schema for a subprocess task in another modularized process.

First simulations based on the ATAPIS prototype show a significantly improved performance of our modularization-based approach compared to the "classical approach" where each subprocess task has to be replaced by it respective (temporal) components. Overall, the prototype demonstrates the applicability of our approach.

6   Conclusions

Time and modular design constitute two fundamental aspects for properly supporting business processes by PAIS. So far, these aspects have only been considered in isolation, although the overall temporal behaviour of a (sub-)process significantly differs from the one of simple tasks. This paper closes this gap by considering modularization and time-awareness of processes in conjunction with each other. In particular, we propose a novel approach for determining and representing the overall temporal behavior of a process, called *guarded range with contingency*. Using this representation, we can specify the possible durations of a (sub-)process as well as any permissible restriction that may be applied to it, while still ensuring the executability of the process. Moreover, we show how this may be used in the context of process repositories and multilayered process hierarchies.

We are currently extending STNPSU to consider conditional aspects as well. In future work, we want to study the integration of (modularized) time-aware processes in PAISs, specifically focusing on aspects like scalability and usability.

References

1. Combi, C., Gambini, M., Migliorini, S., Posenato, R.: Representing business processes through a temporal data-centric workflow modeling language: an application to the management of clinical pathways. IEEE Trans. Syst. Man Cybern.: Syst. **44**(9), 1182–1203 (2014)
2. Combi, C., Hunsberger, L., Posenato, R.: An algorithm for checking the dynamic controllability of a conditional simple temporal network with uncertainty. In: Filipe, J., Fred, A.L.N. (eds.) Proceedings of the 5th International Conference on Agents and Artificial Intelligence, ICAART 2013, vol. 2, pp. 144–156. SciTePress, February 2013
3. Combi, C., Posenato, R.: Towards temporal controllabilities for workflow schemata. In: Markey, N., Wijsen, J. (eds.) 17th International Symposium on Temporal Representation and Reasoning, TIME 2010, pp. 129–136. IEEE Computer Society, September 2010
4. Dechter, R., Meiri, I., Pearl, J.: Temporal constraint networks. Artif. Intell. **49**(1–3), 61–95 (1991)
5. Ede, J., Gruber, W., Panagos, E.: Temporal modeling of workflows with conditional execution paths. In: Ibrahim, M., Küng, J., Revell, N. (eds.) DEXA 2000. LNCS, vol. 1873, pp. 243–253. Springer, Heidelberg (2000)

6. Eder, J., Panagos, E., Rabinovich, M.: Workflow time management revisited. In: Seminal Contributions to Information Systems Engineering, pp. 207–213. Springer, Heidelberg (2013)

7. Hochberg, M.C., et al.: American college of rheumatology 2012 recommendations for the use of nonpharmacologic and pharmacologic therapies in osteoarthritis of the hand, hip, and knee. Arthritis Care Res. **64**(4), 465–474 (2012)

8. Lanz, A., Posenato, R., Combi, C., Reichert, M.: Controllability of time-aware processes at run time. In: Meersman, R., Panetto, H., Dillon, T., Eder, J., Bellahsene, Z., Ritter, N., De Leenheer, P., Dou, D. (eds.) ODBASE 2013. LNCS, vol. 8185, pp. 39–56. Springer, Heidelberg (2013)

9. Lanz, A., Posenato, R., Combi, C., Reichert, M.: Controlling time-awareness in modularized processes (extended version). Tech. Rep. UIB-2015-01, Ulm University, March 2015. http://dbis.eprints.uni-ulm.de/1133/

10. Lanz, A., Posenato, R., Combi, C., Reichert, M.: Simple temporal networks with partially shrinkable uncertainty. In: Proceedings of the International Conference on Agents and Artificial Intelligence, ICAART 2015, vol. 2, pp. 370–381. SciTePress (2015)

11. Lanz, A., Reichert, M.: Dealing with changes of time-aware processes. In: Sadiq, S., Soffer, P., Völzer, H. (eds.) BPM 2014. LNCS, vol. 8659, pp. 217–233. Springer, Heidelberg (2014)

12. Lanz, A., Weber, B., Reichert, M.: Time patterns for process-aware information systems. Requirements Eng. **19**(2), 113–141 (2014)

13. Morris, P.H., Muscettola, N.: Temporal dynamic controllability revisited. In: National Conference on Artificial Intelligence (AAAI 2005), pp. 1193–1198 (2005)

14. Reichert, M., Weber, B.: Enabling Flexibility in Process-aware Information Systems: Challenges, Methods, Technologies. Springer, Heidelberg (2012)

Consistency, Correctness
and Compliance

A Systematic Literature Review of Consistency Among Business Process Models

Afef Awadid$^{(\boxtimes)}$ and Selmin Nurcan

University of Paris 1 Pantheon-Sorbonne, CRI, Paris, France
afef.awadid@malix.univ-parisl.fr,
nurcan@univ-parisl.fr

Abstract. The field of business process modeling has been beset by inter-model consistency problems which are mainly due to the existence of multiple variants of the same business process, for instance when models have been produced by different actors, or through the time by a same (or different) actor(s), as well as the possibility of its modeling from discrete and complementary perspectives (using different lenses). The aim of our research is manifold. First we aim to develop a framework (i) enabling situating new research activities as well as the existing approaches and (ii) targeting to master the inter-model consistency issue. Second, this framework shall offer the capability of handling business process models coherence issue (i) having in mind various modeling goals and targets/products and (ii) having in hand a wide range of problem statements and project situations requiring the use of a large catalogue of business process meta-models. Third, we have the ambition of determining gaps in current research with the aim of suggesting areas for further investigations in the area of inter-models consistency. In order to do so, this paper presents a systematic literature review (SLR) of consistency among business process models, where a total of 982 published papers extracted from the most relevant scientific sources, were considered, of which 41 papers, were ultimately included.

Keywords: Business process models · Modeling perspectives · Inter-model consistency · Systematic literature review

1 Introduction

Business process modeling is chiefly a convergence of two connected modeling disciplines: process modeling [1–3], which aims at providing "an abstract representation of a process architecture, design or definition" [4] and enterprise modeling, which seeks to provide a full and holistic understanding of the enterprise [5, 70]. Reasons for this convergence might be (i) the key role played by business process (BP) models in both enterprise information systems development [6], and organizational management [7, 8], (ii) the similarity between these disciplines in that both may focus on business processes as subject of investigation by capturing the relevant ones [5], and (iii) both have been beset by inter-model consistency problems.

In the field of process modeling, these problems are mainly due to (a) the existence of multiple models or views, which take part in the information systems engineering

© Springer International Publishing Switzerland 2016
R. Schmidt et al. (Eds.): BPMDS/EMMSAD 2016, LNBIP 248, pp. 175–195, 2016.
DOI: 10.1007/978-3-319-39429-9_12

[69] and (b) the existence of many variants of BP models, which capture the occurrences of the same BP. The inconsistencies caused by (a) are the root causes of many errors in the resulting software applications [9], while those caused by (b) constitute a serious obstacle "to dynamically switch process execution from one variant to another if required" [10]. The importance of the first family (a) of inter-model consistency problems is reported in a systematic review of UML model consistency management [11], and a survey on inconsistency management in software engineering [12]. The necessity of dealing with the second family (b) of consistency problems is proved by a large amount of work within this scope.

Similarly, in the field of enterprise modeling, consistency problems are of a great interest to both practitioners and researchers. This interest has emerged from the advent of multi-perspective or multi-view modeling methods where a complex system (e.g., the enterprise architecture, a BP) is captured from different perspectives (views) in order to master its complexity.

Although a SLR [11] and a survey [12] on the inconsistency management of software process models were already carried out and even though software processes are considered as business processes [13, 14], the existing work is mostly related to a particular kind of modeling approaches [15] (mainly object-oriented approaches). This is not the case in BP modeling for which none of the modeling notations is predominant [16] until 2006. This broader extent of the notion of inter-model consistency requires the capability of positioning the great amount of research works in this scope with respect to a reference framework that facilitates identifying the emerging/ unresolved problems in the area of inter-model consistency in BP modeling.

Carrying out a Systematic Literature Review (SLR) in this area seems to be appropriate to set up such a framework. In fact a SLR is defined as a means of identifying, evaluating and interpreting all available research relevant to a particular research question or topic area with the aim, amongst others, of providing a framework (background in order to properly position new research activities [17]).

In this paper, we undertake the first SLR for inter-model consistency in the field of BP modeling. We aim at providing a generic framework enabling positioning existing approaches and determining gaps in the current research. The remainder of the paper is structured as follows. Section 2 outlines the key terms and concepts with regard to the topic of inter-model consistency. We describe our methodology in Sect. 3 and present the results and answer our research questions in Sect. 4. We present the framework in Sect. 5. Section 6 concludes the paper.

2   Inter Model Consistency: Key Terms and Concepts

Consistency issues have been raised in various domains such as databases, information systems development, enterprise modeling and software engineering. Thereby, manifold are the approaches proposing definitions for concepts in this area. Hence, in order to establish a common understanding of the terminology used in this paper, we start by defining key concepts on which this SLR is grounded.

- Diagram: a graphical representation of real world using a particular modeling language.
- Perspective: refers to the notion of view defined as a representation of a system (e.g. a BP) from the angle of a related set of concerns or aspects [18]. For instance informational, functional, behavioral, organizational, operational and intentional in [71].
- Consistency among models: refers to the fact that the information covered in each model should not contradict each other [19]. For instance, if the concept actor appears in more than one model, its instances in all corresponding models have to be syntactically and semantically equivalent.
- Multi-perspective modeling: refers to the notion of multi-view modeling defined as the construction of distinct and separate models of the same system in order to model different aspects of it [20]. For instance, in the domain of business processes, multi-perspective modeling allows us to depict the same BP using distinct and complementary representations adopting distinct modeling languages.
- Projective multi-perspective modeling (commonly referred to as projective multi-view modeling): one comprehensive overarching meta-model is given. All perspectives captured by all concerned modeling languages are defined as projections onto this central meta-model [21]. One example of this approach is the UML, which has, in its current version the Meta Object Facility (MOF) as a common meta-model. All UML diagram types (e.g. activity diagrams, sequence diagrams) are specified by projections onto that MOF meta-model.
- Selective multi-perspective modeling (commonly referred to as selective multi-view modeling): no central meta-model is given. Each perspective is captured by a distinct meta-model and the overall system is obtained as synthesis of the information carried out by the different meta-models [21]. Hence, if one concept (e.g. activity) is used in multiple perspectives, the dependencies between them need to be specified manually.
- Horizontal consistency: refers to the consistency between models at the same phase or abstraction level [22]. For instance, the consistency between two BPMN models produced during the analysis phase.
- Vertical consistency: refers to the consistency between models at different development phases or abstraction levels [22]. For example, the consistency between a BPMN analysis model and the associated BPMN implementation model.
- Syntactic consistency: refers to ensure that a model conforms to its abstract syntax specified by its meta-model [23]. For instance, the roles in the actor-role model should appear in the corresponding role-activity model.
- Semantic consistency: refers to the fact that models behavior should be semantically compatible [23]. For example, actors in the actor-role model have to be defined as business objects in the corresponding business objects model.

3 Method Applied for the SLR

In order to conduct this study as a SLR, we have relied on the review protocol used in [24], since it was based on the original guidelines as proposed by Kitchenham [25]. Two key concepts are mainly associated with the notion of SLR namely (i) the primary study which refers to an empirical study investigating a specific research question and (ii) the secondary study referring to the study that reviews all the primary studies relating to a specific research issue with the aim of integrating/synthetizing evidence related to that issue [25]. The present study is then categorized as a secondary study and involves the steps cited below.

3.1   Research Questions and Search Process

This SLR raises the research questions listed below resulting from our understanding of the key points after the study of the literature.

RQ1. What can be a source of inconsistency among BP models?

RQ2/RQ3. What type(s) of diagram(s) are being tackled? (i) activity-driven diagrams describing a BP as a sequence of activities, (ii) role-driven ones specifying the roles and the organization related issues involved in the BP and/or (iii) product-driven ones that represent a BP through its products/results (or resources) and their evolution. And how many diagrams have been used?

RQ4/RQ5. On which type of inter-model consistency problem focuses the study? Horizontal or vertical and what is the nature of the targeted consistency? Syntactic, semantic or both.

RQ6. What is the main methodological activity [11] on which the consistency management process relies?

RQ7/RQ8. What is the scope of business process models under study? Intra-enterprise or inter-enterprise models and what kinds of multi-perspective modeling are being addressed? Selective or projective multi-perspective modeling.

To perform the manual search process for primary studies, we based on a set of sources that were recommended in [26] as relevant within the research community and that were appropriate for the present study. These sources along with the search fields are presented in Table 1. In the aforementioned sources, we tested with different search string criteria. That which ultimately allowed obtaining the highest number of relevant results was:

("business process model" AND ("consistency" OR "inconsistency"))

In the search process, we also took into account the synonyms and terms related to each of the three concepts, as shown in Table 2.

Table 1. Selected sources along with research fields

Source	Search field
Google Scholar	Title, abstract and full text
ACM Digital Library	Title, abstract and full text
Science Direct	Title, abstract and full text
SCOPUS Database	Title, abstract and keywords
IEEE Computer Society	Title, abstract and full text

Table 2. Other synonyms and terms used in the search process

Concept	Synonym and/or related term
Business process model	Process model; process variant; enterprise modeling or enterprise modeling; multi-perspective modeling or multi-perspective modeling; multi-view modeling or multi-view modeling
Inconsistency	Inconsistencies: incoherence: incohesion

3.2   Inclusion/Exclusion Criteria and Quality Assessment

In this study, peer reviewed papers with the following concerns were included:

- Papers proposing approaches that favor or evaluate consistency between BP models; each included paper raise one of the following questions: how to check consistency between BP models or how to maintain consistency between BP models.
- Papers where the proposed approach was based on the comparison of two or more BP models depicting the same BP, since such comparison is the cornerstone of each inconsistency management activity. For instance, papers dealing with the verification of similarity between BP models are included.
- Papers dealing with the issue of consistency in the context of multi-perspective modeling with a particular focus on the consistency among BP models or with a wider focus towards enterprise modeling.

Articles with the following concerns were excluded:

- Papers focusing on the issue of compliance defined as "a relationship between two sets of specifications: the specifications for executing a BP and the specifications regulating a business" [27]. Thus only papers where the models in question depict the same BP as subject of modeling have been considered.
- Papers dealing with the topic of inter-model consistency, where the subject under study is the software process. This means that our study is not concerned with the inter-model consistency in the field of software engineering.

We also excluded books, doctoral dissertations and non-English papers focusing on the topic of inter-model consistency.

The activity of assessing the "quality" of primary studies is generally viewed as important mainly in guiding the interpretation of findings and determining the strength of inferences as well as in guiding recommendations for further research [25]. The main criteria on which we based the quality assessment of the primary study were (QA1) "is the inter-model consistency the main purpose of the paper in such a way that the issue is studied in a thorough manner, contextualized and validated?", and (QA2) "is the proposed approach generalizable and to which extent is it applicable in another context?" The questions were scored as follows:

QA1. Y (yes), the inter-model consistency problem was contextualized, a well-defined approach was proposed in order to solve it, and a validation of the approach was provided and supported with a tool; P (Partly), the problem was contextualized, a well-defined approach was proposed, a first manual validation was given, but no support tool was offered; N (No), the approach was defined in a general and a succinct way and no validation was given.

QA2. Y (yes), the proposed solution is likely to be applicable outside of the primary study; P (Partly), the proposed solution needs to be slightly altered to meet other requirement outside of the study; N (No), the proposed solution is not likely to be applicable outside of the study (i.e. it is limited to a narrow context).

The scoring procedure was Y = 1; P = 0.5; N = 0. In the coordination between the two authors with regard to the stages of the data collection as well as the quality assessment, each author played a particular role. The one applied inclusion and exclusion criteria during data collection, assessed the quality of primary studies and checked manually the excluded papers based on the abstracts and introduction sections. The other checked all included papers and their score. In case of doubt of the former and lack of availability of the latter to perform a deeper verification, we contacted the authors of the paper.

Data collection and data analysis: The data we extracted from each primary study are: (i) the source, where the paper was found; (i) the data related to the research questions we have raised in Sect. 3.1; (iii) quality evaluation. The data was tabulated in order to put emphasis on the research questions listed in Sect. 3.1.

4   Results

In this section, we summarize and analyze the results of our SLR. We discuss the answers to our research questions and provide recommendations.

4.1   Search Results

Table 3 shows the results of the search procedure respectively before and after applying inclusion and exclusion criteria, along with the selected papers. Before applying the exclusion criterion for eliminating papers that deal with inter-model consistency in the field of software engineering, the number of studies in the first round was very large (982). Applying the aforementioned criterion has considerably decreased this number. This implies that the inter-model consistency is a widely-tackled topic in the field of software engineering. In order to avoid biasing the results of the data analysis, it was essential to ensure that papers appeared in multiple sources were taken into account only once (leading to 982). Also, among the 41 resulting papers the ones describing the same approach were grouped together (leading to 36 as shown in Table 4).

Table 3. Summary of results before and after applying inclusion and exclusion criteria.

Science direct	IEEE	Scopus	Google scholar	ACM	Science direct	Total
Results before	192	12	678	24	76	982
Results after	9	0	24	2	6	41
Selected papers	[3, 28, 34, 38, 47–49, 59, 68]	–	[29–32, 35–37, 41–46, 51, 52, 55–58, 60, 61, 65–67]	[50, 53]	[39, 40, 54, 62–64]	

Thereby, the total number of approaches considered during the data analysis and evinced in Table 4 is 36. The aforesaid Table 4 puts emphasis on the first part of data extracted from each primary study. It includes data related to the source of inconsistency (RQ1), the type (RQ4) and nature (RQ5) of consistency, the type of multi-view modeling (RQ8), the type of diagrams (RQ2), the number of modeling techniques used (RQ3), the scope of BP models (RQ7), and the main activity (purpose) on which the consistency management relies (RQ6). Regarding the latter we identified six fundamental activities in consistency management:

(i) Detect common concepts refers to determining the concepts shared between several models;
(ii) Establish correspondences between elements of models refers to making correspondence between pairs of elements (mainly activities) between two models;
(iii) Evaluate consistency between models refers to checking whether two models are consistent (they do not contradict each other);
(iv) Generate views dependency model refers to generating an intermediate model, which captures the common concepts between multi-perspective models;
(v) Evaluate views dependency model with regard to consistency rules refers to verifying whether the view dependency model complies with the defined consistency rules;
(vi) Generate model from another refers to transforming one model to another.

Table 4. Summary: The first part of extracted data related to the inter-model consistency.

ID	Reference	Inconsistency source (RQ1)	Consistency type (RQ4)	Consistency nature (RQ5)	Type of multi-view modeling (RQ8)	Diagram type (RQ2)	Number of modeling techniques (RQ3)	BP models scope (RQ7)	The main purpose of inconsistency management (RQ6)
S1	Bork and Karagiannis [28]	Multi-persp. modeling	Horizontal	Syntactic & semantic	Projective	DSML	4	Intra-	Generate intermediate model (iv)
S2	Bork et al. [44]	Multi-persp. modeling	Horizontal	Syntactic & semantic	Projective	DSML	4	Intra-	Detect common concepts (i)
S3	Yan et al. [29]	Multi-persp. modeling	Horizontal	Syntactic & semantic	Projective	DSML	>1 (unknown)	Inter-	Detect common concepts (i)
S4	Hallerbach et al. [30]	BP models variants	Vertical	Semantic	–	Activity-	1	Intra-	Establish correspondences between models (ii)
S5	Smirnov et al [31]	BP models merging	Vertical	Semantic (behavioral)	–	Activity-	1	Intra-	Generate model from another (vi)
S6	Koliadis et al. [32, 38]	Multi-persp. modeling	Horizontal	Semantic	–	Activity-Role-	2	Intra-	Evaluate intermediate model (v)
S7	Gerth et al. [33]	BP models merging	Vertical	Semantic (behavioral)	–	Activity-	1	Intra-	Generate model from another (vi)
S8	Zemni et al. [34]	BP models merging	Vertical	Semantic (behavioral)	–	Activity-	1	Intra-	Generate model from another (vi)
S9	Pascalau et al. [35]	BP models variants	Vertical	Semantic	–	Activity-	1	Intra-	Generate model from another (vi)

(*Continued*)

Table 4. (*Continued*)

ID	Reference	Inconsistency source (RQ1)	Consistency type (RQ4)	Consistency nature (RQ5)	Type of multi-view modeling (RQ8)	Diagram type (RQ2)	Number of modeling techniques (RQ3)	BP models scope (RQ7)	The main purpose of inconsistency management (RQ6)
S10	Pascalau and Rath [43]	BP models variants	Vertical	Semantic	–	Activity-	1	Intra-	Establish correspondences between models (ii)
S11	Weidlich et al. [36, 56, 63]	BP models variants	Vertical	Semantic	–	Activity-	1	Intra-	Generate model from another (vi)
S12	Koschmider and Blanchard [37]	BP models variants	Vertical	Semantic	–	Activity-	1	Intra-	Generate model from another (vi)
S13	Milani et al. [39]	BP models variants	Vertical	Semantic	–	Activity-	1	Intra-	Generate model from another (vi)
S14	Dijkman et al. [40, 49, 66]	BP models variants	Horizontal	Semantic	–	Activity-	1	Intra-	Establish correspondences between models (ii)
S15	Lu et al. [41]	BP models variants	Vertical	Semantic	–	Activity-	1	Intra-	Evaluate consistency between models (iii)
S16	Rastrepkina [42]	BP models variants	Vertical	Semantic	–	Activity-	1	Intra-	Establish correspondences between models (ii)

(*Continued*)

Table 4. (*Continued*)

ID	Reference	Inconsistency source (RQ1)	Consistency type (RQ4)	Consistency nature (RQ5)	Type of multi-view modeling (RQ8)	Diagram type (RQ2)	Number of modeling techniques (RQ3)	BP models scope (RQ7)	The main purpose of inconsistency management (RQ6)
S17	Cheng-Leong et al [45]	BP models variants	Vertical	Semantic	–	Activity-	1	Intra-	Generate model from another (vi)
S18	Chen-Burger [46]	Multi-persp. modeling	Horizontal	Syntactic & semantic	Selective	Activity-Product-	2	Intra-	Evaluate intermediate model (v)
S19	Koehler et al. [47]	BP models variants	Vertical	Semantic	–	Activity-	1	Intra-	Evaluate consistency between models (iii)
S20	Worzberger et al. [48]	BP models variants	Vertical	Semantic	–	Activity-	1	Intra-	Evaluate consistency between models (iii)
S21	Lu and Sadiq [50]	BP models variants	Vertical	Semantic	–	Activity-	1	Intra-	Establish correspondences between models (ii)
S22	Gulden and Frank [51]	Multi-persp. modeling	Horizontal	Syntactic & semantic	Projective	DSML	3	Intra-	Detect common concepts (i)
S23	Delen and Benjamin [52]	Multi-persp. modeling	Horizontal	Syntactic & semantic	Projective	unspecified	unspecified	Intra-	Detect common concepts (i)
S24	Leist and Zellner [53]	Multi-persp. modeling	Horizontal	Syntactic & semantic	Projective	Activity-	1	Intra-	Detect common concepts (i)

(*Continued*)

Table 4. (*Continued*)

ID	Reference	Inconsistency source (RQ1)	Consistency type (RQ4)	Consistency nature (RQ5)	Type of multi-view modeling (RQ8)	Diagram type (RQ2)	Number of modeling techniques (RQ3)	BP models scope (RQ7)	The main purpose of inconsistency management (RQ6)
S25	Shunk et al. [54]	Multi-persp. modeling	Horizontal	Syntactic & semantic	Projective	unspecified	unspecified	Intra-	Detect common concepts (i)
S26	Fang et al. [55]	BP models variants	Vertical	Semantic	–	Activity-	1	Intra-	Evaluate consistency between models (iii)
S27	Koubarakis and Plexousakis [57]	Multi-persp. modeling	Horizontal	Syntactic & semantic	Projective	unspecified	unspecified	Intra-	Evaluate intermediate model (v)
S28	Vanderfeesten et al. [58]	BP models variants	Vertical	Semantic	–	Activity-	1	Intra-	Evaluate consistency between models (iii)
S29	Martens [59]	BP models variants	Vertical	Semantic	–	Activity-	1	Intra-	Generate model from another (vi)
S30	Decker and Weske [60]	BP models variants	Vertical	Semantic	–	Activity-	1	Intra-	Evaluate consistency between models (iii)
S31	Fang et al. [61]	BP models variants	Vertical	Semantic	–	Activity-	1	Intra-	Establish correspondences between models (ii)

(*Continued*)

Table 4. (*Continued*)

ID	Reference	Inconsistency source (RQ1)	Consistency type (RQ4)	Consistency nature (RQ5)	Type of multi-view modeling (RQ8)	Diagram type (RQ2)	Number of modeling techniques (RQ3)	BP models scope (RQ7)	The main purpose of inconsistency management (RQ6)
S32	De Medeiros et al. [62]	BP models variants	Vertical	Semantic (behavioral)	–	Activity-	1	Intra-	Evaluate consistency between models (iii)
S33	Niemann et al. [64]	BP models variants	Vertical	Semantic	–	Activity-	1	Intra-	Evaluate consistency between models (iii)
S34	Küster et al. [65]	BP models variants	Vertical	Syntactic & semantic	–	Activity-	1	Intra-	Evaluate consistency between models (iii)
S35	Van der Aalst et al. [67]	BP models variants	Vertical	Semantic	–	Activity-	1	Intra-	Evaluate consistency between models (iii)
S36	Li et al. [68]	BP models variants	Vertical	Semantic	–	Activity-	1	Intra-	Evaluate consistency between models (iii)

DSML: Domain specific modeling language

4.2   Quality Evaluation of Primary Studies

We assessed the studies for quality based on the two quality assessment questions defined in the Sect. 3.2. The score for each study is shown in Table 5.

Table 5. Summary: quality evaluation of studies.

Study	QA1	QA2	Total score	Study	QA1	QA2	Total score
S1	Y	Y	2	S19	P	N	0.5
S2	P	Y	1.5	S20	Y	P	1.5
S3	P	P	1	S21	P	N	0.5
S4	P	P	1	S22	P	N	0.5
S5	P	P	1	S23	P	N	0.5
S6	P	P	1	S24	P	N	0.5
S7	P	P	1	S25	P	Y	1.5
S8	P	Y	1.5	S26	Y	P	1.5
S9	P	P	1	S27	P	N	0.5
S10	Y	P	1.5	S28	Y	P	1.5
S11	Y	Y	2	S29	Y	P	1.5
S12	P	P	1	S30	P	Y	1.5
S13	P	P	1	S31	Y	P	1.5
S14	P	Y	1.5	S32	P	Y	1.5
S15	Y	P	1.5	S33	P	P	1
S16	P	P	1	S34	P	P	1
S17	P	N	0.5	S35	P	Y	1.5
S18	Y	P	1.5	S36	Y	P	1.5

4.3   Analysis of Results and Discussion

The column named "Inconsistency source" in the above Table 4, along with Fig. 1 reveal that mainly three sources (the multi-perspective modeling, the existence of many BP models variants depicting the same BP, and the merging of BP models) prompted researchers to deal with the issue of inter-model consistency.

Figure 1 shows that the BP models variants as source of inconsistency is tackled by 64 % of the studies (23 of 36), whereas the other sources are somehow overlooked. When focusing on this source in relation with the columns named respectively "Consistency type" and "Number of modeling techniques", a strong dependency can be deduced between them as shown in Table 6.

Table 6 reveals on one side that consistency problems caused by the existence of many variants of BP models are mainly vertical consistency problems (95.7 %), whereas those arising from the

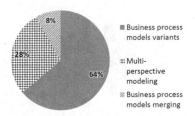

Fig. 1. Sources of inconsistencies addressed in literature (Color figure online)

Table 6. Consistency type and number of techniques in relation with inconsistency source.

Inconsistency source	Consistency type		Number of modeling techniques	
	Horizontal	Vertical	=1	>1
BP models variants	4.3 % (1 of 23)	95.7 % (22 of 23)	100 % (23 of 23)	0 %
BP models merging	0 % (0 of 3)	100 % (3 of 3)	100 % (3 of 3)	0 %
Multi-perspective modeling	100 % (10 of 10)	0 % (0 of 10)	10 % (1 of 10)	90 % (9 of 10)

Table 7. Consistency nature in relation with the inconsistency source.

Inconsistency source	Consistency nature		
	Syntactic	Semantic	Both
BP models merging	0 %	100 % (3 of 3)	0 %
BP models variants	0 %	95.7 % (22 of 23)	4.3 % (1 of 23)
Multi-perspective modeling	0 %	10 % (1 of 10)	90 % (9 of 10)

multi-perspective modeling as inconsistency source refer usually to horizontal consistency problems (100 %). On the other side, all primary studies dealing with consistency across BP models variants rely on one single modeling technique, whereas often more than one technique are (90 % of studies) used when the cause of inconsistency between models is the multi-perspective modeling.

Similarly, the nature of consistency can be strongly linked to the inconsistency source. Table 7 puts forward this link.

Furthermore, the three approaches (S5, S7 and S8) seeking to tackle the consistency when merging two fragments of BP models chiefly target consistency of a semantic nature as depicted in Table 7. This refers to consistency problems related to behavioral aspects of a BP like the exclusiveness of a pair of activity (i.e. the execution logic such as AND, OR, XOR) or their order of potential occurrence. Figures 2 and 3 summarize the relation between the six fundamental activities in inconsistency management, and the three sources of inconsistency.

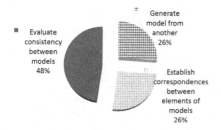

Fig. 2. Activities in inconsistency management when inconsistency source is *BP models variants*

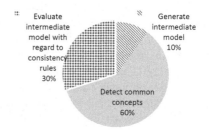

Fig. 3. Activities in inconsistency management when inconsistency source is *Multi-perspective modeling*

Figure 2 highlights that the most common activity (48 %) in the literature with respect to the inter-model consistency is its evaluation, known as *consistency checking* between two or more BP models variants. Figure 3 shows that managing inconsistency in case of multi-perspective modeling consists mainly in detecting common concepts among multiple views (60 %). Evaluating an intermediate model with regard to consistency rules (30 %) is also referred as *consistency checking* in the literature. We also observed that only few concepts are shared between the models when they depict the same system from complementary perspectives. Hence, a partial dependency exists between the BP models. An inter-model consistency problem occurs when a partial or a total (strong) dependency exists between BP models. The former happens in case of multi-perspective modeling, whereas the latter appears when many variants of the same BP exist. The source of inconsistency among BP models seems to be the cornerstone of each attempt for categorizing research works dealing with the inter-model consistency issue.

In the following, we attempt to answer the research questions set in Sect. 3.1.

RQ1: We identified three sources of inconsistency among BP models (see Fig. 1). The majority of approaches focus on the variants of the same BP model (64 %). During our analysis of multi-perspective modeling approaches (28 %), we noted that among the 10 primary studies, only one, S27, has focused on the BP as a subject of modeling, i.e. according to multiple perspectives. Unfortunately, the proposed solution is not applicable outside of this primary study (QA2 = N, see Table 5). Four others (S1, S2, S3, S6) analyze the enterprise as a whole, offering the BP models among the multiple perspectives. Finally, the five latters (S18, S22, S23, S24, S25) offer multiple perspectives in enterprise modeling, excluding BP models. Hence, it will be promising to overcome the lack of approaches dealing with the consistency among multi-perspective BP models.

RQ2/RQ3: The majority of approaches presented (75 %) focused in the activity-driven diagrams, where a BP is modeled as a sequence of activities by using a single modeling technique. This does not allow capturing all facets of a BP in a comprehensive manner. The need to resorting to different types of diagrams emerges, especially for modeling knowledge intensive BPs. Thereby, mastering the consistency between BP models produced using a variety of modeling techniques becomes essential to guarantee a complete and coherent picture of a BP.

RQ4/RQ5: Only 30.6 % of the studied approaches handle the horizontal consistency. The percentage of approaches seeking for both syntactic and semantic consistency among BP models is limited to 27, 8 % (10 of 36). These results reveal the need for enhancing the other approaches by similar capabilities, when the causes of the inconsistencies call for such capabilities.

RQ6: The most recurrent activity in consistency management applied to BP models variants (Fig. 2) is the evaluation of consistency between models (48 %), also called consistency checking. Hence, it will be beneficial if the inconsistencies between models can be prevented (i.e. managing inconsistencies in early steps of modeling) rather than corrected (i.e. managing inconsistencies at late steps of exploitation).

RQ7/RQ8: 97.2 % of approaches focus on BP models within the same enterprise. 80 % of approaches dealing with the consistency issues in the context of multi-perspective modeling are concerned with a projective type, and hence with a particular enterprise modeling method. It may seem obvious that approaches aiming to master the consistency between inter-enterprise BP models, which often implies heterogonous modeling techniques, are still lacking.

5   Towards Categorizing Approaches Related to Inter-model Consistency: A Reference Framework

In the light of the results of the SLR and their analysis summarized in the above Sect. 4, the inconsistency source is considered as the basic factor on which we can rely in order to categorize the approaches dealing with the consistency among BP models. For each class of approaches, related to a particular source of inconsistency amongst the three sources (the variants of BP models, the multi-perspective modeling and the merging of BP models), we consider in turn other factors which may characterize approaches placed in the same class.

Figure 4 shows the proposed framework in the form of a tree. We aim that each research work dealing with the issue of consistency among BP models takes place in

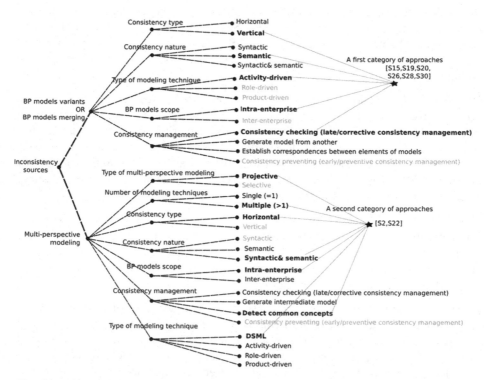

Fig. 4. A Framework towards categorizing approaches focusing on inter-model consistency

this framework (which is a first attempt and is candidate to evolution). Values in bold are the most common ones in the literature regarding the corresponding characteristics. The less common values among the studied approaches are shown in gray; together with the values in simple black, they suggest us research challenges. In Fig. 4, we also illustrated the use of this framework, by *requesting* the approaches offering the indicated values for the *search parameters (shown with the gray lines)*.

6 Conclusion

A problem of inter-model consistency can occur when a partial or strong dependency exists between BP models. A partial dependency arises when few concepts are shared between models; this is the case for multi-perspective modeling. A total dependency occurs when it is possible to establish correspondences between all elements involved in the models. Therefore, challenges related to the decomposition (vertical coherence) or the similarity issues between BP models might be also considered as consistency issues.

In this paper, we presented the results gained from undertaking a SLR on consistency among BP models with the aim of proposing a framework that facilitates (i) categorizing the plethora of existing approaches and (ii) providing directions for promising new research activities with regard to this topic. We considered a total of 982 papers and extracted from the most relevant scientific sources, of which 41 papers were ultimately analyzed in depth by referring to the Kitchenham's guidelines. The results mainly showed that a special attention must be given to the consistency between multi-perspective BP models, where a lack of approaches has been noticed. The results also revealed that the majority of the existing approaches tackle the consistency checking and thus handle the detected inconsistencies between models (i.e. late/corrective consistency management) rather than preventing them (i.e. early/preventive consistency management). The latter seems to be a promising line of research.

References

1. Curtis, B., Kellner, M., Over, J.: Process modelling. Commun. ACM **35**(9), 75–90 (1992)
2. Rolland, C.: A primer for method engineering. In: Proceedings of INFormatique des ORganisations et Systemes d'Information et de Décision, Toulouse, France (1997)
3. Rolland, C.: A comprehensive view of process engineering. In: Pernici, B., Thanos, C. (eds.) CAiSE 1998. LNCS, vol. 1413, pp. 1–24. Springer, Heidelberg (1998)
4. Feiler, P.H., Humphrey, W.S.: Software process development and enactment: concepts and definitions. In: Continuous Software Process Improvement, 2nd International Conference on the Software Process, pp. 28–40. IEEE (1993)
5. Sandkuhl, K., Stirna, J., Persson, A., Wiβotzki, M.: Enterprise Modeling. Tackling Business Challenges with the 4EM Method, pp. 1–309. Springer, Heidelberg (2014)
6. Dumas, M., Van der Aalst, W.M., Ter Hofstede, A.H.: Process-Aware Information Systems: Bridging People and Software Through Process Technology. Wiley, New York (2005)

7. Hammer, M., Champy, J.: Le Reengineering, vol. 93. Dunod, Paris (1993)
8. Peterson, R.S., Smith, D.B., Martorana, P.V., Owens, P.D.: The impact of chief executive officer personality on top management team dynamics: one mechanism by which leadership affects organizational performance. J. Appl. Psychol. **88**(5), 795 (2003)
9. Muskens, J., Bril, R.J., Chaudron, M.R.: Generalizing consistency checking between software views. In: 5th Working IEEE/IFIP Conference on Software Architecture, pp. 169–180. IEEE (2005)
10. Hallerbach, A., Bauer, T., Reichert, M.: Configuration and management of process variants. In: Handbook on Business Process Management 1, pp. 237–255. Springer, Heidelberg (2010)
11. Lucas, F.J., Molina, F., Toval, A.: A systematic review of UML model consistency management. Inf. Softw. Technol. **51**(12), 1631–1645 (2009)
12. Spanoudakis, G., Zisman, A.: Inconsistency management in software engineering: Survey and open research issues. Handb. Softw. Eng. Knowl. Eng. **1**, 329–380 (2001)
13. Nurcan, S.: Ingénierie et Architecture d'Entreprise et des Systèmes d'Information-Concepts, Fondements et Méthodes. Habilitation dissertation, Université Paris 1 Panthéon-Sorbonne (2012)
14. Chapurlat, V.: Vérification et validation de modèles de systèmes complexes: application à la Modélisation d'Entreprise. Habilitation dissertation, Université Montpellier II (2007)
15. Russell, N., van der Aalst, W.M., Ter Hofstede, A.H., Wohed, P.: On the suitability of UML 2.0 activity diagrams for business process modelling. In: Proceedings of 3rd Asia-Pacific Conference on Conceptual Modelling, vol. 53, pp. 95–104 (2006)
16. Mendling, J., Neumann, G., Nüttgens, M.: A comparison of XML interchange formats for business process modelling. In: Workflow Handbook. Future Strategies, Lighthouse Point, pp. 185–198 (2005)
17. Akobeng, A.K.: Understanding systematic reviews and meta-analysis. Arch. Dis. Child. **90** (8), 845–848 (2005)
18. IEEE standard 42010. IEEE (2011)
19. Persson, M., Törngren, M., Qamar, A., Westman, J., Biehl, M., Tripakis, S., Denil, J.: A characterization of integrated multi-view modeling in the context of embedded and cyber-physical systems. In: 11th ACM Int. Conference on Embedded Software, p. 10. IEEE Press (2013)
20. Brooks, C., Cheng, C.P., Feng, T.H., Lee, E.A., Von Hanxleden, R.: Model Engineering Using Multimodeling (No. UCB/EECS-2008-39). California University, Berkeley (2008)
21. Cicchetti, A., Ciccozzi, F., Leveque, T.: A hybrid approach for multi-view modeling. In: Electronic Communications of the European Association for the Study of Science and Technology, vol. 50 (2012)
22. Broy, M., Feilkas, M., Herrmannsdoerfer, M., Merenda, S., Ratiu, D.: Seamless model-based development: from isolated tools to integrated model engineering environments. Proc. IEEE **98**(4), 526–545 (2010)
23. Engels, G., Küster, J.M., Heckel, R., Groenewegen, L.: A methodology for specifying and analyzing consistency of object-oriented behavioral models. ACM SIGSOFT Softw. Eng. Notes **26**(5), 186–195 (2001). ACM
24. Kitchenham, B., Brereton, O.P., Budgen, D., Turner, M., Bailey, J., Linkman, S.: Systematic literature reviews in software engineering–a systematic literature review. Inf. Softw. Technol. **51**(1), 7–15 (2009)
25. Kitchenham, B.: Procedures for Performing Systematic Reviews, vol. 33, pp. 1–26. Keele University, Keele, UK (2004)
26. Keele, S.: Guidelines for performing systematic literature reviews in software engineering. Technical report, version 2.3, EBSE Technical report, EBSE (2007)

27. Governatori, G., Sadiq, S.: The journey to business process compliance (2008)
28. Bork, D., Karagiannis, D.: Model-driven development of multi-view modeling tools the MuVieMOT approach. In: ICSOFT-PT 2014, pp. IS-11. IEEE (2014)
29. Yan, Z., Dijkman, R., Grefen, P.: Business process model repositories–Framework and survey. Inf. Softw. Technol. **54**(4), 380–395 (2012)
30. Hallerbach, A., Bauer, T., Reichert, M.: Capturing variability in business process models: the Provop approach. J. Softw. Maint. Evol.: Res. Pract. **22**(6–7), 519–546 (2010)
31. Smirnov, S., Weidlich, M., Mendling, J.: Business process model abstraction based on behavioral profiles. In: Maglio, P.P., Weske, M., Yang, J., Fantinato, M. (eds.) ICSOC 2010. LNCS, vol. 6470, pp. 1–16. Springer, Heidelberg (2010)
32. Koliadis, G., Vranesevic, A., Bhuiyan, M., Krishna, A., Ghose, A.K.: A combined approach for supporting the business process model lifecycle (2006)
33. Gerth, C., Luckey, M., Kuster, J.M., Engels, G.: Detection of semantically equivalent fragments for business process model change management. In: IEEE International Conference on Services Computing, pp. 57–64. IEEE (2010)
34. Zemni, M.A., Mammar, A., Ben Hadj Alouane, N.: A behavior-aware systematic approach for merging business process fragments. In: 19th International Conference on Engineering of Complex Computer Systems, pp. 194–197. IEEE (2014)
35. Pascalau, E., Awad, A., Sakr, S., Weske, M.: On maintaining consistency of process model variants. In: Muehlen, M., Su, J. (eds.) BPM 2010 Workshops. LNBIP, vol. 66, pp. 289–300. Springer, Heidelberg (2011)
36. Weidlich, M., Mendling, J., Weske, M.: Efficient consistency measurement based on behavioral profiles of process models. IEEE Trans. Softw. Eng. **37**(3), 410–429 (2011)
37. Koschmider, A., Blanchard, E.: User assistance for business process model decomposition. In: 1st IEEE International Conference on Research Challenges in Information Science, pp. 445–454 (2007)
38. Koliadis, G., Ghose, A.: Verifying semantic business process models in inter-operation. In: 2007 IEEE International Conference on Services Computing, pp. 731–738 (2007)
39. Milani, F., Dumas, M., Ahmed, N., Matulevičius, R.: Modelling families of business process variants: a decomposition driven method (2013). arXiv preprint arXiv:1311.1322
40. Dijkman, R., Gfeller, B., Küster, J., Völzer, H.: Identifying refactoring opportunities in process model repositories. Inf. Softw. Technol. **53**(9), 937–948 (2011)
41. Lu, R., Sadiq, S., Governatori, G., Yang, X.: Defining adaptation constraints for business process variants. In: Abramowicz, W. (ed.) Business Information Systems. LNBIP, vol. 21, pp. 145–156. Springer, Heidelberg (2009)
42. Rastrepkina, M.: Managing variability in process models by structural decomposition. In: Mendling, J., Weidlich, M., Weske, M. (eds.) BPMN 2010. LNBIP, vol. 67, pp. 106–113. Springer, Heidelberg (2010)
43. Pascalau, E., Rath, C.: Managing business process variants at eBay. In: Mendling, J., Weidlich, M., Weske, M. (eds.) BPMN 2010. LNBIP, vol. 67, pp. 91–105. Springer, Heidelberg (2010)
44. Bork, D., Buchmann, R., Karagiannis, D.: Preserving multi-view consistency in diagrammatic knowledge representation. In: Zhang, S., Wirsing, M., Zhang, Z. (eds.) KSEM 2015. LNCS, vol. 9403, pp. 177–182. Springer, Heidelberg (2015)
45. Cheng-Leong, A., Li Pheng, K., Keng Leng, G.R.: IDEF*: a comprehensive modelling methodology for the development of manufacturing enterprise systems. Int. J. Prod. Res. **37**(17), 3839–3858 (1999)
46. Chen-Burger, Y.H.: Knowledge sharing and inconsistency checking on multiple enterprise models. In: International Joint Conference on Artificial Intelligence, Knowledge Management and Organizational Memories Workshop, Seattle, Washington, USA (2001)

47. Koehler, J., Tirenni, G., Kumaran, S.: From business process model to consistent implementation: a case for formal verification methods. In: Proceedings of Sixth International Enterprise Distributed Object Computing Conference, pp. 96–106. IEEE (2002)
48. Worzberger, R., Kurpick, T., Heer, T.: On correctness, compliance and consistency of process models. In: IEEE 17th Workshop on Enabling Technologies: Infrastructure for Collaborative Enterprises, pp. 251–252. IEEE (2008)
49. Dijkman, R., Dumas, M., Garcia-Banuelos, L., Käärik, R.: Aligning business process models. In: IEEE International EDOC Conference, pp. 45–53. IEEE (2009)
50. Lu, R., Sadiq, S.K.: On the discovery of preferred work practice through business process variants. In: Parent, C., Schewe, K.-D., Storey, V.C., Thalheim, B. (eds.) ER 2007. LNCS, vol. 4801, pp. 165–180. Springer, Heidelberg (2007)
51. Gulden, J., Frank, U.: MEMOCenterNG–a full-featured modeling environment for organization modeling and model-driven software development. In: 22nd International Conference on Advanced Information Systems Engineering, Hammamet, Tunisia (2010)
52. Delen, D., Benjamin, P.C.: Towards a truly integrated enterprise modeling and analysis environment. Comput. Ind. **51**(3), 257–268 (2003)
53. Leist, S., Zellner, G.: Evaluation of current architecture frameworks. In: Proceedings of 2006 ACM symposium on Applied computing, pp. 1546–1553. ACM (2006)
54. Shunk, D.L., Kim, J.I., Nam, H.Y.: The application of an integrated enterprise modeling methodology—FIDO—to supply chain integration modeling. Comput. Ind. Eng. **45**(1), 167–193 (2003)
55. Fang, X., Liu, L., Liu, X.: Analyzing the consistency of business process based on behavioral Petri Net. Int. J. u- e-Serv. Sci. Technol. **8**(2), 25–34 (2015)
56. Weidlich, M., Dijkman, R., Mendling, J.: The ICoP framework: identification of correspondences between process models. In: 22nd International Conference on Advanced Information Systems Engineering, Hammamet, Tunisia (2010)
57. Koubarakis, M., Plexousakis, D.: A formal framework for business process modelling and design. Inf. Syst. **27**(5), 299–319 (2002)
58. Vanderfeesten, I., Reijers, H.A., Mendling, J., van der Aalst, W.M., Cardoso, J.: On a quest for good process models: the cross-connectivity metric. In: Bellahsène, Z., Léonard, M. (eds.) CAiSE 2008. LNCS, vol. 5074, pp. 480–494. Springer, Heidelberg (2008)
59. Martens, A.: Consistency between executable and abstract processes. In: Proceedings of 2005 IEEE International Conference on e-Technology, e-Commerce and e-Service, pp. 60–67. IEEE (2005)
60. Decker, G., Weske, M.: Behavioral consistency for B2B process integration. In: Krogstie, J., Opdahl, A.L., Sindre, G. (eds.) CAiSE 2007 and WES 2007. LNCS, vol. 4495, pp. 81–95. Springer, Heidelberg (2007)
61. Fang, X., Wang, M., Yin, Z.: Behavior consistency analysis based on the behavior profile about transition multi-set of Petri Net. Przegląd Elektrotechniczny **89**(1b), 171–173 (2013)
62. De Medeiros, A.A., van der Aalst, W.M., Weijters, A.J.M.M.: Quantifying process equivalence based on observed behavior. Data Knowl. Eng. **64**(1), 55–74 (2008)
63. Weidlich, M., Mendling, J.: Perceived consistency between process models. Inf. Syst. **37**(2), 80–98 (2012)
64. Niemann, M., Siebenhaar, M., Schulte, S., Steinmetz, R.: Comparison and retrieval of process models using related cluster pairs. Comput. Ind. **63**(2), 168–180 (2012)
65. Küster, J.M., Gerth, C., Förster, A., Engels, G.: Detecting and resolving process model differences in the absence of a change log. In: Dumas, M., Reichert, M., Shan, M.-C. (eds.) BPM 2008. LNCS, vol. 5240, pp. 244–260. Springer, Heidelberg (2008)

66. Dijkman, R., Dumas, M., García-Bañuelos, L.: Graph matching algorithms for business process model similarity search. In: Dayal, U., Eder, J., Koehler, J., Reijers, H.A. (eds.) BPM 2009. LNCS, vol. 5701, pp. 48–63. Springer, Heidelberg (2009)

67. Van der Aalst, W.M., De Medeiros, A.A., Weijters, A.J.M.M.: Graph matching algorithms for business process model similarity search. In: Dustdar, S., Fiadeiro, J.L., Sheth, J., AP, H. A. (eds.) BPM 2006. LNCS, vol. 4102, pp. 129–144. Springer, Heidelberg (2006)

68. Li, C., Reichert, M., Wombacher, A.: Discovering reference process models by mining process variants. In: IEEE International Conference on Web Services, pp. 45–53. IEEE (2008)

69. Zachman, J.A.: A framework for information systems architecture. IBM Syst. J. **26**(3), 276–292 (1987)

70. Nurcan, S., Rolland, C.: Using EKD-CMM electronic guide book for managing change in organisations. In: Proceedings of 9th European-Japanese Conference on Information Modeling and Knowledge Bases, Iwate, Japan, pp. 105–123 (1999)

71. Daoudi, F., Nurcan, S.: A benchmarking framework for methods to design flexible business processes. Spec. Issue Softw. Process: Improv. Pract. (J. Bus. Process Manag. Dev. Support) **12**(1), 51–63 (2007)

Automatic Signature Generation for Anomaly Detection in Business Process Instance Data

Kristof Böhmer[✉] and Stefanie Rinderle-Ma

Faculty of Computer Science, University of Vienna, Vienna, Austria
{kristof.boehmer,stefanie.rinderle-ma}@univie.ac.at

Abstract. Implementing and automating business processes often means to connect and integrate a diverse set of potentially flawed services and applications. This makes them an attractive target for attackers. Here anomaly detection is one of the last defense lines against unknown vulnerabilities. Whereas anomaly detection for process behavior has been researched, anomalies in process instance data have been neglected so far, even though the data is exchanged with external services and hence might be a major sources for attacks. Deriving the required anomaly detection signatures can be a complex, work intensive, and error-prone task, specifically at the presence of a multitude of process versions and instances. Hence, this paper proposes a novel automatic signature generation approach for textual business process instance data while respecting its contextual attributes. Its efficiency is shown by an comprehensive evaluation that applies the approach on thousands of realistic data entries and 240,000 anomalous data entries.

Keywords: Anomaly detection · Process instance · Regex · Textual data

1 Introduction

Business processes have risen to important and deeply integrated solutions which spawn over various organizations and interconnect a multitude of different services and applications [4]. Hence, ensuring business process security is a crucial challenge [8]. To address this challenge, process models can be interpreted as networks. They connect, for example, legacy applications [13] that were originally not intended to be globally linked or services that are not controlled by the process owner and, therefore, should not be trusted. Although the vulnerability of IT supported business process models is generally accepted [8], we found that the business process security monitoring area is still underdeveloped compared to "classic" IT network security [9].

This surprises because the importance and widespread automated execution of processes makes them an attractive target for attackers [10]. Two different scenarios can occur: In the *first*, a *targeted attack* is executed whereby the attacker has in-depth knowledge of the attacked process model. Such attacks are difficult

© Springer International Publishing Switzerland 2016
R. Schmidt et al. (Eds.): BPMDS/EMMSAD 2016, LNBIP 248, pp. 196–211, 2016.
DOI: 10.1007/978-3-319-39429-9_13

Table 1. Generate signatures automatically from recorded data.

No	Data Type	Analyzed Process Instance Data	Matches Signature
1	Normal	'a=mapred:8_set_PCMA/8000'	— (Signature not yet generated)
2	Normal	'a=mapred:3_startFAPCM_GSM/8000'	— (Signature not yet generated)
3	Normal	'a=mapred:0_test_PCMU/8000'	— (Signature not yet generated)
4	Normal	'a=mapred:3_hello_GSM/8000'	— (Signature not yet generated)
5	Normal	'a=mapred:5_stop_GSM/8000'	— (Signature not yet generated)
6	*Generated Signature*	'^(a=mapred:)(\d\s\w\w\w)(.){0,7}(_PCMA\/8000|_GSM\/8000|_\/8000)$'	*Generated by the presented approach.*
7	Normal	'a=mapred:4_reject_PCMA/8000'	Yes
8	Attacker	'a=attack:0_8D_14_03_PCMU/8000'	*No* → Anomaly detected

to prevent. However, to prepare a targeted attack, the attacker must *probe* the process to identify vulnerabilities—which is the *second* scenario.

We assume that probing and attacks deviate from *normal use*. For example, under normal use data values that trigger potential buffer overflow vulnerabilities, remain, likely, unobserved. Hence, such data values can be detected as anomalies, which are, events with relatively small probabilities of occurrence [1]. If an anomaly, indicating probing or attacks, can be detected in advance, then the affected process instances could, for example, be halted or migrated to a honey pot and the attack, thereby, be prevented.

A common approach to detect anomalies is to define signatures which represent the expected typical behavior [16]. Hence, if the signatures do not match the observed behavior then an anomaly was detected. Compare with Table 1—which is also employed as a *running example*. Multiple normal (i.e., expected structure/type/content) variable values[1] (*No. 1–5*, i.e., Analyzed Process Instance Data) are used to create a signature *(No. 6)*. This signature identifies a probing message from the attacker *(No. 8)* as an anomaly and, therefore, as a potential attack.

However, creating such signatures manually is time-consuming and error-prone because an enormous amount of frequently changing complex business processes [4] is currently in use in large scale systems [12]. Hence, an *automatic signature generation approach* is proposed in the following.

Additionally, we found that existing process anomaly detection work, cf. [1–3,11,16], is not capable of analyzing arbitrary *textual process instance data* values (i.e., string variables holding, e.g., XML, JSON, or EDIFACT data formats along with dates, booleans, or exchanged messages). This limitation is critical when considering that today's business processes frequently utilize textual variables to flexibly store and process various kinds of information. Moreover, existing work, considers *contextual attributes*, such as time, only partly and the generated signatures can *hardly be read or manually adapted*.

Hence, existing work isn't suitable to answer the following research questions:

RQ1 How can anomalies in textual process instance data be detected?

---

[1] Note, those can be extracted from recorded process execution logs which are frequently automatically generated by process execution engines.

RQ2 How can contextual attributes be used to improve anomaly detection?

RQ3 How can signatures be automatically generated and described in a human readable and adaptable way?

Therefore, we propose a novel automatic signature generation approach which enables to exploit contextual attributes to improve anomaly detection. The signatures are defined as human readable regular expressions (regex). The applicability of the proposed approach is shown based on a proof-of-concept implementation which analyzes 240,000 anomalous data entries.

This paper is organized as follows. The process instance data signature syntax and the integration of contextual attributes are discussed in Sect. 2. The proposed signature generation approach is defined in Sect. 3. Evaluation, corresponding results and their discussion are presented in Sect. 4. Section 5 discusses related work. Conclusions and future work is given in Sect. 6.

2   Signatures on Textual Process Instance Data

To protect process instances from unknown attacks we propose an automatic signature generation approach which enables anomaly detection during process execution. The signatures are generated from recorded process execution logs that are created automatically by process execution engines during runtime [14]. Such logs hold, for example, all variables—including their values—which are used by a process model during its execution [5,14]. Hence, a signature can be created that matches the recorded variable data (e.g., variable values exchanged between process activities or received from external partners, such as, other processes/ web services) and therefore allows to distinguish between the recorded—expected and typical—behavior and anomalous behavior that would be observable, e.g., during attack preparations. Detecting such anomalies allows to apply various counter measures, such as stopping the execution of affected process instances.

We assume that especially processes and process execution engines are a *worthwhile application area* for anomaly detection, because, today's processes integrate and share the data of a wide range of services and applications. Hence, integrating signature generation and anomaly detection directly into process execution engines enables to secure a huge amount of potential attack areas at once and provides a direct access to all the required data.

We propose that the signatures should be created automatically to meet the complexity and flexibility of today's business processes [4]. Process repositories frequently contain hundreds of individual process models, which are executed in a versatile service landscape [12]. Hence, a large amount of signatures must be created and constantly updated. Additionally, we assume that the documentation of each service/application that is integrated in the processes is frequently outdated or missing because changes are often implemented in a rapid pace leaving less time to update a constantly increasing amount of documentation. This leads to high signature generation costs or incorrect signatures. To address this challenge we propose to extract signatures (i.e., expected behavior) from real process executions described in recorded process execution logs. Doing so ensures that the signatures match the real behavior and not, e.g., some incorrect behavior based on an outdated manual. Additionally, the log data contains all the information that is required to create context specific signatures, which simplifies the generated signatures and increases their anomaly detection.

Contextual Process Instance Data. Behavior that is anomalous only in a specific context, but not otherwise, is termed *contextual anomaly* [3]. Taking *contextual attributes* into account enables the creation of more focused signatures and hereby increases the anomaly detection rate. So far, contextual attributes were neglected for anomaly detection in the business process domain.

Imagine that signatures should be created for data that is received by two process activities. The first activity is always receiving characters while the second one is always receiving digits. Without contextual attributes only a single signature would be created to check if only characters *or* digits are received. However, when the second activity surprisingly starts to receive characters, then this could not be detected because the signature only checks for characters *or* digits. But when taking contextual attributes into account—here the context is defined as the activity which receives the data, e.g., from an external service—two independent signatures are created for each activity and the behavioral change is detected as an anomaly. Hence, taking the context into account increases the performance (i.e., anomaly detection rate) of the generated signatures and also simplifies the signatures itself. Moreover, two specialized separate signatures— one to match digits and one to match characters—are shorter and easier to read and to maintain than a single signature that has to fulfill multiple tasks at once.

Two kinds of contextual attributes are exploited in the following:

General attributes such as time (by creating unique signatures for specific time periods, such as individual months), because, we assume that process instance data values can be time dependent.

Process instance specific attributes for example, which process or activity has created the analyzed instance data values because we assume that the data can greatly differ between different activities/processes and even between multiple activity data fields/variables.

Further kinds of contextual attributes, for example, which user has defined the observed variable values, are left for future work.

The signature generation starts with a pre-processing step that groups the analyzed data based on the discussed contexts to create an unique signature for each contextual attribute combination. In the following, an anomaly detection system can select the appropriate signature based on the observed context.

Signature Definition. The signatures are defined as regular expressions, for the following reasons: Regular expressions are well-suited to analyze textual data, are supported by many programming languages, human readable and adaptable, and used by existing intrusion detection systems (such as Snort or l7-filter). Further on, high speed matching algorithms are available [15]. We assume that these

Table 2. Simple regular expression signature syntax.

Character	Description
'^'	Matches the start of the compared data.
'$'	Matches the end of the compared data.
'.'	Matches any possible character (including digits and control characters).
'{n}'	Matches the preceding expression, exactly n times.
'{a,b}'	Matches the preceding expression, between a and b times.
'|'	Matches either the expression before or after the vertical bar
'()'	*Capturing group* which concatenates expressions or groups of expressions. Note, a capturing group can be defined in combination with '{n}', '{a,b}', or '|'. For example, '(ab|cd)' matches 'ab' or 'bc' while '(.){5}' matches any possible character exactly five times

advantages ease the integration of the presented anomaly detection approach into existing process execution engines and anomaly detection systems.

The signatures (i.e., regular expressions), defined in the following, consist of a choice operator, groups, and multiple metacharacters which are listed in Table 2. Note, that the '^' and '$' character get added to the start ('^')/end ('$') of each signature to enforce that the whole observed content matches the signature.

The regular expression syntax presented in Table 2 enables the definition of *simple* signatures which match the *structural components* of the observed data (e.g., for XML data this would be the XML tags). Additionally, we propose a novel approach to increase the anomaly detection performance by taking the content of the observed data into account (e.g., for XML this would be the data which is placed between the XML tags, i.e., a XML node value). Accordingly, the presented syntax (cf. Table 2) is extended with character classes (cf. Table 3) to define *complex* signatures.

Character classes allow to differentiate if an observed character is, e.g., a number or a letter. Hence, it becomes possible, for example, to define a signature which checks if the observed bank account number always starts with two letters and ends with at least four numbers. This enables the generated signatures to ensure that structural and content-related properties comply with the expected behavior. An example for a signature that was defined using the described syntax can be found in the running example in Table 1.

3   Generating Signatures

This section presents a novel automatic process instance data signature generation approach. From a given set of training data (e.g., recorded process model instance executions) it generates signatures that allow to detect if the currently observed behavior (i.e., values of textual variables that are exchanged/used during ongoing process instance executions) are anomalous (i.e., do not match the expected common behavior specified at the signature) or not.

Table 3. Extensions for the simple regular expression signature syntax.

Character	Description
'\w'	A character class that matches any word character.
'\s'	A character class that matches any formating character such as tabs or spaces.
'\d'	A character class that matches any digit.
'[^\s\d\w]'	A negated combination of multiple characters which matches any character that is not matched by '\w','\s', or '\d', e.g., a minus sign ('-').

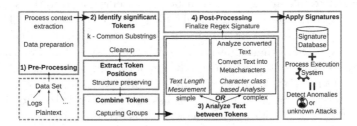

Fig. 1. Abstract signature generation approach overview.

Each signature is generated by four main components (cf. Fig. 1). *(1)* the *pre-processing module* extracts the relevant data from process execution logs. Subsequently, the textual variables—strings—are grouped based on various contextual attributes and each group is individually forwarded to the tokenization module. *(2)* the *tokenization module* identifies tokens (i.e., substrings) which commonly occur in the recorded variables. Next, the position and order of the tokens in the analyzed data is extracted and used to group and combine the tokens. *(3)* is a module that analyzes the text that is placed *between tokens*. The *simple signatures* only utilize the length of the text that is placed between the tokens. Alternatively, *complex signatures* are constructed by converting the text between the tokens into character classes and injecting the result into module *(2)* to check if the converted text reveals previously hidden structures. *(4)* the *post-processing module* takes all the extracted and prepared information and constructs a valid regular expression which can be stored in a signature database and used to detect anomalous behavior during process model instance executions.

Pre-Processing. The purpose of the pre-processing step is twofold: First, recorded process execution logs are prepared so that their content can be processed by the following steps. For this, the logs are analyzed and all variables with textual information and their metadata (i.e., important contextual attributes, e.g., which activity a variable belongs to or when it was created) are extracted.

Secondly, the prepared data is *grouped* based on the *identified contextual attributes* (e.g., time, activity, or process model, hence, the same variable value can be contained in multiple groups). For example, all textual data that belongs

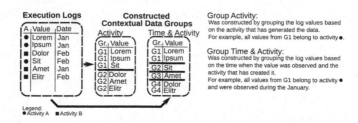

Fig. 2. Exemplary contextual group construction.

to one specific variable in one activity/process model is added to the same group and processed at once. Hereby, individual signatures for each contextual attribute (e.g., time or activity) and their combinations are created. Hence, contextual differences are respected during signature generation to ensure that a variable's content behaves as expected in specific situations, for example, in a specific month, process, or activity. For example, we found—while evaluating this paper—that processes contain activities which always assign the same values to their variables depending on the month when the activity is executed (e.g., each January), while the same activity variables contain a diverse set of values when comparing different months (e.g., January vs. June). Hence, creating an independent signature for each contextual attribute (e.g., time of the year) increases the signature anomaly detection performance, because more fine granular signatures are generated which focus on specific situations and contexts. Additionally, the signatures become simpler (e.g., because only a subset of the available data must be covered by a single signature) and therefore easier to read and maintain.

An example for creating contextual groups is depicted in Fig. 2. For the sake of brevity the execution log data only contains a single process with two activities. Two contextual groups are constructed—based on activity only and the combination of activity and time.

The following steps, starting with identifying significant tokens, are then applied on the textual variable data stored in each generated contextual group.

Identifying Significant Tokens. *Significant tokens* are substrings which commonly occur in the recorded and analyzed process instance textual (string) variable values. Those tokens are used to construct signatures that detect anomalies (i.e., detect that a significant token that is present at all the analyzed data is surprisingly missing). The problem of finding the significant tokens is defined as:

INPUT: A list of strings to analyze S, a minimum token length of $tl_{min} \in \mathbb{N}$ characters, and a minimum occurrence of $to_{min} \in (0,1]$, e.g., the token must occur in at least 10 percent of the analyzed strings in S to be significant.

OUTPUT: A list of distinct substrings D and therefore significant tokens which fulfill the minimum length (tl_{min}) and occurrence requirement (to_{min}).

The *k-common substring algorithm* [6] is applied to this problem to identify the longest substrings which occur in at least k (i.e., to_{min}) strings. For this, a *generalized suffix tree* [6] is generated from S. We have extended the suffix tree data structure so that each tree node contains the list of strings in S that are represented by this node (cf. to_{min}) and the respective length of each sub path (cf. tl_{min}). This enables to ensure the compliance with to_{min} and tl_{min} when extracting significant tokens. The extraction itself starts from *terminator nodes* (i.e., tree nodes that hold the last character/substring of a string in S). For each terminator node that fulfills tl_{min} the algorithm traverses towards the root of the tree until it finds a node that fulfills to_{min}. Then it starts recording the data that is stored in each node until it reaches the root node and therefore has identified a *potentially significant token* which is then stored in D. This is repeated for each terminator node to extract all potentially significant tokens.

Finally, all the potentially significant tokens are cleaned up. First, duplicates are removed so that each potentially significant token only appears once in D. Secondly, each remaining potentially significant token in D is analyzed to identify if it can be completely replaced by one of the longer[2] tokens in D. Why? Because a longer token provides a more strict representation of the analyzed data because it enforces more characters. Imagine that S, inter alia, contains the words '*performance*' and '*performed*' and that, among others, '*perform*' and '*for*' are commonly occurring tokens hold by D. S and D are now evaluated by checking, for each token and analyzed string, if a shorter token could be completely replaced by a longer token. Hence, for the analyzed words '*performance*' and '*performed*' it is checked if for all positions were the token '*for*' occurs also the token '*perform*' occurs. This is the case and so '*for*' will be removed from D to replace it with the longer token '*perform*'.

Applying the described token identification approach to the running example (cf. Table 1) results in following significant tokens (when defining tl_{min} as 5 and to_{min} as 0.2): '*a=mapred:*','*_PCMA/8000*','*_GSM/8000*', and '*/8000*'

Extract Token Positions. The position, order, and occurrence of each significant token is determined for each string in S. Hereby, tokens that are placed on related positions are identified. Subsequently, these tokens are used to form regex groups. The problem of finding token positions/order is defined as:

INPUT: A list of strings S and a list of cleaned up significant tokens D.

OUTPUT: A list P were each $p \in P$ is a list of significant tokens which occur in the respective $s \in S$ ordered based on their position in s.

For each $s \in S$ the left most positioned token[3] $d \in D$ is identified. If such a token was found then it is stored in p, and s is trimmed to remove all characters left from the position where d ends. Subsequently, the search for the left most significant token restarts on the trimmed version of s. This repeats until no more significant tokens can be found in the trimmed s. Note, for each $s \in S$ an respective $p \in P$ is created and utilized/filled.

The following P is generated for the running example's tokens and strings (cf. Table 1) (the list entries are separated using semicolons for P and commas for p). This allows to deduce, for example, that '*a=mapred:*' is present in all $s \in S$ (i.e., all analyzed strings) and that it is always the left most significant

---

[2] Measured based on the number of characters.
[3] If two tokens start on the same position then the longer one is chosen because it enforces more characters during signature checking than a shorter one.

token: '*a=mapred:,_PCMA/8000*'; '*a=mapred:,_GSM/8000*'; '*a=mapred:,/8000*'; '*a=mapred:,_GSM/8000*'; '*a=mapred:,_GSM/8000*'

The ordered tokens and their positions are used during the next step to start with the creation of regular expressions (i.e., signatures).

Grouping Tokens Based on their Order. We propose that the generated signatures should represent the *structural components* (represented by significant tokens) of the analyzed data (e.g., for XML data this would, likely, be the XML tags). However, we assume that most likely not each analyzed string will contain the exact same significant tokens. Hence, *regex groups* are created to enable the signature to choose from multiple token *alternatives*, for example to specify that token A or B should occur. Additionally, we expect that the analyzed textual data is of *variable length* so that the structural components are most likely not overlapping (i.e., use the same absolute positions) for each string in S. Hence, it is not possible to decide which tokens should be grouped solely based on the absolute position of the tokens. Accordingly, we propose to group the identified significant tokens based on the order of their occurrences rather than on their absolute positions. The problem of grouping the tokens is therefore defined as:

INPUT: P, as defined in the previous step.

OUTPUT: A list G were each $g \in G$—for each $p \in P$ an associated $g \in G$ is generated—holds a list of significant tokens that are combined into regex groups.

To combine the tokens the algorithm identifies the shortest entry $p \in P$ (i.e., it is containing the least amount of significant tokens) and extracts its length as $y \in \mathbb{N}$. y is then used as the amount of tokens which should be grouped. To group the tokens a list of indexes ranging from—if y is even—0 to $(\lfloor y/2 \rfloor - 1)$ is created. Subsequently, from each token list $p \in P$ the tokens with the respective indexes are taken and stored in a new list $g \in G$ (first to last, an independent list g is generated for each p). A similar approach is applied on the second half of the indexes (i.e., $(|p| - \lfloor y/2 \rfloor) \cdots (|p| - 1)$). However, this time the algorithm iterates from the last token in each $p \in P$ towards the first token (last to first) and adds the tokens (in reversed order) to the already existing $g \in G$ that belongs to the respective p. If y is uneven then an additional iteration is executed to cover the token index which would else be ignored (i.e., $0 \cdots \lfloor y/2 \rfloor$ is used at first to last).

The approach described above ensures that the generated signatures cover a wide area of the analyzed data. Imagine, that the approach would only incorporate a single direction (e.g., first to last) then an attacker could attach the vulnerable information to the end of the data—especially if the amount of tokens in each p fluctuates. Secondly, we found a positive impact of this two direction approach during the preliminary evaluation, especially, when analyzing XML data because the two direction approach more frequently preserved matching XML start/end tags and therefore more likely recognized missing XML nodes.

Finally from each $g \in G$ the tokens with equal indexes (e.g., all first tokens, all second tokens, and so on) are combined into distinct *regex groups* using the *or* operator ('|'). For the running example (cf. Table 1) the following regex groups are generated: '*(a=mapred:)*'; '*(_PCMA/8000|_GSM/8000|/8000)*'

The significant tokens likely do not represent all the analyzed data, for example, data which is not occurring frequently enough to become a significant token (e.g., varying content that is placed between XML tags). Hence, a novel approach to integrate the remaining data into the generated signatures is presented.

Analyze Textual Data Between Tokens. Until now the textual data which is placed between the identified significant tokens was not yet addressed. This data mainly consists of application data, such as addresses or names, which frequently do not contain stable structural components. However, this data is processed by the process activities and should, therefore, also be checked for anomalies to prevent attackers from injecting vulnerable—anomalous—data. Hence we propose two novel approaches called *simple* and *complex*.

INPUT: A list S and a list G, as defined in the previous step.

OUTPUT: Regex artifacts that represent the textual data between the tokens. Hence, the simple approach utilizes the length of the respective strings between the tokens to represent them. For the complex approach the representations are generated from a mixture of length information and character classes.

Both, the complex and the simple approach, analyze the textual data that is positioned between the identified significant tokens (e.g., this is, for XML data, likely, the data between XML tags). So, this data must first be extracted. Therefore, for each string $s \in S$ the respective list of significant grouped tokens $g \in G$ that occur in s is exploited. Hence, g is used to identify the position of each significant grouped token in s. Further on, the text between each identified token position and its predecessor token is extracted and stored for future analysis. A similar approach is used to extract the text between the first/last token and the start/end of s. Hence, all text that is placed, for example, between the second and the third token (for each $g \in G$) is, in the following, processed at once.

For the running example (cf. Table 1) the following strings are identified as text that is placed between the two generated groups of significant tokens: '8_set'; '3_startFAPCM'; '0_test_PCMU'; '3_hello'; '5_stop'. Subsequently these strings are processed by a complex or a simple approach.

Complex: The *complex approach* converts the textual data into a format that makes it more likely to identify structural information. Imagine, that some bank account numbers should be analyzed (e.g., '*AB12345*', '*GH56521*', and '*UJ56122*'). Initially the token based analysis is not able to detect significant tokens and therefore structure, because, each bank account number is unique and substrings which occur at multiple account numbers can, therefore, not be identified. However, a close analysis reveals that each account number starts with two letters, continued by five digits. To enable the presented complex approach to recognize this pattern the data is converted in an *abstract representation*.

Therefore, each letter is converted into a '*w*', each formating character (e.g., a space) into a '*s*', each digit into a '*d*', and any other character is converted into an '*r*'. Hence, each account number is then represented as '*wwddddd*'. Subsequently, the presented signature generation approach is applied on the prepared data (three times '*wwddddd*', one for each abstracted account number), starting from the "Identifying Significant Tokens" step. Hereby the regex group '*(wwddddd)*' is generated to represent the fact that each analyzed string contains two letters and five digits. Finally, the characters ('w','s','d', and 'r') are replaced with regular expression character classes ('w' → '\w', 'd' → '\d', 's' → '\s', 'r' → '[^\s\d\w]'), cf. Table 3, which enforce, during signature checking, the specified order and occurrence of digits, letters, formating characters, and so on. Hence, '(wwddddd)' becomes '(\w\w\d\d\d\d\d)'. Note, that the complex approach falls back to the simple approach for parts of data where no structure (even when applying the discussed abstraction approach) could be identified.

Simple: The *simple approach* deals with the textual data in a more abstract way than the complex one. Hence, it analyzes the respective data and identifies the shortest and the longest string. Subsequently, the length of these two strings is used to add minimum/maximum length limits to the signatures. Hence, when applying it on the running example (cf. Table 1) the following signature artifacts are generated: The shortest identified textual information is '8_set' and the longest is '3_startFAPCM'. So the content is described as '(.){4,13}' which indicates that any possible text is valid, but, it must be between 4 to 13 characters long. A shorter definition is used if each string is of equal length, for example, '(.){4}' if each analyzed string is exactly 4 characters long.

For the running example (cf. Table 1) the complex approach generates the following regex artifact: First, the substrings which are placed between the two identified token groups are abstracted: '8_set' → 'dswww', '3_startFAPCM' → 'dswwwwwwwwww', '0_test_PCMU' → 'dswwwwswwww', '3_hello' → 'dswwwww', '5_ stop' → 'dswwww'. Then, the complex approach identifies '*dswww*' as a structural component (i.e., significant token). Why not use '*dswwwwwww*'? Because '*dswww*' is the only substring that fulfills the minimum length requirement and occurs frequently enough in S (when using $tl_{min} = 5$ and $to_{min} = 0.2$). However, '*dswww*' is not able to represent all the strings (e.g., '*dswwwwwww*' contains more characters than '*dswww*'). Hence, also the simple approach is applied as a fall back. Altogether, the following result is generated to represent the data which is placed between the two identified significant token groups: '(\d\s\w\w\w)(.){0,7}'

Post-Processing. All the components (e.g., token groups) are now combined to create a signature that is a valid regular expression. Subsequently, the signature can be stored in a signature database and used by process execution engines to detect anomalies and, therefore, potential attacks or attack preparations.

During post-processing three objectives are fulfilled. First, all generated components (e.g., the token groups) are combined to generate a *raw signature*. It is called raw signature because it is not yet ready to be stored in a signature database. Secondly, the characters which have a special meaning in regular expressions are, if necessary, escaped. For example, the plus sign ('+') typically indicates that some character should be matched at least once. However, if a plus sign should be treated as a normal character (e.g., because it is a part of a significant token) it must be escaped by placing a backslash ('\') in front of it. Thirdly, a circumflex ('^') is placed at the start of the signature and a dollar sign ('$') is placed at the end. Why? Because this enforces that the signature must match the whole observed data from the start to the end and not only a part of it. Hence, it increases the anomaly detection performance of the generated signatures because an attacker can no longer send some valid data and then attach the vulnerable data to the end of it, which would otherwise be possible.

For the running example (cf. Table 1) the finalized signature is defined as:
'^(a=mapred:)(\d\s\w\w\w)(.){0,7}(_PCMA\/8000|_GSM\/8000|\/8000)$'

4   Evaluation

The evaluation combines *realistic artificial data* and *real life process instance execution data* to assess the impact of contextual attributes on signature generation and the anomaly detection performance of the presented approach.

Test Problems. The test data which was used for the evaluation consists of (a) *artificially generated test data*[4] (using three different formats, namely, XML, JSON, and EDIFACT) and (b) *real life process execution logs* from the Business Processing Intelligence Challenge 2015[5] (provided by five Dutch municipalities).

For (a) the artificially generated data consists of three different data formats (XML, JSON, and EDIFACT—wide spread data exchange formats, used in various disciples, such as banking or manufacturing) that we found are frequently used in business processes. For each of the three formats a thousand test data entries were generated and randomly separated into *signature generation data* and *test data*. Three hundred (100 for each data format) randomly selected entries from the test data were also used to construct anomalous data to evaluate the anomaly detection performance of the presented approach. Moreover, the generated XML, JSON, and EDIFACT data entries contain realistic data as payload (e.g., realistic e-mail addresses, phone numbers, or company names) along with the required structural components (e.g., XML tags). Each generated XML and JSON data entry holds 4 payload values (e.g., names), while each generated EDIFACT data entry represents a purchase order message with 14 payload values.

For (b) the realistic log data consists of 262,628 independent events from 27 process models and 356 activities—recorded over a period of six years.

Real life and artificial data were combined because the identified real life data only contains simple textual variable values (i.e., textual variables typed as strings [5] that hold dates, booleans, or numbers) which can easily be addressed by the presented approach. Hence, we opted to include complex artificially generated data to assess the performance of the presented anomaly detection approach in situations where the data is complex and, therefore, more challenging. Note, despite the prototypical implementation, the signature generation could be conducted quickly (5 min to generate signatures for all test data items, fractions of a section to decide if a value is anomalous or not – on a 2.6 Ghz Intel Q6700).

Metrics and Evaluation. *Quantitative* and *qualitative metrics* were combined.

---

[4] http://cs.univie.ac.at/wst/research/projects/project/infproj/1057/.
[5] http://www.win.tue.nl/bpi/2015/challenge.

Quantitative: Realistic artificially generated data (i.e., signature generation data) was used to generate signatures, one for each data format. Subsequently, each signature had to match the respective test data to ensure that the signature was not *over-fitted* [7]. An over-fitted signature can lead to many false positives which would reduce the applicability of the presented approach. Note, each generated signature successfully evaluated the test data as non-anomalous. So, no over-fitting occurred. Finally, each signature was applied to anomalies that were generated from the test data to assess its anomaly detection performance.

Qualitative: Real life process execution logs were analyzed to check if contextual attributes, such as time, have an effect on the variables and data fields, used by the process, that would allow to improve anomaly detection. For example, it was evaluated if the variable values of an activity show similarities for specific times of the year (e.g., each April, for multiple years). If this is the case, then respecting contextual attributes (e.g., time) and therefore creating an independent signature for each month is beneficial because less data must be represented by each single signature which improves the anomaly detection performance.

Results. The results were generated by applying the signatures on randomly selected test data entries which were altered to represent 8 *anomaly classes*.

The following anomaly classes were evaluated: *(a)* The length of the data entry was extended by 4–10 random characters, *(b)* The data entry was completely replaced by random characters, *(c)* Content (e.g., for XML data this is the value of a XML node) was replaced by random characters, *(d)* Content was duplicated and attached to the original value, *(e)* Between 4–10 characters of the content were randomly selected and flipped (e.g., a letter was replaced with a random digit), *(f)* An element (e.g., a complete XML node) of the data entry was completely removed, *(g)* An element (e.g., a complete XML node) of the data entry was duplicated, *(h)* A structural element (e.g., a XML tag) was replaced with random data. Note, that the anomaly classes *(b)*, *(c)*, and *(h)* replaced data with a randomly generated equivalent that has the exact same length as the replaced data.

We assume that the generated anomalous data entries realistically represent data that can be observed during attacks. For example, the anomaly class *(d)* can be used to check for potential buffer overflows or the anomaly *(c)* represents the attempt to inject machine code into a process model instance. Overall 240,000 anomalous data entries were generated and evaluated. The evaluation was executed a hundred times to even out the random behavior of the anomaly class adaptation approach. During each execution a hundred test data entries were individually adapted by 8 anomaly classes, for three different data formats.

Primary tests were executed to identify appropriate configuration values for the discussed signature generation approach, resulting in $tl_{min} = 4$ and $to_{min} = 0.75$. The average results of the evaluation are shown in Fig. 3.

The results show that the presented approach is capable of detecting a wide range of anomalies. Already the simple approach generates reasonable results for most anomaly classes. However, the simple approach is not able to detect anomalies that only affect the content (e.g., XML node values) of the analyzed

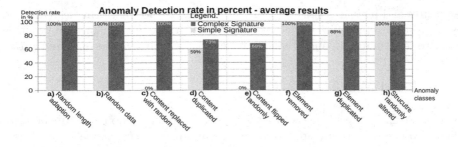

Fig. 3. Anomaly detection performance of simple and complex signatures.

data without changing its length (e.g., only specific characters are replaced, cf. anomaly *(c)* and *(e)*). This is not surprising because the simple approach only enforces length restrictions on the content. Here, the complex approach comes into play. By analyzing the content and its internal structure it can, for example detect flipped characters (e.g., anomaly *(e)*). Hence, we conclude that the presented novel complex signature generation approach is capable of providing remarkable strict signatures while the simple signatures are easier to read. They are shorter, and are already able to detect important length based vulnerabilities (e.g., buffer overflows). Why are anomalies *(d)* and *(e)* not always detected? This can occur, for *(d)*, if the duplicated value is still shorter then other representations of this value in the signature generation data or, for *(e)*, if the flipped character value is also present at the same place at data entries in the signature generation data.

Table 4. Influence assessment of respecting contextual attributes.

No	Context. Attribute	Beneficial	No	Context. Attribute	Beneficial
1	Process activity	Yes	3	Time	Yes
2	Process model	Yes	4	Combination	Yes

The importance of contextual attributes for process signature generation was evaluated using process execution logs provided by the Business Processing Intelligence Challenge 2015. It was checked, for three different contextual attributes (activity, process, time, along with their combinations), if the generated signatures *benefit* from respecting these contextual attributes during signature generation (e.g., by generating an independent signature for each activity and month). We found clear indications that the recorded data is influenced by the described attributes, cf. Table 4. For example, some activities always used the same variable values during specific times of the year or when integrated into specific process models. Moreover, we found that activities, despite equal variable names, store vastly different data formats. We conclude that respecting contextual attributes during the signature generation allows to generate simpler signatures and increases the signature anomaly detection performance (because less diverse data must be covered by a single signature, so the signature becomes easier to read/maintain and it can be more strictly represent the analyzed data).

5   Related Work

Related work, in the business process anomaly detection domain, can be classified into two categories: process instance data and process model control flow anomaly detection. The existing *data anomaly* detection approaches concentrate on integer variables and apply statistical regression analysis to identify outliers and, therefore, anomalies [11]. *Control flow anomaly* detection approaches mine process logs to extract control flows which are then, for example, compared with a reference process model. Alternative approaches check how frequently each control flow is found, infrequent flows are then marked as anomalies, cf. [1,2].

We conclude that textual business process instance data is currently not addressed by existing process anomaly detection approaches. Moreover, we found that contextual attributes are currently not exploited in the business process domain. In general, in the *security domain*, anomaly detection in textual data is currently mainly applied to detect novel topics in a collection of documents [3] or on highly standardized network protocols [16], such as SIP, neglecting the security critical aspects of arbitrary textual data. This circumstances reduce the protection gained from today's, process, anomaly detection solutions.

6   Conclusion

This paper provides process instance anomaly detection and signature generation approaches ($\mapsto$ **RQ1** to **RQ3**) which will be integrated in our "ProTest" project which focuses on creating automatic process behavior verification. Future work will exploit the generated signatures as a foundation to construct realistic test data to improve process model testing. In addition, we are confident that the described approach can also be applied to related domains (e.g., web services) that process textual data and, even, other data types (e.g., binary data).

The evaluation results show the flexibility and applicability of the presented approach for complex data formats ($\mapsto$ **RQ1**). Additionally, we found that contextual attributes affect the analyzed business process instance data and conclude that contextual attributes can be exploited to improve the signature quality (i.e., anomaly detection performance; $\mapsto$ **RQ2**). Overall, this work provides the first process instance anomaly detection approach that addresses textual data and enables to replace error prone manual signature generation ($\mapsto$ **RQ3**).

Future work will strive to enhance the performance of the generated signatures, and to identify ways which enable to measure how much the observed behavior deviates from the expected one. Hereby multiple anomalies and their effects can be aggregated to decrease the risk of improperly assessing small, probably harmless, anomalies, as large, probably harmful, anomalies (i.e., attacks).

Acknowledgment. The research was partly funded by COMET K1, FFG - Austrian Research Promotion Agency.

References

1. Bezerra, F., Wainer, J.: Algorithms for anomaly detection of traces in logs of process aware information systems. Inf. Syst. **38**, 33–44 (2013)
2. Bezerra, F., Wainer, J., van der Aalst, W.M.: Anomaly detection using process mining. In: Halpin, T., Krogstie, J., Nurcan, S., Proper, E., Schmidt, R., Soffer, P., Ukor, R. (eds.) Enterprise, Business-Process and Information Systems Modeling. LNBIP, vol. 29, pp. 149–161. Springer, Heidelberg (2009)
3. Chandola, V., Banerjee, A., Kumar, V.: Anomaly detection: a survey. ACM Comput. Surv. **41**, 15–87 (2009)
4. Fdhila, W., Rinderle-Ma, S., Indiono, C.: Change propagation analysis and prediction in process choreographies. Coop. Inf. Syst. **24**, 47–62 (2015)
5. Günther, W.C., Verbeek, E.: XES – Standard. Technical report, TU Eindhove (2014)
6. Gusfield, D.: Algorithms on Strings, Trees and Sequences: Computer Science and Computational Biology. Cambridge University Press, New York (1997)
7. Hawkins, D.M.: The problem of overfitting. Chem. Inf. Comput. Sci. **44**, 1–12 (2004)
8. Herrmann, P., Herrmann, G.: Security requirement analysis of business processes. Electron. Commer. Res. **6**, 305–335 (2006)
9. Liao, H.J., Lin, C.H.R., Lin, Y.C., Tung, K.Y.: Intrusion detection system: a comprehensive review. Netw. Comput. Appl. **36**, 16–24 (2013)
10. Müller, G., Accorsi, R.: Why are business processes not secure? In: Fischlin, M., Katzenbeisser, S. (eds.) Number Theory and Cryptography. LNCS, vol. 8260, pp. 240–254. Springer, Heidelberg (2013)
11. Quan, L., Tian, G.s.: Outlier detection of business process based on support vector data description. In: Computing, Communication, Control, and Management. pp. 571–574. IEEE (2009)
12. Rosemann, M.: Potential pitfalls of process modeling: part B. Bus. Process Manag. **12**, 377–384 (2006)
13. Sneed, H.M.: Integrating legacy software into a service oriented architecture. In: Software Maintenance and Reengineering. pp. 11–22. IEEE (2006)
14. Van Der Aalst, W.: Process Mining: Discovery, Conformance and Enhancement of Business Processes. Springer, Heidelberg (2011)
15. Yamagaki, N., Sidhu, R., Kamiya, S.: High-speed regular expression matching engine using multi-character nfa. In: Field Programmable Logic and Applications. pp. 131–136. IEEE (2008)
16. Zuech, R., Khoshgoftaar, T.M., Wald, R.: Intrusion detection and big heterogeneous data: a survey. Big Data **2**, 1–41 (2015)

Hybrid Diagnosis Applied to Multiple Instances in Business Processes

Rafael Ceballos$^{(\boxtimes)}$, Diana Borrego, María Teresa Gómez-López,
and Rafael M. Gasca

Universidad de Sevilla, Seville, Spain
{ceball,dianabn,maytegomez,gasca}@us.es

Abstract. Business Process compliance is an important issue in control-flow and data-flow perspectives. Control-flow correctness can be analysed at design time, whereas data-flow accuracy should be verified at runtime, since data is accessed and modified during execution. Compliance validation should consider the conformance of data to business rules. Business compliance rules are policies or statements that govern corporate behaviour. Since business compliance rules and data change during process execution, faults can appear due to the erroneous inclusion of rules and/or data in the process. A hybrid diagnosis therefore needs to be performed regarding the likelihood of faults in data vs. business rules. In order to achieve the correct diagnosis, it is fundamental to attain the best assumption concerning the degree of likelihood. In this paper, we present an automatic process to diagnose possible faults that simultaneously combines business rules and data of multiple process instances. This process is based on Constraint Programming paradigm to efficiently ascertain a minimal diagnosis. Furthermore, a methodology for calculation of the most appropriate degree of likelihood of faults in data vs. business rules is proposed.

Keywords: Business process analysis · Diagnosis · Business rules · Business data constraints · Constraint programming · Databases

1 Introduction

Business processes (BPs) permit the description of the activities necessary to achieve an objective in a company. This description includes, among other things: a workflow model, a set of business rules or policies, and the data interchanged during the execution. The correctness of an execution implies the correctness of these three aspects. Frequently, BPs are supported by Process Aware Information Systems (PAISs) [23]. A PAIS is a software system that manages and executes operational processes involving people, applications, and/or information sources on the basis of process models. This type of system provides a way to manage data stored in a repository layer that is read and written by a BP. The diagnosis of the workflow tends to be performed at design time to prevent errors after process deployment. However, the updating of business rules, such as

© Springer International Publishing Switzerland 2016
R. Schmidt et al. (Eds.): BPMDS/EMMSAD 2016, LNBIP 248, pp. 212–227, 2016.
DOI: 10.1007/978-3-319-39429-9_14

compliance rules, and the management of data at runtime is common practice. Since data and compliance rules may be modified at runtime, their diagnosis cannot be included in only the design phase.

The development an efficient diagnosis of possible faults is essential, since these faults appear at runtime and a great quantity of data and business compliance rules are probably involved. The special problem faced in BP diagnosis is that the data is not involved in only one instance, and it is not isolated from other instances executed in the past or in the future. The data written or read in an instance can be shared with other instances, or even with another process, such as when the data involved is stored in a repository, typically in a relational database. These relations must be used in the diagnosis process, since the isolation of a fault that explains a failure of an instance cannot contradict the diagnoses found for other instances.

In previous work [8], the importance of data correctness in BPs is studied, including relational databases as the main source of data. This previous work, however, only includes the possibility of faults in data, but fails to consider defects in business compliance rules. Derived from the modification of business compliance rules and data, certain faults can be produced due to the erroneous inclusion of the rules and/or data in the process. The importance of verifying the correctness of data in PAISs is known [12], although how to combine faults in business compliance rules and data at the same time remains a challenge. In this paper, we propose an automatic model-based diagnosis methodology to verify the compliance to the business rules by data in multiple instances, and to isolate the origin of the faults. Since data is more numerous and even more frequently updated, it is more likely a fault appearing due to incorrect data than due to an erroneous business compliance rule. In order to obtain the best assumption about this degree of likelihood, we propose a methodology for calculating the most appropriate degree of likelihood of faults between data vs. business compliance rules.

The paper is organised as follows. Section 2 presents a motivating example to illustrate the concepts. Section 3 introduces the adaptation of model-based diagnosis methodology. Section 4 presents the Constraints Programming paradigm used to perform the diagnosis. In Sect. 5, the diagnosis of the motivating example is performed. Section 6 presents an overview of related work found in the literature. And finally, conclusions are drawn and future work is proposed.

2   Using Business Data Constraints. A Motivating Example

In this paper, a real example of a financial economic application is used to illustrate the hybrid diagnosis. The activities of the company are oriented towards negotiating collaborative projects developed over a number of years. The process consists of the management of costs of projects, during their execution, as detailed in [8], and is represented in Fig. 1. All these tasks are carried out by a total of 25 employees, who modify the stored information for more than

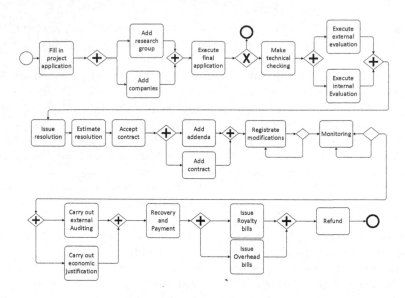

Fig. 1. Business Process example

300 projects. Each employee is responsible for certain activities of the process. The persistence layer that supports this business process is formed of a database with 86 tables. Each employee can introduce an average of 200 items of data per project, during the 4 years that a project can last.

BPs with a high level of human interaction demand a more frequent data validation and diagnosis, since humans are more likely to introduce intermittent faults. The intermittence of faults complicates the detection and diagnosis of model violations, since an inconsistency detected during the execution of an activity does not necessarily imply a failure in the activity, neither does it imply that this fault may appear again in the future. Our paper focuses on the concept of Business Data Constraints (BDCs), which was introduced in [8].

Definition 1. *Business Data Constraints are a subset of business compliance rules that represent the compliance relation between the values of data during a business process instance.*

In this paper, we assume that the BDCs specification can be incorrect, and therefore inconsistent with the introduced data, where the data is correct. The use of BDCs hugely facilitates the data consistency analysis, and the diagnosis of the origin of an inconsistency. The diagnosis methodology must consider the following characteristics:

- Data involved in the diagnosis is not strictly flowing through the process, some of them may also be stored in databases. This implies that the quantity of data involved in the diagnosis of each instance can be very large.
- The data managed in an instance is not independent from the data of other instances, since data can be shared between instances.

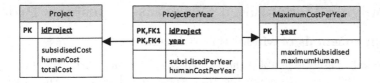

Fig. 2. Subset of relational model of the example

Project					ProjectPerYear					MaximumCostPerYear		
idProject	subsidised Cost	human Cost	total Cost		idProject	year	subsidised PerYear	humanCost PerYear		year	maximum Subsidised	maximum Human
223	55000	25000	80000		223	2015	7000	8000		2015	10000	10000
224	60000	26000	88000		223	2016	16000	9000		2016	10000	9000
...	...	...	...		224	2015	8000	9000		2017	15000	12000
					224	2016	18000	6000		...	...	...
					224	2017	9000	8000				
					...	...	...	...				

Fig. 3. Subset of tables of the example

- The BDCs tend to be updated in order to represent new conditions. Therefore, it is necessary to include these modified rules as possible faults.

The activities of a BP can modify certain data from a relational database, thereby making it necessary to evaluate certain BDCs. For the example, the BDCs must be satisfied with the various values of the project data. In order to express the BDCs, we use the grammar proposed in [8], which is based on numerical constraints over natural, integer, and float variables. A subset of BDCs for the activities is presented below:

1. Execute final application:
   (a) humanCost + subsidisedCost = totalCost   (BDC$_1$)
   (b) 3 · humanCost ≤ totalCost   (BDC$_6$)
2. Accept contract:
   (a) subsidisedCost ≥ 2 · subsidisedPerYear   (BDC$_2$)
   (b) humanCost ≥ 4 · humanCostPerYear   (BDC$_3$)
3. Recovery and payment:
   (a) subsidisedPerYear ≤ maximumSubsidised   (BDC$_4$)
   (b) humanCostPerYear ≤ maximumHuman   (BDC$_5$)

The variables in the previous BDCs are stored in a database whose relational model is shown in Fig. 2. The three tables represent the information about the project (*Project*), the details for each project in each year *ProjectPerYear*, and the maximum spending limit allowed in each year and for each cost item (*MaximumCostPerYear*). Examples of the stored data are shown in Fig. 3.

In order to show our diagnosis methodology, in this example, we introduce 3 defects: 2 erroneous items of data, and 1 incorrect BDC. The following section shows how to adapt model-based diagnosis to the hybrid problem.

3   Applying Model-Based Diagnosis to Business Processes

3.1   Fundamentals of Model-Based Diagnosis

Model-based diagnosis enables the identification of the parts that fail in a system. It is performed by comparing the expected behaviour of the system with real behaviour. The expected behaviour is modelled using the knowledge of the system to diagnose, whereas real behaviour is known by analysing the events produced. This implies that model-based diagnosis is considered by the pair $\{SD, OM\}$, where SD is the System Description and OM is the Observational Model. The SD is a set of constraints, and the OM is a set of values of the observable data. A fault is visible when a discrepancy between the expected behaviour (SD) and the observed behaviour (OM) is found.

Model-based diagnosis is based on the parsimony principle [17], in order to attain a minimal diagnosis that explains the conflicts in an efficient way. This principle states that among competing hypotheses, the one with the fewest assumptions should be selected. For example, a conflict is detected in the following SD and OM: $\{a + b = c, a + 2 \cdot b = d\}$, $\{a = 7, b = 5, c = 9, d = 14\}$. If the assumption $\{a = 7\}$ is false (i.e. it is modified, $a = 4$), then the remaining assumptions are satisfied between them. Model-based diagnosis identifies the smallest assumption that causes conflicts. In this example, there are other possible hypotheses, but these imply the modification of more than one assumption.

The following subsections analyse: (1) how to design the model to be diagnosed (SD); and (2) how to obtain the observational model (OM).

3.2   System Description: Relational Database Model and Business Data Constraints

Since the stored data participates in the BDCs, it is necessary to include the BDCs and the relational database scheme into the model to be diagnosed. Business Data Constraints describe the semantic relation between the data values that are introduced, read and modified during the BP instances. It should be borne in mind that the variables participating in the constraints can come from the database or from the data-flow.

A Relational Database is a collection of predicates over a finite set of variables described by means of a set of relations. A relation R is a data structure which consists of a heading and an unordered set of tuples which share the same type, where $A_1, A_2, \ldots, A_n$ are attributes of the domains $D_1, D_2, \ldots, D_n$. A number of the attributes of a relation can be described as *Primary Key Attributes*. The relation between two tables is described by a referential integrity. Two tables can be related by means of their *Primary* and *Foreign Key Attributes*, described in the literature as the relational model.

3.3   Observational Model: Tuples of the Database

In a PAIS, the information is typically stored in a relational database, and therefore the tuples of the tables compose the OM. Since the variables involved in

BDCs:		(1,2)		(1,3,6)		(1,6)			(2,4)		(3,5)		(4)		(5)	
tuple	idProj.	Subs. Cost Var.	Subs. Cost	Hum. Cost Var.	Hum. Cost	Tot. Cost Var.	Tot. Cost	Year	Subs. PerYear Var.	Subs. Per Year	Hum. PerYear Var.	Hum. Per Year	Max. Subs. Var.	Max. Subs.	Max. Hum. Var.	Max. Hum.
1	223	Subs. Cost1	55000	Hum. Cost1	25000	Tot. Co1	80000	2015	Subs. PerYear1	7000	Hum. PerYear1	8000	Max. Subs1	10000	Max. Hu1	10000
2	223	Subs. Cost1	55000	Hum. Cost1	25000	Tot. Co1	80000	2016	Subs. PerYear2	16000	Hum. PerYear2	9000	Max. Subs2	10000	Max. Hu2	9000
3	224	Subs. Cost2	60000	Hum. Cost2	26000	Tot. Co2	88000	2015	Subs. PerYear3	8000	Hum. PerYear3	9000	Max. Subs1	10000	Max. Hu1	10000
4	224	Subs. Cost2	60000	Hum. Cost2	26000	Tot. Co2	88000	2016	Subs. PerYear4	18000	Hum. PerYear4	6000	Max. Subs2	10000	Max. Hu2	9000
5	224	Subs. Cost2	60000	Hum. Cost2	26000	Tot. Co2	88000	2017	Subs. PerYear5	9000	Hum. PerYear5	8000	Max. Subs3	15000	Max. Hu3	12000
...	...	...	...	...	...	...	...	...	...	...	...	...	...	...	...	...

Fig. 4. Denormalized tuples

a BDC have different origins, it is possible that attributes from various tables are related in a single BDC. The location of the data in various tables is due to the necessity to follow the *Normal Forms* defined in relational database theory. The normalization rules are designed to prevent update anomalies and data inconsistencies. Since data is stored in various tables, to ascertain the full tuple of values for an OM, a *denormalization process* needs to be carried out. This denormalization process is only used for the purpose of diagnosing; the relational database undergoes no changes at all, only a new join relation is obtained with all related attributes together used for the diagnosis in a temporal way. Although normalization methods are applied at schema level, it is possible to apply the opposite methods to ascertain the denormalized relations between the data. The related attributes are those that appear in the same BDC together with the primary-foreign key attributes necessary to join the related attributes that belong to various tables. For the example in Fig. 2, the obtained join-table is shown in Fig. 4. In the figure, the related BDCs are shown at the top of each column. Although the details of how the join-table is created are described in [8], the general idea can be understood by analysing the relational model in Fig. 2, and by observing the specific values for the example introduced in Fig. 3. Since the attribute *idProject* of the table *ProjectPerYear* is a foreign key of table *Project*, then each tuple of *ProjectPerYear* is related to the tuple of *Project* whose foreign and primary keys are equal. In a similar way, these are relation between the variables *MaximumPerYear* and *ProjectPerYear*.

In the denormalization process, a column is included for each attribute to distinguish between the provenance of the values. Distinction is made due to the fact that it is necessary to know whether two values correspond to the same attribute after the denormalization, since two equal values in a column do not imply they represent the same variable. In the denormalization process, the same value can appear in various tuples, derived from the $1..n$ relation between the tables, such as the value associated to the human cost of project 223 (Variable *humanCost1*) that appears in the two first tuples of Fig. 4, since this project was developed over two years.

4 Constraint Programming for Hybrid Business Process Diagnosis Models

4.1 Fundamentals of Constraint Programming

In order to find the minimal incorrect part of the SD and OM in an automatic and efficient way, we propose using Constraint Programming, since the definition of BDCs is very close to the definition of the logic and arithmetic constraint modelled in a Constraint Satisfaction Problem (CSP). A CSP [18] represents a reasoning framework consisting of variables, domains and constraints $\langle V, D, C \rangle$, where V is a set of n variables $v_1, v_2, ..., v_n$ whose values are taken from finite domains $D_{v1}, D_{v2}, ..., D_{vn}$ respectively, and C is a set of constraints on their values. The constraint c_k $(x_{k1}, ..., x_{kn})$ is a predicate that is defined on the Cartesian product $D_{k1} \times ... \times D_{kj}$. This predicate is true iff the value assignment of these variables satisfies the constraint c_k.

If the model represented by $\{SD, OM\}$ is satisfiable, then the OM conforms to the BDCs that describe the SD. However, if no solution is found, the minimal non-conformance parts of the model should be determined. In order to ascertain the minimal explanation regarding faults, it is necessary to find a minimal subset $ss \subset \{OM \cup BDCs\}$ that satisfies $\{SD \cup OM\} - ss$. Since BDCs are applied by various tuples with different values, all instances of BDC_i must be included in the search. This search is performed using a Constraint Optimization Problem (COP), which is solved as a Min-CSP. A Min-CSP is a COP, where the goal is to minimize an optimization function. The application of a Min-COP to a model-based diagnosis problem implies defining an optimization function in order to minimize the subset ss.

4.2 Compliance Verification by Means of the Observational Model

In order to verify the compliance of the BDCs, we have applied the obtained tuples to instantiate the BDCs. For the tuples shown in Fig. 4, the BDCs of the example have been instantiated, and the results are shown in Table 1. Analysing each BDC according to the tuples:

Table 1. Results of the compliance verification of the BDCs

BDC	Tuple				
	1	2	3	4	5
1	BDC_1^1 ✓			BDC_1^2 ✗	
2	BDC_2^1 ✓	BDC_2^2 ✓	BDC_2^3 ✓	BDC_2^4 ✓	BDC_2^5 ✓
3	BDC_3^1 ✗	BDC_3^2 ✗	BDC_3^3 ✗	BDC_3^4 ✓	BDC_3^5 ✗
4	BDC_4^1 ✓	BDC_4^2 ✗	BDC_4^3 ✓	BDC_4^4 ✗	BDC_4^5 ✓
5	BDC_5^1 ✓	BDC_5^2 ✓	BDC_5^3 ✓	BDC_5^4 ✓	BDC_5^5 ✓
6	BDC_6^1 ✓			BDC_6^2 ✓	

- BDC_1: two different instances of this BDC can be obtained (BDC_1^1 and BDC_1^2) for the five tuples ($\{1, 2, 3\}$ and $\{4, 5\}$). BDC_1^1 is satisfiable (tuples 1, 2 and 3), but BDC_1^2 is non-compliant (tuples 4 and 5). An error in one single input: *humanCost2*, *subsidisedCost2* or *totalCost2*, can explain this abnormal behaviour. It is less likely that there is a fault in a BDC (usually written by a business expert), than in an input (usually typed in by an user).
- BDC_2, BDC_5 and BDC_6 are consistent for all tuples.
- BDC_3: Only BDC_3^4 is satisfiable. There are several explanations for this behaviour: (a) related to the data values, and (b) related to the BDC. An error committed on writing BDC_3 could provide a single explanation. As mentioned earlier, it is less likely that a BDC is erroneous than data. The question is, in what percentage should this likelihood be described? The analysis of this percentage of likelihood is part of our methodology and is detailed in the following subsection.
- BDC_4: BDC_4^2 and BDC_4^4 are not satisfied. The failure can be explained with a single variable,$\{maximumSubsidised2\}$, whose value is 10000 and is shared by these two instances.

Although the possible diagnosis has been explained separately for each BDC_i, it is necessary to ascertain the minimal diagnosis that explains all discrepancies for the whole problem. Our methodology is able to obtain the minimal diagnosis that explains all this non-compliant behaviour.

Variables of the problem:
integer subsidisedCost1, subsidisedCost2, humanCost1, humanCost2, ...
integer[0,1] rSc1, rSc2, rHc1, rHc2, rTc1, rTc2, rHp1, rHp2, rHp3, rHp4, rHp5, ..., $rBDC_1$, $rBDC_1^1$, $rBDC_1^2$, $rBDC_2$, $rBDC_2^1$, ..., $rBDC_5^5$, $rBDC_6$, $rBDC_6^1$, $rBDC_6^2$
Constraints to represent the instantiation of Variables:
rSc1 = ¬(subsidisedCost1 = 55000)   rSc2 = ¬(subsidisedCost2 = 60000) rHc1 = ¬(humanCost1 = 25000)   ...
Constraints to represent the instantiation of BDCs:

$rBDC_1^1$ = ¬(humanCost1 + subsidisedCost1 = totalCost1)
$rBDC_1^2$ = ¬(humanCost2 + subsidisedCost2 = totalCost2)
$rBDC_2^1$ = ¬(subsidisedCost1 ≥ 2 · subsidisedPerYear1)
 	
               ...
$rBDC_1$ = ($rBDC_1^1$ + $rBDC_1^2$ ≥ minLik$_1$)
$rBDC_2$ = ($rBDC_2^1$ + $rBDC_2^2$ + $rBDC_2^3$ + $rBDC_2^4$ + $rBDC_2^5$ ≥ minLik$_2$)
 	
               ...
(($rBDC_1^1$ + $rBDC_1^2$ = 0) ∨ ($rBDC_1^1$ + $rBDC_1^2$ ≥ minLik$_1$)
($rBDC_2^1$ + $rBDC_2^2$ + $rBDC_2^3$ + $rBDC_2^4$ + $rBDC_2^5$ = 0) ∨
               ($rBDC_2^1$ + $rBDC_2^2$ + $rBDC_2^3$ + $rBDC_2^4$ + $rBDC_2^5$ ≥ minLik$_2$)
 	
               ...

Objective function:
minimize(rSc1 + rSc2 + rHc1 + rHc2 + rTc1 + rTc2 + ... + $rBDC_1$·minLik$_1$ + $rBDC_2$·minLik$_2$ + ... + $rBDC_6$·minLik$_6$)

Fig. 5. Min-CSP example

4.3 Min-CSP Applied to Model-Based Diagnosis

The execution of the diagnosis entails the translation of the problem into a CSP, including BDCs and tuples for each execution instance. Figure 5 shows the Min-CSP created to diagnose the example. Below, the modelling of the parts of the CSP are detailed (variables, domains, constraints and objective function).

In order to declare the **Variables of the problem**, a new variable is added to the Min-CSP for each variable obtained in BDC_j^i as explained in Subsect. 3.3 following the syntax: $type\ var_k^1,...,var_k^m$.

Furthermore, in order to provide the Min-CSP with the ability to distinguish between different sources of faults (i.e. data and/or BDCs), we use reified constraints. A reified constraint relies on a variable that denotes its truth value. It is therefore necessary to add new variables to the CSP, whose domain is reduced to values 0 (false value) and 1 (true value). These variables are associated to each BDC_i, BDC_i^j (**Constraints to represent the instantiation of BDCs**) and assignments of value to an input (**Constraints to represent the instantiation of Variables**), in order to denote whether they are satisfiable. Both $rBDC^i$ and $rBDC_i^j$ are included to differentiate the BDC from its application for each tuple. The proposed syntax is:

//Reified variables to ascertain the satisfiability of the BDCs
integer[0,1] $rVar_k^1,...,rVar_k^m$
integer[0,1] $rBDC_i, rBDC_i^1,...,rBDC_i^n$
//Constraints to represent the reified variables assignment
$rVar_k^j = \neg(var_k^j = value_k^j)$
//Constraints to represent the reified BDCs
$rBDC_i^1 = \neg(BusinessRule_i$ instantiated by tuple 1)
...
$rBDC_i^n = \neg(BusinessRule_i$ instantiated by tuple n)

The reified variables are equalized to the negated constraints since the objective function is to minimize the number of elements with abnormal behaviour (non-compliant). In order to ascertain when a defect in a BDC is less likely than in data errors, the following constraint is added for each BDC:

$$rBDC_i = rBDC_i^1 + ... + rBDC_i^n = \sum_j^n rBDC_i^j \geq minLik_i$$

These constraints incorporate the likelihood concept into the CSP, by using the parameter $minLik_i$.

Definition 2. *The parameter $minLik_i$ is the minimum number of faults (non-compliant instances of a BDC_i) that is set as the threshold to indicate that there is a defect in a BDC_i.*

For example, if there are 5 tuples where BDC_i is involved, $minLik_i$ can take a value between 1 and 5. If at least the $minLik_i$ threshold number of instances are not satisfiable, then BDC_i is considered as a part of the minimal diagnosis. How the values of each $minLik_i$ is determined is detailed in the following subsection.

The **Objective function** is defined as:

$$\text{minimize}(\text{rVar}_k^1 + ... + \text{rVar}_k^m + ... + \text{rBDC}_1 \cdot \text{minLik}_1 + ... + \text{rBDC}_q \cdot \text{minLik}_q)$$

Each $rBDC_i$ has a weighting that is proportional to each parameter $minLik_i$. This objetive implies finding the minimal hybrid diagnosis.

Finally, it is important to add this constraint for each BDC_i:

$$(\text{rBDC}_i^1 + ... + \text{rBDC}_i^n = 0) \vee (\text{rBDC}_i^1 + ... + \text{rBDC}_i^n \geq \text{minLik}_i)$$

This constraint permits two options: (1) $\sum_j^n \text{rBDC}_i^j$ is equal to 0, and therefore the BDC_i is correct; or (2) $\sum_j^n \text{rBDC}_i^j$ is equal to or greater than $minLik_i$, and therefore the BDC_i has a defect. Intermediate values between 0 and $minLik_i$ are not allowed, and therefore, if there are inconsistencies in a BDC_i^j, it can only be avoided by relaxing the variables rVar_q related to BDC_i^j. In other words, the defects are only in input data and the BDC_i is correct.

4.4   Calculation of the MinLik Parameter

The appropriate value of the minLik parameter for each BDC_i depends on several factors. It is necessary to take into account the number of tuples and variables affected by each BDC, therefore $minLik$ should be calculated at runtime, when the diagnosis process is performed. The calculations for the example are shown in Table 2. The meaning of each column is as follows:

Table 2. *minLik* parameter calculation

BDC	nVar	nInst	%errors	nErrors	domain	media	rep?	cover	reduced	minLik
1	3	2	20 %	$1.2 \approx 1$	[1]	1	No			2
2	2	5	20 %	2	[1, 2]	2	Yes	2	1	1
3	2	5	20 %	2	[1, 2]	2	Yes	2	1	1
4	2	5	20 %	2	[1, 2]	2	Yes	3	2	2
5	2	5	20 %	2	[1, 2]	2	Yes	3	2	2
6	2	2	20 %	$0.8 \approx 1$	[1]	1	No			2

- *BDC*: number that identifies a BDC_i.
- *nVar*: number of variables involved in a BDC_i.
- *nInst*: number of instances (BDC_i^j) of the BDC_i.
- *%errors*: average percentage of data errors estimated by the expert.
- *nErrors*: probable number of data errors that can appear in all instances of a BDC_i. This is obtained as the product of: $nErrors = nVar \cdot nInst \cdot \%errors$, and it is rounded to the nearest integer.

- *domain*: interval between the minimum and maximum number of instances that can be influenced by a number of data errors equal to *nErrors*.
- *media*: most likely case within the range obtained in the previous *domain* column. It is calculated as the weighted average of all possible cases (integers of the *domain* column) and probability, rounded to the nearest integer.
- *rep?*: column set to Yes if there is at least one variable that appears in different instances of the same BDC_i. For example, humanCost2 appears in BDC_3^3, BDC_3^4, BDC_3^5.
- *cover*: minimum number of items of erroneous data such that all BDC_i^j of a BDC_i are incorrect. This depends on the existence of variables that participate in different instances of the same BDC_i (previous column *rep?*). For example, for BDC_3, the minimal number is two: variables humanCost1 and humanCost2.
- *reduced*: the diagnosis process tries to explain the anomalous behaviour with the minimum number of errors, but the variables that are repeated in different instances can explain errors in several instances at the same time. To counteract this effect, this column is calculated as $1+(media \cdot cover)/nInst$, and is rounded to the nearest integer.
- *minLik*: calculation performed as the minimum between the values $media+1$ and *reduced*.

4.5   Diagnosing the Example

The minimal diagnosis is obtained solving the Min-CSP presented in Fig. 5. The minimal diagnosis is a set of three elements: maximumSubsidised2, BDC_3, and {humanCost2, subsidisedCost2, or totalCost2}. In order to satisfy all BDCs and the compliance verification presented in Table 1, three modifications could be made:

- The input data associated to the *maximumSubsidised2* variable must be changed. This change solves the compliance problems in BDC_4.
- The BDC_3 must be changed. This change solves the compliance problems in BDC_3.
- Finally, the input data associated to, *humanCost2* or *subsidisedCost2* or *totalCost2*, must be changed. Only one of three variables. This change solves the compliance problems in BDC_1.

The minimal diagnosis found with our methodology includes the three introduced faults.

5   Evaluation

In order to evaluate our proposal, we have designed a set of tests where the possible single and double faults in data and BDCs are simulated. With these tests, we can confirm the validity of our methodology to achieve the minimal diagnosis for various cases. To cover the most relevant cases, variables and BDCs of the example have been divided into different sets according to the number of tuples and BDCs that are affected by each of them. Regarding the 22 variables in Fig. 4, we have divided them into 7 different sets, where each set is formed of the variables that appear in the same number of tuples and involved in the same BDCs. Regarding BDCs, the 6 BDCs of the example are divided into 4 sets according to the number of variables affected by them. In greater detail, the two sets of variables and BDCs are:

- **Set of Variables:** {subsidisedCost1, 2, 3, 4, 5, maximumHuman3, maximum-Subsidised3} are in one tuple and affected by one BDC; {humanPerYear1, 2, 3, 4, 5} are in one tuple and affected by three BDCs; {maximumSubsidised1, 2, maximumHuman1, 2} are in two tuples and affected by one BDC; {subsidised-Cost1, totalCost1} in two tuples and affected by two BDCs; {HumanCost1} is in two tuples and affected by three BDCs; {subsidisedCost2, totalCost2} are in three tuples and affected by two BDCs; {humanCost2} is in three tuples and affected by three BDCs.
- **Set of BDCs:** {BDC_6} involves four variables, {BDC_1} involves six variables, {BDC_2, BDC_3} involve seven variables, {BDC_4, BDC_5} involve eight variables.

Figure 6 depicts the execution time (ms) for the proposed hybrid diagnosis. Each possible fault is simulated for an example of each set of variables and BDCs, since the study for the elements within the same set is equivalent. In the figure, different symbols are used to represent the results of the attained diagnosis: (1) green diamonds, the introduced fault is the minimal and single fault found with our methodology; (2) blue squares, the introduced fault is one of the minimal faults found by our methodology; and (3) red triangles, the introduced fault is not minimal, and hence our approach found another minimal explanation.

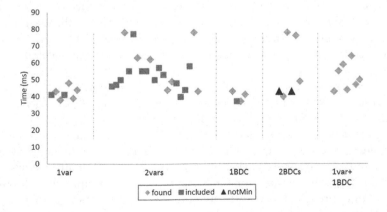

Fig. 6. Execution time of hybrid diagnosis test cases

In these forty-five tests, only in two cases did our approach find another minimal explanation. In these two cases, the correct diagnosis is not the minimal diagnosis, but is proposed as the second minimal diagnosis.

6   Related Works

Papers where data is involved in model-based diagnosis in BPs are divided into two types: model analysis at design time, and analysis of instances at runtime.

Regarding the analysis of the data model at design time, one of the main focuses is on the detection of possible faults in the data flow, such as missing, redundant, and conflicting data [21]. This research has been extended to deal with the analysis of process models that contain both control flow and data flow, and with artifact-centric orientation instead of activity-centric models [7,10,20]. Although a variety of mechanisms have been developed to prevent errors at the structural level (deadlocks, livelocks, ...) [22], they also have to comply with business level rules and policies. In [5,6], the activities are attributed with pre-conditions and post-conditions that describe the data behaviour to verify the correctness of the model at design time. Artifact-centric orientation has also been used to support consistent specifications [26].

Regarding the importance of data for the runtime conformity of BPs, both stored data and data flow are objects of this study. In relation to the persistence layer and data-flow, relational databases have been used in BPs, for example in [2], which presents a solution where data is audited and stored in a relational database. However, no validation of the semantics is performed for this persistence layer and the business rules.

The analysis of the correctness of BPs is typically related to the activity executed according to the value of a data variable in each case, by verifying whether the model and the log conform to each other [1]. Although some authors have noticed that relational databases are the typical repository where the changed data is stored instead of log events [3], the stored data itself does not represent the objective of the diagnosis. In [15,16], business constraint monitoring is presented based on Event Calculus. In [12], the importance of validating the correctness of data in PAISs is highlighted, although the challenge remains of how to find a fault in data instead of in a decision related to data. Therefore, conformance-checking analysis on log events is insufficient [19] to claim correctness in a BP.

In this paper, we present the necessity to study the correctness of rules and data themselves, and define data-aware compliance rules (BDCs). Related to how to model data-aware compliance rules, studies such as [4,11,12,14,24] define graphical notations to represent the relationship between data and compliance rules by means of data conditions. In [25], a method for monitoring control-flow deviations during process execution is proposed. In [13], "semantic constraints" and the SeaFlows framework for enabling integrated compliance support are proposed. An approach for semantically annotating activities with preconditions and effects that may refer to data objects is introduced in [9], and an efficient algorithm for compliance verification using propagation is also discussed.

Summarizing, to the best of our knowledge, this is the first contribution that addresses a hybrid approximation, where faults in rules and data are considered at the same time. Previous studies can be found in the literature about the *Possible Minimal Set of Incorrect Data* or *Possible Minimal Set of Incorrect BDCs*, but this work is centred on both types of errors at the same time. A preliminary study [8] diagnoses data stored according to the model, but not combined with possible faults in business rules.

7 Conclusions and Future Work

A diagnosis methodology that considers both business data constraints and data (either flowing or stored) as possibly being responsible for incorrect behaviour is presented. The combination of types of faults (i.e. in BDCs and/or data) necessitates a hybrid diagnosis, which is performed regarding the likelihood of faults in data vs. those in BDCs. To this end, Constraint Programming is used by modelling the problem as a CSP. Moreover, this proposal takes into account that data may be shared by various instances of the BP, and deals with it accordingly, for example by diagnosing faults in an instance that were caused by the updating of data by another running instance.

As future work, we plan to offer an easier and customized way to determine the likelihood between data and BDC malfunction. In order to improve our approach, we would like to consider roles or the organization view. Moreover, we intend to perform the diagnosis even when certain data still remains unknown, in order to allow the detection of potential errors in advance.

Furthermore, we would like to extend the idea to include those BPs that manage aggregate data. Another interesting line would be to manage a log of diagnoses, whereby the cause of a malfunction is ascertained by analysing previous diagnoses.

Acknowledgement. This work has been partially funded by the Ministry of Science and Technology of Spain (TIN2015-63502) and the European Regional Development Fund (ERDF/FEDER). Thanks to Lesley Burridge for revision of the English version of the manuscript.

References

1. van der Aalst, W., Adriansyah, A., van Dongen, B.: Replaying history on process models for conformance checking and performance analysis. Wiley Int. Rev. Data Min. Knowl. Disc. **2**(2), 182–192 (2012)
2. van der Aalst, W., van Hee, K., van der Werf, J.M., Kumar, A., Verdonk, M.: Conceptual model for online auditing. Decis. Support Syst. **50**(3), 636–647 (2011)
3. van der Aalst, W.M.P.: Extracting event data from databases to unleash process mining. In: vom Brocke, J., Schmiedel, T. (eds.) BPM - Driving Innovation in a Digital World. Management for Professionals, pp. 105–128. Springer, Switzerland (2015)

4. Awad, A., Weidlich, M., Weske, M.: Visually specifying compliance rules and explaining their violations for business processes. J. Vis. Lang. Comput. **22**(1), 30–55 (2011)
5. Borrego, D., Eshuis, R., Gómez-López, M.T., Gasca, R.M.: Diagnosing correctness of semantic workflow models. Data Knowl. Eng. **87**, 167–184 (2013)
6. Borrego, D., Gasca, R.M., Gómez-López, M.T.: Automating correctness verification of artifact-centric business process models. Inf. Softw. Technol. **62**, 187–197 (2015)
7. Eshuis, R., Kumar, A.: An integer programming based approach for verification and diagnosis of workflows. Data Knowl. Eng. **69**(8), 816–835 (2010)
8. Gómez-López, M.T., Gasca, R.M., Pérez-Álvarez, J.: Compliance validation and diagnosis of business data constraints in business processes at runtime. Inf. Syst. **48**, 26–43 (2015)
9. Governatori, G., Hoffmann, J., Sadiq, S., Weber, I.: Detecting regulatory compliance for business process models through semantic annotations. In: Ardagna, D., Mecella, M., Yang, J. (eds.) BPM 2008 Workshops. LNBIP, vol. 17, pp. 5–17. Springer, Heidelberg (2009)
10. Hull, R.: Artifact-centric business process models: brief survey of research results and challenges. In: Tari, Z., Meersman, R. (eds.) OTM 2008, Part II. LNCS, vol. 5332, pp. 1152–1163. Springer, Heidelberg (2008)
11. Liu, Y., Müler, S., Xu, K.: A static compliance-checking framework for business process models. IBM Syst. J. **46**(2), 335–362 (2007)
12. Ly, L.T., Rinderle-Ma, S., Dadam, P.: Design and verification of instantiable compliance rule graphs in process-aware information systems. In: Pernici, B. (ed.) CAiSE 2010. LNCS, vol. 6051, pp. 9–23. Springer, Heidelberg (2010)
13. Ly, L.T., Rinderle-Ma, S., Göser, K., Dadam, P.: On enabling integrated process compliance with semantic constraints in process management systems. Inf. Syst. Front., 1–25 (2009)
14. Ly, L.T., Rinderle-Ma, S., Knuplesch, D., Dadam, P.: Monitoring business process compliance using compliance rule graphs. In: Meersman, R., et al. (eds.) OTM 2011, Part I. LNCS, vol. 7044, pp. 82–99. Springer, Heidelberg (2011)
15. Maggi, F.M., Montali, M., van der Aalst, W.M.P.: An operational decision support framework for monitoring business constraints. In: de Lara, J., Zisman, A. (eds.) FASE 2012. LNCS, vol. 7212, pp. 146–162. Springer, Heidelberg (2012)
16. Montali, M., Maggi, F.M., Chesani, F., Mello, P., van der Aalst, W.M.P.: Monitoring business constraints with the event calculus. ACM TIST **5**(1), 17 (2013)
17. Peng, Y., Reggia, J.: Abductive Inference Models for Diagnostic Problem-Solving. Symbolic Computation. Springer, New York (1990)
18. Rossi, F., van Beek, P., Walsh, T.: Handbook of Constraint Programming. Elsevier, Amsterdam (2006)
19. Rozinat, A., van der Aalst, W.: Conformance checking of processes based on monitoring real behavior. Inf. Syst. **33**(1), 64–95 (2008)
20. Sidorova, N., Stahl, C., Trcka, N.: Soundness verification for conceptual workflow nets with data: Early detection of errors with the most precision possible. Inf. Syst. **36**(7), 1026–1043 (2011)
21. Sun, S.X., Zhao, J.L., Nunamaker, J.F., Sheng, O.R.L.: Formulating the data-flow perspective for business process management. Inf. Syst. Res. **17**(4), 374–391 (2006)
22. Trčka, N., van der Aalst, W.M.P., Sidorova, N.: Data-flow anti-patterns: discovering data-flow errors in workflows. In: van Eck, P., Gordijn, J., Wieringa, R. (eds.) CAiSE 2009. LNCS, vol. 5565, pp. 425–439. Springer, Heidelberg (2009)

23. Weber, B., Sadiq, S.W., Reichert, M.: Beyond rigidity - dynamic process lifecycle support. Comput. Sci. - R&D **23**(2), 47–65 (2009)
24. Weber, I., Hoffmann, J., Mendling, J.: Semantic business process validation. In: SBPM 2008: 3rd International Workshop on Semantic Business Process Management at ESWC 2008, June 2008
25. Weidlich, M., Ziekow, H., Mendling, J., Günther, O., Weske, M., Desai, N.: Event-based monitoring of process execution violations. In: Rinderle-Ma, S., Toumani, F., Wolf, K. (eds.) BPM 2011. LNCS, vol. 6896, pp. 182–198. Springer, Heidelberg (2011)
26. Yongchareon, S., Liu, C., Zhao, X.: A framework for behavior-consistent specialization of artifact-centric business processes. In: Barros, A., Gal, A., Kindler, E. (eds.) BPM 2012. LNCS, vol. 7481, pp. 285–301. Springer, Heidelberg (2012)

Process and Data Mining

Connecting Databases with Process Mining: A Meta Model and Toolset

E. González López de Murillas[1,3]([✉]), Hajo A. Reijers[1,2],
and Wil M.P. van der Aalst[1]

[1] Department of Mathematics and Computer Science,
Eindhoven University of Technology, Eindhoven, The Netherlands
{e.gonzalez,h.a.reijers,w.m.p.v.d.aalst}@tue.nl
[2] Department of Computer Science, VU University Amsterdam,
Amsterdam, The Netherlands
[3] Lexmark Enterprise Software, Gooimeer 12, 1411DE Naarden, The Netherlands

Abstract. Process Mining techniques require event logs which, in many cases, are obtained from databases. Obtaining these event logs is not a trivial task and requires substantial domain knowledge. In addition, the result is a single view on the database in the form of a specific event log. If we desire to change our view, e.g. to focus on another business process, and generate another event log, it is necessary to go back to the source of data. This paper proposes a meta model to integrate both process and data perspectives, relating one to the other and allowing to generate different views from it at any moment in a highly flexible way. This approach decouples the data extraction from the application of analysis techniques, enabling its use in different contexts.

Keywords: Process mining · Database · Data schema · Meta model · Event extraction

1   Introduction

The field of process mining offers a wide variety of techniques to analyze event data. Process discovery, conformance and compliance checking, performance analysis, process monitoring and prediction, and operational support are some of the techniques that process mining provides in order to better understand and improve business processes. However, most of these techniques rely on the existence of an event log.

Anyone who has dealt with obtaining event logs in real-life scenarios knows that this is not a trivial task. It is not common to find logs exactly in the right form. In many occasions, they simply do not exist and need to be extracted from some sort of storage, like databases. In these situations, when a database exists, several approaches are available to extract events. The most general is the classical extraction in which events are manually obtained from the tables in the database. To do so, a lot of domain knowledge is required in order to

© Springer International Publishing Switzerland 2016
R. Schmidt et al. (Eds.): BPMDS/EMMSAD 2016, LNBIP 248, pp. 231–249, 2016.
DOI: 10.1007/978-3-319-39429-9_15

select the right data. Some work has been done in this field to assist in the
extraction and log generation task [2]. Also, studies have been performed on
how to extract events in very specific environments like SAP [5,6,13] or other
ERP systems [8]. A more general solution to extract events from databases,
regardless of the application under study, is presented in [10], which describes
how to automatically obtain events from database systems that generate *redo logs*
as a way to recover from failure. The mentioned approaches aim at, eventually,
generating an *event log*, i.e. a set of traces, each of them containing a set of
events. These events represent operations or actions performed in the system
under study, and are grouped in traces following some kind of criteria. However,
there are multiple ways in which events can be selected and grouped into traces.
Depending on the perspective we want to have on the data, we need to extract
event logs differently. Also, a database contains a lot more information than
just events. The extraction of events and its representation as a plain event log
can be seen as a "lossy" process during which valuable information can get lost.
Considering the prevalence of databases as a source for event logs, it makes sense
to gather as much information as possible, combining the process view with the
actual data.

We see that process mining techniques grow more and more sophisticated.
Yet, the most time-consuming activity, event log extraction, is hardly supported.
This paper provides mature support to tackle the problem of obtaining, transforming, organizing and deriving data and process information from databases.
This makes easier to connect the registration system of enterprises with analysis
tools, generating different views on the data in a flexible way. Also, this work
presents a comprehensive integration of process and data information in a consistent and unified format. All of this is formally supported with automated
techniques. Moreover, the provided solution has the benefit of being universal,
being applicable regardless of the specific system in use. Figure 1 depicts an
environment in which the information is scattered over several systems from a
different nature, like ERPs, CRMs, BPM managers, database systems, redo logs,
etc. In such a heterogeneous environment, the goal is to extract, transform and
derive data from all sources to a common representation that is able to connect
all the pieces together such that analysis techniques like process mining can be
readily applied.

The remainder of this paper is structured as follows: Sect. 2 presents a running
example used through the paper. Section 3 explains the proposed meta model and

Fig. 1. Data gathering from several systems to a meta model

the formalization. Implementation details are presented in Sect. 4. The approach
is evaluated in Sects. 5 and 6 shows the related work. Finally, Sect. 7 presents
the conclusions and future work.

2   Running Example

In this section we propose a running example to explain and illustrate our approach. Assume we want to analyze a setting where concerts are organized and
concert tickets are sold. To do so, a database is used to store all the information related to concerts, concert halls (*hall*), seats, tickets, bands, performance
of bands in concerts (*band_playing*), customers and bookings. Figure 2 shows
the data schema of the database. In it we see many different elements of the
involved process represented. Let us consider now a complex question that can
be of interest from a business point of view: *What is the interaction with the*
process of customers between 18 and 25 years old who bought tickets for concerts
of band X? This question represents a challenge starting from the given database
for several reasons:

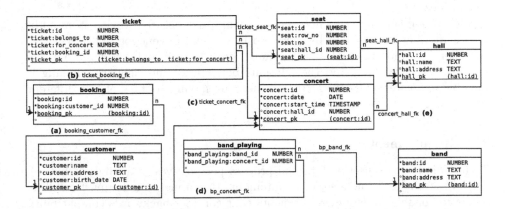

Fig. 2. Data schema of the example database

1. The database does not provide an integrated view of process and data. Therefore, questions related to the execution of the underlying process cannot be
   answered with a query.
2. The current database schema fits the purpose of storing the information in
   this specific setting, but it does not have enough flexibility to extend its functionality allocating new kinds of data such as events or objects of a different
   nature.
3. The setting lacks execution information in an accessible way (events, traces
   and logs are missing so one cannot apply process mining without a lot of extra
   work), and there is no assistance on how to extract or derive this information
   from the given data.

4. If we plan to use the data as it is, it requires to adapt to the way it is stored for every question we want to answer.

All these reasons make the analysis complex, if not impossible, in many settings. At best, such an analysis can only be carried out by either the extraction of a highly specialized event log or the creation of a complex ad hoc query.

3   Meta Model

As has been shown before, a need exists for a way to store execution information in a structured way, something that accepts data from different sources and allows to build further analysis techniques independently from the origin of this data. Efforts in this field have already been made as can be observed in [14] with the OpenXES standard. This standard defines structure to manage and manipulate logs, containing events and traces and the corresponding attributes. Therefore, XES is a good target format to represent behavior. However, a XES file is just one view on the data and, despite being an extensible format, it does not provide a predefined structure to store all the linked information we want to consider.

Because of this, it seems necessary to define a structured way to store additional information that can be linked to the classical event log. This new way to generalize and store information must provide sufficient details about the process, the data types and the relations between all the elements, making it possible to answer questions at the business level, while looking at two different perspectives: data and process.

3.1   Requirements

To be able to combine the data and process perspectives in a single structure, it is important to define a set of requirements that a meta model must fulfill. It seems reasonable to define requirements that consider backwards-compatibility with well established standards, support of additional information, its structure and the correlation between process and data views:

1. The meta model must be compatible with the current meta model of XES, i.e. any XES log can be transformed to the new meta model and back without loss of information,
2. It must be possible to store several logs in the new meta model, avoiding event duplication,
3. Logs stored in the same meta model can share events and belong to different processes,
4. It must be possible to store some notion of process in the meta model,
5. The meta model must allow to store additional information, like database objects, together with the events, traces and processes, and the correlation between all these elements,

6. The structure of additional data must be precisely modeled,
7. All information mentioned must be self contained in a single storage format, easy to share and exchange, similarly to the way that XES logs are handled.

The following section describes the proposed meta model which complies with these requirements, providing a formalization of the concepts along with explanations.

3.2   Formalization

Considering the typical environments subject to study in the process mining field, we can say that it is common to find systems backed up by some sort of database storage system. Regardless of the specific technology behind these databases, all of them have in common some kind of structure for data. We can describe our meta model as a way to integrate process and data perspectives, providing flexibility on its inspection and assistance to reconstruct the missing parts. Figure 3 shows a high level representation of the meta model. On the right hand side, the data perspective is considered, while the left models the process view. Assuming that the starting point of our approach is data, we see that the less abstract elements of the meta model, *events* and *versions*, are related, providing the connection between the process and data view. These are the basic blocks of the whole structure and, usually, the rest can be derived from them. However, in Sect. 5 we will see that, given enough information, we can also derive any of these two basic blocks from the other.

The data side considers three elements: **data model**, **objects** and **versions**. The data model provides a schema describing the objects of the database. The objects represent the unique entities of data that ever existed or will exist in our database, while the versions represent the specific values of the attributes of an object during a period of time. Versions represent the evolution of objects through time. The process side considers **events**, **instances** and **processes**. Processes describe the behavior of the system. Instances are traces of execution for a given process, being sets of events ordered through time. These events

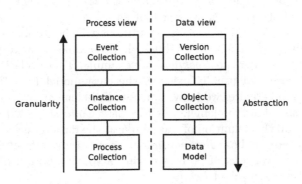

Fig. 3. Diagram of the meta model at a high level

represent the most granular kind of execution data, denoting the occurrence of an activity or action at a certain point in time.

The remainder of this section proposes a formalization of the elements in this meta model, starting from the data and continuing with the process side. As has been mentioned before, we can assume a way to classify elements in types or *Classes* exists. Looking at our running example, we can distinguish between a *ticket* class and a *customer* class. This leads to the definition of data model as a way to describe the schema of our data.

Definition 1 (Data Model). *A data model is a tuple* $DM = (CL, AT,$ *classOfAttribute, RS, sourceClass, targetClass) such that*

- *CL is a set of class names,*
- *AT is a set of attribute names,*
- *classOfAttribute* $\in AT \rightarrow CL$ *is a function that maps each attribute to a class,*
- *RS is a set of relationship names,*
- *sourceClass* $\in RS \rightarrow CL$ *is a function that maps each relationship to its source class,*
- *targetClass* $\in RS \rightarrow CL$ *is a function that maps each relationship to its target class*

Each of these elements belonging to a *Class* represents a unique entity, something that can be differentiated from the other elements of the same class, e.g. *Customer A* and *Customer B*. We will call them *Objects*, being unique entities according to our meta model.

Definition 2 (Object Collection). *Assume OBJ to be the set of all possible objects. An object collection OC is a set of objects such that* $OC \subseteq OBJ$.

Something we know as well is that, during the execution of a process, the nature of these elements can change over time. Modifications can be made on the attributes of these *objects*. Each of these represents *mutations* of an object, modifying the values of some of its attributes, e.g. modifying the address of a customer. As a result, despite being the same object, we will be looking at a different *version* of it. The notion of *Object Version* is therefore introduced to show the different stages in the life-cycle of an *Object*.

During the execution of a process, operations will be performed and, many times, links between elements are established. These links allow to relate *Tickets* to *Concerts*, or *Customers* to *Bookings*, for example. These relationships are of a structured nature and usually exist at the data model level, being defined between *Classes*. Therefore, we know upfront that elements of the class *Ticket* can be related somehow to elements of the class *Concert*. *Relationships* is the name we use to call the definition of these links at the data model level. However, the actual instances of these *Relationships* appear at the *Object Version* level, connecting specific versions of objects during a specific period of time. These specific connections are called *Relations*.

Definition 3 (Version Collection). *Assume V to be some universe of values, TS a universe of timestamps and DM = (CL, AT, classOfAttribute, RS, sourceClass, targetClass) a data model. A version collection is a tuple OVC = (OV, attValue, startTimestamp, endTimestamp, REL) such that*

- *OV is a set of object versions,*
- *attValue ∈ (AT × OV) ↛ V is a function that maps a pair of object version and attribute to a value,*
- *startTimestamp ∈ OV → TS is a function that maps each object version to a start timestamp,*
- *endTimestamp ∈ OV → TS is a function that maps each object version to an end timestamp such that $\forall ov \in OVC : endTimestamp(ov) \geq startTimestamp(ov)$,*
- *REL ⊆ RS × OV × OV is a set of triples relating pairs of object versions through a specific relationship.*

At this point, it is time to consider the *process* side of the meta model. The most basic piece of information we can find in a process event log is an event. These are defined by some attributes, among which we find a few typical ones like *timestamp, resource* or *lifecycle*.

Definition 4 (Event Collection). *Assume V to be some universe of values and TS a universe of timestamps. An event collection is a tuple EC = (EV, EVAT, eventAttributeValue, eventTimestamp, eventLifecycle, eventResource) such that*

- *EV is a set of events,*
- *EVAT is a set of event attribute names,*
- *eventAttributeValue ∈ EV × EVAT ↛ V is a function that maps a pair of an event and event attribute name to a value,*
- *eventTimestamp ∈ EV → TS is a function that maps each event to a timestamp,*
- *eventLifecycle ∈ EV → {start, complete, . . . } is a function that maps each event to a value for its life-cycle attribute,*
- *eventResource ∈ EV → V is a function that maps each event to a value for its resource attribute.*

When we consider events of the same activity but relating to a different lifecycle, we gather them under the same *activity instance*. For example, two events that belong to the activity *make booking* could have different lifecycle values, being *start* the one denoting the beginning of the operation (first event) and *complete* the one denoting the finalization of the operation (second event). Therefore, both events belong to the same *activity instance*. Each of these activity instances can belong to different *cases* or *traces*. At the same time, *cases* can belong to different *logs*, that represent a whole set of *traces* on the behavior of a process.

Definition 5 (Instance Collection). *An instance collection is a tuple IC = (AI, CS, LG, aisOfCase, casesOfLog) such that*

- AI is a set of activity instances,
- CS is a set of cases,
- LG is a set of logs,
- $aisOfCase \in CS \rightarrow \mathcal{P}(AI)$ is a function that maps each case to a set of activity instances,
- $casesOfLog \in LG \rightarrow \mathcal{P}(CS)$ is a function that maps each log to a set of cases.

The last piece of the our meta model is the *process model collection*. This part stores *process models* on an abstract level, i.e. as sets of *activities*. An *activity* can belong to different *processes* at the same time.

Definition 6 (Process Model Collection). *A process model collection is a tuple $PMC = (PM, AC, actOfProc)$ such that*

- PM is a set of processes,
- AC is a set of activities,
- $actOfProc \in PM \rightarrow \mathcal{P}(AC)$ is a function that maps each process to a set of activities.

Now we have all the pieces of our meta model, but it is still necessary to wire them together. A *connected meta model* defines the connections between these blocks. Therefore, we see that *versions* belong to *objects* (*objectOfVersion*) and *objects* belong to a class (*classOfObject*). In the same way, *events* belong to *activity instances* (*eventAI*), *activity instances* belong to *activities* (*activityOfAI*) and can belong to different *cases* (*aisOfCase*), *cases* to different *logs* (*casesOfLog*) and *logs* to process (*processOfLog*). Connecting both data and process views, we find *events* and *versions*. They are related (*eventToOVLabel*) in a way that can be interpreted as a causal relation between *events* and *versions*, i.e. when *events* happen they trigger the creation of *versions* as a result of modifications on data (the update of an attribute for instance). Another possibility is that the event represents a read access or query of the values of a *version*.

Definition 7 (Connected Meta Model). *Assume V to be some universe of values, $DM = (CL, AT, classOfAttribute, RS, sourceClass, targetClass)$ a data model, OC an object collection, $OVC = (OV, attValue, startTimestamp, endTimestamp, REL)$ a version collection, $EC = (EV, EVAT, event AttributeValue, eventTimestamp, eventLifecycle, eventResource)$ an event collection, $IC = (AI, CS, LG, aisOfCase, casesOfLog)$ an instance collection and $PMC = (PM, AC, actOfProc)$ a process model collection. A connected meta model is a tuple $CMM = (DM, OC, classOfObject, objectOfVersion, OVC, EC, eventToOVLabel, IC, eventAI, PMC, activityOfAI, processOfLog)$ such that*

- $classOfObject \in OC \rightarrow CL$ is a function that maps each object to a class,
- $objectOfVersion \in OV \rightarrow OC$ is a function that maps each object version to an object,
- $eventToOVLabel \in EV \times OV \nrightarrow V$ is a function mapping pairs of an event and an object version to a label. If a pair $(ev, ov) \in domain(eventToOVLabel)$, this means that both event and object version are linked. The label itself defines the nature of such link, e.g. "insert", "update", "read", "delete", etc.,

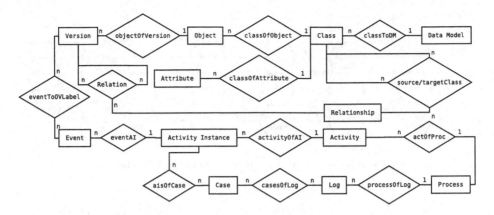

Fig. 4. ER diagram of the meta model

- $eventAI \in EV \rightarrow AI$ is a function that maps each event to an activity instance,
- $activityOfAI \in AI \rightarrow AC$ is a function that maps each activity instance to an activity,
- $processOfLog \in LG \rightarrow PM$ is a function that maps each log to a process.

An instantiation of this meta model fulfills the requirements set in Sect. 3.1 in terms of storage of data and process view. Some characteristics of this meta model that enable full compatibility with the XES standard have been omitted in this formalization for the sake of brevity. In addition to this formalization, an implementation has been made. This was required in order to provide tools that assist in the exploration of the information contained within the meta model. More details on this implementation are explained in the following section.

4   Implementation

The library OpenSLEX, based on the meta model proposed in this work, has been implemented in Java[1]. This library provides an interface to insert data in the meta model, and to access it in a similar way to how XES Logs are managed by OpenXES [14]. However, under the hood it relies on SQL technology. Specifically, the meta model is stored in an SQLite[2] file. This provides some advantages like an SQL query engine, a standardized format as well as storage in self contained single data files that benefits its exchange and portability. Figure 4 shows an ER diagram of the internal structure of the meta model. However, it represents a simplified version to make it more understandable and easy to visualize. The complete class diagram of the meta model can be accessed in the

---

[1] http://www.win.tue.nl/~egonzale/projects/openslex/.

[2] http://www.sqlite.org/.

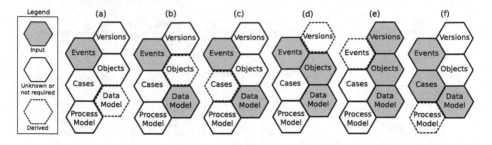

Fig. 5. Input scenarios to complete meta model elements

tool's website[3]. In addition to the library mentioned earlier, an inspection tool has been developed. This tool allows to explore the content of OpenSLEX files by means of a GUI in an exploratory fashion, which lets the user dig into the data and apply some basic filters on each element of the structure. The tool presents a series of blocks that contain the *activities, cases, activity instances, events, event attribute values, data model, objects, object versions, object version attribute values* and *relations* in the meta model. Some of the lists in the inspector (*cases, activity instances, events* and *objects*) have tabs that allow to filter the content they show. For instance, if the tab *"Per Activity"* in the *cases* list is clicked, only cases that contain events of such activity will be shown. In the same way, if the tab *"Per Case"* in the *events* list is clicked, only events contained in the selected case will be displayed. An additional block in the tool displays the attributes of the selected event.

5   Evaluation

The development of the meta model presented in this paper has been partly motivated by the need of a general way to capture the information contained in different systems combining the data and process view. These systems, usually backed by a database, use very different ways to internally store their data. Therefore, in order to extract this data, it is necessary to define a translation mechanism tailored to the wide variety of such environments. Because of this, *the evaluation aims at demonstrating the possibility of transforming information from different environments to the proposed meta model.* Specifically, three source environments are analyzed:

1. Database Redo Logs: files generated by the DBMS in order to maintain the consistency of the database in case of failure or rollback operations.
2. In-table version storage: Application-specific schema to store new versions of objects as a new row in each table.
3. SAP-style change table: changes on tables are recorded in a "redo log" style as a separate table, the way it is done in SAP systems.

---

[3] http://www.win.tue.nl/~egonzale/projects/meta-model/.

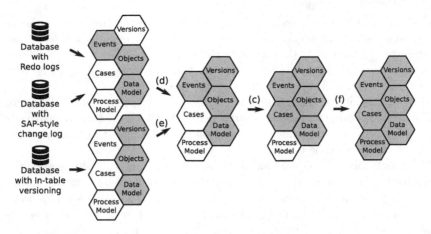

Fig. 6. Meta model completion in the three evaluated environments

The benefit of transforming the data to a common representation is that it allows for decoupling the application of techniques for the analysis from the sources of data. In addition, a centralized representation allows to link data from different sources. However, the source of data may be incomplete. In this case, when transforming the data to fit in our meta model is not enough, we need to apply some inference techniques. This allows to derive the missing information and create a complete and fully integrated view.

The first part of this evaluation (Sect. 5.1) presents the different scenarios that we can find when transforming data. Each of these scenarios start from data that corresponds to different parts of the meta model. Then, it shows how to derive the missing elements from the given starting point. Sections 5.2, 5.3 and 5.4 analyze the three realistic environments mentioned before. We will demonstrate that data extraction is possible and that the meta model can be used to apply process mining instantly. Section 5.5 shows an example of the corresponding output meta model for the three environments.

5.1   Meta Model Completion Scenarios

It is rare to find an environment that explicitly provides the information to fill every *cell* of our meta model. This means that additional steps need to be taken to evolve from an incomplete meta model to a complete one. To do so, Fig. 5 presents several scenarios in which, starting from a certain input, it is possible to infer other elements. Applying these steps consecutively will lead us, in the end, to a completely filled meta model:

a) One of the most basic elements we require in our meta model to be able to infer other elements is the *event collection*. Starting from this input and applying schema, primary key and foreign key discovery techniques [12,15], it is possible to obtain a data model describing the structure of the original data.

b) The events, when combined with a data model, constitute one of the basic components and allows to infer objects. To do so, it is necessary to know the attributes of each class that identify the objects (primary keys). Finding the unique values for such attributes in the events corresponding to each class results in the list of unique objects of the meta model.

c) Also, we can derive cases from the combination of events and a data model. The event splitting technique described in [10], which uses the transitive relations between events defined by the data model, allows to generate different sets of cases, or event logs.

d) The events of each object can be processed to infer the *object versions* as results of the execution of each event. To do so, events must contain the values of the attributes of the object they relate to at a certain point in time or, at least, the values of the attributes that were affected (modified) by the event. Then, ordering the events by (ascending) timestamp allows to reconstruct the versions of each object.

e) The inverse of scenario *d* is the one in which events are inferred from object versions. Looking at the attributes that differ between consecutive versions it is possible to create the corresponding event for the modification.

f) Finally, a set of cases is required to discover a process model using any of the multiple miners available in the process mining field.

The following three sections analyze realistic environments and relate them to these scenarios to demonstrate that the complete meta model structure can be derived in each of them. The goal is to create an integrated view of data and process, specially when event logs are not directly available.

5.2   Database Redo Logs

The first environment focuses on database *redo logs*, a mechanism present in many DBMSs to guarantee consistency, as well as providing additional features such as rollback, point-in-time recovery, etc. Redo logs have already been considered in previous works [1,10] as a source of event data for process mining. Table 1 shows an example of a redo log obtained from an Oracle DBMS. After its processing, explained in [10], these records are transformed into events.

Table 1. Fragment of a redo log: each line corresponds to the occurrence of an event

#	Time + Op + Table	Redo	Undo
1	2014-11-27 15:57:08.0 + INSERT+CUSTOMER	insert into "SAMPLEDB". "CUSTOMER" ("ID", "NAME", "ADDRESS", "BIRTH_DATE") values ('17299', 'Name1', 'Address1', TO_DATE('01-AUG-06', 'DD-MON-RR'));	delete from "SAMPLEDB". "CUSTOMER" where "ID" = '17299' and "NAME" = 'Name1' and "ADDRESS" = 'Address1' and "BIRTH_DATE" = TO_DATE('01-AUG-06', 'DD-MON-RR') and ROWID = '1';
2	2014-11-27 16:07:02.0 + UPDATE+CUSTOMER	update "SAMPLEDB". "CUSTOMER" set "NAME" = 'Name2' where "NAME" = 'Name1' and ROWID = '1';	update "SAMPLEDB". "CUSTOMER" set "NAME" = 'Name1' where "NAME" = 'Name2' and ROWID = '1';
3	2014-11-27 16:07:16.0 + INSERT+BOOKING	insert into "SAMPLEDB". "BOOKING" ("ID", "CUSTOMER_ID") values ('36846', '17299');	delete from "SAMPLEDB". "BOOKING" where "ID" = '36846' and "CUSTOMER_ID" = '17299' and ROWID = '2';
4	2014-11-27 16:07:16.0 + UPDATE+TICKET	update "SAMPLEDB". "TICKET" set "BOOKING_ID" = '36846' where "BOOKING_ID" IS NULL and ROWID = '3';	update "SAMPLEDB". "TICKET" set "BOOKING_ID" = NULL where "BOOKING_ID" = '36846' and ROWID = '3';
5	2014-11-27 16:07:17.0 + INSERT+BOOKING	insert into "SAMPLEDB". "BOOKING" ("ID", "CUSTOMER_ID") values ('36876', '17299');	delete from "SAMPLEDB". "BOOKING" where "ID" = '36876' and "CUSTOMER_ID" = '17299' and ROWID = '4';
6	2014-11-27 16:07:17.0 + TICKET+UPDATE	update "SAMPLEDB". "TICKET" set "ID" = '36876' where "BOOKING_ID" IS NULL and ROWID = '5';	update "SAMPLEDB". "TICKET" set "ID" = NULL where "BOOKING_ID" = '36876' and ROWID = '5';

Figure 6 shows a general overview of how the meta model elements are completed according to the starting input data and the steps taken to derive the missing ones. In this case, the analysis of database redo logs allows to obtain a set of events, together with the objects they belong to and the data model of the database. These elements alone are not sufficient to do process mining without the existence of an event log (a set of traces). In addition, the versions of the objects of the database need to be inferred from the events as well.

Fortunately, a technique to build logs using different perspectives (Trace ID Patterns) is presented in [10]. The existence or definition of a data model is required for this technique to work. Figure 6 shows a diagram of the data transformation performed by the technique, and how it fits in the proposed meta model structure. The data model is automatically extracted from the database schema and is the one included in the meta model. This data model, together with the extracted events, allows to generate both cases (c) and object versions (d). Then, process discovery completes the meta model with a process (f). Once the meta model structure is filled with data, we can make queries on it taking advantage of the established connections between all the elements and apply process mining to do the analysis.

5.3   In-Table Versioning

It is not always possible to get redo logs from databases. Sometimes they are disabled or not supported by the DBMS. Also, we simply may not be able to obtain credentials to access them. Whatever the reason, we often face a situation in which events are not explicitly stored. This enormously limits the analysis that can be performed on the data. The challenge in this environment is to obtain, somehow, an event log to complete our data.

It can be the case that we encounter an environment such that, despite of lacking events, versioning of objects is kept in the database, i.e. it is possible to retrieve the old value for any attribute of an object at a certain point in time. This is achieved by means of duplication of the modified versions of rows. The table at the bottom left corner of Fig. 7 shows an example of an in-table versioning of objects. We see that the primary key is formed by the fields *id* and *load_timestamp*. Each row represents a version of an object and every new reference to the same *id* at a later *load_timestamp* represents an update. Therefore, if we order rows (ascending) by *id* and *load_timestamp*, we get sets of versions for each object. The first one (with older *load_timestamp*) represents an insertion, and the rest updates on the values.

Looking at Fig. 7 it is clear that, ordering by timestamp the versions in the original set (bottom left), we can reconstruct the different states of the database (right). Each new row in the original table represents a change in the state of the database. Performing this process for all the tables, allows to infer the events in a setting where they were not explicitly stored. Figure 6 shows that, thanks to the meta model proposed it is possible to derive events starting from a data model, a set of objects, and their versions as input (Fig. 5.e). The next step is to obtain cases from the events and data model applying the technique from [10]

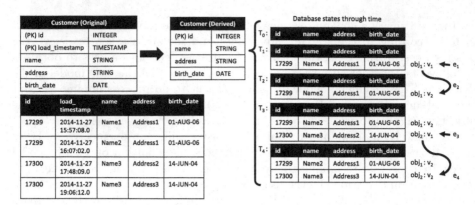

Fig. 7. Example of in-table versioning and its transformation into objects and versions

to split event collections into cases selecting an appropriate *Trace ID Pattern* (scenario *c*). Finally, process discovery will allow us to obtain a process model to complete the meta model structure (scenario *f*).

As a result of the whole procedure, we have a meta model completely filled with data (original and derived) that enables any kind of analysis available nowadays in the process mining field. Moreover, it allows for extended analysis combining data and process perspectives.

5.4 SAP-style Change Table

The last environment we will consider is related to very widespread ERP systems such as SAP. These systems provide a huge amount of functionalities to companies by means of configurable modules. They can run on various platforms and rely on databases to store all their information. However, in order to make them as flexible as possible, the implementation tries to be independent of the specific storage technology running underneath. We see SAP systems running on MSSQL, Oracle or other technologies but they do not make intensive use of the features that the database vendor provides. Therefore, data relations are often not defined in the database schema, but managed at the application level. This makes the life of the analyst who would be interested in obtaining event logs a bit more complicated. Fortunately, SAP implements its own redo log like mechanism to store changes in data, and it represents a valid source of data for our purposes. In this setting we lack event logs, object versions, a complete data model and processes. Without some of these elements, performing any kind of process mining analysis becomes very complicated, if not impossible. For instance, the lack of an event log does not allow for the discovery of a process and, without it, performance or conformance analysis are not possible. To overcome this problem, we need to infer the lacking elements from the available information in the SAP database.

First, it must be noted that, despite the absence of an explicitly defined data model, SAP uses a consistent naming system for their tables and columns, and

CDHDR	
OBJECTCLAS	STRING
CHANGENR	INTEGER
OBJECTID	INTEGER
USERNAME	STRING
UDATE	DATE
UTIME	TIME
TCODE	INTEGER

CDPOS	
OBJECTCLAS	STRING
CHANGENR	INTEGER
TABNAME	STRING
TABKEY	STRING
FNAME	STRING
VALUE_NEW	STRING
VALUE_OLD	STRING

OBJECTCLAS	CHANGENR	OBJECTID	USERNAME	UDATE	UTIME	TCODE
CUST	0001	0000001	USER1	2014-11-27	15:57:08.0	0001
CUST	0002	0000001	USER2	2014-11-27	16:07:02.0	0002
CUST	0003	0000002	USER2	2014-11-27	17:48:09.0	0003
CUST	0004	0000002	USER1	2014-11-27	19:06:12.0	0004
...	...	...	...	...	...	...

Customer	
(PK) id	INTEGER
name	STRING
address	STRING
birth_date	DATE

OBJECTCLAS	CHANGENR	TABNAME	TABKEY	FNAME	VALUE_NEW	VALUE_OLD
CUST	0001	CUSTOMER	17299	name	Name1	
CUST	0001	CUSTOMER	17299	address	Address1	
CUST	0001	CUSTOMER	17299	birth_date	01-AUG-06	
CUST	0002	CUSTOMER	17299	name	Name2	Name1
CUST	0003	CUSTOMER	17300	name	Name3	
CUST	0003	CUSTOMER	17300	address	Address2	
CUST	0003	CUSTOMER	17300	birth_date	14-JUN-04	
CUST	0004	CUSTOMER	17300	address	Address3	Address2
...	...	...	...	...	...	...

Fig. 8. Example of SAP change tables CDHDR and CDPOS

there is lots of documentation available that describes the data model of the whole SAP table landscape. On the other hand, to extract the events we need to process the change log. This SAP-style change log, as can be observed in Fig. 8, is based on two change tables: *CDHDR* and *CDPOS*. The first table (*CDHDR*) stores one entry per change performed on the data with a unique change id (*CHANGENR*) and other additional details. The second table (*CDPOS*) stores one entry per field changed. Several fields in a data object can be changed at the same time and will share the same *CHANGENR*. For each field changed, the table name is recorded (*TABNAME*) together with the field name (*FNAME*), the key of the row affected by the change (*TABKEY*) and the old and new values of the field (*VALUE_OLD*, *VALUE_NEW*).

As can be seen in Fig. 6, after processing the change log and providing an SAP data model, we are in a situation in which the events, objects and data model are known. Then, we can infer the versions of each object (d), split the events in cases (c) and finally, discover a process model (f). With all these ingredients it becomes possible to perform any process mining analysis and answer complex questions combining process and data perspectives.

5.5   Resulting Meta Model

In the three previous sections we have explored different environments in which, from a given starting input, we can derive the missing blocks to completely fill our meta model. The three environments are different, but based on the running example presented in Sect. 2. Therefore, we assume that the same information can be accessed in all three of them. As a result, the resulting meta model will contain the same information, but different parts have been derived through a different procedure depending on the starting input data. Figure 9 provides a simplified view on the final content of the meta model for any of the three environments. However, a full meta model considering all the tables presented

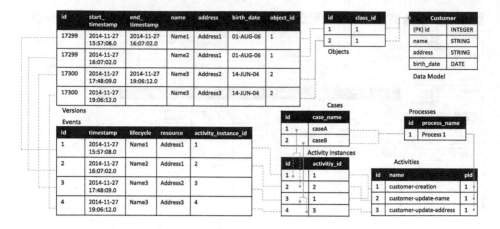

Fig. 9. Sumarized example of resulting meta model

in the running example is available online[4] and can be explored using our tool shown in Sect. 4. Now we have all the information we need in a centralized storage. However, it cannot be used yet to do process mining analysis. Most of these techniques require an event log in XES format. Fortunately, our meta model allows full compatibility with this target format. In addition, our meta model has the advantage of allowing to generate many different event logs depending on the view we want to obtain of our data. Let us reconsider the question proposed in Sect. 2: *What is the interaction with the process of customers between 18 and 25 years old who bought tickets for concerts of band X?*

In order to answer this question, we need to make a proper selection on the data and transform it to a XES event log. The OpenSLEX library provides the functionality required to programmatically select the desired view of the data in our meta model and instantly export it to XES. The great advantage is that, regardless of the source environment from which we obtained the data, the structure in which to make this selection is always the same. Figure 10 shows the result of mining the selected event log. We see that customers make two kind of operations: ticket booking and customer updates. In order to buy a ticket, customers make a booking (*BOOKING+INSERT*) and, afterward, the corresponding ticket is updated (*TICKET+UPDATE*). This means that the ticket is assigned to the booking to avoid it from being sold twice. The other operation, customer updates (*CUSTOMER+UPDATE*), is performed quite often and it corresponds with changes in the details of the customer, e.g. a change of address. The dashed lines denote deviations of the log respect to the discovered model. In this image we see that, in 5 cases, the *TICKET+UPDATE* operation was skipped after executing *BOOKING+INSERT*. This could mean that the booking process does not finish or it is empty. Therefore, no ticket is ever booked. Also, we see some deviations represented by a dashed loop line on top of

---

[4] http://www.win.tue.nl/~egonzale/projects/meta-model/mm-01.zip.

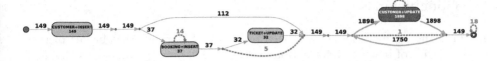

Fig. 10. Process model of the selected events

an activity or place, for example the one in activity *BOOKING+INSERT* that reads a counter with the value 14. This means that a certain amount of "move on log" operations took place, i.e. an event happened in the log that cannot be aligned with the model.

At this point we have shown that, from the proposed meta model, it is possible to generate views on the data. Then, exporting them to the target XES format allows to perform process mining analysis to get more insights on the behavior of the system under study.

6   Related Work

Several efforts have been made trying to provide structure to the representation of execution data in the information systems field. Workflow management systems are an example of environment in which the community has focused on providing models to describe their functioning and allow for the analysis of their behavior. Papers like [9,11] provide meta models to give structure to audit trails on workflows. However, they focus mainly on the workflow or process perspective. Process mining has different needs and the desire to store event data in a unified way is obvious. In [3], the authors provide a meta model to define event logs, which would evolve later in the OpenXES format [14]. This format represents a great achievement from the point of view of standardization, and allows to exchange logs and develop mining techniques assuming a common representation of the data. The XES format is, in fact, a target format from our perspective and not a source of information, i.e. we aim at, from a richer source, generate different views on data in XES format to enable process mining.

The main flaw of these approaches resides in the way they force the representation of complex systems by means of a flat event log. The data perspective is missing, only allowing to add attributes at the event, trace or log level. More recent works try to improve the situation, analyzing data dependencies [7] in business models with the purpose of improving them, or even observing changes on object states to improve their analysis [4]. However, none of the existing approaches provides a generic and standard way of gathering, classifying, storing and connecting process and data perspectives on information systems, specially when dealing with databases where the concept of structured process can be fuzzy or nonexistent.

7   Conclusion

In this paper, a meta model has been proposed that provides a bigger picture of the reality of business information systems. This meta model aligns the data and process perspectives and enables the application of existing process mining techniques (at the same time that unleashes a new way to query data and historical information). This is possible thanks to the combination of data and process perspectives on the analysis of information systems. The applicability of the technique has been demonstrated by means of the analysis of several real-life environments. Also, an implementation of the proposed solution has been developed and tested. However, from the authors' point of view, the main contribution of this work is not only on what new features it enables, but on the universality of the proposed solution. Its applicability to so many different environments provides a common ground to separate data extraction and analysis as different problems, creating an interface that is much richer and powerful than the current existing standards. Several challenges remain open. For instance, a system that enhances the query building experience, allowing for a more natural and user friendly way is desirable. Also, mechanisms to exploit the benefits of the process part of the structure when combined to the data will make the solution really beneficial in comparison to the regular queries that can be performed in the source database systems. In addition, the development of techniques to incorporate data from more varied sources and systems will certainly make the proposed meta model a real candidate for extended use and standardization.

References

1. van der Aalst, W.M.P.: Extracting event data from databases to unleash process mining. In: vom Brocke, J., Schmiedel, T. (eds.) BPM - Driving Innovation in a Digital World. Management for Professionals, pp. 105–128. Springer International Publishing, Switzerland (2015)
2. Buijs, J.: Mapping Data Sources to XES in a Generic Way. Master's thesis, Technische Universiteit Eindhoven, The Netherlands (2010)
3. van Dongen, B.F., van der Aalst, W.M.P.: A meta model for process mining data. EMOI-INTEROP 160, 30 (2005)
4. Herzberg, N., Meyer, A., Weske, M.: Improving business process intelligence by observing object state transitions. Data Knowl. Eng. 98, 144–164 (2015)
5. Ingvaldsen, J.E., Gulla, J.A.: Preprocessing support for large scale process mining of SAP transactions. In: ter Hofstede, A., Benatallah, B., Paik, H.-Y. (eds.) BPM 2007 Workshops. LNCS, vol. 4928, pp. 30–41. Springer, Heidelberg (2008)
6. ER, M., Astuti, H.M., Wardhani, I.R.K.: Material movement analysis for warehouse business process improvement with process mining: a case study. In: Bae, J., Suriadi, S., Wen, L. (eds.) AP-BPM 2015. LNBIP, vol. 219, pp. 115–127. Springer, Heidelberg (2015)
7. Meyer, A., Pufahl, L., Fahland, D., Weske, M.: Modeling and enacting complex data dependencies in business processes. In: Daniel, F., Wang, J., Weber, B. (eds.) BPM 2013. LNCS, vol. 8094, pp. 171–186. Springer, Heidelberg (2013)

8. Mueller-Wickop, N., Schultz, M.: ERP event log preprocessing: timestamps vs. accounting logic. In: vom Brocke, J., Hekkala, R., Ram, S., Rossi, M. (eds.) DESRIST 2013. LNCS, vol. 7939, pp. 105–119. Springer, Heidelberg (2013)
9. zur Muhlen, M.: Evaluation of workflow management systems using meta models. In: Proceedings of the 32nd Annual Hawaii International Conference on Systems Sciences, HICSS-32 (1999)
10. González-López de Murillas, E., van der Aalst, W.M.P., Reijers, H.A.: Process mining on databases: Unearthing historical data from redo logs. In: Motahari-Nezhad, H.R., Recker, J., Weidlich, M. (eds.) BPM 2015. LNCS, vol. 9253, pp. 367–385. Springer, Switzerland (2015)
11. Rosemann, M., Zur Muehlen, M.: Evaluation of workflow management systems-a meta model approach. Aust. J. Inf. Syst. **6**(1), 103–116 (1998)
12. Sismanis, Y., Brown, P., Haas, P.J., Reinwald, B.: Gordian: efficient and scalable discovery of composite keys. In: Proceedings of the 32nd International Conference on Very Large Data Bases, pp. 691–702. VLDB Endowment (2006)
13. Štolfa, J., Kopka, M., Štolfa, S., Koběrský, O., Snášel, V.: An application of process mining to invoice verification process in SAP. In: Abraham, A., Krömer, P., Snášel, V. (eds.) Innovations in Bio-inspired Computing and Applications. AISC, vol. 237, pp. 61–74. Springer, Heidelberg (2014)
14. Verbeek, H.M.W., Buijs, J.C.A.M., van Dongen, B.F., van der Aalst, W.M.P.: XES, XESame, and ProM 6. In: Soffer, P., Proper, E. (eds.) CAiSE Forum 2010. LNBIP, vol. 72, pp. 60–75. Springer, Heidelberg (2011)
15. Zhang, M., Hadjieleftheriou, M., Ooi, B.C., Procopiuc, C.M., Srivastava, D.: On multi-column foreign key discovery. Proc. VLDB Endowment **3**(1–2), 805–814 (2010)

Towards a Formal Framework for Business Process Re-Design Based on Data Mining

Thai-Minh Truong$^{(\boxtimes)}$ and Lam-Son Lê

Faculty of Computer Science and Engineering,
Ho Chi Minh City University of Technology, Ho Chi Minh City, Vietnam
thaiminh@cse.hcmut.edu.vn, lam-son.le@alumni.epfl.ch

Abstract. In today's ever changing world, business processes need to be dynamic. Data accumulated as the processes operate capture the meaning of transactions in the past, which opens a door for the dynamics of the business processes in question. Mining the operational data to explicitly represent this meaning could lead to process re-design to make the business processes more efficient. In this paper, we propose a formal framework for redesigning business processes taking data mining rules and business rules as the driver. We formally represent business processes using the artifact-centric approach put forward by the IBM Research. We devise redesigning algorithms that take classification rules extracted from data mining together with business rules and transform the business process in question by eliminating redundant tasks and/or re-ordering inefficiently placed tasks. We illustrate our algorithms and report experiments that were conducted using a proof-of-concept case-study.

Keywords: Artifact-centric processes · Process redesign · Process modeling · Data mining · Formal methods

1 Introduction

In today's ever changing world, business processes need to be dynamic, e.g. flexible process structure, dynamic resource allocation, tasks assigned alternatively to business actors. Data collected as the processes operate might capture the meaning of the dynamics of the business processes in question. Mining the operational data to explicitly represent this meaning could lead to process re-design to make the processes more efficient. For example, in insurance processing, data mining techniques could suggest changes made to the business process in order to improve its fraud detection, i.e. better chance to spot frauds from claimants [1]. Data mining may also help speed up insurance-claiming processes. In finance, reviewing and approving loan applications is a time-consuming, risk-prone activity in which multiple roles could participate. Classification in data mining might suggest changes to be made, e.g. re-positioning it, breaking it into sub tasks some of which could be moved/suppressed in order to speed up the process as a whole [2]. In hospitals, data mining might lead to a business process being

© Springer International Publishing Switzerland 2016
R. Schmidt et al. (Eds.): BPMDS/EMMSAD 2016, LNBIP 248, pp. 250–265, 2016.
DOI: 10.1007/978-3-319-39429-9_16

redesigned to change the order in which patients are admitted, queued and later on treated [3].

Business process redesign [2,4,5] has been researched on from different disciplines including Management Information Systems, Process Management and Information Technology. Technically, redesign could be performed based on patterns [6], based on Petri Net [7], by means of formalization [8]. While it is not a new topic of research, redesigning processes based on data mining techniques had not received much attention from the community (of business processes). Conceptually, redesign performed based on data mining is considered changes to be made to business processes at the *design-time*[1].

The remaining of this paper is structured as follows. Section 2 gives the background of our work. Section 3 formulates our research questions by means of a running example. Section 4 is dedicated to the main contribution of our work including algorithms and implementation. We survey related work in Sect. 5. Section 6 concludes the paper and outlines our further work.

2   Background

2.1   Data Mining

"We are living in the information age" is a popular saying nowadays [9]. This saying goes around within reasons. Over time, data are getting bigger and play an increasingly important role in our daily "information age" life. In the context of enterprise computing, enterprises usually have data warehouse and transactional data of their business processes both of which have hidden values and are crucial to their competitiveness. The data in question typically fall into activities, habits, tendencies, characteristics and the like of individuals or organizations who took part in the business processes. Mining the data in this case is a powerful tool for predicting the outcome of up-coming transactions.

Data mining started to gain momentum in late 1980s. In the context of our work, the primary concern is to leverage data mining in order to change business processes of the enterprise in ways that it becomes more competitive. Several pieces of work have put fort the concept of data mining in business process management. For instance, Wegener proposed integrating data mining into business process based on CRISP-DM [1]. Wegener also considered reusing data mining in business process [6]. Rupnik outlined the deployment of data mining into operational business processes [10].

2.2   Artifact-Centric Business Process Modeling

Business process modeling traditionally put a lot of effort on the behavioral aspect. The so-called activity-centric modeling community became very crowded, resulting in a plenty of initiatives on both imperative and declarative modeling

---

[1] As opposed to changes made at *runtime* that are commonly referred to as process adaptation or process reconfiguration.

of the control-flow of business processes. Making an exhaustive list of them is not the goal of this paper. To name just a few, we have BPEL, YAWL, UML activity diagram, Petri net, etc.

Artifact-centric modeling takes a different approach to modeling business processes [11,12]. Conceptually, both business objects and the process control are taken into account in models, resulting in explicit representation of the transformation of business artifact in contract-like description[2] of business tasks [13]. This approach has been further researched in connection to the UML [14] and BPMN [15].

2.3   Semantically Annotated Business Processes

We based our work on a technique called ProcessSEER, involves doing two stages of computation [16]. In the first stage, we derive a set of possible *scenario label*(s) for the given point in the process view. Each scenario label is a precise list of steps that define a path leading from the start point to a the point being considered. In the second stage, the contiguous sequence of steps in each scenario label is taken into account to accumulate effects annotated to steps along this scenario in a pair-wise fashion.

A functionality annotation states the effect of having functionality delivered at a specified task. The effect can be textual. Alternatively, it could be written in first-order logic (FOL) or some computer-interpretable form. The total functionality delivered up to a certain task is the accumulation of all effects of the precedent tasks. The cumulative effect of tasks can inductively be defined as follows. The cumulative effect of the very first task is equal to its delivery annotation. Let $\langle Tk_i, Tk_j \rangle$ be an ordered pair of consecutive tasks such that Tk_i precedes Tk_j; let e_i be an effect scenario associated with Tk_i and e_j be the delivery annotation associated with Tk_j. Without loss of generality, we assume that e_i and e_j are sets of clauses. The resulting cumulative effect, denoted by $acc(e_i, e_j)$ is defined as follows.

- $acc(e_i, e_j) = e_i \cup e_j$ if $e_i \cup e_j$ is logically consistent
- Otherwise $acc(e_i, e_j) = e_i' \cup e_j$ whereby $e_i' \subseteq e_i$ such that $e_i' \cup e_j$ is consistent and we do not have any $e_i'' \subseteq e_i' \subseteq e_i$ such that $e_i'' \cup e_j$ is consistent."

We may extend the notion of accumulation as defined above to the postcondition (of business tasks) and an additional set of clauses. For instance, if we have two consecutive tasks t_1 and t_2 and a domain constraint dom, we combine them as $acc(dom, acc(t_1.\beta, t_2.\beta))$, where the reduction of clauses in case we have inconsistent ones might be applied to the domain constraint. Alternatively, the accumulation could be in the form $acc(acc(t_1.\beta, t_2.\beta), dom)$.

---

[2] Each business task is described in terms of pre-conditions (i.e., conditions that must be held for the task to be invoked) and post-conditions (i.e., conditions that will be help upon the completion of the task).

3   Motivation

3.1   Running Example

Let us consider the business process[3] of reserving, procuring, picking up and
returning rental cars at a car rental company. Initially, a customer books a car for
rent. The receptionist then receives customer's requirements, checks the inventory to ensure the requested car is available (task "Reserve"). The receptionist
opens a contract filling customer's information (task "Procure booking"). Next,
as the contract is created, the customer is requested to register her profile using
her credit card for deposit (task "Registration"). On one hand, the mechanics
of the rental company take care of the rental cars and record it to the contract
before the customers pick them up (task "Service booked vehicles"). On the
other hand, the receptionist also makes sure that the rental cars are cosmetically
ready for the new renters and record it to the contract (task "Pre-rent check").
Next, the contract is considered by the manager (task "Rental approval") in
order to approve her contract, followed by the customer's rental pick-up (task
"Pick-up vehicles"). Otherwise, the customer's contract will be denied, followed by the customer being notified of a cancellation (task "Cancel rent").
The customer is expected to return her rental cars on time. Returned cars are
handled (task "Return vehicles") and checked for any damages done by the customer. Late return fees and liability may apply to the customer's contract (task
"Post-rent check"). The rental contract should be closed at this point of time.

We use the Business Process Model and Notation (BPMN)[4] to represent this
business process (Fig. 2(a)).

3.2   Research Problems

Based on the analysis in Subsect. 3.1, we come down to the following research
problems about changes that could be made to a business process for the sake
of improvement.

- *Multiple tasks doing the same work that might be* **redundant**. If multiple tasks
  share a common outcome that explicitly stated in the rules extracted by data
  mining, can we downgrade (or get rid of) one or more of these tasks without
  changing the overall effect of the business process in question? Relationship
  between candidate tasks will decide on the removal/reduction.
- *A single task could be split into sub tasks some of which could be* **reordered**
  to make the process as a whole function more efficiently.

Our goal is to devise algorithms for spotting the all tasks that fall into the
categories described above, which will lead to a redesigning procedure.

---

[3] The running example will be walked through again in Subsect. 4.4 when we make a
redesign for the business process in question.
[4] BPMN Specification http://www.bpmn.org/.

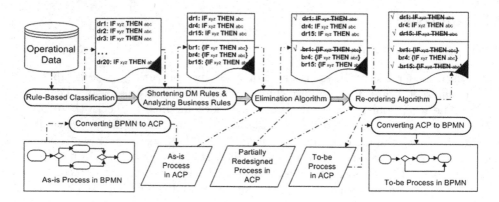

Fig. 1. Overview Framework for Business Process Re-design

4 Framework

4.1 Overview

Figure 1 presents the overall picture of our approach. The operational data is exploited using rule-based classification (i.e., extracting IF-THEN rules from a decision tree) where every non-key attribute of all tables in the operational data will be used as a class label [9]. The resulting classification rules will be assessed and chosen jointly by business analysts and data mining experts, with which they will associate a business rule, e.g., $dr2$ associated with $br2$. Next, the classification rules (associated with the business rules) and the as-is process (represented in ACP) are fed to the elimination step (Algorithm 1), which yields a partially redesigned process (in ACP again) where redundant tasks are either reduced (i.e., their functionality get lessened) or got rid of. Rules that did not take effect in the elimination step will be the input of the next step – Algorithm 2. The output of this step is a final version in ACP that will be converted back to BPMN. In the end, we may have some classification rules and business rules that did not take effect in any step of our framework.

4.2 Definitions

An artifact schema is defined as $M = \{C_1, C_2, ..., C_n\}$ [11], where $C_i \in M(1 \leq i \leq n)$ is an artifact class. We define an artifact class as $C = (A, S)$ where $A = \{a_1, a_2, ..., a_x\}$, each $a_i \in A(1 \leq i \leq x)$ is an attribute of a scalar-typed value (string and real number) or an undefined value. $S = \{s_1, s_2, ..., s_y\} \cup \{s^{init}\}$ where $s_i \in S(1 \leq i \leq y)$ is a state and s^{init} denotes the initial state.

We adopt the concept of pre- and post-condition [11] in formally representing business tasks. Let's formally define a business task $t = (\lambda, \beta, \beta', \nu, \sigma)$, where λ and β are pre-condition and post-condition of the task in question, respectively. β' stands for the post-condition excludes all instate expressions. $\nu \in V$ is a service to be performed. σ is a set of artifact classes are involved in task t: $\sigma = \{C_1, C_2, ..., C_y\}$ where $C_i \in M (1 \le i \le y)$.

Definitions 1 and 2 formally captures the pre- and post-condition of a business task and a classification rule / associated business rule, respectively. $defined(C_i.a_k)$ denotes that attribute a_k of artifact class C_i has some value. $instate(C_i, C_i.s_m)$ denotes that artifact class C_i is in state s_m.

Definition 1. λ or $\beta = \bigwedge\limits_{i=1}^{|M|} \bigwedge\limits_{m=1}^{|C_i.S|} instate(C_i, C_i.s_m) \bigwedge\limits_{j=1}^{|M|} \bigwedge\limits_{k=1}^{|C_j.A|} defined(C_j.a_k)$

$\bigwedge\limits_{h=1}^{|M|} \bigwedge\limits_{l=1}^{|C_h.A|} \neg defined(C_h.a_l) \bigwedge\limits_{q=1}^{|M|} \bigwedge\limits_{p=1}^{|C_q.A|} \bigwedge\limits_{r=1}^{|C_q.a_p.X|} C_q.a_p = x_r$

where: $C_i, C_j, C_h, C_q \in M$ is an artifact class; $C_i.s_m \in C_i.S$ is a state of class C_i; $C_j.a_k \in C_j.A$ is an attribute of class C_j; $C_h.a_l \in C_h.A$ is an attribute of class C_h; $C_q.a_p \in C_q.A$ is an attribute of class C_q; $x_r \in C_q.a_p.X$ is an attribute value of attribute a_p in class C_q; and $C_j.a_k \ne C_h.a_l \ne C_q.a_p$.

Definition 2. Let's consider a set of classification rules $DR = \{dr_1, dr_2, ..., dr_n\}$. Each classification rule dr_l is optionally associated with a business rule br_l. Both classification rules and business rules are given in the form of IF Cond THEN Consqnt, where Cond and Consqnt are the condition and the consequent of the rule being represented, respectively. Formally, we have

$Cond \equiv \bigwedge\limits_{i=1}^{|M|} \bigwedge\limits_{m=1}^{|C_i.S|} instate(C_i, C_i.s_m) \bigwedge\limits_{j=1}^{|M|} \bigwedge\limits_{k=1}^{|C_j.A|} \bigwedge\limits_{r=1}^{|C_j.a_k.X|} C_j.a_k = x_r$

$Consqnt \equiv C_q.a_p = x_h$ where: $q \in \{1, ..., |M|\}$; $p \in \{1, ..., |C_q.A|\}$; $h \in \{1, ..., |C_q.a_p.X|\}$; $dr_l \in DR (1 \le l \le n)$ is a classification rule; br_l is a business rule associated with the classification rule dr_l; $C_i, C_j, C_q \in M$ is a class; $C_i.s_m \in C_i.S$ is a state of class C_i; $C_j.a_k \in C_j.A$ is an attribute of class C_j; $C_q.a_p \in C_q.A$ is an attribute of class C_q; $x_r \in C_j.a_k.X$ is an attribute value of attribute a_k in class C_j; $x_h \in C_q.a_p.X$ is an attribute value of attribute a_p in class C_q.

4.3 Algorithms for Business Process Re-Design

In this paper, we suppose the algorithms to eliminate or reorder a task in business process based on the classification rule in data mining. We assume that business tasks represented in BPMN diagram are the same as business rules formally defined in ACP, formally as a tuple $t(\lambda, \beta, \beta', \nu, \sigma)$ as presented in Subsect. 4.2.

To navigate back and forth in a scenario label, we define the following denotations.

- $next(L, t_i)$: returns the immediately succeeding task of t_i in scenario label L.
- $next^*(L, t_i)$: denotes the transitive closure of the succeeding tasks of t_i in scenario label L.
- $prev(L, t_i)$: yields the immediately preceding task of t_i in scenario label L.
- $prev^*(L, t_i)$: denotes the transitive closure of the precedent tasks of t_i in scenario label L.

Tasks Elimination. We address the first research problem in Subsect. 3.2. We devise Algorithm 1 to eleminate redundant tasks in a business process with respect to a data mining rule. We first identify a set of candidate tasks for elimination. Out of these tasks, if we can find out a pair of tasks one of which entails the other, the latter should be eliminated. We then figure out the common part of the post-conditions of the remaining tasks. Finally, we keep the common part of the specific remaining task while getting rid of these of the other remaining tasks. In Algorithm 1, we assume the following.

- $instance = Cond(p_i.\sigma)$: is a logic clause which is specified by reflecting condition of classification rule ($Cond$) into a specific set of artifact classes of task ($p_i.\sigma$).
- ν_{comm} has been identified by the user. When we show β_{comm} and request the user specifies the corresponding service that will be reduced.
- $Eliminate(q_k)$: gets rid of task q_k from the original process.
- $Literals(p_i.\beta)$: returns the set of literals of the post-condition of task p_i.
- q_y: is a remaining task which is chosen by a business user.
- $ReduceTask(q_k, \nu_{comm}, \beta_{comm})$: reduces a part of task q_k which has the service likes ν_{comm} and the post-condition likes β_{comm} (i.e., the service and the post-condition of this task is reduced).

Tasks Reordering. We devise Algorithm 2 using the notion of cumulative effect given in Subsect. 2.3. It comes with the following assumption.

- $dom = Cond(\bigcup_{l=1}^{i} p_l.\sigma)$: is a logic clause which is specified by reflecting condition of classification rule ($Cond$) into a specific set of artifact classes of these tasks (from p_1 until p_i).
- $t_{new} = Decompose(t_{target})$: task t_{new} is a specific case of task t_{target} which is recommended by the classification rule.
- $ReorderAfter(L, t_{new}, p_i)$: task t_{new} is re-ordered to the immediately succeeding task of p_i in scenario label L.

4.4   Implementation

This subsection is dedicated to the implementation of our re-design framework. We illustrate it using the running example presented in Subsect. 3.1. Let us begin by giving a primary set of artifact classes in the following.

- `Contract = ({contractID, customerID, carID, depositedType, depositedBudget, insuranceType, preRentCondition, carAge, carMileage, expectedPickedupTime, realPickedupTime, expectedReturnedTime, realReturnedTime, approval, preRentCondition, postRentCondition, totalCost, postRentAssessment}, {initial, opened, deposited, pickedup, returned, canceled, closed})`

Algorithm 1. Algorithm for tasks elimination

Input: $dr < Cond, Consqnt >$; associated business rule br; scenario label L
Data: business process P

1 Let $T_{list} \leftarrow getTasks(L)$; /* gets all tasks of L from P */
2 Let $T_{target} \leftarrow \emptyset$; /* is a set of target tasks   */
3 **foreach** $task\ p_i \in T_{list}$ **do**
4 $\quad$ **if** $(acc(instance, p_i.\lambda) \cup \{br\}) \vdash Cond$ **and** $(\{Consqnt\} \cup \{br\}) \vdash p_i.\beta'$ **then**
5 $\quad\quad$ $T_{target} \leftarrow T_{target} + p_i$;
6 $\quad$ **end**
7 **end**
8 **if** $t_{target} \neq \emptyset$ **then**
9 $\quad$ Let $T_{rem} \leftarrow T_{target}$;
10 $\quad$ **foreach** $target\ task\ p_i \in T_{target}$ **do**
11 $\quad\quad$ **foreach** $target\ task\ q_k \in T_{target} \setminus \{p_i\}$ **do**
12 $\quad\quad\quad$ **if** $p_i.\beta \vdash q_k.\beta$ **and** q_k has not been eliminated **then**
13 $\quad\quad\quad\quad$ $Eliminate(q_k)$;
14 $\quad\quad\quad\quad$ $T_{rem} \leftarrow T_{rem} - q_k$;
15 $\quad\quad\quad$ **end**
16 $\quad\quad$ **end**
17 $\quad$ **end**
18 $\quad$ **if** $|T_{rem}| > 1$ **then**
19 $\quad\quad$ Let $l_{comm} = \bigcap\limits_{p_i \in T_{rem}} Literals(p_i.\beta)$;
20 $\quad\quad$ **if** $l_{comm} \neq \emptyset$ **then**
21 $\quad\quad\quad$ Let $\beta_{comm} = \bigwedge\limits_{i=1...|l_{comm}|} l_{comm_i}$;
22 $\quad\quad\quad$ **foreach** $target\ task\ q_k \in T_{rem} \setminus \{q_y\}$ **do**
23 $\quad\quad\quad\quad$ $ReduceTask(q_k, \nu_{comm}, \beta_{comm})$;
24 $\quad\quad\quad$ **end**
25 $\quad\quad$ **end**
26 $\quad$ **end**
27 **end**

- Customer = ({customerID, name, age, gender, address, nationality, job, licenseNumber, driverLicensePeriod, expiredDate}, $\emptyset$)
- Car = ({carID, level, manufacturer, model, serialNumber, color}, {nonBooked, booked, booked &Serviced, nonBooked &Serviced, readyForRent, pickedup, returned})

After having performed the rule-based classification step (see Fig. 1), we obtain certain classification rules. The business analysts and data mining experts of the car rental company pick up classification rules $dr1$ and $dr2$ by analyzing their coverage and accuracy.

- $dr1$ = IF $Contract.carAge <= 2 \wedge Car.level = "luxury" \wedge Contract.carMileage < 30000$ THEN $Contract.preRentCondition = "soundCondition"$

Algorithm 2. Tasks re-ordering

Input: $dr < Cond, Consqnt >$; associated business rule br; scenario label L
Data: business process P

1 Let $T_{list} \leftarrow getTasks(L)$; /* gets all tasks of L from P */
2 Let $t_{target} \leftarrow \emptyset$; /* is a target task   */
3 **foreach** $task\ p_i \in T_{list}$ **do**
4   **if** $(\{Consqnt\} \cup \{br\}) \vdash p_i.\beta'$ **then**
5          $t_{target} \leftarrow p_i$;
6   break;
7   **end**
8 **end**
9 **if** $t_{target} \neq \emptyset$ **then**
10   **foreach** $task\ p_i \in prev^*(L, t_{target})$ **do**
11   **if** $(acc(dom, acc(p_{i-1}.\beta, p_i.\beta)) \cup \{br\}) \vdash Cond$ **then**
12             $t_{new} = Decompose(t_{target})$;
13             $ReorderAfter(L, t_{new}, p_i)$;
14   break;
15   **end**
16   **end**
17 **end**

- $dr2 =$ IF $Customer.age = [18...30] \wedge Customer.gender = "male" \wedge Customer.job =$ $"unidentified" \wedge Customer.driverLicensePeriod = [0...2] \wedge Contract.insuranceType =$ $"breakdown" \wedge Car.level = "luxury" \wedge Car.model = "X6"$ THEN $Contract.postRent$ $Assessment = "avoided"$

They then associate $dr2$ to business rule $br2$ defined in the following. Note that classification rule $dr1$ is associated to no business rule.

$br2 =$ IF $Contract.postRentAssessment = "avoided"$ THEN $Contract.approval = "denied"$

Now, both $dr1$ and $dr2$ are given to the elimination step. Only $dr1$ takes effect in this step (detailed in Tables 1 and 2) and marked as exploited. Rule $dr2$ is then fed to the re-ordering step. This classification rule and its associated business rule $br2$ take effect in this step (detailed in Tables 3 and 4). Finally, our framework transforms the original business process depicted in Fig. 2(a) into what is presented in Fig. 2(b). The redesigned process differs from the original one primarily on (i) task **Service booked vehicles** is reduced; (ii) task **Rental approval** is split into two pieces of work one of which becomes a new task named **Check rent request** – this new task is placed right next to task **Procure booking** to be able to reject rent requests from "bad" drivers with the aid of a data mining rule textually depicted in the figure. We provide the execution details of our algorithms for this scenario as below.

Tasks Elimination. We consider a scenario label that runs as: **"Reserve"**, $\oplus$, **"Service booked vehicles"**, $\oplus$, **"Pre-rent check"**, **"Rent approval"**, $\otimes$, **"Cancel rent"**, end event – where: $\otimes$ stands for an exclusive gateway, $\oplus$ stands

Table 1. Description of tasks along the scenario label considered for a redesign of the car rental business process using tasks elimination

t_1 Reserve	
Artifact classes	$Car, Contract$
Pre-condition	$instate(Car, nonBooked) \wedge instate(Contract, initial) \wedge \neg defined(Contract.carID) \wedge$ $\quad \neg defined(Contract.contractID)$
Service	$reserveACar(Car, Contract)$
Post-condition	$instate(Car, booked) \wedge instate(Contract, opened) \wedge defined(Contract.carID) \wedge$ $\quad defined(Contract.contractID) \wedge Contract.carID =$ $\quad Car.carID \wedge defined(Car.level) \wedge defined(Car.manufacturer) \wedge$ $\quad defined(Car.model) \wedge defined(Car.serialNumber) \wedge defined(Car.color)$
t_2 Service Booked Vehicles	
Artifact classes	$Car, Contract$
Pre-condition	$instate(Car, booked) \wedge instate(Contract, opened) \wedge$ $\quad \neg defined(Contract.preRentCondition) \wedge Contract.carID = Car.carID$
Service	$serviceBookedVehicles(Car, Contract)$
Post-condition	$defined(Contract.preRentCondition) \wedge instate(Car, booked$ $\quad \& Serviced)$
t_3 Pre-rent Check	
Artifact classes	$Car, Contract$
Pre-condition	$instate(Car, booked \& Serviced) \wedge instate(Contract, deposited) \wedge Contract.carID =$ $\quad Car.carID$
Service	$pre - rentCheck(Car, Contract)$
Post-condition	$defined(Contract.preRentCondition) \wedge instate(Car, readyForRent)$
t_4 Rental Approval	
Artifact classes	$Contract$
Pre-condition	$instate(Contract, opened) \wedge defined(Contract.customerID) \wedge$ $\quad defined(Contract.carID)$
Service	$approveRent(Contract)$
Post-condition	$defined(Contract.approval)$
t_5 Cancel Rent	
Artifact classes	$Contract, Car$
Pre-condition	$instate(Contract, opened) \wedge instate(Car, booked) \wedge$ $\quad Contract.approval = "denied" \wedge Contract.carID = Car.carID$
Service	$cancelRent(Contract, Car)$
Post-condition	$instate(Contract, canceled) \wedge instate(Car, nonBooked) \wedge Contract.approval = "denied"$

for a parallel gateway. Tables 1 and 2 describe the business tasks involved and how Algorithm 1 works for the running example, respectively.

Tasks Decomposition & Reordering. We consider the scenario label includes these tasks: "Reserve", $\oplus$, "Procure booking", "Registration", $\oplus$, "Pre-rent check", "Rent approval", $\otimes$, "Cancel rent", end event – where: $\otimes$ stands for an exclusive gateway, $\oplus$ stands for a parallel gateway. Tables 3 and 4 describes the business tasks involved and how Algorithm 2 works for the running example, respectively.

We implemented Algorithms 1 and 2 using the Orbital[5] library – a Java library supporting logic inference. Our implementation[6] has 662 Java lines of

---

[5] Homepage of Orbital symbolaris.com/orbital.

[6] Source code can be downloaded at www.esp-lab.net/images/src.zip.

Table 2. Execution of Algorithm 1 for the running example

Lines		Variables	Values
1		T_{list}	$\{t_1, t_2, t_3, t_4\}$
2		T_{target}	$\emptyset$
3-8		$dr < Cond >$	$Contract.carAge <= 2 \land Car.level = "luxury" \land Contract.carMileage < 30000$
		$dr < Consqnt >$	$Contract.preRentCondition = "soundCondition"$
		br	$\emptyset$
	t_1	$instance$	$Contract.carAge <= 2 \land Car.level = "luxury" \land Contract.carMileage < 30000$
		$acc(instance, t_1.\lambda)$	$instate(Car, nonBooked) \land instate(Contract, initial) \land \neg defined(Contract.carID)$
		β'	$defined(Contract.carID) \land defined(Contract.contractID) \land Contract.carID = Car.carID \land defined(Car.level) \land defined(Car.manufacturer) \land defined(Car.model) \land defined(Car.serialNumber) \land defined(Car.color)$
	t_2	$instance$	$Contract.carAge <= 2 \land Car.level = "luxury" \land Contract.carMileage < 30000$
		$acc(instance, t_2.\lambda)$	$instate(Car, booked) \land instate(Contract, opened) \land \neg defined(Contract.preRentCondition) \land Contract.carID = Car.carID \land Contract.carAge <= 2 \land Car.level = "luxury" \land Contract.carMileage < 30000$
		β'	$defined(Contract.preRentCondition)$
		T_{target}	$\{t_2\}$
	t_3	$instance$	$Contract.carAge <= 2 \land Car.level = "luxury" \land Contract.carMileage < 30000$
		$acc(instance, t_3.\lambda)$	$instate(Car, booked\&Serviced) \land instate(Contract, deposited) \land Contract.carID = Car.carID \land Contract.carAge <= 2 \land Car.level = "luxury" \land Contract.carMileage < 30000$
		β'	$defined(Contract.preRentCondition)$
		T_{target}	$\{t_2, t_3\}$
	t_4	$instance$	$Contract.carMileage < 30000$
		$acc(instance, t_4.\lambda)$	$instate(Contract, opened) \land defined(Contract.customerID) \land defined(Contract.carID) \land Contract.carMileage < 30000$
		β'	$defined(Contract.approval)$
		T_{target}	$\{t_2, t_3\}$
	t_5	$instance$	$Contract.carAge <= 2 \land Car.level = "luxury" \land Contract.carMileage < 30000$
		$acc(instance, t_5.\lambda)$	$instate(Contract, opened) \land instate(Car, booked) \land Contract.approval = "denied" \land Contract.carID = Car.carID \land Contract.carAge <= 2 \land Car.level = "luxury" \land Contract.carMileage < 30000$
		β'	$Contract.approval = "denied"$
		T_{target}	$\{t_2, t_3\}$
9		T_{rem}	$\{t_2, t_3\}$
10-17		$t_2.\beta$	$defined(Contract.preRentCondition) \land instate(Car, booked\&Serviced)$
		$t_3.\beta$	$defined(Contract.preRentCondition) \land instate(Car, readyForRent)$
			No tasks are eliminated, as $t_2.\beta$ does not entail $t_3.\beta$ and $t_3.\beta$ does not entail $t_2.\beta$
		T_{rem}	$\{t_2, t_3\}$
18-26		l_{comm}	$defined(Contract.preRentCondition)$
		β_{comm}	$defined(Contract.preRentCondition)$
		q_y	t_3 (notation: this task is chosen by a business user)
	Reduce Task	q_k	t_2
		ν_{comm}	note the pre-rent conditon of a car on the contract
		β_{comm}	$defined(Contract.preRentCondition)$

code. It was done in the Eclipse[7] platform that ran on a 64-bit Windows (Core i5, 6 GB of RAM). Taking the car rental example as the input, Algorithm 1 and 2 came out with the re-designed process as depicted in Fig. 2(b) within 1129 milliseconds. Time breakdown is as follows.

- $dr1$ takes effect in Algorithm 1 – 387 miliseconds.
- $dr2$ does not take effect in Algorithm 1 – 126 miliseconds.
- $dr2$ takes effect in Algorithm 2 – 616 miliseconds.

---

[7] Homepage of Eclipse eclipse.org.

Table 3. Description of tasks along the scenario label considered for a redesign of the car rental business process using tasks re-ordering

t_1 Reserve	
Artifact classes	$Car, Contract$
Pre-condition	$instate(Car, nonBooked) \wedge instate(Contract, initial) \wedge \neg defined(Contract.carID) \wedge \neg defined(Contract.contractID)$
Service	$Reserveacar(Car, Contract)$
Post-condition	$instate(Car, booked) \wedge instate(Contract, opened) \wedge defined(Contract.carID) \wedge defined(Contract.contractID) \wedge Contract.carID = Car.carID \wedge defined(Car.level) \wedge defined(Car.manufacturer) \wedge defined(Car.model) \wedge defined(Car.serialNumber) \wedge defined(Car.color)$
t_2 Procure Booking	
Artifact classes	$Contract, Customer$
Pre-condition	$instate(Contract, opened) \wedge \neg defined(Contract.customerID) \wedge defined(Contract.carID)$
Service	$procureBooking(Contract, Customer)$
Post-condition	$defined(Contract.customerID) \wedge defined(Contract.insuranceType) \wedge defined(Contract.expectedPickedupTime) \wedge defined(Contract.expectedReturnedTime) \wedge defined(Customer.customerID) \wedge defined(Customer.name) \wedge defined(Customer.age) \wedge defined(Customer.gender) \wedge defined(Customer.address) \wedge defined(Customer.licenseNumber) \wedge defined(Customer.driverLicensePeriod) \wedge defined(Customer.expiredDate) \wedge Contract.customerID = Customer.customerID$
t_3 Registration	
Artifact classes	$Contract$
Pre-condition	$instate(Contract, opened) \wedge \neg defined(Contract.depositedType) \wedge \neg defined(Contract.depositedBudget)$
Service	$Registration(Contract)$
Post-condition	$instate(Contract, deposited) \wedge defined(Contract.depositedType) \wedge defined(Contract.depositedBudget)$
t_4 Pre-rent Check	
Artifact classes	$Car, Contract$
Pre-condition	$instate(Car, booked\&Serviced) \wedge instate(Contract, deposited) \wedge Contract.carID = Car.carID$
Service	$pre - renCheck(Car, Contract)$
Post-condition	$defined(Contract.preRentCondition) \wedge instate(Car, readyForRent)$
t_5 Rental Approval	
Artifact classes	$Contract$
Pre-condition	$instate(Contract, opened) \wedge defined(Contract.customerID) \wedge defined(Contract.carID)$
Service	$approveRent(Contract)$
Post-condition	$defined(Contract.approval)$
t_6 Cancel Rent	
Artifact classes	$Contract, Car$
Pre-condition	$instate(Contract, opened) \wedge instate(Car, booked) \wedge Contract.approval = "denied" \wedge Contract.carID = Car.carID$
Service	$cancelRent(Contract, Car)$
Post-condition	$instate(Contract, canceled) \wedge instate(Car, nonBooked) \wedge Contract.approval = "denied"$

Table 4. Execution of Algorithm 2 for the running example

Lines	Variables			Values
1	T_{list}			$\{t_1, t_2, t_3, t_4, t_5, t_6\}$
2	T_{target}			$\emptyset$
3-8	$dr < Consqnt >$			$Contract.postRentAssessment = "avoided"$
	br			IF $Contract.postRentAssessment = "avoided"$ THEN $Contract.approval = "denied"$
	t_1	t_1		task "Reserve"
		$t_1.\beta'$		$defined(Contract.carID) \land defined(Contract.contractID) \land Contract.carID = Car.carID \land defined(Car.manufacturer) \land defined(Car.level) \land defined(Car.model) \land defined(Car.serialNumber) \land defined(Car.color)$
	t_2	t_2		task "Procure Booking"
		$t_2.\beta'$		$defined(Contract.customerID) \land defined(Contract.insuranceType) \land defined(Contract.expectedPickedupTime) \land defined(Contract.expectedReturnedTime) \land defined(Customer.customerID) \land defined(Customer.name) \land defined(Customer.age) \land defined(Customer.gender) \land defined(Customer.address) \land defined(Customer.licenseNumber) \land defined(Customer.driverLicensePeriod) \land defined(Customer.expiredDate) \land Contract.customerID = Customer.customerID$
	t_3	t_3		task "Registration"
		$t_3.\beta'$		$defined(Contract.depositedType) \land defined(Contract.depositedBudget)$
	t_4	t_4		task "Pre-rent Check"
		$t_4.\beta'$		$defined(Contract.preRentCondition)$
	t_5	t_5		task "Rental Approval"
		$t_5.\beta'$		$defined(Contract.approval)$
	T_{target}			t_5
9-17	$prev^*(L, t_5)$			$\{t_1, t_2, t_3, t_4\}$
	$dr < Cond >$			$Customer.age = [18...30] \land Customer.gender = "male" \land Customer.job = "unidentified" \land Customer.driverLicensePeriod = [0...2] \land Contract.insuranceType = "breakdown" \land Car.level = "luxury" \land Car.model = "X6"$
		dom		$Customer.age = [18...30] \land Customer.gender = "male" \land Customer.job = "unidentified" \land Customer.driverLicensePeriod = [0...2] \land Contract.insuranceType = "breakdown" \land Car.level = "luxury" \land Car.model = "X6"$
		$acc(dom, acc(t_1.\beta, t_2.\beta))$		$defined(Contract.customerID) \land defined(Contract.insuranceType) \land defined(Contract.expectedPickedupTime) \land defined(Contract.expectedReturnedTime) \land defined(Customer.customerID) \land defined(Customer.name) \land defined(Customer.age) \land defined(Customer.gender) \land defined(Customer.address) \land defined(Customer.licenseNumber) \land defined(Customer.driverLicensePeriod) \land defined(Customer.expiredDate) \land Contract.customerID = Customer.customerID \land instate(Car, booked) \land instate(Car, opened) \land defined(Contract.carID) \land defined(Contract.contractID) \land Contract.carID = Car.carID \land defined(Car.manufacturer) \land defined(Car.level) \land defined(Car.model) \land defined(Car.serialNumber) \land defined(Car.color) \land Customer.age = [18...30] \land Customer.gender = "male" \land Customer.job = "unidentified" \land Customer.driverLicensePeriod = [0...2] \land Contract.insuranceType = "breakdown" \land Car.level = "luxury" \land Car.model = "X6"$
		t_{new}		$Decompose(t_5)$
	$ReorderAfter(L, t_5, t_2)$			

5   Related Work

Bernhard Mitschang's group [2] proposed a (semi-)automated business process optimization platform based on actual execution data of the car loan process by using "best-practice" optimization patterns.

Furthermore, Dennis Wegener and Stefan Rüping [1] discussed the integrating data mining into business processes and its evaluation in BPR (Business Process Re-engineering) context based on BPMN and CRISP-DM (CRoss-Industry Standard Process for Data Mining). Rok Rupnik and Jurij Jaklič [10] proposed a deployment of data mining into operational business processes using JDM API (Java Data Mining Application Interface) and CRISP-DM version 1.0 [17]. They went further with this research line by proposing the reuse of

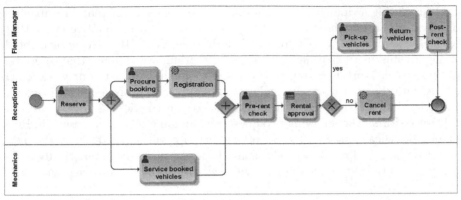

(a) Original process

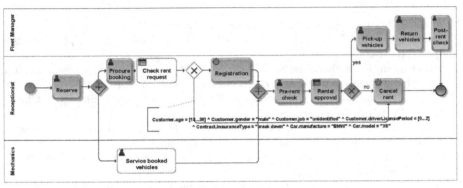

(b) Re-designed process

Fig. 2. Car rental business process represented in BPMN

successful data mining solutions in order to minimize manual coordination and adjustment [6].

In the context of our work, redesign of business processes is considered design-time changes or process re-engineering, which could be grouped as the process flexibility according to van der Aalst [4]. We devised algorithms for the elimination of redundant tasks and re-ordering tasks. We implemented our algorithms and partially validated them on the car rental example.

6   Conclusions

In this paper, we present a formal framework for redesigning business processes. In our work, the driver for process redesign comes from data mining that is performed on the operational data collected. We proposed algorithms for task elimination and task re-ordering, respectively. Pre- and post-conditions of business tasks are coupled with instantiation predicates, which will be checked for entailment against the consequent declared in the data mining rules. We illustrate our algorithms using a running example about the business of car rental companies. To reason about entailment, we rely on the notion of accumulative effect that was made popular in a technique called ProcessSEER [16]. We implemented our algorithms in Java using an research programming library that supports logic inference. Work is currently underway to further validate our work using real-life data. Further work also includes extending the proposed algorithms for additional redesigning mechanisms, e.g. tasks decomposition.

Discussions. The following two questions remain open in our research. How to methodologically shorten the list of classification rules generated by the data mining "engine"? How to precisely define business rules and semantically associate them to the list of classification rules shortened by the business analysts?

References

1. Wegener, D., Rüping, S.: On integrating data mining into business processes. In: Abramowicz, W., Tolksdorf, R. (eds.) BIS 2010. LNBIP, vol. 47, pp. 183–194. Springer, Heidelberg (2010)
2. Niedermann, F., Radeschütz, S., Mitschang, B.: Design-time process optimization through optimization patterns and process model matching. In: Proceedings of the 12th IEEE Conference on Commerce and Enterprise Computing, Shanghai, China, pp. 48–55. IEEE Computer Society (2010)
3. Fayyad, U.M., Piatetsky-Shapiro, G., Smyth, P.: From data mining to knowledge discovery in databases. AI Mag. **17**(3), 37–54 (1996)
4. van der Aalst, W.M.: Business process management: a comprehensive survey. ISRN Softw. Eng. **2013**, 1–37 (2013)
5. Mansar, S.L., Reijers, H.A.: Best practices in business process redesign: use and impact. bus. process manag. j. **13**(2), 193–213 (2007)
6. Wegener, D., Rüping, S.: On reusing data mining in business processes - a pattern-based approach. In: Muehlen, M., Su, J. (eds.) BPM 2010 Workshops. LNBIP, vol. 66, pp. 264–276. Springer, Heidelberg (2011)
7. van der Aalst, W., van Hee, K.: Business process redesign: a Petri-net-based approach. Comput. Ind. **29**(1–2), 15–26 (1996)
8. Koliadis, G., Ghose, A.: A conceptual framework for business process redesign. In: Halpin, H., Krogstie, J., Nurcan, S., Proper, E., Schmidt, R., Soffer, P., Ukor, R. (eds.) BPMDS 2009 and EMMSAD 2009. LNBIP, vol. 29, pp. 14–26. Springer, Heidelberg (2009)
9. Han, J., Kamber, M., Pei, J.: Data Mining: Concepts and Techniques, 3rd edn. Morgan Kaufmann, San Francisco (2011)
10. Rupnik, R., Jaklič, J.: The deployment of data mining into operational business processes. In Ponce, J., Karahoca, A. (eds.) Data Mining and Knowledge Discovery in Real Life Applications. I-Tech Education and Publishing (2009)
11. Yongchareon, S., Liu, C.: A process view framework for artifact-centric business processes. In: Meersman, R., Dillon, T.S., Herrero, P. (eds.) OTM 2010. LNCS, vol. 6426, pp. 26–43. Springer, Heidelberg (2010)
12. Kunchala, J., Yu, J., Yongchareon, S.: A survey on approaches to modeling artifact-centric business processes. In: Benatallah, B., Bestavros, A., Catania, B., Haller, A., Manolopoulos, Y., Vakali, A., Zhang, Y. (eds.) WISE 2014. LNCS, vol. 9051, pp. 117–132. Springer, Heidelberg (2015)

13. Bhattacharya, K., Gerede, C.E., Hull, R., Liu, R., Su, J.: Towards formal analysis of artifact-centric business process models. In: Alonso, G., Dadam, P., Rosemann, M. (eds.) BPM 2007. LNCS, vol. 4714, pp. 288–304. Springer, Heidelberg (2007)
14. Estañol, M., Queralt, A., Sancho, M.R., Teniente, E.: Artifact-centric business process models in UML. In: La Rosa, M., Soffer, P. (eds.) BPM 2012 Workshops. LNBIP, vol. 132, pp. 292–303. Springer, Heidelberg (2013)
15. Lohmann, N., Nyolt, M.: Artifact-centric modeling using BPMN. In: Pallis, G., et al. (eds.) ICSOC 2011 Workshops. LNCS, vol. 7221, pp. 54–65. Springer, Heidelberg (2012)
16. Hinge, K., Ghose, A., Koliadis, G.: Process SEER: a tool for semantic effect annotation of business process models. In: Proceedings of the 13th IEEE International Conference on Enterprise Distributed Object Computing, Auckland, New Zealand, pp. 49–58. IEEE Computer Society, September 2009
17. Shearer, C.: The CRISP-DM model: the new blueprint for data mining. J. Data Warehousing 5(4), 13–22 (2000)

Repairing Alignments: Striking the Right Nerve

Borja Vázquez-Barreiros[1]([✉]), Sebastian J. van Zelst[2], Joos C.A.M. Buijs[2],
Manuel Lama[1], and Manuel Mucientes[1]

[1] Centro de Investigación en Tecnoloxías da Informacíon (CiTIUS),
University of Santiago de Compostela, Santiago de Compostela, Spain
{borja.vazquez,manuel.lama,manuel.mucientes}@usc.es
[2] Department of Mathematics and Computer Science,
Eindhoven University of Technology, Eindhoven, The Netherlands
{s.j.v.zelst,j.c.a.m.buijs}@tue.nl

Abstract. Process Mining is concerned with the analysis, understanding and improvement of business processes. One of the most important branches of process mining is conformance checking, i.e. assessing to what extent a business process model conforms to observed business process execution data. Alignments are the de facto standard instrument to compute conformance statistics. Alignments map elements of an event log onto activities present in a business process model. However, computing them is a combinatorial problem and hence, extremely costly. In this paper we show how to compute an alignment for a given process model, using an existing alignment and an existing process model as a basis. We show that we are able to effectively repair the existing alignment by updating those parts that no longer fit the given process model. Thus, computation time decreases significantly. Moreover, we show that the potential loss of optimality is limited and stays within acceptable bounds.

Keywords: Process mining · Conformance checking · Alignments

1 Introduction

Today's information systems store an overwhelming amount of data related to the execution of business processes. *Process mining* [1] is concerned with the analysis, understanding and improvement of business processes based upon such data in the form of *event logs*. Three main branches form the basis of process mining: *process discovery, conformance checking* and *process enhancement*. Within process discovery the main goal is to discover a business process model based on an event log. Within conformance checking the main goal is to check whether a given process model conforms to the behavior recorded in an event log. Within process enhancement the main goal is to improve business processes, primarily (though not exhaustively) using the two aforementioned fields.

Alignments [2] have proven to be very effective for the purpose of conformance checking. In essence, an alignment aligns an event log to a process model. Based on

© Springer International Publishing Switzerland 2016
R. Schmidt et al. (Eds.): BPMDS/EMMSAD 2016, LNBIP 248, pp. 266–281, 2016.
DOI: 10.1007/978-3-319-39429-9_17

such alignment, a variety of analysis techniques can be applied resulting in different statistics describing the relation between the event log and the process model. A plethora of process mining techniques use alignments internally [3–7]. Replay-fitness and precision, two essential process mining quality dimensions [8], are computed on the basis of alignments [3,4]. In evolutionary process discovery [5,6], replay-fitness and precision are used to judge the quality of a newly generated process model. The Inductive Visual Miner [7] uses replay-fitness measures to visualize the flow of cases through a given process model.

The sheer complexity of computing alignments has its effects on the techniques that internally use them. Using alignments in combination with realistically sized event logs and process models, typically results in poor run time performance. However, some of the aforementioned techniques share an interesting common property, i.e. the potential use of *similar* process models. For example, within evolutionary process discovery a new generation of process models is created based on slight manipulations of the current generation of process models [9]. The Inductive Visual Miner allows the user to apply filtering techniques, resulting in a new, rather similar, process model for which we need to recompute alignments. Hence, the question arises whether we can use previously computed alignments as a basis for computing of new alignments.

In this paper we propose an *alignment repair method* that, given a process model and an existing alignment on a different process model, computes a new alignment for the given process model. The technique identifies fragments of the existing alignment that do not correspond to the given process model and replaces them with new alignment fragments that do correspond. Because the method only focuses on those alignment fragments that do not fit, i.e. the method strikes the right nerve, computation time decreases significantly. Moreover, we show that the loss of optimality is limited and stays within acceptable bounds.

The remainder of this paper is structured as follows. Section 2 explains event logs, process trees and alignments. Section 3 describes the alignment repair method in detail. In Sect. 4 we present an evaluation of the approach. Section 5 discusses related work and Sect. 6 concludes the paper.

2   Preliminaries

In this section we introduce the notion of event logs, process trees and alignments.

2.1   Sequences, Bags and Event Logs

We write a bag as $[e_1^{n_1}, e_2^{n_2}, ..., e_k^{n_k}]$ where element e_1 occurs n_1 times, with $n_1 > 0$. As an example, $B_1 = [a^3, b^5]$ denotes a bag consisting of 3 a's, 5 b's and 0 c's. Sequences are written as $\langle e_1, e_2, ..., e_n \rangle$. Sequence concatenation of sequences σ_1 and σ_2 is written as $\sigma_1 \cdot \sigma_2$. As an example consider the concatenation of sequences $\langle a, b \rangle$ and $\langle c, d \rangle$: $\langle a, b \rangle \cdot \langle c, d \rangle = \langle a, b, c, d \rangle$. The set of all possible sequences over some set of elements X is denoted as X^*, e.g. $\langle a, b \rangle \in \{a, b, c\}^*$. Given a set X and an element $e \notin X$, we write X^e as a shorthand for $X \cup \{e\}$.

Event logs often act as a primary input for process mining techniques and describe the actual execution of activities within a business process. In essence, an event log is a bag of sequences that consist of business process events. Consider Table 1 depicting a snapshot of an event log of a fictional loan application handling process. Let us consider all activities related to the case *3554*. First, John *Checks the application form*, after which Harold *checks the applicant's credit history*. Pete *appraises the property* after which Harold performs a *loan risk assessment*. Finally, Harry *assesses the eligibility of the client for the loan* and in the end he decides to *reject the application*. A sequence of events, e.g. the execution of the activities related to case *3554*, is referred to as a *trace*. Thus, from the *control-flow perspective*, i.e. the sequential ordering of *activities* w.r.t. *cases*, case *3554* can be written as ⟨ *Check application form, Check credit history, Appraise property, Assess loan risk, Assess eligibility, Reject application* ⟩.

Table 1. A fragment of an event log loosely based on a fictional loan application process [10], where each individual line corresponds to an event.

Case-id	Activity	Resource	Time-stamp
⋮	⋮	⋮	⋮
3554	*Check application form*	*John*	*2015-10-08T09:45:37+00:00*
3555	*Check application form*	*Lucy*	*2015-10-08T10:12:37+00:00*
3554	*Check credit history*	*Harold*	*2015-10-08T10:14:25+00:00*
3555	*Check credit history*	*Harold*	*2015-10-08T10:31:02+00:00*
3554	*Appraise property*	*Pete*	*2015-10-08T10:45:22+00:00*
3554	*Assess loan risk*	*Harold*	*2015-10-08T10:49:52+00:00*
3555	*Assess loan risk*	*Harold*	*2015-10-08T11:01:51+00:00*
3553	*Return application to client*	*John*	*2015-10-08T11:03:18+00:00*
3556	*Check application form*	*Lucy*	*2015-10-08T11:05:10+00:00*
3555	*Assess eligibility*	*Harry*	*2015-10-08T11:06:22+00:00*
3554	*Assess eligibility*	*Harry*	*2015-10-08T11:33:42+00:00*
3554	*Reject application*	*Harry*	*2015-10-08T11:45:42+00:00*
3557	*Check application form*	*Lucy*	*2015-10-08T13:48:12+00:00*
3555	*Prepare acceptance pack*	*Sue*	*2015-10-08T14:02:22+00:00*
⋮	⋮	⋮	⋮

2.2   Process Trees

A process tree [5,11] is an abstract hierarchical representation of a block-structured workflow net [12]. The leafs of a process tree are labeled with *activities*. The internal nodes are labeled with *operators*, used to specify the relation between their children. Formally, every node within a process tree describes a language, i.e., a set of sequences of activities. The language of the process tree itself is equal to the language of the root node of the process tree. Thus, the labels of the leafs of the process tree form the *alphabet* of a process tree's language. The operators describe how the languages of their children have to be combined.

There are five standard operator types [5] defined for process trees: the sequential operator ($\rightarrow$), the parallel execution operator ($\land$), the exclusive choice operator ($\times$), the non-exclusive choice operator ($\lor$), and the repeated execution (loop) operator ($\circlearrowright$). Operators can have an arbitrary number of children in any arbitrary order, except for the sequence and loop operators. For the sequence operator ($\rightarrow$), the number of children can be arbitrary, though the order of the children specifies the order in which they must be evaluated, i.e. from left to right. Loop nodes ($\circlearrowright$) always have three children: the left child is the *do* part of the loop, the middle child is the *redo* part, and the right child is the *exit* part. We refer to [5,11] for an exact, formal, language definition of process trees.

Figure 1 shows an example process tree P_1 with all five possible operators. The root node n_1 is labeled with a sequence operator ($\rightarrow$), hence we first evaluate its left-most child, n_2, which is a leaf labeled with activity a. Thereby, every sequence present in the language of P_1 starts with an a-activity. The second child of the root, n_3, is a sub-tree describing the parallel execution ($\land$) of activity b, with a non-exclusive choice ($\lor$) between activities c and d. The third child again refers to a single activity, labeled e. Finally a loop ($\circlearrowright$) will be executed. The *do* part of the loop consists of an exclusive choice ($\times$) between f and g. If we decide to *re-do* the loop, we execute activity h. Executing activity h enforces us to re-execute the exclusive choice between f and g. The *exit* part of the loop is labeled with activity τ. This activity is unobservable, i.e., it is not part of any sequence in the language of P_1.

Instead of recording the activity labels directly, we first record the sequence of leaf-nodes described by the process tree. As a second step, this sequence is projected on the activities associated with the leaf nodes. As an example consider the sequence of leaf nodes $\langle n_2, n_4, n_6, n_8, n_{12}, n_{14} \rangle$. Projected on the activity labels yields $\langle a, b, c, e, g, \tau \rangle$. The final label τ is an unobservable label and hence the sequence becomes $\langle a, b, c, e, g \rangle$. Due to the loop operator n_9, the language of P_1 is infinite. Some other exam-

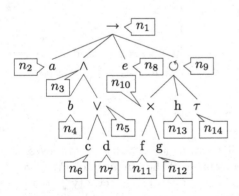

Fig. 1. Process Tree P_1.

ple sequences present in P_1's language are: $\langle n_2, n_7, n_4, n_8, n_{11}, n_{14} \rangle \equiv \langle a, d, b, e, f \rangle, \langle n_2, n_6, n_4, n_7, n_8, n_{11}, n_{13}, n_{12}, n_{14} \rangle \equiv \langle a, c, b, d, e, f, h, g \rangle$, etc.

2.3   Alignments

Alignments map events present in a trace to activities in a process model, i.e., the leaf nodes of a process tree. Alignments allow us to decide to what degree a given trace fits the language of a process tree. Given a trace σ_L and a sequence of leaf nodes σ_P, an alignment is a partial injective function mapping the elements of σ_L to the elements of σ_P. If the function is total and every non-mapped element in the range has a τ label, the trace σ_L *perfectly aligns* σ_P.

Throughout the paper we will use A to represent the activities of a trace and N to represent the set of leafs of some process tree P. Additionally, we use $\gg$ (i.e. $\gg \notin A$, $\gg \notin N$) to represent the *skip move*. The skip move represents an element from the domain (range) that is not part of the alignment, i.e. not part of the partial injective function. We represent an alignment as a sequence of pairs, combining the events in a trace with leaf nodes of a process tree, i.e., an alignment is represented as an element of $(A^{\gg} \times N^{\gg})^*$. A pair $(a, n) \in (A^{\gg} \times N^{\gg})$ is referred to as a *move*. The premise of an alignment is that: *(i)* the elements of the $A^{\gg}$-part respect the ordering of the events within the trace; and *(ii)* the elements of the $N^{\gg}$-part, projected onto their activity labels, form an element of the language of the process tree. We distinguish the following moves (a, n): *(i)* a synchronous move, if $a \in A$ and $n \in N$ s.t. n's label equals a or $a = \gg$ and $n \in N$ s.t. n's label is τ; *(ii)* a model move, if $a = \gg$ and $n \in N$; and *(iii)* a log move, if $a \in A$ and $n = \gg$. Other combinations are considered *illegal*. Given a trace σ and a process tree P, we write $\gamma_{(\sigma,P)}$ to denote an alignment of σ and P. We refer to [2] for a formal, Petri-net based definition of alignments.

Consider the trace $\sigma_1 = \langle a, b, c, e, f \rangle$ and the leaf sequence $\langle n_2, n_4, n_6, n_8, n_{11}, n_{14} \rangle$ of process tree P_1 of Fig. 1. Clearly $\langle n_2, n_4, n_6, n_8, n_{11}, n_{14} \rangle \equiv \langle a, b, c, e, f \rangle$ and thus if we (trivially) map $\sigma_1(1)$ *onto* n_2, $\sigma_1(2)$ *onto* n_4, ..., $\sigma_1(5)$ *onto* n_{11} we find a perfect alignment of σ_1 on the process tree.[1]

$$\gamma^1_{(\sigma_1,P_1)} = \frac{A^{\gg}}{N^{\gg}} \begin{array}{|c|c|c|c|c|c|} \hline a & b & c & e & f & \gg \\ \hline n_2 & n_4 & n_6 & n_8 & n_{11} & n_{14} \\ \hline \end{array}$$

Alignment $\gamma^1_{(\sigma_1,P_1)}$ is not the only possible alignment between σ_1 and P_1. It is also possible to map σ_1 to the leaf sequence $\langle n_2, n_4, n_6, n_8, n_{11}, n_{13}, n_{12}, n_{14} \rangle$:

$$\gamma^2_{(\sigma_1,P_1)} = \frac{A^{\gg}}{N^{\gg}} \begin{array}{|c|c|c|c|c|c|c|c|} \hline a & b & c & e & f & \gg & \gg & \gg \\ \hline n_2 & n_4 & n_6 & n_8 & n_{11} & n_{13} & n_{12} & n_{14} \\ \hline \end{array}$$

However, we favor $\gamma^1_{(\sigma_1,P_1)}$ over $\gamma^2_{(\sigma_1,P_1)}$ as it contains less $(\gg, n)$-typed moves.

In the previous example, the trace is an element of the language of P_1. If we consider the trace $\sigma_2 = \langle a, b, c, d, e, f, g \rangle$, this is not the case. Activity f and g can never co-occur in any sequence present in the language of P_1, unless activity h is in between them (due to the loop operator). For σ_2, we are able to construct (amongst others) these alignments:

---

[1] Note that n_{14} is a leaf with a τ label and maps to synchronous move $(\gg, n_{14})$.

$$\gamma^1_{(\sigma_2,P_1)} = \frac{A^{\gg}}{N^{\gg}} \begin{array}{|c|c|c|c|c|c|c|c|} \hline a & b & c & d & e & f & g & \gg \\ \hline n_2 & n_4 & n_6 & n_7 & n_8 & n_{11} & \gg & n_{14} \\ \hline \end{array}$$

$$\gamma^2_{(\sigma_2,P_1)} = \frac{A^{\gg}}{N^{\gg}} \begin{array}{|c|c|c|c|c|c|c|c|} \hline a & b & c & d & e & f & g & \gg \\ \hline n_2 & n_4 & n_6 & n_7 & n_8 & \gg & n_{12} & n_{14} \\ \hline \end{array}$$

For alignments $\gamma^1_{(\sigma_2,P_1)}$ and $\gamma^2_{(\sigma_2,P_1)}$, it is less obvious which one is favored over the other one, or, if both alignments are equally favorable. In general, given a trace and a process model, a multitude of alignments exist. We are however interested in an *optimal* alignment. In essence, an optimal alignment is an alignment that minimizes some cost function $\kappa : (A^{\gg} \times N^{\gg})^* \to \mathbb{R}^+$. For a given alignment γ, $\kappa(\gamma)$ often is computed deterministically as each type of move gets a cost assigned. A synchronous move typically has cost 0, whereas any illegal move has cost ∞. Costs of model/log moves are usually problem specific though usually greater than 0. Optimal alignments are computed for ordinary Petri nets with an initial marking and a (collection of) final marking(s), e.g. by using the A^* algorithm [2]. Trivially this applies to workflow nets and, as a consequence, process trees as well. Hence, for the purpose of this paper, we assume the availability of an *oracle* function o that, given a trace σ and a model P, produces an (optimal) alignment.

3   Repairing Alignments

As indicated, some process mining techniques share an interesting property, i.e., alignments need to be computed for multiple relatively *similar* process models. Henceforth, the main research question addressed in this paper is formulated as follows. Given a trace σ_L, a process tree P, a process tree P', and an alignment $\gamma_{(\sigma_L,P)}$, are we able to compute an alignment $\gamma_{(\sigma_L,P')}$ by reusing and repairing $\gamma_{(\sigma_L,P)}$?

3.1   Repairing Alignments: A Concrete Example

We illustrate alignment repair by providing an algorithmic sketch based on a running example. We use process trees P_1 and P_2 (Fig. 2) as a running example. Consider trace $\sigma_3 = \langle a, c, d, e, f, g, h, g, f \rangle$ and alignment $\gamma_{(\sigma_3,P_1)}$ of σ_3 on P_1:

$$\gamma_{(\sigma_3,P_1)} = \frac{A^{\gg}}{N^{\gg}} \begin{array}{|c|c|c|c|c|c|c|c|c|c|} \hline a & \gg & c & d & e & f & g & h & g & f & \gg \\ \hline n_2 & n_4 & n_6 & n_7 & n_8 & n_{11} & \gg & n_{13} & n_{12} & \gg & n_{14} \\ \hline \end{array}$$

Trace σ_3 is missing a b activity enforced by n_3 ($\wedge$). Moreover, $n_{10}(\times)$, does not allow for executing both the f- and g-activity without an h-activity in between. We are interested in changing the operator type of n_{10} as it yields two moves that include a $\gg$ symbol: $(g, \gg)$ and $(f, \gg)$. This is fixed by changing the operator type of n_{10} to either $\wedge$ or $\vee$. Consider process tree P_2 depicted on the right-hand side of Fig. 2, in which the operator type of n_{10} is changed to $\wedge$. For convenience, n_{10} and

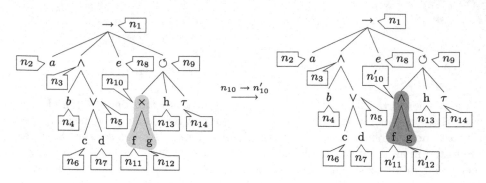

Fig. 2. Modification of node n_{10} in process tree P_1 (left), resulting in process tree P_2 (right). Scope of change S_1 is highlighted in P_1 (light gray), scope S_2 is highlighted in P_2 (dark gray).

its children n_{11} and n_{12} are relabeled to n'_{10}, n'_{11} and n'_{12}, respectively. The process to compute $\gamma_{(\sigma_3,P_2)}$ by reusing $\gamma_{(\sigma_3,P_1)}$ consists of three steps: *(i)* scope of change detection, *(ii)* sub-alignment computation, and *(iii)* alignment reassembly.

Step 1; Scope of Change Detection. The first step in reusing $\gamma_{(\sigma_3,P_1)}$ involves detecting the scope of change of P_1 w.r.t P_2 and vise versa. The scope of change itself is a process tree and is defined by the modified node and its children. For process trees P_1 and P_2 we highlighted scopes of change S_1 and S_2 in light and dark gray in Fig. 2. S_1 consists of nodes n_{10}, n_{11} and n_{12} of P_1. S_2 consists of nodes n'_{10}, n'_{11} and n'_{12} of P_2. Using this information, we need to detect what elements of $\gamma_{(\sigma_3,P_1)}$ belong to the scope of change of P_1, i.e. to leaf nodes of S_1. In this step, we linearly walk through all moves of $\gamma_{(\sigma_3,P_1)}$ checking for each $(a,n) \in \gamma_{(\sigma_3,P_1)}$, whether or not it belongs to scope S_1. We take the following (a,n) moves in consideration for scope S_1:

1. If n is a leaf of scope S_1, (a,n) belongs to S_1.
2. If $n =\gg$, and the previous move $(a',n') \in \gamma_{(\sigma_3,P_1)}$ belongs to S_1, then also (a,n) belongs to S_1.

Using the aforementioned rules, we start constructing the new alignment $\gamma_{(\sigma_3,P_2)}$. Every pair $(a,n) \in \gamma_{(\sigma_3,P_1)}$ *not belonging* to S_1 remains untouched and is copied in the exact same position into the new alignment $\gamma_{(\sigma_3,P_2)}$. On the other

Table 2. Schematic overview of the first step of the alignment repair algorithm.

$A^{\gg}$	a	$\gg$	c	d	e	f	g	h	g	f	$\gg$
$N^{\gg}$	n_2	n_4	n_6	n_7	n_8	n_{11}	$\gg$	n_{13}	n_{12}	$\gg$	n_{14}
	$\downarrow$	$\downarrow$	$\downarrow$	$\downarrow$	$\downarrow$			$\downarrow$			$\downarrow$
$A^{\gg}$	a	$\gg$	c	d	e	$\sigma_3^1 = \langle f,g \rangle$		h	$\sigma_3^2 = \langle g,f \rangle$		$\gg$
$N^{\gg}$	n_2	n_4	n_6	n_7	n_8			n_{13}			n_{14}

hand, we skip every move (a,n) *belonging to* S_1, yet for each of such move we remember the exact position in $\gamma_{(\sigma_3,P_1)}$. Moreover, whenever we encounter the first move (a,n) inside scope S_1, we create an intermediary sequence $\sigma_3^1 = \langle a \rangle$ (or $\sigma_3^1 = \epsilon$ if $a = \gg$). For every subsequent move (a',n') belonging to scope S_1 we update σ_3^1 to $\sigma_3^1 \cdot \langle a' \rangle$ (or $\sigma_3^1 \cdot \epsilon = \sigma_3^1$ if $a' = \gg$). However, if during this process we detect a move (a'',n'') belonging to S_1, which indicates a *new execution* of the process tree described by S_1, we create a new trace $\sigma_3^2 = \langle a'' \rangle$ (or $\sigma_3^2 = \epsilon$ if $a'' = \gg$). Note that this type of behavior might be present in the alignment due to loop structures in P_1. Let us consider Table 2. The first five elements of $\gamma_{(\sigma_3,P_1)}$ are outside of scope S_1, hence they will be copied directly into $\gamma_{(\sigma_3,P_2)}$. The sixth and seventh element, i.e. (f,n_{11}) and $(g,\gg)$, belong to scope S_1 and thus we remember their positions and create σ_3^1 based on them. The eight element is again outside of scope S_1 and will be directly copied into $\gamma_{(\sigma_3,P_2)}$. The ninth and tenth element, i.e. (g,n_{12}) and $(f,\gg)$, again belong to scope S_1. These two moves indicate a new execution of the process tree described by S_1 and hence, we create a new sequence σ_3^2 out of the two elements. Finally the last element of the alignment is again outside of scope S_1.

Step 2; Alignment Calculation. We now constructed a part of the new alignment $\gamma_{(\sigma_3,P_2)}$ together with a set of sequences, i.e. the lower part of Table 2. For each of these sequences, we additionally have a set of pointers referring to the elements of $\gamma_{(\sigma_3,P_1)}$ that generated the sequence. The next step of the repair consists of creating new chunks of alignments for the sub-sequences generated from the elements belonging to S_1. The core idea is that sequences σ_3^1 and σ_3^2 are both referring to behavior related to S_1. However, in P_2, S_2 is the replacement of S_1. Thus in the new alignment, this behavior can no longer be present and needs to be updated in context of S_2. As S_2 itself defines a process tree we use the *oracle* function o to compute two new alignments $\gamma_{(\sigma_3^1,S_2)}$ and $\gamma_{(\sigma_3^1,S_2)}$. The result of computing these alignments, together with S_2, are depicted in Fig. 3.

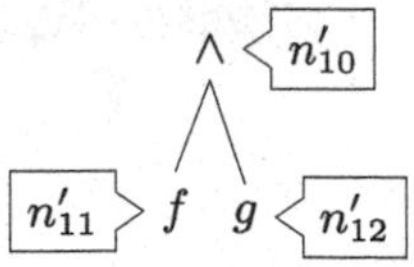

$$\gamma_{(\sigma_3^1,S_2)} = \begin{array}{c||c|c} A\gg & f & g \\ \hline N\gg & n'_{11} & n'_{12} \end{array} \qquad \gamma_{(\sigma_3^2,S_2)} = \begin{array}{c||c|c} A\gg & g & f \\ \hline N\gg & n'_{12} & n'_{11} \end{array}$$

Fig. 3. Process tree S_2 and the two alignments $\gamma_{(\sigma_3^1,S_2)}$ and $\gamma_{(\sigma_3^2,S_2)}$.

Step 3; Alignment Reassembly. The final step of the approach concerns placing back the newly created alignment fragments into the partially finished alignment, i.e. the bottom part of Table 2. Recall that we stored the position of every move in $\gamma_{(\sigma_3,P_1)}$ that belonged to S_1. For each such move (a,n) inside the scope in $\gamma_{(\sigma_3,P_1)}$ with $a \neq \gg$, we know that there is a move (a,n') in either $\gamma_{(\sigma_3^1,S_2)}$ or $\gamma_{(\sigma_3^2,S_2)}$. In our example consider (f,n_{11}) vs. (f,n'_{11}) in $\gamma_{(\sigma_3^1,S_2)}$, $(g,\gg)$ vs. (g,n'_{12}) in $\gamma_{(\sigma_3^1,S_2)}$, (g,n_{12}) vs. (g,n'_{12}) in $\gamma_{(\sigma_3^2,S_2)}$ and $(f,\gg)$ vs. (f,n'_{11}) in $\gamma_{(\sigma_3^2,S_2)}$. These type of moves serve as *anchor points* for placing the new alignment fragments into the partially finished alignment. If an alignment fragment does not contain any anchor point, i.e. caused by replay on an empty sequence, we are still able to place the new alignment back due to the fact that we have a one-to-one mapping between the old alignment moves and the (empty) sequence used in step 2. Once the third step is performed, we obtain an alignment of σ_3 on P_2, constructed by only calculating alignments on S_1 rather than P_1 as a whole. The final step resulting in $\gamma_{(\sigma_3, P_2)}$ is depicted in Table 3.

Table 3. Schematic overview of the third step of the alignment repair algorithm.

$A^{\gg}$	a	$\gg$	c	d	e	$\sigma_3^1 = \langle f, g \rangle$	h	$\sigma_3^2 = \langle g, f \rangle$	$\gg$
$N^{\gg}$	n_2	n_4	n_6	n_7	n_8		n_{13}		n_{14}
	$\downarrow$	$\downarrow$	$\downarrow$	$\downarrow$	$\downarrow$	$\downarrow$	$\downarrow$	$\downarrow$	$\downarrow$
$A^{\gg}$	a	$\gg$	c	d	e	f	g	h	g	f	$\gg$
$N^{\gg}$	n_2	n_4	n_6	n_7	n_8	n'_{11}	n'_{12}	n_{13}	n'_{12}	n'_{11}	n_{14}

In the example, intermediary sequences σ_3^1 and σ_3^2 directly correspond to a consecutive block of elements of $\gamma_{(\sigma_3, P_1)}$. Due to parallelism, either by nodes labeled with an $\wedge$ or an $\vee$ operator, this is not necessarily the case, i.e. the subsequences can be related to a set of alignment moves scattered around the original alignment. Note that due to the use of the *anchor moves*, we are still able to put the elements of the newly created alignments into a correct position.

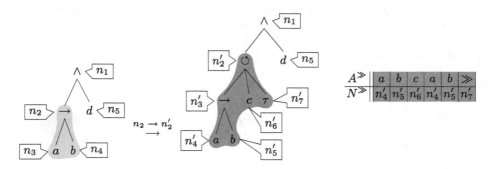

Fig. 4. Process Trees P_3 (left), P_4 (right), scope of change S_3 (light gray), scope of change S_4 (dark gray) and alignment $\gamma_{(\sigma_4^1, S_4)}$.

3.2   Optimality of Repaired Alignments

In the previous example, the repaired alignment is optimal. In general, we are not able to guarantee that the resulting repaired alignment is optimal. Consider the change between process trees P_3 and P_4 depicted in Fig. 4, trace $\sigma_4 = \langle a, b, c, d, a, b, c, a, b \rangle$ and optimal alignment $\gamma_{(\sigma_4, P_3)}$:

$$\gamma_{(\sigma_4, P_3)} = \frac{A^{\gg} \;\Vert\; a \;\vert\; b \;\vert\; c \;\vert\; d \;\vert\; a \;\vert\; b \;\vert\; c \;\vert\; a \;\vert\; b}{N^{\gg} \;\Vert\; \gg \;\vert\; \gg \;\vert\; \gg \;\vert\; n_5 \;\vert\; n_3 \;\vert\; n_4 \;\vert\; \gg \;\vert\; \gg \;\vert\; \gg}$$

Scope S_3 is defined by the sub-tree of P_3 starting from node n_2 whereas scope S_4 is the subtree of P_4 starting at node n_2'. Note that if we apply the algorithm as described in Sect. 3.1, the light gray colored moves belong to scope S_3. The algorithm will subsequently create the intermediary sequence $\sigma_4^1 = \langle a, b, c, a, b \rangle$ and let *oracle* function o compute the alignment $\gamma_{(\sigma_4^1, S_4)}$ as depicted in Fig. 4. Eventually, $\gamma_{(\sigma_4^1, S_4)}$ is combined with the first four moves of $\gamma_{(\sigma_4, P_3)}$, yielding $\gamma_{(\sigma_4, P_4)}$:

$$\gamma_{(\sigma_4, P_4)} = \frac{A^{\gg} \;\Vert\; a \;\vert\; b \;\vert\; c \;\vert\; d \;\vert\; a \;\;\; b \;\;\; c \;\;\; a \;\;\; b \;\; \gg}{N^{\gg} \;\Vert\; \gg \;\vert\; \gg \;\vert\; \gg \;\vert\; n_5 \;\vert\; n_4' \; n_5' \; n_6' \; n_4' \; n_5' \; n_7'}$$

Clearly $\gamma_{(\sigma_4, P_4)}$ is not optimal, i.e. consider $\gamma^*_{(\sigma_4, P_4)}$, which in fact is an optimal alignment of σ_4 on P_4.

$$\gamma^*_{(\sigma_4, P_4)} = \frac{A^{\gg} \;\Vert\; a \;\vert\; b \;\vert\; c \;\vert\; d \;\vert\; a \;\vert\; b \;\vert\; c \;\vert\; a \;\vert\; b \;\vert\; \gg}{N^{\gg} \;\Vert\; n_4' \;\vert\; n_5' \;\vert\; n_6' \;\vert\; n_5 \;\vert\; n_4' \;\vert\; n_5' \;\vert\; n_6' \;\vert\; n_4' \;\vert\; n_5' \;\vert\; n_7'}$$

Unfortunately, we are not able to assign the three log moves at the start of alignment $\gamma_{(\sigma_4, P_3)}$, i.e. $(a, \gg)$, $(b, \gg)$ and $(c, \gg)$ to scope S_3. Hence these moves remain untouched whereas they in fact should be mapped onto elements of the leafs of S_4. Therefore, this leads to a repaired alignment that is not optimal. Nevertheless, in Sect. 4 we show that the potential loss of optimality is limited and stays within acceptable bounds.

3.3   Feasibility of Repaired Alignments

One of the basic requirements of the presented approach is that, after reusing an existing (optimal) alignment, the repaired alignment itself is an alignment. Due to the rather informal nature of this paper, we provide an intuition on the fact that the repaired alignment is indeed an alignment, rather than a formal proof. Recall that the premise of an alignment is that: *(i)* the elements of the $A^{\gg}$-part respect the ordering of the events within the trace and *(ii)* the elements of the $N^{\gg}$-part, projected onto their activity labels, form an element of the language of the process tree. Let P and P' denote two process trees and let S and S' denote the scopes of change of P and P' respectively. Moreover let $\sigma \in A^*$ be a trace and let $\gamma_{(\sigma, P)}$ denote an (optimal) alignment of σ on P. Let $\gamma_{(\sigma, P')}$ denote the sequence of $(A^{\gg} \times N'^{\gg})$ moves resulting from a repair of $\gamma_{(\sigma, P)}$ based on P'.

Observe that by using moves (a, n) with $a \neq \gg$ as *anchor points*, we effectively keep all elements of the $A^{\gg}$-part in place w.r.t. each other. The *oracle* function o by definition respects the order of the elements of the $A^{\gg}$-part w.r.t. the generated intermediary subsequences. Thus we enforce that the elements of the $A^{\gg}$-part of $\gamma_{(\sigma, P')}$ respect the ordering of trace σ. Hence $\gamma_{(\sigma, P')}$ fulfills part *(i)* of the premise.

The intuition of part *(ii)* of the premise is a bit more involved. Let us consider the case in which there is no $\wedge$ or $\vee$ operator on the path from the root of P to the root of S. In this case any valid execution of S always results in a consecutive block of leaf nodes of S in $\gamma_{(\sigma,P)}$. For each of these consecutive blocks we know that at that point in time sub-tree S must have been active. In step two, for each of these consecutive blocks we create a new alignment fragment based on S'. The *oracle* function o guarantees that *(ii)* holds for these fragments. We then place the newly created chunks, corresponding to behavior of S', exactly at the points where S was active. If we assume that in this case *(ii)* does not hold, this contradicts either the *oracle* function or the fact that $\gamma_{(\sigma,P)}$ was an alignment in the first place, hence *(ii)* must hold. In case there is an $\wedge$ or $\vee$ operator on the path from the root of P to the root of S, we know that there might be interference of other parts of the tree w.r.t. S. The intermediate sequences can be potentially build up out of multiple chunks of alignment moves scattered around $\gamma_{(\sigma,P)}$. In this case, S was active throughout the whole span of the first chunk mapping to an intermediate subsequence up until the last chunk mapping to the same intermediate subsequence. Moreover, we know that within the span of S, we can reorder any leaf node of S with any other leaf node not in S, as long as we do not reorder any two leafs of S. The *oracle* function o provides us with the guarantee that *(ii)* holds for the new alignment fragments based on S'. Since we use the *anchor points* we know that we might only reposition leafs of S' w.r.t. leafs of P' that are not an element of S'. Leaf nodes of S' are however never shuffled. We know that at any position where we place the new (chunks of) fragments based on S' back, S' has to be active. Thus, also in this case, if we assume that in this *(ii)* does not hold, this contradicts either the *oracle* function or the fact that $\gamma_{(\sigma,P)}$ was an alignment.

4   Evaluation

To validate the usefulness of the presented technique, we answer two main questions: *(i)* What is the time needed to align a model and a log with the presented technique? and *(ii)* How close/far is the repaired alignment from the optimal alignment? In this section we answer these questions by *(i)* comparing the time needed for alignment repair with the time expended to compute a new, optimal alignment and *(ii)* by measuring the quality of the repaired alignments w.r.t. the new, optimal alignment.

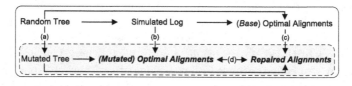

Fig. 5. Process followed during the experimentation.

4.1   Experimental Set-Up

Figure 5 shows a schematic overview of the experimental set-up. We generate an initial *random* process tree with a random size. Based on this model, we simulate a non-fitting event log, i.e. the event log contains noise, consisting of 2000 traces. We then calculate the optimal alignments of all traces in the event log w.r.t. the initial model. As a second step, we perform a set of random changes on the base model (step a in Fig. 5), generating a total of 150 *different* mutated process trees. We enforce that every mutated model is unique. The possible changes applied over the base model are: randomly adding a new node, randomly removing a node and randomly changing a node of the tree. Then, we calculate two different types of alignments for each mutated tree: the optimal alignments based on the simulated log (step b in Fig. 5) and the repaired alignments reusing the optimal alignments previously calculated on the base model (step c in Fig. 5). Finally, we compare both outputs (step d in Fig. 5).

Following this process, we created a set of 50 initial *random* trees with arbitrary sizes between 21 and 47 nodes. Thus, we applied the presented technique over $50 \times 150 \times 2000 \approx 1.5 \cdot 10^6$ alignments. We additionally checked whether the repaired alignment is indeed an alignment, which *was true in all cases*.[2]

4.2   Running Time

As the time needed to compute alignments varies significantly between runs, we grouped the results of the experiments based on the size of the *initial random process trees*. We created a bucket with initial trees of sizes between 21 and 28 nodes (12 trees in total), a bucket with sizes between 29 and 31 nodes (12 trees in total), a bucket with sizes between 32 and 34 nodes (13 trees in total) and a bucket with sizes greater than 35 nodes (13 trees in total).

Figure 6 shows the time comparison, using box plots, for each bucket of experiments. Due to the high dispersion of the data, on the right-hand side of Fig. 6 we also show the box plots zoomed into the domain 0–100 s.

In general we observe there is *no overlap* in the second and third quartiles of computing alignments based on the repair method versus computing an optimal alignment from scratch. This implies that in nearly all cases, the time needed to align a model and an event log by applying alignment repair outperforms computing a new optimal alignment. In this case, the time needed for alignment repair is directly related to the size of the scope of change which explains the rather high range of the right whiskers in the box plots for alignment repair. Clearly, if the change is performed in the root node of a process tree, the scope of change is the process tree as a whole. The time needed to apply the presented approach will be roughly equal to the time needed to compute the optimal alignment as there is no room to repair the old alignment. Thus, we conclude that using the presented technique, guarantees a lower, or, in worst case equal, running time compared with computing the optimal alignments between an event log and a process tree from scratch.

---

[2] All results can be found at https://svn.win.tue.nl/repos/prom/Packages/EvolutionaryTreeMiner/Branches/BorjaImp/experiments/bpmds2016/.

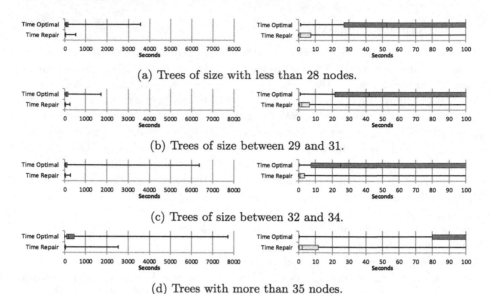

(a) Trees of size with less than 28 nodes.

(b) Trees of size between 29 and 31.

(c) Trees of size between 32 and 34.

(d) Trees with more than 35 nodes.

Fig. 6. Box plots showing the time needed to repair an alignment versus computing the optimal alignments for each bucket of experiments. The right-hand side shows the results zoomed into the domain 0–100 s.

4.3 Alignment Quality

As explained in Sect. 3.2, alignment repair does not guarantee optimality. It is not straightforward to assess how well the repaired alignment scores in terms of optimality. To judge the "rank" of the repaired alignment, i.e. how many other alignments are closer to the optimal alignment, we need to traverse all possible alignments of a trace and a process tree. This is rather involved from a run-time complexity point and hence hard to incorporate within the experiments.

We propose a *grade* measure, that grades the repaired alignment, based on the relative distance of the alignment w.r.t. the optimal alignment. To compute the distance, we first compute the cost of the optimal alignment γ^*. Additionally, we create an alignment γ^w, consisting of only $(a, \gg)$-moves and $(\gg, n)$-moves, such that the a-moves form the trace and the n-moves form a shortest possible valid sequence of leaf nodes. The γ^w alignment represents "the best of the worst" alignment. Finally, we calculate the cost of the repaired alignment γ^r. Based on the difference between the cost of γ^* and γ^w we compute the relative cost of γ^r. Let c^*, c^w and c^r denote the costs for γ^*, γ^w and γ^r. We grade the cost of γ^r as follows: $grade(\gamma^r) = 1 - \frac{c^r - c^*}{c^w - c^*}$. Clearly, $0 \leq grade(\gamma^r) \leq 1$. We used the following cost function $\kappa : (A^{\gg} \times N^{\gg}) \rightarrow \mathbb{R}^+$: $\kappa(a, n) = +\infty$ if a and n's label do not match, $\kappa(a, n) = 5$ if $a \in A$ and $n = \gg$, $\kappa(a, n) = 2$ if $a = \gg$ and $n \in N$, and finally $\kappa(a, n) = 0$ if either $a \in A$ a and n's label match, or, $a = \gg$ and n's label is τ.

Figure 7 shows the box plots for the computed average grades of the repaired alignments. As the figure shows, we always have a grade above 0.84, and in the top 75 % of all experiments is above 0.98. Thus, when the repaired alignments are not optimal, the difference with the optimal alignments is minimal. Hence, the loss of optimality is limited and stays within acceptable bounds.

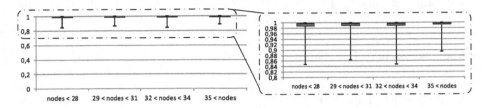

Fig. 7. Normalized *grade* of the repaired alignments.

Again, there is a close relation between the size of the scope of change and the potential loss of optimality. If the change is performed close to the root node, more moves of the previous alignment will belong to the scope of change. Consequently, the probability of retrieving an optimal alignment is higher. If the root of the point of change is the root node, we do have optimality.

5   Related Work

Alignments were introduced in [3]. In [2] an alignment computation approach is presented based on the A^* algorithm. The concept presented in this paper, i.e. solving a sub-problem rather than the whole problem, is similar to methods that aim at decomposition of process mining techniques [13–15]. In [15] the authors present a decomposition technique that partitions process models and event logs into smaller parts that can be analyzed independently. A similar approach for data-aware conformance checking problems is presented in [14]. The main difference compared to these works is the fact that the presented technique results in an alignment for the *whole trace* and the *whole process model*, whereas decomposition techniques typically provide solutions for sub-problems, which in aggregated form provide lower bounds rather than a full solution.

6   Conclusion

We presented a novel approach to compute alignments based on an existing alignment, instead of (re)computing the alignment from scratch. The approach has been validated with a set of random trees and event logs. The evaluation shows that our approach always retrieves an alignment in a significantly lower, or equal, time than computing optimal alignments. Furthermore, we show that the potential loss of optimality is limited and stays within acceptable bounds.

We plan to improve and/or extend the approach as follows. Depending on the type of operators in the tree, it might be possible to extend or shrink the scope of change, allowing to reduce the loss of optimality. Moreover, we plan to develop means to predict optimality, allowing us to decide at which point it would be necessary to compute the optimal alignment instead of reusing and existing one. Based on the achieved results, we plan to apply the presented technique, if applicable, in different process mining domains, e.g. within handling concept drift in stream-based process discovery [16].

Acknowledgments. This research was supported by the Spanish Ministry of Economy and Competitiveness (grant TIN2014-56633-C3-1-R, co-funded by the European Regional Development Fund - FEDER program), the Galician Ministry of Education under the projects EM2014/012, CN2012/151, GRC2014/030, and the DELIBIDA research program supported by NWO.

References

1. van der Aalst, W.M.P.: Process Mining - Discovery, Conformance and Enhancement of Business Processes. Springer (2011)
2. Adriansyah, A.: Aligning Observed and Modeled Behavior. Ph.D. thesis, Eindhoven University of Technology (2014)
3. van der Aalst, W.M.P., Adriansyah, A., van Dongen, B.F.: Replaying history on process models for conformance checking and performance analysis. Wiley Interdisc. Rew. Data Min. Knowl. Disc. **2**(2), 182–192 (2012)
4. Adriansyah, A., Munoz-Gama, J., Carmona, J., van Dongen, B.F., Aalst, W.M.P.: Measuring precision of modeled behavior. Inf. Syst. E-Business Manage. **13**(1), 37–67 (2015)
5. Buijs, J.C.A.M.: Flexible Evolutionary Algorithms for Mining Structured Process Models. Ph.D. thesis, Eindhoven University of Technology (2014)
6. Vázquez-Barreiros, B., Mucientes, M., Lama, M.: ProDiGen: mining complete, precise and minimal structure process models with a genetic algorithm. Inf. Sci. **294**, 315–333 (2015)
7. Leemans, S.J.J., Fahland, D., van der Aalst, W.M.P.: Process and deviation exploration with inductive visual miner. In: Proceedings of the BPM Demo Sessions 2014, Eindhoven, The Netherlands, 10 September 2014, p. 46 (2014)
8. Buijs, J.C.A.M., van Dongen, B.F., van der Aalst, W.M.P.: Quality dimensions in process discovery: the importance of fitness, precision, generalization and simplicity. Int. J. Coop. Inf. Syst. **23**(1), 305–322 (2014)
9. van Eck, M.L., Buijs, J.C.A.M., van Dongen, B.F.: Genetic process mining: alignment-based process model mutation. In: Fournier, F., Mendling, J. (eds.) BPM 2014 Workshops. LNBIP, vol. 202, pp. 291–303. Springer, Heidelberg (2015)
10. Dumas, M., Rosa, M.L., Mendling, J., Reijers, H.A.: Fundamentals of Business Process Management. Springer (2013)
11. Leemans, S.J.J., Fahland, D., van der Aalst, W.M.P.: Discovering Block-Structured Process Models from Event Logs - A Constructive Approach. In: Colom, J.-M., Desel, J. (eds.) PETRI NETS 2013. LNCS, vol. 7927, pp. 311–329. Springer, Heidelberg (2013)

12. van der Aalst, W.M.P.: The application of petri nets to workflow management. J. Circuits Syst. Comput. **8**(1), 21–66 (1998)

13. van der Aalst, W.M.P.: Decomposing petri nets for process mining: a generic approach. Distrib. Parallel Databases **31**(4), 471–507 (2013)

14. de Leoni, M., Munoz-Gama, J., Carmona, J., van der Aalst, W.M.P.: Decomposing alignment-based conformance checking of data-aware process models. In: Meersman, R., Panetto, H., Dillon, T., Missikoff, M., Liu, L., Pastor, O., Cuzzocrea, A., Sellis, T. (eds.) OTM 2014. LNCS, vol. 8841, pp. 3–20. Springer, Heidelberg (2014)

15. Munoz-Gama, J., Carmona, J., van der Aalst, W.M.P.: Single-entry single-exit decomposed conformance checking. Inf. Syst. **46**, 102–122 (2014)

16. Burattin, A., Sperduti, A., van der Aalst, W.M.P.: Heuristics Miners for Streaming Event Data. CoRR **abs/1212.6383** (2012)

Process Variability

A Process Variant Modeling Method Comparison: Experience Report

Banu Aysolmaz[1(✉)], Ali Yaldiz[2], and Hajo Reijers[1]

[1] Vrije Universiteit Amsterdam,
De Boelelaan 1105, 1081 HV Amsterdam, The Netherlands
{b.e.aysolmaz,h.a.reijers}@vu.nl
[2] 4S Information Technologies, Ankara, Turkey
ali.yaldiz@4s.com.tr

Abstract. Various process variant modeling methods have been introduced in the literature to manage process diversity in a business context. In industrial settings, it is difficult to select a method suitable for the needs and limitations of the organization due to the limited number of examples and guidelines. In this paper, we report our experiences on variant modeling in a process management consultancy company. The company experienced difficulties in maintaining and reusing process definitions of their customers and decided to evaluate variant modeling methods as a solution. We selected two methods, the Decomposition Driven and the Provop, to develop variant models of seven software project management processes from five customers. We evaluated the results together with company experts. This study contributes to the field by providing real-life examples of two variant modeling methods, a comparison of the results with these methods and a guideline for choosing a method under comparable conditions.

Keywords: Business process modeling · Process variant modeling · Decomposition driven method · Provop

1 Introduction

In enterprises, business process modeling (or process modeling for short) is of great importance to reveal processes and develop business process management systems (BPMS). In process modeling, one of the problems that analysts encounter is the need to deal with process variability. Due to the diversity in business contexts, variants of the same process may be modelled and used in multiple cases in the same organization [1]. This diversity may be caused by various factors such as differences in delivered products, customer types, and divergent business requirements in countries. When such factors are present, consideration of process variants during process modeling is inevitable [2]. However, in the design of a process model, it is a challenging task to either maintain variants of the same process separately while managing the relations between them or integrate the process variants into a single model while preventing complexity and redundancy [2].

© Springer International Publishing Switzerland 2016
R. Schmidt et al. (Eds.): BPMDS/EMMSAD 2016, LNBIP 248, pp. 285–300, 2016.
DOI: 10.1007/978-3-319-39429-9_18

To overcome such difficulties, various methods have been proposed to incorporate variant management into the phases of the business process management (BPM) life cycle [3]. These methods provide solutions for different cases. However, in real-life settings, it is difficult for an organization to make the proper choice between variant modeling methods. It is hard to find studies from the literature on the evaluation and comparison of methods in practice as well as any guidelines to make a method selection [3]. The study presented in this paper stems from the difficulties observed within 4S Information Technologies (4S for short), a company that provides consultancy services to its customers to analyze and improve their processes and develop BPMSs using HP PPM tools [4]. For each customer, 4S defines a new variant of a process, such as software project management, demand management, software change request management, risk and issue management, etc. 4S maintains separate process definitions and artefacts for each variant, yet the interrelations between the variants are not tracked. As a result, 4S cannot systematically reuse its process knowledge for creating a new variant for a new customer. The same problem applies for maintenance, as they need to update each variant independently without the opportunity to reuse the effort. For these reasons, 4S was motivated to implement a variant management method to more efficiently apply its knowledge in process analysis, design and improvement activities.

In accordance with the needs of 4S, the aim of this study is to implement and compare process variant modeling methods in a real-life setting. For this purpose, we selected two different, well-accepted process variant modeling methods focusing on the analysis and design of process variants: the Decomposition Driven method [5] and the Provop Method [1]. We applied these methods to 7 software project management processes of 5 4S customers. A team of 6 employees from the company participated in the study that was led by one of the authors. The team evaluated the application of the two methods in terms of effort spent, structure and flexibility of the outputs for maintenance and utilization in new projects. On the basis of this evaluation, we present a guideline that companies may follow when they face a similar situation.

The rest of the paper is structured as follows. Section 2 presents the design of this study describing the organization, the need, method selection, the purpose and the plan. In Sect. 3, we explain how we applied the Decomposition Driven method and in Sect. 4, the Provop method. Section 5 includes an evaluation of the results together with guidelines for method selection. Section 6 concludes the study.

2 The Design of the Process Variant Modeling Study

2.1 The Organization and the Need for Variant Modeling

4S is a consultancy company that provides process analysis, improvement and automation services to its customers using HP PPM product [4]. HP PPM provides a flexible workflow development environment specializing in project and demand management processes. 4S has customers from various countries and industries focusing on different process areas. Usually, 4S analysts need to rely on their own expertise to discover other activities and improve the existing process. They cannot systematically exploit process knowledge obtained from previous similar companies for new customers. Based on the problem, the need for using a process variant modeling method for 4S can be summarized as follows:

- When they start to work with a new customer, 4S analysts need to combine their knowledge on previous customers as a baseline for understanding the new as-is process and suggesting improvements. Analysts would be better off if they would have an integrated model, which they can practically use as a jump start in the project initiation phase.
- Through the steps of process analysis, improvement and design, 4S analysts design various processes for customers. Even when they start developing a process based on a previously encountered process, the knowledge of such related processes and the connections hereto are soon lost. Analysts cannot benefit from one another's experiences as it is hard for them to go over each process to find out if it is relevant for a new case. The same problem persists through process enactment phase; as developers cannot easily find out similar automated processes and activities for example, to reuse their form design and flow logic.
- When an improvement or update is needed, 4S needs to go over each customer's processes to find out which ones are affected and where updates are needed. This requires a lot of effort and can introduce errors due to manual review process.

2.2   Process Variant Modeling Method Selection

4S needs a process variant modeling method to manage customer models in an integrated way, utilize the knowledge in the following projects and enhance maintainability of multiple process models. Process variant modelling approaches have been proposed in the literature over a spectrum of *single* to *multi* model solutions [5]. On the one end of the spectrum, multi-model approaches capture every possible variant of a process as a separate model. Using such an approach, redundancy and maintainability problems are introduced, which is basically the problem that comes with not managing process variants at all [6]. On the other end of the spectrum, methods that model all variants in an integrated single model produce integrated models for multiple variants. The resulting decrease in the total number of process elements and the improvement of maintainability is balanced against increased complexity and comprehension problems. As a result of benefits, single model approaches are more popular in the industry. Considering the situation at hand, we focused on single-model approaches as well.

Single model approaches apply different techniques to integrate multiple process variants into a single source and use the single source to configure a specific process variant. Such techniques include questionnaire-based models, feature models, goal models, and decision tables [3]. Some single model approaches decrease complexity via providing only delete and condition selection operations based on a comprehensive base model. In 4S case, best practice model developed based on PMBOK guide was used as the starting point as it is common for various domains [3]. PMBOK is a book that provides a set of guidelines to define and implement project management processes such as scope, time and quality management. However, the best practice model of 4S is not inclusive of all activities – it is rather a brief process model including must-have activities. For this reason, we needed a method that has more flexibility to define process variants. Another criteria for method selection was on the need of a variant modeling tool. For some approaches, a tool that has specific process variant modeling features are required to properly benefit from the approach [7]. Due to the concern of increase in effort by adding a new tool to company repository and training needs, 4S eliminated the methods that need a specific tool.

Considering these needs and limitations, we identified an initial list of process variant modeling methods based on existing literature reviews [3] and our review of the related work. We discussed potential pros and cons of these methods with the 4S team and made a joint selection. As stated, we selected the Decomposition Driven method and the Provop method.

The Decomposition Driven method was selected because it provides flexibility for certain parts of the model. By means of step-wise decomposition, users can choose to model some of the sub-processes together and some others separately [5]. Moreover, the team specifically appreciated how the method does not only approach identification of variants mechanically but it considers the wider business environment via business drivers and syntactic drivers inherent to the processes. In turn, the Provop method was selected due to its robust mechanism to treat all variants equally and create a big model. The usage of the list of options to mechanically end up in new variants in a plug and play logo-like feeling was seen as another advantage. The team focused on selecting methods that have a different approach for variant modeling. In this way, a comparison of the benefits of different process variant modeling approaches for future use would be feasible.

2.3   Purpose of the Study

Based on the identified problems and the needs of 4S, we formulated the following questions:

- How can we develop process variant models for a process where different process models are developed for diverse companies although they share the same best practices?
- How does the application of two variant modeling methods, the Decomposition Driven and the Provop methods, compare for flexibility in terms of reusing the knowledge to define processes for new customers and maintain all variants in case of a change in one process?
- What factors are to be considered for an organization to select a proper variant modeling method based on its setting, needs and constraints?

2.4   Process Variant Modeling Plan

4S decided to use software project management processes of five customers for this study. Four of the companies are from Turkey and other is a Turkish branch of an international company. For all of them, their software project management processes are defined based on PMI's PMBOK guide [8]. One of the authors of this paper who is affiliated with 4S, worked as a leader of the team in the company that implements the methods and evaluates the results.

Although 4S uses the PMBOK guide as the baseline, the best practices provided just the essential steps of a project management process. The processes were defined as workflow definitions on HP PPM, but process models were not developed for analysis purposes. We converted the low level workflow models to process models in BPMN notation through discussion sessions with the experts. We aggregated workflow tasks to higher level activities in BPMN. The experts found it easier to observe and define relations between process variants using the BPMN models. For this reason, we decided to use these models in variant modeling activities. For each variant, we developed a high level software project management process. We created a relation table for the corresponding workflow tasks for each BPMN activity. In this way, we achieved more comprehensible process definitions where the experts could better observe the relations between the variants. Still, we are able to map process model activities to workflow tasks via the relation table. This enabled the experts to analyze workflow definitions together with variant models after the study is completed. A summary of the companies and their process metrics can be found in Table 1. Process models for all process variants can be seen in [9].

Table 1. Metrics for software project management processes of 4S customers

Process	Field	Number of workflow tasks	Number of workflow gateways	Number of BPMN activities	Number of BPMN gateways
Company 1	Annuity Insurance	15	7	10	4
Company 2	Insurance	40	9	12	6
Company 3	Banking	21	14	9	0
Company 4-1	Banking	48	7	14	6
Company 4-2	Banking	8	2	9	2
Company 5-1	Telecom	46	11	11	2
Company 5-2	Telecom	44	8	11	2
Average		31.7	8.3	10.9	3
PMBOK best practice		13	0		

The described research project was initiated based on the need in 4S as defined in previous sections. After the analysis of related work, elimination took place of multi-model approaches and approaches that require usage of a specific variant modeling tool and that conduct automated process discovery. Subsequently, the approaches mentioned earlier were selected. Upon the selection of the methods, the following steps were planned:

- Identify the process to apply selected process variant modeling methods: The team selected the software project management process, which is the most frequent process that they provide consultancy for their customers.
- Identify the context for application: Five customers were identified that are representative for different industries. Two of the customers implement two variants of software project management process.
- Define process models for each variant: We developed process models in BPMN notation for each customer as described at the beginning of this section.
- Apply the Decomposition Driven and the Provop methods to develop process variant models: The team conducted the relevant steps for applying the two methods as described in the following sections. Two methods were applied in parallel to prevent the effect of the learning curve.
- Evaluate the process of method application and outputs: The team collected data on the effort on each method and compared the outputs. In addition to comparing the outputs and facts, we conducted interviews to understand how the experts interpret the usability, complexity and efficiency.
- List the guidelines to choose proper method: The team identified the benefits and disadvantages brought by the two methods and how can one select the proper method with respect to priorities and benefits expected.

3   Applying the Decomposition Driven Method

The method starts with the definition of a main top-level process [5]. Then, each activity in the main process is defined in detail in a sub-process. Later, the sub-processes is further decomposed into sub-processes until there is no meaningful decomposition possible. At every level, the so-called variation map is created which contains activities and relations necessary to configure every variant. In the following sections, we describe the conduct of each step as prescribed by the method [5].

3.1   Step 1 – Model the Main Process

We started to apply the method by developing a main software project management process that acts as a process map applicable for all variants. The high level process can be seen in Fig. 1. Only one activity, "Plan Resources" was added and the remaining activities were directly used from the best practice. While modeling the main process, we also investigated and summarized each company's existing processes in order to point out how they add value to the process.

3.2   Step 2 – Identify Variation Drivers

An outstanding feature of the Decomposition Driven method is the consideration of business and syntactic drivers to understand the emergence of variations and using them to flexibly develop the models [5]. Business drivers are determined based on factors such as: resources used, products and services produced, customers, countries. In our case, we focused on how the high level activities in the main process are performed and possible causes of variation. We observed that the main cause of variation is the variety of customers. Another driver is identified as location of the services. This driver is used to differentiate the processes of Company 4, leading to variations of national and international services.

Syntactic drivers are the second type of drivers which diversify the way multiple variants produce their outcomes. They are defined based on the similarity of the process models of the variants. The method allows consolidation or separation of variant models due to syntactic drivers. In 4S, we manually assessed the similarity of process variant models with respect to the main process modeled in Fig. 1 [9]. We conclude that there is no explicit syntactic driver, as the main process can be used to reach the variant models by mostly adding nodes and alternative paths to the main process.

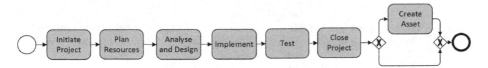

Fig. 1. Software project management high level process

3.3 Step 3 – Assess the Relative Strength of Variation Drivers

In this step, variation drivers are analyzed to specify their priority as well as their effect on defining variants. The business driver with high priority is the variety of customers in our case. Additionally, we have the driver for location of services. This driver is used to define variations for Customer 4: International vs. national services. Although Company 5 had two processes, only one variant is specified as there is no applicable, distinguishing driver.

3.4 Step 4 – Identify the Variations in Each Subprocess

In this step, we populated a variation matrix for each subprocess of the main process, as seen in Table 2. To generate this matrix for every activity in the high level process, we identified subprocesses for each driver. We then named each different subprocess and marked the subprocesses used by every driver in the matrix. For example in Table 2, *Simple Initiation* variant of *Initiate Project* subprocess is used by Company 3 and Company 4-2. This subprocess includes simple project definition activities on the system. *Complex Initiation* subprocess used by Company 1 includes a wider extent of activities such as approval of scope, project manager assignment and quality control initiation. The activities in subprocesses and all similarity decisions can be seen in [9].

Table 2. Variation matrix showing varying activities of first level subprocesses

	Initiate project	Plan resources	Analyze & design	Implement	Test	Close project	Create asset
Company 1	Complex Initiation	Moderate Planning	Basic Analyze and Design	Basic Implementation	Basic Test	Detailed Closure	
Company 2	Moderate Initiation	Complex Planning	Detailed Analyze and Design	Detailed Implementation	Detailed Test	Complex Closure	Asset Creation
Company 3	Simple Initiation	Basic Planning	Basic Analyze and Design	Basic Implementation	Basic Test	Basic Closure	Asset Creation
Company 4-1	Detailed Initiation	Simple Planning	Detailed Analyze and Design	Detailed Implementation	Detailed Test	Fast Closure	
Company 4-2	Simple Initiation	Fast Planning	Basic Analyze and Design	Basic Implementation	Basic Test	Simple Closure	Asset Creation
Company 5	Basic Initiation	Detailed Planning	Detailed Analyze and Design	Detailed Implementation	Detailed UAT	Moderate Closure	

3.5   Step 5 – Perform Similarity Assessment of Variants for Each Subprocess

In this step, we performed a similarity assessment by analyzing each subprocess of the variation matrix in Table 2. We asked the experts to identify the similarity of activities in the subprocesses for each driver. To evaluate the similarity, the experts focused on how those activities are performed. For this, they investigated the information on data used and produced while performing activities, the number of workflow steps, and the role of performer to investigate the similarities between subprocesses. As a result, activities in different subprocesses that have high similarity were marked. For example, *initiation approval* activity in *Moderate Initiation* variant was indicated to have high similarity with *initiation announcement* of *Detailed Initiation* variant.

3.6   Step 6 – Construct the Variation Map

As outputs from Step 4 and Step 5, we have the variants of subprocesses for each activity in the high level process and a list of similar activities in the subprocesses. We mapped these variants in the variation map as seen in Fig. 2. We used the decision framework of the Decomposition Driven method to decide merging of activities in the variant map [5]. For example, *Moderate Initiation* and *Detailed Initiation* variants were merged as they were assessed to be similar. For both *Plan Resources* and *Close Project* activities, a different subprocess was defined for each variant. Only two variants among six were assessed to be similar for both activities. Rest of the subprocesses had very strong drivers and were assessed to be not similar. Thus, there were five variations of these activities as seen on the variation map. The details on activities in the subprocesses, the similarity decisions for activities and the merging of subprocesses can be seen in [9].

3.7   Step 7 – Configure a Specific Process Variant

The generated variation map acts as a reference model to observe both the process map and help the experts to arrive at possible variations by means of the flow defined by gateways. This model does not include knowledge of a specific variant. Thus, if one wants to configure a process variant, she needs to understand that specific variant and go through the variation map to select relevant activities. This selection is done for Company 4 as shown with darker colored activities in Fig. 2. We manually verified that we can generate all our variants as syntactically correct and sound.

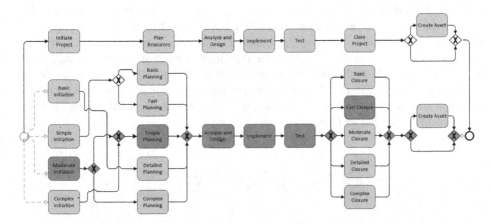

Fig. 2. Variation map for software project management process

After this step, the Decomposition Driven method suggests the iteration of all steps for the subprocesses of the main process. We applied the Decomposition Driven method completely in the first level of decomposition in 4S. Moreover, we identified the activities to be placed in each subprocess and discussed a sketch of the variation maps with the experts. In this way, the experts were able to observe how the Decomposition Driven method provided a flexible way of variant modeling in different granularity levels. For example. For "Implement" process, variation in the high level is not found necessary. However, it is observed that variants of this subprocess need to be handled considering other business drivers such as project type.

4   Applying the Provop Method

The Provop method focuses on creating a single base process model which includes adjustment points and their related sets of options [6]. The options include a set of atomic operations such as insert, delete, move and modify; which are used to configure the base model to reach a certain variant. Defining the set of operations options provides a reusable mechanism to define common operations for multiple variants. This mechanism decreases the complexity and increases controllability to configure a variant. Moreover, the Provop method can support automated variant configuration by defining context-aware configuration options.

4.1   Step 1 – Design a Base Process

The Provop method offers different policies to identify the base process on which the process variants are configured. One can either use the standard reference process used within the particular industry, use the most frequent process variant, design a version that has minimal average distance to all variants, or create a superset or intersection of all process variants. In our case, we applied a combination of these policies. First, the standard PMBOK reference model is taken as the starting point. Next, we extended this model by consideration of the policy 2, that is the variant of Company 1 which is the most frequent process worked on in 4S. We utilized policy 3 to identify process elements so that it will require the least number of operations in total to reach process variants while we also included activities at the intersection of all variants as suggested by policy 5. As a result, the base model evolved from the initial best practice model in Fig. 3 to final version in Fig. 4. Here we can indicate that it was relatively easy for us to design the base process, as we already know variant processes beforehand and we had a relatively simple and linear high level process.

Fig. 3. Best practice process model

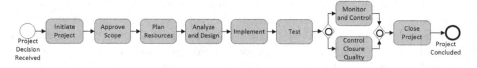

Fig. 4. Final base process model

4.2 Step 2 – Define Adjustment Points

The next action is to determine the explicit positions of the adjustment points that specify where the options can be applied on the base model. In this step, we analyzed the base process model and identified the adjustment points necessary to be able to generate all process variants provided in [9]. The final model with adjustment points can be seen in Fig. 5.

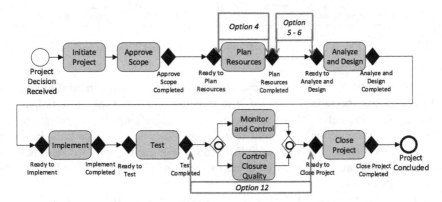

Fig. 5. Base process model with adjustment points (Color figure online)

4.3 Step 3 – Design and Model the Options

In this step, options of the process should be designed and modelled. To this end, the possible change operations for generating the variants based on the base process are investigated. Then, the conditional branches in the model are examined in order to determine that they are only variant-specific or included in all variant models. Granularity of options, the number of operations combined to define options, is important to enhance reusability and maintainability of options while keeping the number of options minimal [1]. As an example we can consider Option 1 (that contains the operation of "Delete Approve Scope") and Option 7 (that contains the operation of "Insert Control Plan Quality") seen in Fig. 6. To configure company 3 variant, we could create an integrated option including the operations in Option 1 and 7. However, in that case, we would need to define another option for the operation of "Delete Approve Scope" for company 5 variant. This would increase the number of operations and decrease reusability and understandability. Therefore, we divided this option into two as Option 1 and Option 7. 14 options are identified in total, which include 17 operations. An example set of operations are shown in Fig. 6.

4.4 Step 4 – Configure Variants

For variant configuration, the Provop suggests the usage of three substeps. First, relevant options need to be selected to configure the relevant process variant. This can be

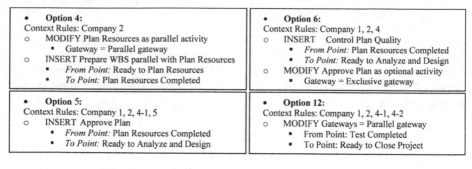

• **Option 4:**
Context Rules: Company 2
o   MODIFY Plan Resources as parallel activity
    ▪   Gateway = Parallel gateway
o   INSERT Prepare WBS parallel with Plan Resources
    ▪   *From Point:* Ready to Plan Resources
    ▪   *To Point:* Plan Resources Completed
	• **Option 6:**
	Context Rules: Company 1, 2, 4
	o INSERT Control Plan Quality
	    ▪ *From Point:* Plan Resources Completed
	    ▪ *To Point:* Ready to Analyze and Design
	o MODIFY Approve Plan as optional activity
	    ▪ Gateway = Exclusive gateway

• **Option 5:**
Context Rules: Company 1, 2, 4-1, 5
o   INSERT Approve Plan
    ▪   *From Point:* Plan Resources Completed
    ▪   *To Point:* Ready to Analyze and Design
	• **Option 12:**
	Context Rules: Company 1, 2, 4-1, 4-2
	o MODIFY Gateways = Parallel gateway
	    ▪ From Point: Test Completed
	    ▪ To Point: Ready to Close Project

Fig. 6. Example set of options for company 2 variant

done by asking users to manually choose specific variants, which is hard if there are a lot of options and specialized knowledge is required. To overcome the problem, the Provop suggests the definition of context rules by identifying, for each option, the context in which the options are applicable. In our case, the available knowledge on business drivers became useful to define the context. For each option, we identified the set of variants that are to be configured via this option. This can be seen in Fig. 6 as context rules.

Another point to be considered while applying the options is the possible constraints with the options. For example, there may be implication relation between options, an option implying the usage of another one [1]. We had an order constraint for options 5 and 6, as option 5 always needs to be applied before 6. We observed that the modelers need to pay special attention for constraints especially for options effective on the same adjustment point pairs.

In conformance with the constraints, we manually apply the set of options shown in Fig. 6 to the base process as indicated with red markers on Fig. 5 to achieve the variant process of company 2 as in Fig. 7. In the following section, we evaluate the results of applying the two methods and provide a guideline for selection.

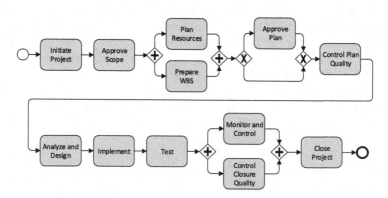

Fig. 7. Company 2 process model after configuration

After the definition of the base process model with adjustment points for the software project management process, we needed to analyze variants for subprocesses of the activities in the base process model. We were not able to identify specific guidelines for applying the Provop method in a hierarchical process structure. We plan to define subprocesses for each variant and conduct the same set of steps to develop base process models of each subprocess. However, we need to consider that new activities may be added to the high level process via options. In this case, we plan to define base process maps for the subprocesses under those activities as well. This will introduce problems in reading, as the user is not able to see and associate such a subprocess in the base process model. We also observed that special attention is needed to prevent conflicts among options for different levels of granularity.

5   Evaluating the Results

As 4S is evaluating two variant modeling methods to implement in all its projects in the future, it is important to identify the method that is practical to apply and meets the needs for reusing process knowledge and maintaining multiple variants. For practical reasons, we evaluated the effort spent on applying the two methods, structure of the outputs and flexibility of using the outputs in new projects and maintaining them when there is an update in one of the variants.

22 h were spent in 5 sessions for the Decomposition Driven method, whereas 15 h were spent in 4 sessions with the Provop method. The experts appreciated the idea of incorporating the business context to identify sources of variation. However, the variety of customers was already an explicit business driver for 4S from the start of the study. The experts think that their extra effort for the Decomposition Driven method will pay off when they implement the method for low level subprocesses and other process types with potentially more varied business drivers.

Comparing the structure of the resultant models, variation map of the Decomposition Driven method has 25 activities, 10 gateways and 50 edges. The Provop method produced a simpler model with only 9 activities, 2 gateways, 13 edges and 11 adjustment points as customized elements. The Decomposition Driven method seems to produce a bigger and more complex model (due to edges/activities ratio). However there is an extra artefact, list of options, required to read and customize the Provop base process model. The experts indicated that it was easier for them to read the Provop's base process model and "picturize" how the adjustments may be conducted even without seeing the option list. They found it non-intuitive to interpret the variation map of the Decomposition Driven method, e.g. in particular with respect to finding out where to start reading the process and how to configure a specific variant. This point reduces the flexibility to maintain existing variants. On the other hand, the experts found it more flexible to use the variation map for defining a new process, as they can see all options together with constraints on the map.

The experts appreciated the flexibility of the Decomposition Driven method for modeling variants in different granularity levels. It is conventional to develop separate models when there are variants of subprocesses which are very different from each other although the higher level process is similar. In this way, it is possible to balance complexity and comprehensibility on multiple process models. For example, no variation was needed at this level for "Initiate Project" and "Close Project" activities. The variations in lower level activities are to be evaluated in variant modeling of their subprocesses.

The 4S experts found it easier to use the Provop base model for configuration of process variants. Similar change operations grouped in the Provop decreased the complexity to generate a variant and made it easier to configure a variant without much knowledge of the customer. Variation map of the Decomposition Driven method does not provide any information on variants, one needs to have specialized knowledge. Lastly, in case organizations need automated configuration of variants, the Provop method provides functionality to integrate variant management with some modeling tools [10]. In the following section, we provide a list of guidelines to help organizations to choose the relevant method for their setting, needs and constraints.

5.1   Guidelines for Process Variant Modeling Method Selection

Based on our evaluation described in the previous sections and the feedback from the 4S experts, we identified the guidelines in Table 3 to make a selection between the two methods.

Table 3. Variant modeling method selection guidelines

Needs and constraints of the organization	Suggested method
You want all information to be embedded in your models and your main purpose for using variant models is to reuse knowledge when you need to create a new variant	Decomposition Driven Method
You need a stepwise description of the operations to configure all your existing variants	Provop Method
You cannot use extra constructs in your modeling tool	Decomposition Driven Method
You plan to use automated process variant configuration	Provop Method
You need a hierarchical approach to analyze your process variants and you have different levels of variability in your subprocesses	Decomposition Driven Method
You prefer a lean model where your domain experts can easily understand the base model and your analysts can go deeper in configuring variants using extra information	Provop Method

We observe that the experts can benefit from the merits of the two methods even when they use another method specifically in the following ways:

- The guidelines of the Decomposition Driven method may be used to extensively evaluate why process variants emerge in your business context.
- The approach of the Decomposition Driven method for modeling variants of hierarchical process models may be implemented in other methods as well.

- The policies of the Provop method to define a base model, such as usage of reference models, most frequent variant or minimal distance, may be used while defining a main process in any single-model variant modeling approach.
- When the generated single-model does not include the steps of configuring a specific process variant, option list approach can be used.

6   Conclusion

In this study, we implemented two different single-model approach process variant modeling methods in a real-life setting. The involved company has various process definitions of the same process as they provide BPM services in similar areas to their customers. The company experienced problems in reusing their existing knowledge to define a new process for a customer, and finding out the related process variants and maintaining them properly when there is an update in an existing process. To explore the solutions to their problems, they wanted to employ process variant modeling methods to evaluate their benefits and compare with each other. For this purpose, we selected the Decomposition Driven and the Provop methods. We observed that both of these methods can be applied with a reasonable effort and will bring benefits by providing a single integrated model to configure models. We observed benefits of both methods in 4S from different aspects.

As is the case for many things in life, there is not a single answer for the question which method to select. Both methods we analyzed here have their merits while they still introduce complexity due to new analysis techniques and notations to be applied. Even when professionals decide to use another variant modeling method or no method at all, learning about variant analysis through these methods will bring benefits. For example, when organizations explore business drivers causing variations, they can use this information to evaluate root-causes and deal with this variation on a strategical level. Another point is that every organization can adopt the idea of using policies to define its base models. Considering this fact, we prepared a list of guidelines to help organizations to select a proper method and to utilize the insights provided by these methods when they have process variation in their organization.

In future work, we will completely apply the methods for low level processes of software project management as already initiated in current work. This will enable a thorough evaluation of the methods for hierarchical processes. Also, we plan to apply the methods to the demand request process, which even shows more variation with respect to customers and other factors. In parallel, 4S plans to start a gradual usage of variant modeling in its company. For this, new experts will be trained. Then, prototypes will be identified from the projects where the experts will use the outputs of this study to define processes of the new customers.

Acknowledgements. This study has been conducted in 4S Information Systems in collaboration with process management team of the company. This study received funding from the European Union's Horizon 2020 research and innovation programme under the Marie Sklodowska-Curie grant agreement No. 660646.

References

1. Hallerbach, A., Bauer, T., Reichert, M.: Capturing variability in business process models: the Provop approach. J. Softw. Maint. Evol. Res. Pract. **22**, 519–546 (2010)
2. Döhring, M., Reijers, H., Smirnov, S.: Configuration vs. adaptation for business process variant maintenance: an empirical study. Inf. Syst. **39**, 108–133 (2014)
3. Ayora, C., Torres, V., Weber, B., Reichert, M., Pelechano, V.: VIVACE: a framework for the systematic evaluation of variability support in process-aware information systems. Inf. Softw. Technol. **57**, 248–276 (2015)
4. Enterprise, H.P.: Project and Portfolio Management – PPM. http://www8.hp.com/us/en/software-solutions/ppm-it-project-portfolio-management/
5. Milani, F., Dumas, M., Ahmed, N., Matulevicius, R.: Modelling families of business process variants: a decomposition driven method. CoRR abs/1311.1 (2013)
6. Hallerbach, A., Bauer, T., Reichert, M.: Configuration and management of process variants. In: Brocke, J., Rosemann, M. (eds.) Handbook on Business Process Management 1 SE - 11, pp. 237–255. Springer, Heidelberg (2010)
7. Conforti, R., Dumas, M., Rosa, M.La., Maaradji, A., Nguyen, H.H., Ostovar, A., Raboczi, S.: Analysis of Business Process Variants in Apromore (2015)
8. Project Management Institute Inc: A guide to the project management body of knowledge (PMBOK® guide) (2000)
9. Yaldiz, A.: Evaluation of process variant modeling approaches: a case study, Ankara, Turkey (2016). http://expertjudgment.com/publications/METU_II_TR_2016_YILDIZ.pdf
10. Reichert, M., Rechtenbach, S., Hallerbach, A., Bauer, T.: Extending a business process modeling tool with process configuration facilities: the Provop demonstrator. In: Proceedings of BPM 2009 Demonstration Track (2009)

Fundamental Issues in Modeling

Representing Dynamic Invariants in Ontologically Well-Founded Conceptual Models

John Guerson$^{(\boxtimes)}$ and João Paulo Almeida

Ontology and Conceptual Modeling Research Group (NEMO),
Federal University of Espírito Santo (UFES), Vitória, ES, Brazil
{jguerson,jpalmeida}@inf.ufes.br

Abstract. Conceptual models often capture the invariant aspects of the phenomena we perceive. These invariants may be considered static when they refer to structures we perceive in phenomena *at a particular point in time* or dynamic/temporal when they refer to regularities *across different points in time*. While static invariants have received significant attention, dynamics enjoy marginal support in widely-employed techniques such as UML and OCL. This paper aims at addressing this gap by proposing a technique for the representation of dynamic invariants of subject domains in UML-based conceptual models. For that purpose, a temporal extension of OCL is proposed. It enriches the ontologically well-founded OntoUML profile and enables the expression of a variety of (arbitrary) temporal constraints. The extension is fully implemented in the tool for specification, verification and simulation of enriched OntoUML models.

Keywords: Conceptual modeling · OntoUML · Temporal OCL

1 Introduction

In a broad perspective, conceptual modeling has been characterized as "the activity of formally describing some aspects of the physical and social world around us for purposes of understanding and communication" [20]. Many of the efforts in conceptual modeling attempt to represent a conceptualization about a given subject domain [15], which is often accomplished by capturing in a model the invariant aspects of the phenomena we perceive. These invariants may be considered *static* when they refer to structures we perceive in phenomena at a particular point in time or *dynamic* when they refer to regularities across different points in time.

Take for instance a domain about persons, their stages in life and their marriages. At a particular point in time, a number of persons will exist, each of which may be male or female, may be a child, a teenager or an adult, and may be related to someone else by marriage. The *static invariants* that may be represented in a conceptual model of this domain include the various categories of entities in a domain (in our example, "person", "man", "woman", "child", "teenager", "adult", "elder", "marriage") as well as their relations (a "child" is a "person", "marriage" may be established between two "persons", etc.). The *dynamic invariants* in turn reflect the fact that across different points in time entities of the domain undergo change. In our example, persons are born and die, become

© Springer International Publishing Switzerland 2016
R. Schmidt et al. (Eds.): BPMDS/EMMSAD 2016, LNBIP 248, pp. 303–317, 2016.
DOI: 10.1007/978-3-319-39429-9_19

teenagers and adults, marry, divorce, etc. Dynamic invariants represent what may change and what must remain constant in time. For example, children cannot suddenly become adults, adults cannot later in life become teenagers and elders cannot become children, teenagers or adults.

Much attention has been given to the representation of static invariants in a number of modeling notations including ER diagrams, ORM diagrams [16], and UML Class Diagrams [23]. The UML for example has been enriched with the Object Constraint Language (OCL) to capture invariant expressions [22]. With respect to the dynamic invariants, these have been mostly confined to the representation of pre- and post-conditions for operations or simple meta-attributes for features such as "read only" [23, pp. 125, 129]. Further, due to the strict correspondence that is often established between modeling languages and programming languages, many UML-based approaches lack support for dynamic classification (e.g. USE [13], HOL-OCL [5], UML2Alloy [1, 8]). While this facilitates the mapping to specific programming languages or formalisms, this renders these approaches less suitable to enable the expression of important conceptual structures that rely on dynamic classification (e.g., the classification of persons into life phases: child, teenager, and adult, the classification of persons into roles they play contingently such as husband and wife)[1].

In order to address the shortcomings of the UML and OCL specifications, many approaches have been proposed to extend UML and OCL with dynamic aspects. Some of these approaches address dynamic aspects as part of an overall approach to handle temporal/time aspects [3, 4, 6, 7, 9, 12, 15, 18, 19, 27]. The OntoUML [15], for example, introduces various dynamic aspects through stereotypes referring to meta-attributes of classes and properties such as rigidity and immutability. Similarly, [6] extends UML with stereotypes, augmenting it with dynamic notions of durability and frequency. Others have aimed at enriching OCL with extensions in order to cope with dynamic/ temporal properties of systems. For example, some have extended OCL with Linear-Temporal Logic and Computational-Tree logic (LTL/CTL) operators [3, 7, 19, 27], created new logic formalisms [4, 9], extended OCL with temporal patterns [18], defined a Real-Time extension for OCL with a temporalized CTL [12], etc.

Despite these recent advances, a comprehensive approach to the representation of dynamic aspects in UML-based conceptual models is still lacking. This gap is addressed in this paper, in which we extend OCL with the capability to express rich dynamic invariants in OntoUML. Our approach is based on a reification of world states, with no specialized knowledge in tense logic required to use the approach.

This paper is structured as follows. Section 2 presents our running example, illustrating the various requirements for the representation of dynamics aspects. Section 3 introduces the proposed OCL extension. Section 4 revisits our running example to show how the approach meets the requirements. Section 5 discusses related work while Sect. 6 discusses the implementation and concluding remarks.

---

[1] Note that while dynamic classification is supported in principle by UML diagrams, this is not reflected in tool support and language usage, with little mention in the UML specification.

2   Requirements for the Representation of Dynamic Aspects

According to [6], the UML is a non-temporal conceptual modeling language. Thus, a UML class diagram represents the *actual* state of a system assuming that the "information base" contains only the current instances of classes and relationships. Figure 1 depicts a conceptual model in UML about a domain involving persons, their stages in life and marriages. UML multiplicities in the model define the allowable number of individuals to which a particular individual may be linked in any given state of the system. For example, a partner (husband or wife) can only be married to one other partner at a time. The model is silent with respect to the number of persons a person can marry in time, i.e., whether they may or may not divorce and remarry.

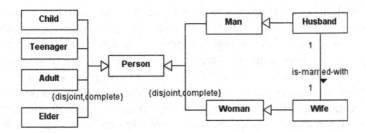

Fig. 1. UML structural conceptual model example

The model of Fig. 2 revisits the model of Fig. 1 employing the OntoUML profile. The profile uses class stereotypes to determine which ontological category from the UFO applies to each class [15]. This means that OntoUML can address some of dynamic aspects of this domain that are not addressed in plain UML. For example, the class Person is stereotyped as ≪kind≫, meaning that it applies necessarily to its instances. Thus, a person cannot cease to be a person without ceasing to exist. This modal notion corresponds to what is called **Rigidity** in UFO. The consequence of rigidity in terms of dynamic aspects is that an individual of a rigid class instantiates this class throughout its life. A kind can be used in a taxonomic structure with rigid subtypes known as subkinds (e.g., Man and Woman).

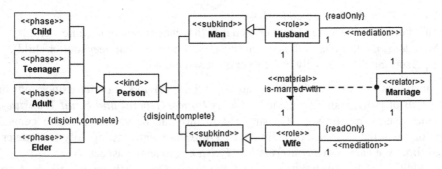

Fig. 2. OntoUML structural conceptual model example

Other examples of dynamic aspects expressed in Fig. 2 include the stereotypes of ≪roles≫ and ≪phase≫. Husband and Wife are stereotyped as ≪role≫ and Child, Teenager, Adult and Elder as ≪phase≫. Roles and phases are both anti-rigid concepts (e.g. a wife can cease to be a wife without ceasing to exist). **Anti-Rigidity** states that a class C is anti-rigid iff for all its individuals, there will be a point in time w in which they exist but do not instantiate C, at w. The difference between the two is that the former defines contingent properties of an individual exhibited in a relational context (e.g. a person is a wife contingently and only in the context of a marriage) while the later through an intrinsic change of an individual's property (e.g. a child has the intrinsic property of being a child).

The class Marriage is stereotyped as ≪relator≫. Relators can be viewed as objectified properties, as entities that "connect" other entities. They are the truthmakers of the so-called ≪material≫ relationships. For example, it is the existence of a particular marriage connecting man X and wife Y that makes true the relation *is-married-with*(X, Y). A derivation relationship on the other hand holds between a relator and a material relationship and exemplifies the truth-maker relation. Relators are rigid concepts and existentially dependent on the instances they connect through ≪*mediation*≫ relationships. A mediation is a dependency relationship that defines existential dependence from their source entity, e.g. Marriage, to their target entities, e.g., a Wife and a Husband. This means that a marriage only exists at some point in time, if wife and husband also exist at that point in time. A particular marriage then depends specifically on two "fixed" persons, i.e., the marriage between Bob and *Alice* cannot ever become the marriage between Bob and *Anna*. Mediations are thus always defined as *readOnly* at their target-side by default. From a logical point of view, this dynamic aspect of existential dependence can be viewed as a type of *immutability* (a marriage never changes their participating wife and husband). **Immutability** states that if an individual (e.g. marriage) exists at a point in time w, then at every subsequent time w' from w that the individual exists, that individual will have in w' the same property (e.g. same wife and husband) as it had in w. Finally, the classes Husband and Wife are related through exactly 1 Marriage, meaning that we represent monogamous heterosexual marriage, i.e., a partner can only be married to one partner at a time.

While some distinctions in OntoUML enable the representation of dynamic aspects of the domain, a number of other dynamic invariants cannot be expressed:

1. A person is created in the child phase.
2. An adult cannot become a child or teenager (a teenager cannot became a child).
3. An elder is the last phase of a person (it cannot become adult, teenager or child).
4. A person should eventually cease to exist at some point.

Constraints *1–3* can be viewed as more general behaviors about *classifications* or *transitions* of individuals. We named these as **Initial, General and Final Transition** dynamic aspects, respectively. Constraint *4* in turn can be viewed as more general behavior about the *existence* of individuals. *Transient* means lasting only for a determined time. We named this as a **Transient Existence** dynamic aspect.

In addition to the constraints specified above, the OntoUML model still does not represent how many times a person can marry throughout his/her life, and the model is silent about this aspect of the conceptualization. In this model, a marriage could still be transient (when it ceases to exist eventually) or permanent (when it never ceases to exist once created). The permanence view of marriages could refer, for example, to a religious conceptualization, where marriages are divine "contracts" between two people and cannot be undone. If committed to this view, a desired dynamic invariant would be:

5. If a marriage is established then it can no longer be destroyed.

We named this as a *Permanent Existence* dynamic aspect. On the other hand, if a marriage is transient and ceases to exist, it may be desirable in a given conceptualization to refer to ex-spouses, i.e., people who participated in a past marriage, which no longer exists in the present:

6. A person will only be an ex-husband or ex-wife if he/she was a husband/wife in a marriage in the past which no longer exists.

This is a common dynamic invariant that has been called *Derivation by Past Specialization* in [21].

The kinds of limitations in the expressiveness of diagrammatic languages (e.g. OntoUML, UML) we identify here are often an explicit language design decision, in order to manage the complexity of graphical representation. In order to complement the graphical representation and address the expressiveness needs, textual representations such as OCL have been proposed to enrich the model as captured in a diagrammatic language. We also take that approach and employ OCL in order to enrich OntoUML in other to support the representation of dynamic aspects of the domain.

The modeling approach is required to:

- support dynamic classification (i.e., allow for individuals to change types throughout their existence);
- enable the expression of modal constraints on types (e.g., rigidity, anti-rigidity);
- enable the expression of transition rules (constraining the order in which individuals instantiate types), transient/permanent existence and past specializations; and
- finally, enable the expression of arbitrary dynamic invariants, i.e., *invariants whose satisfaction is determined by examining the world at more than one point in time.*

Transition rules include: (i) initial type rules (determining the type that is instantiated necessarily upon creation); (ii) final type rules (determining the types that once instantiated, classify the individual); (iii) other general transition rules (constraining arbitrary order of instantiation). Transition rules may be expressed by behavior models such as with UML state chart diagrams. However, we are aiming here for a general approach to define *arbitrary* types of dynamic constraints. Further, the modeling approach must not rely on operations of classes, as these are not employed by OntoUML [15], and also not employ specialized tense/temporal logic-based operators as [3, 7, 19, 27], in order to retain its ease of use for UML/OCL modelers.

3   A Standard OCL Temporal Extension

A standard OCL invariant is a static condition that should hold for each single state of the model's instances. As a consequence, the so-called "context" of a standard invariant is a single state, and no notion of "state" is manipulated in standard OCL invariants. In order to enable the manipulation of states and consequently the representation of dynamic aspects, we *reify* the notion of "world states" (or simply "worlds") (Sect. 3.2). *Reification* gives the ability of referencing, quantifying and qualifying over an objectified entity (in this case, the domain's states). We use the "world" as an index to refer to the properties at a particular point in time (Sect. 3.3). We propose a branching world structure, which can be used to enable arbitrary reference to worlds and world paths in invariants (Sect. 3.4). We adjust a few standard OCL pre-defined operations in order to support world indexing (Sect. 3.5). The use of the resulting temporal extension of OCL described in this section is shown in Sect. 4.

3.1   Temporal Extension Approach

Our approach for extending OCL with dynamic invariants consists of using a temporalized UML/OCL model in the background with the notion of world states reified, such as illustrated by Fig. 3. The OntoUML model is translated into a world-reified model in plain UML. This model is enriched with constraints in standard OCL to ensure the former OntoUML model semantics. Arbitrary standard OCL constraints can then be bound to this temporalized enriched model. With this binding, only few adjustments in standard OCL are required in order for standard OCL to behave as a temporal language. Our extension employs these adjustments: (i) defining built-in operations for world-indexed navigations (Sect. 3.3) and (ii) creating support for world states in some pre-defined OCL object and classifier operations (Sect. 3.5). Using the binding with a world-reified model in background we are able to use standard OCL as if it was a temporal language.

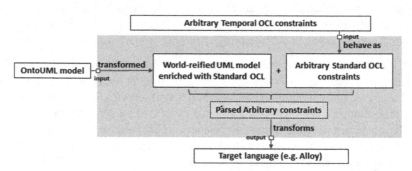

Fig. 3.   Temporal extension approach

Temporal OCL constraints are then treated as standard OCL (with only few adjustments), and verified syntactically in order to be transformed to other languages such as Alloy [17]. The modeler expresses a conceptual model in OntoUML and Temporal OCL and is shielded completely from this underlying support, which ultimately generates an Alloy model for simulation and validation of constraints.

3.2 Underlying the World-Reified Model of Background

The idea behind world states reification is to treat the world states (or "worlds") as entities. Thus we introduce the class "World" in this reification step. The OntoUML model example about people, their stages in life and marriages, previously depicted in Fig. 1, is translated into a world-reified UML model. Figure 4 depicts only a fragment of that resulting UML model. In this model, UML is employed as a temporal model and therefore UML classes represent individuals existing at all possible states of the world. Every former OntoUML class (e.g. the kind Person, the relator universal Marriage) now specializes UML class *Individual*, in order to support the *existsIn* relation, which holds for the worlds in which an individual exists. In this manner, all OntoUML classes can be indexed in time through this relation of existence. Note that all UML relationships in this model are *readOnly* by default since time was reified and each property change is now characterized by a change in the world states.

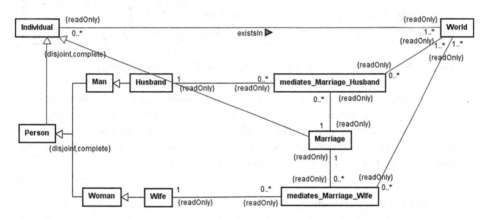

Fig. 4. A fragment of the world-reified UML model of background

In order to capture the dynamics of OntoUML relationships in this reified UML model, all former relationships are reified (translated) to a UML class, with three UML binary relationships and OCL constraints to maintain the semantics of the original OntoUML relationship. The class representing that relationship defines the worlds in which the relationship exists. For instance, the UML class *mediates_Marriage_Wife* represents the former OntoUML mediation relation between Marriage and Wife. Since all wives must be related to a marriage, it *existsIn* a non-empty set of Worlds. In each world, there may be a set of relationships of this type.

Finally, OntoUML multiplicities define *actual* multiplicity constraints (i.e. they restrict how many individuals an individual may be linked to at a single world state). We chose not to represent OntoUML's actual multiplicity constraints in this reified model using UML because only the lower actual cardinality can be represented using a

UML lifetime multiplicity (e.g. a wife has exactly one marriage at a time, which means that she has also at least one marriage in her lifetime). For this reason, we represented all OntoUML's actual multiplicity in this reified model of Fig. 4 as additional OCL constraints. Thus, the multiplicities from Wife and Marriage to the reified mediation in the reified model are defined as simply 0..*. Figure 5 exemplifies these additional constraints specifying OntoUML's actual multiplicity of the mediation between Marriage and Wife (they are embedded in the world-reified model and transparent to the modeler). The first constraint states that a marriage mediates exactly one wife at a world i.e. for every world (*self*), for all marriages at *self*, the number of *mediates_Marriage_Wife* linked (at *self*) to that marriage is equal to 1. Conversely, the second constraint states that a wife is mediated by exactly one marriage at a world.

```
context World
inv marriage_mediates_one_wife_at_a_time:
    self.individual->select(i | i.oclIsKindOf(Marriage))->forAll(m |
    m.mediates_marriage_wife->select(r | r.world = self)->size() = 1)
inv wife_is_mediated_by_one_marriage_at_a_time:
    self.individual->select(i| i.oclIsKindOf(Wife))->forAll(h |
    h.mediates_marriage_wife->select(r | r.world = self)->size() = 1)
```

Fig. 5. World-Reified UML Model: Reflecting OntoUML Multiplicities using OCL

In order to maintain the actual semantics of OntoUML, additional constraints are required in our world-reified UML model (e.g., to capture the fact that relationships, relators and relata co-exist in all worlds in which they exist, to reflect the immutability of relata on which a relator depends, etc.). They are all represented in plain OCL in the world-reified model. We omit them here due to space constraints.

3.3 Temporal OCL Navigations

Usually, OCL navigations on ternary relationships can proceed in three stages: (i) navigating from a ternary relationship to each class it relates, (ii) from each related class to the ternary relationship itself, and (iii) navigating from a first related class to a second related class, filtering the result to a third related class. In our previous world-reified UML model, only (iii) is allowable filtering the result of navigation with respect to world states. (i) and (ii) are not supported because the reified ternary relationship (i.e. the UML class acting as the relationship in our reified model) is hidden from the modeler (they are an implicit construct generated only in the background), and secondly because we want to refer to properties at a specific point in time, making explicit the world state.

Figure 6 specifies the definition of the allowable temporal OCL navigations, as (iii). The first navigation *Wife::marriage(w)* is defined from Wife to Marriage filtered by a specific world state. It returns all marriages of a wife at world *w*. The second navigation is defined from Marriage to Wife *Marriage::wife(w)*, returning the wife related to a specific marriage at *w*. These world-indexed navigations are available to the modeler in order to refer to the relation in a particular state. Furthermore, we also enabled a temporal OCL navigation without a world parameter which returns all individuals of a property, at all possible worlds. For example, if *self* is a wife, then *self.marriage* returns all marriages of a wife in her entire life.

```
context Wife
def: marriage(w: World): Set(Marriage) =
  self.mediates_Marriage_Wife->select(m | m.world=w)->collect(marriage)->asSet()
context Marriage
def: wife(w: World): Set(Wife) =
  self.mediates_Marriage_Wife->select(m | m.world=w)->collect(wife)->asSet()
```

Fig. 6. World-reified UML model: definition of temporal OCL navigations

3.4 World Structures

An ordered structure of world states models how the subject domain behaves in time. We adopt a structure of possible worlds inspired in Kripke structures of modal logic semantics; more specifically, we assume the branching structure previously defined in [2]. Each world has a set of immediate next worlds and at most one previous world (it is a tree with branches towards the future, capturing the notion that the future may unfold in different ways). For each world state, there is only one sequence of worlds to a future state of the world (meaning that branches do not join again). A history, i.e., a path, is comprised by a non-empty set of worlds while a world must be included in at least one history, such as depicted by Fig. 7 using UML. This structure of worlds is a built-in part of every world-reified UML model, dictating how worlds are accessible from each other and specifying a number of pre-defined temporal operations for Worlds and Paths.

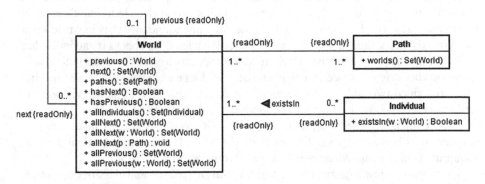

Fig. 7. World-reified UML model: world structure and temporal operations

Differently from [2], we have reified the notion of "paths". These are useful to express constraints which are usually expressed in the CTL tense logic, quantifying not only over states but also over paths of states, both universally and existentially. Quantifying universally and existentially over paths is an important feature to some dynamic properties of systems. Since Path is also an entity as World, several additional constraints are defined in standard OCL to complement this world-reified UML model in order to enforce properly the semantics of histories (paths). A history must contain exactly one initial and one terminal world, no two histories should have the same terminal world and every terminal world must be in exactly one history. Additionally, the worlds contained in a history should be derived from all previous worlds of that history's terminal world. We validated our branching world structure using the lightweight formal method of validation based on Alloy simulation and analysis [17], as a means to check the correct semantics of the reified histories (paths) that we introduced in the world structure. The following set of temporal operations are pre-defined: next() and previous(), which return a world's immediate next and immediate previous world; hasNext() and hasPrevious(), which checks whether a world has an immediate previous or immediate next world; allIndividuals(), which returns all existing individuals at a specific world; existsIn(), which checks the existence of an individual at a specific world; and allNext(), which returns all subsequent worlds of a particular world. This operation has two variants, allNext(w), which returns all subsequent worlds until a particular world w is reached (not including w) and allNext(p), which returns all subsequent worlds from a world, contained in a given path p. Analogously, allPrevious() returns all precedent worlds of a particular world. Finally, p.worlds() returns all worlds of a path p, and w.paths() returns all paths in which the world w is contained. These pre-defined temporal operations were all implemented using standard OCL (body conditions) over the world-reified UML model. For the sake of brevity, we omit here the Alloy code of our branching world structure and the implementation of these pre-defined temporal operations.

3.5   Standard OCL Operations Revisited

oclIsNew() is only used in post-conditions [22, p. 154]. As our subset of OCL does not consider pre- and post- conditions (OntoUML disallows operations) oclIsNew() is not supported. Instead, we defined two temporal operations to check an individual's creation and deletion at a world w. oclIsCreated(w) checks if an individual exists in world w but does not exist in its immediate previous world. oclIsDeleted(w) checks if and individual does not exist in w but does exist in its immediate previous world. These are Individual's operations because existence is a characteristic of domain objects that persist in time. We also defined two additional operations for the classification of an individual at w. oclBecomes(C, w) checks whether an individual is classified as C at w but is not classified as C in w's immediate previous world. oclCeasesToBe(C, w) in turn checks whether an individual ceases to be classified as C at w. That is, the individual does not instantiate C in w's immediate previous world, but instantiate C at w.

There are only few adjustments to some built-in OCL object and classifier operations that need to be established due to our world reification approach. Type conformance operations must explicit the point in time in which the types are checked. Since standard OCL does not natively support world states, we include a world state parameter at oclIsKindOf(T, w), oclIsTypeOf(T, w), oclAsType(T, w) and oclType(w). The allInstances() operation is still allowed and it returns the extension of a class at all possible worlds i.e. the set of all instances of a class independent of their actual existence in a particular point in time. Expressions such as (i) World.allInstances(), (ii) Path.allInstances(), or (iii) Individual.allInstances() are then all valid. They return respectively, the set of all possible worlds, all histories and all individuals at all worlds. Additionally, we assume a temporal operation allInstances(w) for every UML domain class. allInstances(w) returns all instances of a class at a world w (expressions such as World.allInstances(w) or Path.allInstances(w) are not valid constructions since worlds were reified and neither worlds nor paths exist within worlds).

4   Representing Dynamic Invariants (the Modeler's View)

In this section, we represent the dynamic aspects which were required to model as accurately as possible the conceptualization presented initially in Sect. 2, thereby showing how the approach satisfies the requirements. These dynamic aspects include transition rules, existence of individuals and past specializations.

Transition rules constrain the change from one (antecedent) state to another (subsequent) state. The *Initial Transition* rule is a peculiar type of transition rule that mentions to antecedent state. The condition holds at the first world of an individual's existence. Figure 8 exemplifies this rule in Temporal OCL, stating that every time a person is created, he/she is a child at that moment.

```
context Person
temp initial_transition: World.allInstances()->forAll(w |
    self.oclIsCreated(w) implies self.oclIsKindOf(Child, w))
```

Fig. 8. Temporal OCL: initial transition rule

The keyword *temp* defines a temporal invariant. The "context" defines a class extension at all worlds e.g. all individuals that at some point will instantiate Person. The condition then must hold for each of these individuals. Figure 9 exemplifies the *Final Transition* in Temporal OCL. It states that for every individual that will eventually be an Elder, for every world, if that individual is an Elder at that world, then for every world after that, if the individual exists, then it instantiates Elder (i.e. there is no other possible and allowed transition for that individual before ceasing to exist).

```
Context Elder
temp final: World.allInstances()->forAll(w | self.oclBecomes(Elder, w) implies
    w.allNext()->forAll(n| self.existsIn(n) implies self.oclIsKindOf(Elder,n)))
```

Fig. 9. Temporal OCL: final transition rule

Differently from initial transitions, final transition can be viewed as a more specific case of a general rule for arbitrary transitions. This general transition rule states that an antecedent type A1 is transitioned to one or more types S1 + … + SN. This means that there is no allowed transition for the instance of A1, before ceasing to exist, if not being A1 itself, or being one of the subsequent types S1, S2…SN. The final transition is just a special case of this general rule where there is no subsequent type only the antecedent type A1. Figure 10 exemplifies the general transition in temporal OCL stating that a teenager can only transit to teenager (i.e. A1) or to adult (i.e. S1) phases.

```
context Teenager
temp transition: World.allInstances()->forAll(w | self.oclBecomes(Teenager, w)
    implies w.allNext()->forAll(n | self.existsIn(n) implies
    self.oclIsKindOf(Teenager, n) or self.oclIsKindOf(Adult, n)))
```

Fig. 10. Temporal OCL: general transition rule

Figure 11 exemplifies the ***Transient and Permanent Existence*** in Temporal OCL. The first invariant states that for every person that comes into existence, there will be at least one world after that in which that person will cease to exist. The second invariant states that every marriage, once created, will exist at all possible worlds after that. Since a marriage is existentially dependent on a husband and a wife, by implication, the roles Husband and Wife are final transitions of a person and married persons are permanent. While the individual constraints are meaningful, they are inconsistent with each other, which can be checked using our support for Alloy. The analyzer would show that there is no valid instantiation of the model with these two constraints.

```
context Person
temp transient: World.allInstances()->exists(w | self.oclIsCreated(w) and
    w.allNext()->exists(n | not self.existsIn(n)))
context Marriage
temp permanent: World.allInstances()->exists(w | self.oclIsCreated(w) and
    w.allNext()->forAll(n | self.existsIn(n)))
```

Fig. 11. Temporal OCL: transient and permanent existence

Finally, Fig. 12 exemplifies a case where ex-husbands and ex-wives are required as cases of a ***Derivation by Past Specialization***. The invariant states that for every eventual ex-wife, for every world, if an ex-wife exists at a world then there exists a set of previous worlds from w in which she was a wife and her related past marriage does not exist in w.

```
context ExWife
temp past_spec: World.allInstances()->forAll(w | self.oclIsKindOf(ExWife, w)
    implies w.allPrevious()->exists(p | self.oclIsKindOf(Wife, p) and not
    self.oclAsType(Wife, p).marriage(p).existsIn(w)))
```

Fig. 12. Temporal OCL: derivations by past specializations

5 Related Work

There have been many proposals in literature that aimed at extending OCL in order to cope with dynamics/temporal aspects of systems [3, 4, 7, 9, 12, 18, 19, 27]. *Gogolla and Ziemman*'s extension of OCL [27] is based on a set of Linear Temporal Logic (LTL) operators. They introduced an environment's index to characterize the temporal evolution of the system and its current state. *Conrad and Turowski* [7] extended OCL with LTL operators to specify software contracts for business components, where contracts are represented as pre- and post-conditions. *Bill et al.* [3] presented an OCL extension named cOCL, based on Computational Tree Logic (CTL). Their verification framework consists of cOCL specifications and a model checker called MocOCL that can verify cOCL constraints. *Flake and Mueller* [12] defined a state-oriented Real-Time extension of OCL whose semantics is given through a mapping to clocked CTL logics (CCTL). They focus on the specification of real-time systems. Differently from these approaches, we do not use tense logic operators explicitly, choosing to use reification of world states to obtain the expressiveness that would be obtained with tense operators. Extensions based on modal/tense logic operators require a level of logic expertise that most modelers are not expected to have. *Distefano et al.* [9] defined an object-based extension of CTL called BOTL (Object-Based Temporal Logics), a logic formalism inspired by OCL to define specifications of static and dynamic properties in object-oriented systems. BOTL looks syntactically very similar to CTL and although BOTL's concepts are defined clearly and precisely, no tool support is actually provided. *Mullins and Oarga* [19] extended OCL with CTL operators and some first-order features. Their extension termed EOCL is largely inspired by BOTL [9] and based on the framework of *Bradfield et al.* [4]. Their SOCLe tool translates exactly one UML class diagram, one state-chart and one object diagram into an Abstract State Model (ASM) specification, which in turn is translated into an execution graph that can verify on-the-fly EOCL constraints. *Bradfield et al.* [4] proposed a formalism, termed $O\mu(OCL)$ which requires such understanding of temporal logics (as stated by the authors) that is unrealistic to expect most developers to acquire it [4, p. 2]. *Kanso and Taha* extended OCL [18] according to the set of Dwyer's temporal property patterns [10] with the explicit inclusion of events. They have fully implemented the OCL extension in an Eclipse/MDT OCL Plugin, which allows OCL temporal constraints to be defined with Ecore/UML models. However, the set of temporal patterns are not suitable to OntoUML's set of requirements, such as the initial transition dynamic aspect, usually, due to the pattern's closed/open edges of intervals. Finally, *Cabot et al.* [6] extended OCL with instant reification but solely to retrieve immediate past values of UML model properties.

6 Concluding Remarks

In this paper, we have defined a temporal extension for standard OCL to cope with dynamics in ontologically well-founded conceptual models with OntoUML. The temporal OCL extension developed requires only few adjustments to standard OCL; in particular, to four OCL type conformance operations and the allInstances() operation.

Our temporal OCL is expressive not only to represent the implicit dynamics of OntoUML (e.g. rigidity, anti-rigidity, immutability), but also to incorporate *user-defined* dynamics aspects into conceptual models, such as transitions, transience, permanence, past derivations, etc.

The extension is fully incorporated into the OLED[2] tool, which is an editor for the creation, development and validation of OntoUML structural conceptual models. We have thus extended the previous work of [14] with the support for a temporal OCL extension, which includes: (i) a temporal OCL editor with syntax highlighting and code-completion, (ii) a parser for temporal OCL constraints using Eclipse's OCL support [11] and (iii) a transformation from temporal-enriched OntoUML models into the Alloy logic-based language, enabling simulation and verification of dynamic constraints written in our temporal extension.

In the future, we plan to compare our approach with other approaches such as *Kanso and Taha*'s temporal OCL extension and their set of temporal patterns [18] and the ontology-based behavioral specification language (OBSL) [26]. These approaches trade expressiveness for ease of use, so we expect that all of the constraints that can be expressed in these approaches can be expressed with our OCL extension. We also plan to represent *Sales*' simulations scenarios for semantic anti-patterns detection [24] as a means to further demonstrate the expressivity of our extension of OCL.

Finally, we should investigate whether some of the dynamic aspects discussed here (e.g., transience and permanence) can be introduced in the graphical notation (e.g., as additional stereotypes) to improve the language's usability. We should also investigate a combination of the approach with other notations such as state diagrams which could support the specification of some of the rules (e.g., transition rules). These diagrams would ultimately be transformed into temporal OCL constraints.

Acknowledgements. This research is funded by the Brazilian Research Funding Agencies FAPES (grant number 59971509/12) and CNPq (grants number 310634/2011-3, 485368/2013-7 and 461777/2014-2).

References

1. Anastasakis, K., Bordbar, B., Georg, G., Ray, I.: On challenges of model transformation from UML to Alloy. Softw. Syst. Model **9**(1), 69–86 (2010)
2. Benevides, A.B., Guizzardi, G., Braga, B.F.B., Almeida, J.P.A.: Validating modal aspects of OntoUML conceptual models using automatically generated visual world structures. J. Univers. Comput. Sci. **16**, 2904–2933 (2011)
3. Bill, R., Gabmeyer, S., Kaufmann, P., Seidl, M.: OCL meets CTL - towards CTL-extended OCL model checking. In: MoDELS, vol. 1092, pp. 13–22 (2013)
4. Bradfield, J.C., Küster Filipe, J., Stevens, P.: Enriching OCL using observational mu-calculus. In: Kutsche, R.-D., Weber, H. (eds.) FASE 2002. LNCS, vol. 2306, pp. 203–217. Springer, Heidelberg (2002)

---

[2] https://code.google.com/p/ontouml-lightweight-editor/.

5. Brucker, A.D., Wolff, B.: HOL-OCL: a formal proof environment for UML/OCL. In: Fiadeiro, J.L., Inverardi, P. (eds.) FASE 2008. LNCS, vol. 4961, pp. 97–100. Springer, Heidelberg (2008)
6. Cabot, J., Olivé, À., Teniente, E.: Representing temporal information in UML. In: Stevens, P., Whittle, J., Booch, G. (eds.) UML 2003. LNCS, vol. 2863, pp. 44–59. Springer, Heidelberg (2003)
7. Conrad, S., Turowski, K.: Temporal OCL meeting specification demands for business components. In: UML 2001, vol. 2185, pp. 151–165 (2001)
8. Cunha, A., Garis, A., Riesco, D.: Translating between Alloy specifications and UML class diagrams annotated with OCL. Softw. Syst. Model **14**, 5–25 (2013)
9. Distefano, S., Katoen, J.P., Rensink, A.: On a temporal logic for object-based systems. In: Fourth International Conference on Formal Methods for Open Object-Based Distributed Systems IV, vol. 49, pp. 305–325 (2000)
10. Dwyer, M.B., Avrunin, G.S., Corbett J.C.: Patterns in property specifications for finite-state verification. In: Proceedings of the 21st International Conference on Software Programming, pp. 411–420 (1999)
11. Eclipse MDT OCL. http://www.eclipse.org/modeling/mdt/
12. Flake, S., Muller, W.: Formal semantics of static and temporal state-oriented OCL constraints. Softw. Syst. Model. **2**(3), 164–186 (2003)
13. Gogolla, M., Bohling, J., Richters, M.: Validating UML and OCL models in USE by automatic snapshot generation. Softw. Syst. Model. **4**(4), 386–398 (2005)
14. Guerson, J., Almeida, J.P.A., Guizzardi, G.: Support for domain constraints in the validation of ontologically well-founded conceptual models. In: Bider, I., Gaaloul, K., Krogstie, J., Nurcan, S., Proper, H.A., Schmidt, R., Soffer, P. (eds.) BPMDS 2014 and EMMSAD 2014. LNBIP, vol. 175, pp. 302–316. Springer, Heidelberg (2014)
15. Guizzardi, G.: Ontological Foundations for Structural Conceptual Models. Telematica Instituut, The Netherlands (2005)
16. Halpin, T., Morgan, T.: Information Modeling and Relational Databases. Morgan Kaufmann, Los Altos (2010)
17. Jackson, D.: Software Abstractions-Logic, Language, and Analysis, Revised edn. The MIT Press, Cambridge (2012)
18. Kanso, B., Taha, S.: Specification of temporal properties with OCL. Sci. Comput. Program. **96**, 527–551 (2014)
19. Mullins, J., Oarga, R.: Model checking of extended OCL constraints on UML models in SOCLe. In: Bonsangue, M.M., Johnsen, E.B. (eds.) FMOODS 2007. LNCS, vol. 4468, pp. 59–75. Springer, Heidelberg (2007)
20. Mylopoulos, J.: Conceptual modeling and telos. In: Conceptual Modeling, Databases, and CASE: an Integrated View of Information Systems Development. Wiley, Chichester (1992)
21. Olivé, A., Teniente, E.: Derived types and taxonomic constraints in conceptual modeling. Inf. Syst. **27**(6), 391–409 (2002)
22. OMG: OCL Specification v2.4.1 (2014)
23. OMG: UML Superstructure v2.4.1 (2012)
24. Sales T.P.: Ontology validation for managers. MSc thesis, Federal University of Espírito Santo, UFES (2014)
25. Sider, T.: Quantifiers and temporal ontology. Mind **115**(457), 75–97 (2006)
26. Wiegers, R.: Behaviour specification for ontologically grounded conceptual models. M.Sc thesis, University of Twente (2014)
27. Ziemann, P., Gogolla, M.: OCL extended with temporal logic. In: 5th International Andrei Ershov Memorial Conference, PSI, vol. 2890, 351–357 (2003)

Requirements and Regulations

MATRA: A Framework for Assessing Model-Based Approaches on the Transformation Between Requirements and Architecture

Carlos E. Salgado[1(✉)], Ricardo J. Machado[1], and Rita S.P. Maciel[2]

[1] Centro ALGORITMI, Universidade do Minho, Guimarães, Portugal
carlos.salgado@algoritmi.uminho.pt, rmac@dsi.uminho.pt
[2] Departamento de Ciência da Computação, Universidade Federal da Bahia, Salvador, Brazil
ritasuzana@dcc.ufba.br

Abstract. The activity of linking requirements and software engineering, as described by the Twin Peaks model, has set the standard for the transformation between business requirements and system architectures. Still, much is left to do regarding model-based activities in this topic where numerous proposals occur. Although counting on a set of common issues vital for their success, analysing or comparing any of these approaches remains a challenging task. Following previous work on their systematic review and comparison, and supported in a set of selected proposals focused in model-based approaches, we present a framework covering their involved key issues which allows classifying and assessing the different approaches. Accordingly, besides proposing the conceptual design of the framework we demonstrate its use by applying it to the selected transformation approaches, in order to validate this solution. Furthermore, the pros and cons of each approach are further discussed, and future steps on this work analysed.

Keywords: Requirements elicitation · System architecture · Transformation · Model-based · Alignment and traceability

1 Introduction

Software Engineering (SE) is inherently a modelling activity, in which abstract models of information systems are derived from Requirements Engineering (RE) and then systematically developed from problem to solution space [1]. The proficient use of models has been helping traditional engineering in achieving success and boosting product quality, nevertheless, in RE and SE that is not so perceptible. This is due to technical issues but also to a number of complex social and economical ones [2]. Even though, the use of standards and models in RE and SE is noticeable, as researchers try to communicate clearly and contribute on common grounds, while practitioners struggle to adopt existing proposals [3].

Model-driven development research aims to play a role in establishing and spreading the use of model-based approaches in RE and SE. In this context, the early step of transforming requirements into an analysis model is a crucial and challenging one. The

© Springer International Publishing Switzerland 2016
R. Schmidt et al. (Eds.): BPMDS/EMMSAD 2016, LNBIP 248, pp. 321–332, 2016.
DOI: 10.1007/978-3-319-39429-9_20

Twin Peaks model [4], with its spiral life-cycle archetypal, stands as a reference in linking the RE and SE fields, setting the track for the transformation, traceability and alignment between the elicited business requirements and the candidate system architectures. Also, issues as quality characteristics, architecture evaluation and development of supporting tools have become essential elements for any proposed approach in this domain [5].

Inspired by the transformation process between these two worlds, numerous model-based proposals have been presented and evolved along the years, raising several issues in this subject [6]. Among them, a well-grounded choice of an elicitation technique [7], accompanied by a clear description of the transformation, alignment and traceability mechanisms between requirements and its associated analysis model elements are seen as vital. Moreover, questions about representational models for both requirements and architectures, evolution of the relations between business and system requirements, detailed process specifications and design, case study evaluations and support tool design and development, have been explored. The continued evolution and evaluation of all these issues, namely through empirical studies, is ever necessary in order to advance research and influence practitioners adoption [8].

Grounded on the foundations of these previous works of comparisons [6] and systematic reviews [8] for existing transformation approaches, while supported on a set of recent, research relevant, model-based proposals, we now aim to present a framework covering all their involved key issues. This framework will allow classifying and categorizing the individual details of each approach, further facilitating the assessment and comparison of methods, and the respective artefacts involved in the different approaches. It also details and updates previous proposals results, again according to the latest research evolutions in this topic, with a focus on structural transformation approaches and in line with the influence of the Twin Peaks model.

In order to assess our proposal and validate the proposed framework, we demonstrate its use by applying it to the selected transformation approaches. Accordingly, they are framed and compared by performing a review on their transformation issues from requirements to architectures. The selected approaches to take part on this study, a choice based on their associated research relevance and alignment with the Twin Peaks model, are the CBSP technique [5], the ATRIUM methodology [9], the STREAM process [10], and our method for the derivation of a logical architecture from process, goals and rules requirements, the 4SRS-SoaML [11]. These last two count on some associated research variants included, adaptations aimed at covering recent research trends either in the requirements and in the architectural side.

In this paper we first contextualize the background problems and trends faced by the requirements to architecture model-based transformations, in Sect. 2. Then, throughout Sect. 3, we go over a set of selected approaches representing them and reviewing all contemporary issues to involve in the study. Next, Sect. 4 presents the proposed framework with the coverage for each relevant issue analysed, while applying it to the referred approaches in each of its parts. Section 5 discusses the current status and future steps of this work and finally some conclusions are drawn on this paper.

2 Background

The presentation of the Twin Peaks model [4] allowed for a better understanding and articulation of the conceptual differences between requirements and design, and inspired much research in both domains. Nevertheless, although its spiral life-cycle model stands as a reference in the requirements engineering and information systems architecture fields, the process of moving between the problem and solution worlds is not as well recognized.

On one hand, the approaches inspired in the Twin Peaks model are better prepared to avoid the danger of merely focusing on functionality while ignoring quality concerns, and also to fulfil the need for early prototypes and architecture evaluation in order to meet their goals. Also, although this guidance can influence and constrain development independently of the method adopted, it allows for a better comparison of the different approaches and the assessment of all the steps and artefacts involved. On the other hand, many ongoing challenges still remain, whether related to its iterative nature, or to the validation and measurability of its current practices, among others [12].

To begin with, the relation between architecturally significant requirements and architectural design decisions is not always free of controversial. In fact, there seems to be no fundamental distinction between them, as they can be perceived as being observed from different perspectives [13]. A certain amount of creativity is always involved and there are different levels of perspective from diverse stakeholders, leading to ambiguity and whether to call something a requirement or a design decision. Nevertheless, as complementary and aligned approaches, one cannot do without the other.

Regarding the transformation process between the two worlds, traditional solutions as the CBSP approach [5] try to solve this issue in diverse ways. As requirements and architectures use different terms and concepts to capture the model elements relevant to each other, one solution is to relate and reconcile those using intermediate models. The process of reconciling is always a difficult task, much based on intuition and experience, where some automation through tool support is desirable, but not a full one, as human intervention is ever decisive.

Also, gradually, quality issues have been the target of increased attention, with system architecture as a major determinant of system quality [14]. While functional properties determine what the software is able to do, the non-functional (quality) properties determine how well the software performs, where explicit architectural decisions can facilitate optimization among quality attributes. Standards like ISO/IEC 25010 [9] help define quality attributes from both an internal and an in use perspective, addressing architectural design and system realization, respectively.

Moreover, architectural evaluation is becoming a familiar practice for developing quality software, as it reduces development efforts and costs. By verifying the addressability of quality requirements and identifying potential risks, it provides assurance to developers that their chosen architecture will meet both functional and non-functional quality requirements [15]. Standing as a reference in the field, the Architecture Tradeoff Analysis Method (ATAM) technique [16] supports the evaluation of architectures and architectural decision alternatives in light of quality attribute requirements. It takes proposed architectural approaches, analyses them, and identifies sensitivity and trade-off points, describing stakeholders' interaction with the system.

Alongside quality, the realities and necessities of modern software development acknowledge the need to develop architectures that are stable, yet adaptable, in the presence of changing requirements. The question of software evolvability, which describes a software system's ability to easily accommodate future changes, makes evolvability a strong quality requirement in an ever-changing world. As business and technology progress and software becomes more complex, development teams face the challenge on how to evolve the systems in their operationally changing contexts [17]. Concurrently, there is also the implicit occurrence of traceability in formalisms, elements and structures through the transformation steps of model-based development approaches. Although seamlessly easy to produce and follow, there is still a lack of support regarding the research and practical use of traceability links [18].

3   Model-Based Transformation Approaches

Within the domain of the transformation from business requirements to system architectures, a number of diverse approaches have been proposed, becoming the target of comparisons [6] and systematic reviews [8]. Among them, the CBSP technique [5] stands as one early, well-cited reference, directly following the Twin Peaks model. Following those studies and in line with this influence, we identified three other recent and research relevant, model-based approaches. The early ATRIUM methodology [9] and the more recent STREAM process [10] present heavy-modelled solutions, counting on diverse views, which have sparked interesting discussions in this research domain. Also, our latest proposal derived from the original 4SRS method, the 4SRS-SoaML method [11], continues to improve on the V-Model solution by adding coverage on services and quality characteristics concerns. Next, these four approaches are presented in more detail.

3.1   The CBSP Technique

Directly following the Twin Peaks model, the CBSP technique [5] stood as an early reference for the transformation from requirements to architecture. Its taxonomies for both requirements and architecture representational models, counts additionally with an intermediate model to iteratively evolve them. Also, its clearly defined transformation process supports the iterative, concurrent development of requirements, architectures and the intermediate CBSP model. With the associated case study and tool definition presented in the paper, it represented a fairly complete solution at that time for a scalable and human intensive problem.

The work on refining requirements complements its process with a structured transformation technique and tool support, emphasizing a multi-perspective of requirements engineering and also on conflict detection and resolution. Nevertheless, it lacked a more formal treatment of requirements. Its support for traceability eases capturing and tracing links, by narrowing the gap between informal or semi-formal requirements and architecture models, as the intermediate CBSP model also helps to relate architectural issues and requirements.

More recently, its use was reported in an App project, where the method was tailored according to the projects needs and even an extension was developed, although the tool support was considered inadequate [19]. With regards to recent trends, it lacks a deeper integration with standards reference models and evaluation of the resulting architecture. Although presenting itself as a simple approach, it already bridges different levels of formality, models non-functional requirements, maintains evolutionary consistency, incomplete models and iterative development, and also handles scale and complexity.

3.2 The ATRIUM Methodology

ATRIUM [9] is a methodology for developing interactive systems, considering both functional and non-functional requirements in different levels of abstraction, and using ISO/IEC 25010 as one of the inputs for its process. It counts on a supporting tool (MORPHEUS [20]) to aid with each proposed model and activity, with a strong focus on goals and scenarios definition, followed by the generation of a proto-architecture within a synthetizing and transformation procedure. Also, it can be iterated over in order to define and refine the different models.

It is an entirely model-based methodology, guiding the concurrent development from system requirements and software architecture, and deals with quality issues, as they are considered from the very beginning of its application. Also it is defined as standard quality compliance, especially with SQuaRE, providing the advantage of a proper separation of the concerns of the system-to-be. It bridges the gap between requirements and software solutions, using jointly interaction and design patterns, exploiting these last as solutions for their implementation.

Overall, it allows understanding flaws in the application architecture, before code is even written. By using the associated tool, each of its models can be easily described and the traceability throughout its elements maintained, allowing for a continued evaluation of its application. Its recent work continues, more directed to the architectural side on design decisions and anti-patterns.

3.3 The STREAM Process

The STREAM process originated from an initial approach, based on model transformations [21], to generate architectural models from requirements models, where the source and target languages are respectively the i* modelling language and the Acme architectural description language (ADL). It counted on activities as the analysis of internal elements, and the application of horizontal and vertical rules.

Later, this approach to derive architectural structural specifications from system goals, had the added development of an activity for selection of an architectural design solution to better achieve non-functional criteria, and the possibility to refine an architecture, inspired by architectural patterns [10]. It presents important work on heuristics which always require experience and know-how from the analysts involved. It has no current tool support, but integration with the iStarTool is planned.

Associated to this base approach there were a number of proposed extensions, namely the STREAM-ADD, supporting the documentation of architectural design decisions, the F-STREAM, presenting a more flexible and systematic process to derive architectural models from requirements, and the STREAM-AP, devoted to improve the choice of architectural patterns from non-functional requirements. Besides these, there is a wealth of other studies surrounding the research community associated to this approach.

3.4 The 4SRS-SoaML Method

Our own 4SRS-SoaML method [11], an evolution on the initial 4SRS method [22] integrated in a V-Model approach, supports and guides the design of information systems architectures. By successive models derivation based on domain specific needs, it promotes the alignment and traceability between the solution logical architecture and the requirements supporting models. It begins in a domain-specific perspective, at a very high level in the chosen domain, and ends with a technological view of the system, with a context for product design. The generated models and the alignment between both problem and solution specific domains, as well as the inherent traceability, are represented by a V-Model.

The heart of the V-Model comprises a complete business trio of processes, goals and rules (PGR), with functional and non-functional requirements handling, serving as input for the 4SRS-SoaML method, to iterate and derive a logical architecture, built on SoaML participants of the future system to-be. As previously referred, since all transformations are model-based, each elicited trio aligns directly with one or more elements in the derived logical architecture. These, in turn, can be traced back to their originating requirements. The use of the SoaML notation is believed to be more adequate for relating business and system information, by leveraging the business requirements and transporting them to implementation phase.

The added non-functional alignment between the problem side PGR requirements and the solution side service architecture participants stands as the latest extension to our work [23]. It aims to generate the quality information associated to architectural services, from business requirements, where the quality information attributes choice and representation are in line with the CISQ Software Quality Characteristics. Besides the natural, model-based, connection between the PGR requirements and the architecture participants, the need for a tighter non-functional information integration led to a detailed proposal of a quality-oriented alignment, by further specifying each participant service with CISQ quality characteristics, and linking them to PGR goals and rules.

4   The MATRA Framework

The number of proposed solutions for the transformation between requirements and architecture is wide and diverse, either in the way they try to answer the different challenges involved as in the different contexts where they are developed and applied. The need to assess and compare the different approaches, justifying each decision and the origin of the proposed issues, led to a couple of noticeable studies. These focused either in the comparison of different perspectives with a criteria checklist [6], mainly composed of a boolean classification of their identified features, and in the systematic review of transformation approaches [8], presenting a taxonomy on its constituents in order to present a conceptual framework for their analysis and comparison.

These previously published results already considered the central issues of requirements elicitation techniques, artefacts on requirements variability and candidate architectures, heuristics and human intervention, iterations and traceability, abstraction and views, and research maturity and quality. Also, in what regards the problem of assessing or comparing approaches in this domain, we too opt for analysing and evaluating each aspect of the reviewed approaches, in opposite to trying to analysing the final results of the different methods [8], as this later is particularly challenging to realize in practice. Following on these studies and also on the earlier analysed solutions from our four selected approaches, which focus on model-based proposals inspired in the Twin Peaks model, we then propose MATRA, a framework for assessing Model-based Approaches on the Transformation between Requirements and Architecture (MATRA).

In order to assess and compare the different approaches, our proposed twofold framework covers the diverse issues within the transformation from requirements to architectures, as needed. First, it centres on the issues that represent the core of a model-based transformation approach from requirements to architecture, counting with the representation models of both the requirement and architecture elements, as well as the transformation process itself (Table 1). In a second view, the issues around traceability, heuristics use and evaluation of each proposed solution are presented (Table 2). Accordingly, the framework is alongside demonstrated by applying it to the four earlier selected approaches.

Table 1. Comparison framework for representation models and transformation process.

	RE model (F + NF)	Transformation method/ notation	Architectural model
CBSP	CBSP metamodel	Architectural styles/ CBSP	CBSP metamodel
ATRIUM	UML scenarios, KAOS and NFR	Architectural style and patterns/QVT	Architectural elements (UML)
STREAM	i*	Horiz./vert. rules and archit. patterns/ATL	Acme
4SRS-SoaML	Processes (use cases), goals and rules (PGR)	4-step rules set/UML	SoaML participants

Table 2. Comparison framework for transformation and evaluation issues.

	Iterative traceable	Heuristics	Tool	Case study	Architecture evaluation
CBSP	Yes/yes	Architectural styles	Rational rose	Cargo router	No
ATRIUM	Yes/yes	PRISMA models	MORPHEUS	Teach mover	No
STREAM	No/yes	Software engineer	iStarTool (*dev.*)	BTW-UFPE	Suite of metrics
4SRS-SoaML	Yes/yes	BP-analyst/ architect	OutSystems studio (*dev.*)	AAL4ALL	ATAM (*dev.*)

4.1   Representation Models and Transformation Method

As a high-level view on the transformation method, we first envision it in a black-box style, focusing on what goes in and what comes out. The first issue to analyse is the elicited requirements representation model, where the handling of functional and non-functional requirements is transversal to all studies and proposals. While STREAM is based on the widely used i* language and the 4SRS-SoaML in UML based standards, both handling together functional and non-functional requirements, ATRIUM uses a mixture of both UML standards regarding the functional side, and KAOS and NFR for the non-functional aspects. The CBSP technique defines its own metamodel for the representation of the base requirements, allowing to identify and isolate non-functional requirements at the system level (SP, system properties) and architectural-element level (CP and BP, component and bus properties respectively).

Considering the representation model of the architectural side, the common deliverable is the presentation of an architecture model. Correspondingly in CBSP, the same own-defined metamodel is used to represent the obtained architectural building blocks required to architect a given system, as this method only identifies architecturally relevant information. ATRIUM uses architectural elements based on UML, with the possibility to further generate a proto-architecture, and the 4SRS-SoaML uses SoaML participants, both OMG-related specifications, while STREAM uses the also well-known Acme language. All these three present different proposals, but all highly-modelled and standard-oriented solutions.

In what regards to the transformation method itself, it is usually classified as rule, ontological or pattern-based, or even as an identity transformation [8]. In the four selected approaches, mostly rely on architectural styles or patterns, but there is also a combination of rule and pattern-based, one pure rule-based, and none recurring to ontological-based or identity transformations. Additionally, all of these three more recent proposals work with standard model-based notations, counting with QVT (ATRIUM), ATL (STREAM, this one still in development) and UML (4SRS-SoaML) supported transformations. The CBSP solution stands a little aside with its close-defined, proprietary process, although it is well defined and modelled. All this information is summarized below in Table 1.

4.2   Transformation and Evaluation

Looking inside the transformation 'black-box', the approaches are normally organized in a series of steps depicted in a workflow-style graph and then described in a step-by-step fashion. Hopefully, they also depict the work products involved, as in the case of ATRUM and STREAM, and preferably the method is specified and tailor-ready in a standard model-based language like SPEM [24], as in the future plans of the 4SRS-SoaML. Although not in a so standard fashion, CBSP uses ETVX (Entry, Task, Verification, and eXit) to document all its steps in a detailed and clear way.

Regarding other issues related to the transformation method and the evaluation of the obtained architecture, according to the issues raised in past research and in the four selected approaches, we summarize the results in Table 2. These issues are around the desired iterative and traceability support that any proposal following the Twin Peaks model should support, and also the definition of heuristics and a supporting tool to aid practitioners in applying the transformation process, due to their human-intensive nature. Lastly, evaluation issues as the execution of a case study or a similar demonstration, and the existence of a final review on the resulting architecture are also checked for.

All approaches support traceability between the requirements and architectures elements, accordingly to the Twin Peaks vision, nevertheless, regarding iteration (as in the example depicted in Fig. 1 for the 4SRS-SoaML associated V-Model), the STREAM process does not seem to be adequate for it. The other approaches support increasingly detailed iterations to further refine the requirements and architecture elements. This also means that STREAM could not initiate from the architectural side, being a purely transformation from requirements to architecture [12].

Heuristics are also an essential part of any transformation process, as human participation is always required (although tool support can accompany its application). While CBSP and ATRIUM impose stricter rules, the STREAM and 4SRS-SoaML approaches allow for more human intervention, namely from the software engineer and the business-process analyst/architect, respectively.

All the analysed approaches involve a Case Study in the evaluation of their proposals, although the one for the 4SRS-SoaML is still in an early phase of the project involved. Also regarding the tool usage, all adhere to it, counting with open-access development platforms, but the STREAM and 4SRS-SoaML solutions are still in an initial development phase.

As referred previously, the plain evaluation of a transformation proposal is no longer enough, an added evaluation of the quality of the resulting architecture is also important nowadays. The STREAM approach already performed an evaluation based on a previously defined suite of metrics, but plans to advance for deeper evaluations. Similarly, the 4SRS-SoaML intends to realize an ATAM evaluation over their resulting architecture.

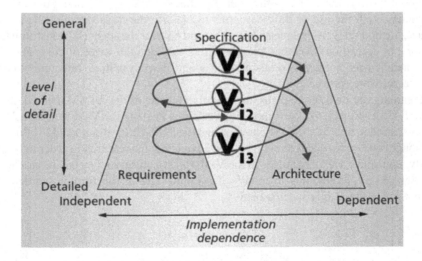

Fig. 1. V-Model integration with the Twin Peaks model.

5 Discussion

Diverse approaches are currently available, using different strategies to handle the connection between requirements and architecture, according to each intervention setting. Although structurally alike, each approach has its proper insights. Our 4SRS-SoaML tries to follow on the same structure of the other similar approaches, but also presenting its added values. The complete alignment between the business requirements and quality characteristics of logical architectures is our latest proposal, with a complete ATAM evaluation being prepared [23].

Although there are several other similar approaches in the research universe, these four we selected were the ones that comply with most of the issues being analysed and that can set the trends for future research in this area. The usage of model-based languages and tools as well as standard-based reference models is increasing, being almost omnipresent in recent research proposals.

Questions as the inclusion of non-functional requirements, added to the traditional functional requirements, and increased balance between hard transformation rules and loose user heuristics are also ubiquitous. Although several heuristics have been proposed for each existing process, some are more creativity and knowledgeability-dependable, and sometimes much is left to the responsibility and decision of the analysts. Even so, the need to further refine existing heuristics and develop new heuristics is ever-present.

Case studies remain the preferred way to evaluate and validate any proposal, but architecture evaluation is on the mind of researchers already. Even so, further evaluation is always recommended, especially when dealing with quality issues and complex scenarios.

Relative to alignment and traceability consistency, all proposals seem full-proof as all its steps and elements are modelled. Nevertheless, there is still some work to do, relating to a quicker access in obtaining related elements, especially involving manual tasks performed by analysts, as not all the proceedings are tool supported. So, additional tool support development is an interesting asset for any proposal.

Open questions to be dealt with are also in the poor formalism among the different approaches, with the use of different terms to classify the transformation (approach, method, methodology, technique and process) and also for the act itself (transformation, transition or derivation). Another question refers to widen the scope of the approaches from modelled requirements to business ones and from system architectures to ones closer to software specification.

Regarding our own future plans, they involve the use of the ATAM method, inside the AAL4ALL project, to assess the new extensions made to the V-Model, namely the PGR metamodel, the 4SRS-SoaML method and the PGR-CISQ alignment [24]. Another long sought objective relates to evolvability, advancing this research in order to be able to fully align and trace all elements in the V-Model structure, so they can be stored and used during design, enabling to modify the architecture at run time in configurable systems, according to any specific quality.

6   Conclusion

The Twin Peaks model has set the foundations for much of the research around the transformation of requirements engineering to information system architectures. In the meantime, research has advanced those basis and several new trends, in diverse research directions, have been proposed, so there is a need to classify and compare them within their constituents.

This study proposed a framework to assess our proposal, while framing and comparing our 4SRS-SoaML method relatively to other similar and relevant model-based approaches in this domain. By performing a review on their transformation issues from requirements to architectures, it covered each of the previously identified research issues and then classified each of the considered approaches accordingly.

Furthermore, we identify open issues and make recommendations for future work. Besides the need for a better specification of the steps involved in the different methods, clearly stating the tasks, work products and roles involved, there is also a necessity to improve the formalism involved in the research work.

Acknowledgments. This work has been supported by FCT – *Fundação para a Ciência e Tecnologia* in the scope of projects: PEst-OE/EEI/UI0319/2013 and FCT/MITP-TB/CS/ 0026/2013.

References

1. France, R., Rumpe, B.: Model-driven development of complex software: a research roadmap. In: Future of Software Engineering, pp. 37–54 (2007)
2. Selic, B.: What will it take? A view on adoption of model-based methods in practice. Softw. Syst. Model. **11**, 513–526 (2012)
3. Whittle, J., Hutchinson, J., Rouncefield, M.: The state of practice in model-driven engineering. IEEE Softw. **31**, 79–85 (2014)
4. Nuseibeh, B.: Weaving together requirements and architectures. Computer **34**, 115–119 (2001)
5. Grünbacher, P., Egyed, A., Medvidovic, N.: Reconciling software requirements and architectures with intermediate models. Softw. Syst. Model. **3**, 235–253 (2004)
6. Galster, M., Eberlein, A., Moussavi, M.: Comparing methodologies for the transition between software requirements and architectures. In: IEEE International Conference on Systems, Man and Cybernetics, pp. 2380–2385 (2009)
7. Loniewski, G., Insfran, E., Abrahão, S.: A systematic review of the use of requirements engineering techniques in model-driven development. In: Petriu, D.C., Rouquette, N., Haugen, Ø. (eds.) MODELS 2010, Part II. LNCS, vol. 6395, pp. 213–227. Springer, Heidelberg (2010)
8. Yue, T., Briand, L.C., Labiche, Y.: A systematic review of transformation approaches between user requirements and analysis models. Requir. Eng. **16**, 75–99 (2011)
9. Montero, F., Navarro, E.: ATRIUM: software architecture driven by requirements. In: 14th IEEE International Conference on Engineering of Complex Computer Systems, pp. 230–240 (2009)

10. Castro, J., Lucena, M., Silva, C., Alencar, F., Santos, E., Pimentel, J.: Changing attitudes towards the generation of architectural models. J. Syst. Softw. **85**, 463–479 (2012)
11. Salgado, C.E., Teixeira, J., Santos, N., Machado, R.J., Maciel, R.S.P.: A SoaML approach for derivation of a process-oriented logical architecture from use cases. In: Nóvoa, H., Drăgoicea, M. (eds.) IESS 2015. LNBIP, vol. 201, pp. 80–94. Springer, Heidelberg (2015)
12. Cleland-Huang, J., Hanmer, R.S., Supakkul, S., Mirakhorli, M.: The twin peaks of requirements and architecture. IEEE Softw. **30**, 24–29 (2013)
13. de Boer, R.C., van Vliet, H.: On the similarity between requirements and architecture. J. Syst. Softw. **82**, 544–550 (2009)
14. Croll, P.R.: Quality attributes - architecting systems to meet customer expectations. In: 2nd Annual IEEE Systems Conference, pp. 1–8 (2008)
15. Shanmugapriya, P., Suresh, R.M.: Software architecture evaluation methods - a survey. Int. J. Comput. Appl. **49**, 19–26 (2012)
16. Kazman, R., Klein, M., Barbacci, M., Longstaff, T., Lipson, H., Carriere, J.: The architecture tradeoff analysis method. In: 4th IEEE International Conference on Engineering of Complex Computer Systems, pp. 68–78 (1998)
17. Breivold, H.P., Crnkovic, I., Larsson, M.: A systematic review of software architecture evolution research. Inf. Softw. Technol. **54**, 16–40 (2012)
18. Winkler, S., von Pilgrim, J.: A survey of traceability in requirements engineering and model-driven development. Softw. Syst. Model. **9**, 529–565 (2010)
19. Vogl, H., Lehner, K., Grünbacher, P., Egyed, A.: Reconciling requirements and architectures with the CBSP approach in an iPhone app project. In: 19th IEEE International Requirements Engineering Conference (RE 2011), pp. 273–278 (2011)
20. Navarro, E., Gómez, A., Letelier, P., Ramos, I.: MORPHEUS: a supporting tool for MDD. In: Song, W.W., Xu, S., Wan, C., Zhong, Y., Wojtkowski, W., Wojtkowski, G., Linger, H. (eds.) Information Systems Development, pp. 255–267. Springer, New York (2011)
21. Lucena, M., Castro, J., Silva, C., Alencar, F., Santos, E., Pimentel, J.: A model transformation approach to derive architectural models from goal-oriented requirements models. In: On the Move to Meaningful Internet Systems: OTM 2009 Workshops, pp. 370–380 (2009)
22. Machado, R.J., Fernandes, J.M., Monteiro, P., Rodrigues, H.: Refinement of software architectures by recursive model transformations. In: Münch, J., Vierimaa, M. (eds.) PROFES 2006. LNCS, vol. 4034, pp. 422–428. Springer, Heidelberg (2006)
23. Salgado, C.E., Machado, R.J., Maciel, R.S.P.: Aligning business requirements with services quality characteristics by using logical architectures. In: 3rd World Conference on Information Systems and Technologies, pp. 1–10 (2015)
24. Salgado, C.E., Machado, R.J., Maciel, R.S.P.: A three-dimensional approach for a quality-based alignment between requirements and architecture. In: Borangiu, T., Dragoicea, M., Nóvoa, H. (eds.) IESS 2016. LNBIP, vol. 247, pp. 112–125. Springer, Heidelberg (2016). doi:10.1007/978-3-319-32689-4_9

Enterprise and Software Ecosystem Modeling

Data Accountability in Socio-Technical Systems

Kristian Beckers$^{(\boxtimes)}$, Jörg Landthaler, Florian Matthes,
Alexander Pretschner, and Bernhard Waltl

Faculty of Informatics, Technical University of Munich, Munich, Germany
{kristian.beckers,joerg.landthaler,matthes,
alexander.pretschner,b.waltl}@tum.de

Abstract. Data-accountability encompasses responsibility for data and the traceability of data flows. This is becoming increasingly important for Socio-Technical Systems (STS). Determining root causes for unwanted events after their occurrence is often not possible, e.g. because of missing logs. A better traceability of root causes can be supported by the integration of accountability mechanisms at design time.

We contribute a structured method for designing an accountability architecture for STS at design time. Therefore, we propose the elicitation of accountability goals to answer why an unwanted event happened and who is responsible for it. We also identify four different interaction types in STS. Additionally, we derive accountability graphs from a generic accountability model for STS that serve as a baseline for designing accountability mechanisms for all relevant entities in an STS. The resulting architecture is adjusted to legal requirements, regulations and contracts. We demonstrate the applicability of our approach with an eHealth case study.

Keywords: Data accountability · eHealth · Socio-technical systems · Accountability architecture · Interaction types · Accountability method · Accountability graph

1 Introduction

An important aspect of information systems (IS) is compliance to legal standards and organizational policies. In case of a violation of a statute it becomes more and more important to identify responsible parties, i.e. to hold someone accountable, assuming that a system can not be responsible on it's own. Hence, an IS is embedded in a so-called Socio-Technical System (STS, (c.f. [1])) encompassing juristic persons and technical systems. Currently, the impact of technical systems to prior analogue world definitions of accountability is unclear. Only few attempts have been made to include the ability of tracing root causes of unwanted events by design, such as [2]. Current efforts for data accountability IS lack either the social aspect of IS or a solution by design or both. Feigenbaum et al. [3] adopts the concept of accountability that is known well in the analogue world to ensure security in information systems and proposes that the ability to

© Springer International Publishing Switzerland 2016
R. Schmidt et al. (Eds.): BPMDS/EMMSAD 2016, LNBIP 248, pp. 335–348, 2016.
DOI: 10.1007/978-3-319-39429-9_21

punish people will prevent them from doing illegal actions. The question, how to enable IS with this ability remains open. One much-noticed approach is proposed by Weitzner et al. [4] by demanding that each subsystem should be responsible to ensure accountability on its own using an appropriate accountability mechanism, e.g. by policy-aware transaction logs.

However, it is unclear how the interaction between humans and machines affect accountability. Within an organization it is important to identify (and define) responsible roles. Another important task is to identify relevant regulations and policies. We analyze possible interaction types in STS, use Data Governance principles to allocate responsibilities and provide a model of STS. Afterwards, we propose a structured method empowering engineers to enhance an STS with accountability by design, which can *a posteriori* determine causalities within such a system. Accountability in this work is a capability of an STS to answer questions regarding the cause of occurred unwanted events (e.g. privacy or security violations).

We limit ourselves to data accountability, i.e. unwanted events whose causes are data-related. Consider e.g. software malfunctions producing wrong data, hardware failures due to faulty interpretation of data or wrong usage of technical systems (intended or not) caused by wrong instructions. The system should be enabled to determine the causality and identify responsible parties. Accountability mechanisms (which could be, but is not restricted to, logging) need to answer the question why an unwanted event happened.

The creation of an accountability solution is related to legal compliance in two ways. Firstly, an accountability solution can support legal compliance. Companies have to ensure transparency and provenance of their data, e.g. when adherence to regulatory frameworks such as HIPAA in the U.S., [5], demands specific needs for the confidentiality and security of healthcare information that describe specific principles regarding transparency and provenance of data. Secondly, the introduction of an accountability solution itself can cause additional legal compliance demands. For example, the gathering and storage of personal data requires compliance to data protection acts, such as the EU Data Protection Directive (95/46/EC). Our structured method explicitly identifies and considers legal compliance as part of designing an accountability solution.

This paper is structured as follows: Sect. 2 reviews related work on data governance and accountability. Section 3 describes our structured method for creating an accountability architecture, Sect. 4 illustrates our accountability STS model, and in Sect. 5, we inherit STS relation types. We apply our method in Sect. 6 and conclude our research in Sect. 7.

2   Background and Related Work

2.1   Data Governance

Weill and Ross [6] presented a framework and structuring mechanism for IT Governance. For a management perspective, they identified five key decision domains: IT architecture, IT Infrastructure, IT investment and prioritization decisions and business applications needs. Based on this research Khatri and Brown [7] focused on Data Governance and identified five decision domains, namely data principles, data quality, data access, metadata and data lifecycle. The authors use these domains as a blueprint for identifying and assigning data responsibilities, i.e. roles. However, their work does not include the causal aspect of accountability nor the relations between different roles. Our method shows how to use their framework to identify relations between roles to determine responsible persons for an unwanted event.

An important part of data governance, is to determine the origin of a certain datum by means of its source, often called Data Provenance, e.g. Buneman et al. [8]. Moreover, the manipulation history of a datum and potentially the person or other technical components or even a chain of components that led to the manipulation of the datum can be useful to answer accountability questions. We propose to design an accountability solution for a STS that allows to address challenges of Data Provenance.

2.2 Accountability

Weitzner et al. [4] proposed an understanding of accountability such that it reflects the ability of a system to answer questions regarding the why of occurred events. For example, why was personal data released to unauthorized staff?

Accountability is the subject of active research in different areas of computer science such as network engineering, see Bechtold and Perrig [9]. Fundamental for accountability is an understanding of causality in general (c.f. Gössler and Le Métayer [10], Halpern and Pearl [11]). In our research, it is essential to consider the efforts of Data Governance as explained above, which provide the essential information of responsible parties and rules of how data should be treated. In particular, effective Data Provenance is necessary to enable the traceability of data through an STS. Without these traces, answering the accountability question is impossible. A straight forward accountability mechanism is the employment of logs at every node of a system, which follows Weitzner et al. [4]. The authors proposed policy-aware transaction logs, which are created by logging of relevant information by individual entities in a system. Identifying fundamental accountability concepts of who and how are discussed in Eriksn [12].

We illustrate our approach using an eHealth example. Gajanayake et al. [13,14] address a similar problem by applying information accountability to systems in the eHealth domain. However, they neither provide a structured method to design a general accountability architecture, nor incorporate data governance structures, nor examine causality chains.

3   A Structured Method for Designing an Accountability Architecture

We contribute a structured method for the design of an accountability architecture (see Fig. 1). Our method is presented in a sequential fashion for simplicity's sake; iterations between different steps are possible during its application.

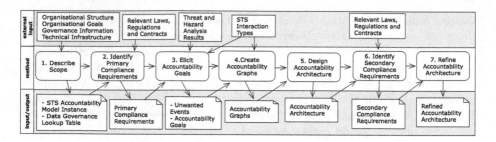

Fig. 1. A method for designing an accountability architecture

Step 1. Describe Scope Initially. We have to understand the STS for which we want to elicit accountability. We propose an accountability model that contains abstract descriptions of all technical components and roles in the scope of an accountability system. Thereby, we describe interactions between these STS elements in order to be able to trace unwanted events to their sources in later steps of our method. We describe in detail how to model the STS and its relation in Sects. 4 and 5.

Step 2. Identify Primary Compliance Requirements. We identify relevant compliance requirements arising from regulative texts for the STS as motivation for designing an accountability architecture. For example, if we process personal information in our STS in Germany the Federal Data Protection Act (BDSG) is relevant.

Step 3. Elicit Accountability Goals. We elicit unwanted events by considering the STS model and the organizational goals of the customer. Note that in our work an unwanted event is any occurrence within an STS that concerns violations of safety, security, or privacy requirements. We distinguish unwanted events regarding these software qualities as follows. Safety analysis focuses on hazards caused by the engineers of soft- and hardware and random faults in these systems. Hazards are situations that lead to accidents that harm humans. Hence, in safety unwanted events are accidents. Security is about protecting an asset, an item of value for a stakeholder from threats caused by malicious attackers or unintentional acts of stakeholders. Realized threats are attacks. Thus, in security analysis unwanted events are successful attacks. Privacy concerns the protection of personal information of stakeholders. An unwanted event in privacy is a data leak of personal information to unauthorized stakeholders. Our work does not restrict the techniques for hazard or threat analysis. For space reasons, we exclude these analyses in this paper. Afterwards each unwanted event is mapped to an accountability goal that describes the abilities the system shall have to identify causes for this particular unwanted event (see examples in Sect. 6). After having identified the cause for the unwanted events in the STS, we need to derive the responsible actor for that element. The information can be derived from existing governance data of the organization. For this purpose, we create a Data Governance Lookup Table mapping STS elements to responsible actors.

Step 4. Create Accountability Graphs. We provide a divide and conquer approach for accountability to reduce the complexity of the overall design problem. For each accountability goal, we create a separate accountability graph as follows. Nodes in our graph are STS elements e.g. humans or machines, while edges are communication channels between these STS elements. The first node of the graph is the STS element where the unwanted event occurs. Afterwards we include all STS elements and relations that are part of the normal operations of the first node of the graph. Using the accountability graph, we have the capability to identify the potential loci of the root cause of the unwanted event by conducting a search along all relevant nodes of the information flow between involved entities. The reasoning for creating the graph is documented for later analysis and should be checked by independent experts, which shall prevent that our accountability graphs have incomplete information.

Step 5. Design Accountability Architecture. We need to ensure that all elements in the accountability graph have the capability to support the information needed for our causal reasoning from the unwanted event to its source. In particular, each node has to have a mechanism to monitor and document relevant events. The result is an accountability architecture that ensures the satisfaction of the accountability goal.

Step 6. Identify Secondary Compliance Requirements. we identify relevant secondary compliance requirements that can be identified for our resulting accountability architecture. For example, an intensive logging of personal information may conflict with a given privacy legislation. This step results in a set of compliance documents that are refined into precise compliance requirements for our proposed accountability architecture.

Step 7. Refine Accountability Architecture. The planned accountability architecture is revised according to the compliance requirements. The result of this step is a compliant and precise description of an accountability architecture that satisfies the initial accountability goals, as well as the elicited compliance requirements.

4 A Generic Model for Socio-Technical Systems

As already stated, socio technical systems become increasingly complex. This has several reasons and up to certain extent this is due to the contained entities and their tight interconnectedness among each other. Results from Enterprise Architecture Management (EAM) have significantly improved the understanding of business, their capabilities and the interconnectedness to technical and physical entities Lankhorst [15], Jonkers et. al [16]. Based on the insights gained from EAM, we are able to transfer those results into a generic STS model. The interconnectedness of the STS elements provides insights into the different communication channels between those elements, which we show in our generic STS model (depicted in Fig. 2).

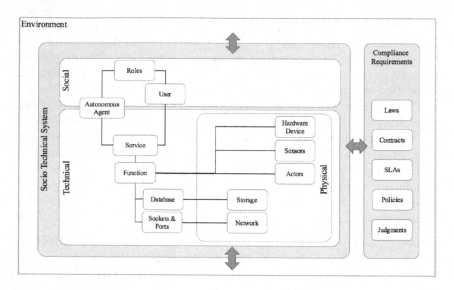

Fig. 2. A generic accountability model for STS differentiating social and technical layers.

The model shown in Fig. 2 consists of three different layers: social, technical and physical. Accountability in STS has several dimensions, which we are now going to examine. Thereby, we are able to assign each dimension to at least one of those three layers. Consequently, this approach provides a constructive and structured way of differentiating the term accountability into sub-problems, which can then be investigated separately. Trivially, the model shown in Fig. 2 serves as a base line to answer the question regarding the dimensions of accountability; therefore, it remains - as every model - an abstraction from a real world STS. Nevertheless, it is comprehensive in the sense that we can use it to differentiate between the dimensions of accountability in interacting STS. Most relevant for the accountability dimensions are of course the obligations of an STS arising through legislation, contracts, SLAs and other policies governing the flow and management of data and information assets in general.

Social. The social dimension of an enterprise covers all organizational units and human actors, which are interacting among each other and with the technical systems. Commonly they are organized into roles aggregating them according to responsibilities, tasks and goals. In addition to the users, autonomously acting users, i.e. agents, get more and more in the focus of investigations regarding accountability.

Technical and Physical. The technical dimension of an enterprise covers the application landscape with its services and functions. Hereby, "functions" can be understood in a technical sense, such that concrete functionalities, such as network communication and persisting data in databases, are subsumed. Those technical functions are aggregated to more complex services, which are later on consumed by users or agents to fulfil their needs. The physical layer is part of the technical layer and covers hardware devices and interactions at hardware level. Every device that is either measuring physical phenomena or states, i.e. sensors, or changing the physical environment, i.e. actors, belongs to the physical layer. This differentiation in layers briefly shows how the components of an STS are interacting among each other. These interactions have to be investigated to fully understand the challenges and drawbacks of data accountability (see Sect. 5).

5   Interaction Types in STS

Based on the model for an interacting STS (see Fig. 2) we distinguish the different types of interactions. We identified four different types of interactions, namely human-human, human-machine, machine-human, machine-machine, which we describe in Table 1. Those interaction types have an impact on the design and implementation of the accountability mechanism.

Accountability mechanisms in STS have to consider these interaction types, otherwise no comprehensive reconstruction of behavior and explanations can be performed. This has consequences for the design of such accountability mechanisms. Those mechanisms heavily influence the way in which data has to be tracked and logged in the overall STS and how this data can be stored. Based on this stored data it is possible to automatically derive accountability information and to reconstruct the root cause of an event, c.f. Waltl et al. [17].

Table 1. Interaction Types in STS between Humans and Machines

Human → Human	In order to reconstruct the behavior of a human, and the reasons for it, it might be necessary to understand and retrace the information (e.g. instructions) he got from another human. This reconstruction requires information which might not be codified properly (e.g. burden of proof), such that an explanation cannot be given.
Human → Machine	Human interact with machines, i.e. services, sensors, etc. provided by the technical and physical layer of the STS. However, if the human interacts with machines, such as insertion, update or deletion of information, this has effects on the technical layer. Consequently, to reconstruct the actions done by machines it is required to understand the triggers that caused the machine to perform a certain action.
Machine → Human	Machines can offer information to human, such as notifications about an event. The provision of information by a technical system causes the human to perform a follow-up action or hinders him from doing some actions. Keeping trace of the information that was offered to humans is not trivial, but essential in order to reconstruct the behavior of the STS.
Machine → Machine	The interaction between machines, such as retrieving and aggregating data from sensor networks or forwarding commands to an actor that changes the physical environment, is the fourth interaction type in STS. Which service, respectively function, has processed which data and forwarded it to which instance, is a critical question that has to be answered by accountability mechanisms in STS.

6   Case Study

We illustrate our approach with the case study *eHealth Record (EHR)* adapted from the NESSoS[1] project. EHRs contain any information created by health care professionals or devices in the context of the care of a patient. Examples are laboratory reports, X-ray images, and data from monitoring equipment. The method will be executed by a dedicated accountability officer in coordination with lawyers, domain experts, and software engineers.

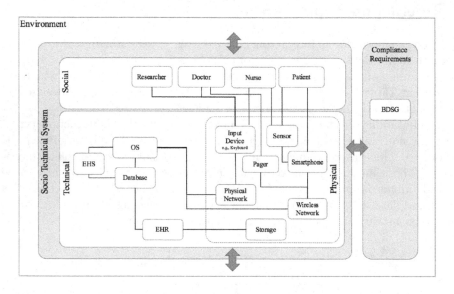

Fig. 3. Instance of our accountability model for STS for the eHealth scenario

Step 1. Describe Scope EHRs are part of an eHealth System (EHS) owned by a hospital. The overall organizational goal of a hospital is to fulfil the societal goal to provide health care for patients. An EHS with its EHRs shall help to treat patients more efficiently and effectively. For example, the nurse does not need to take the vital signs for specific time intervals and deliver them to the doctor manually, because the EHS fulfils these tasks automatically, hence saving working time. We illustrate our example in Fig. 3. The EHS is a software that stores medical information in EHRs. Further, it interacts with different users and communicates with various devices and serves as the example of an STS. In Germany, an EHS has to be compliant with the Federal Data Protection Act (BDSG). Hence, the information stored in the EHR shall only be accessed with the patient's informed consent. An exception to this rule is a medical emergency, in which case the patient's physical status may prevent her from giving the consent. In addition, the information in the EHR supports clinical research, which

---

[1] Network of Excellence on Engineering Secure Future Internet Software Services and Systems (NESSoS), http://www.nessos-project.eu, last access on 03/23/2016.

is represented by a researcher in this scenario. The patient wears a sensor that is monitoring her vital signs and communicates them to the patient's smartphone. The smartphone is transmitting the data via a wireless network to the EHS. Doctor, nurse and researcher use a terminal that is connected to the EHS vial a physical network. The doctor carries a pager in order to receive emergency calls from the EHS. The EHS is embedded in the organizational structure of a hospital. In the following, we design a runtime accountability mechanism for this example scenario. We start by defining the data governance roles for our EHR scenario. Hospitals often host a large IT landscape for various purposes. The EHS supports different user roles and is also embedded in the organizational structure of a hospital. A Data Owner is essential to our method. Note that we refer to a data owner in the sense of Data Governance according to Khatri and Brown [7]. A Data Owner is the trustee responsible for data and its uses. In our example the doctor is the Data Owner being in charge of the health data in general, be cause on the one hand he is allowed to access the data while on the other hand he often assumes managerial tasks regarding EHRs in a hospital. The nurse is assigned a Data Steward role that enters certain data about patients into the EHS. A Data Steward according to Khatri and Brown [7] is responsible for what is stored in a set of data. This is a delegation of responsibilities from the doctor as Data Owner to the nurse as Data Steward. Furthermore, the nurse is also assigned the Data Custodian role. A Data Custodian takes care of a working technical infrastructure for collecting or transporting data. In our case, the nurse has to take care of the maintenance of the sensor e.g. exchanging its batteries. The patient assumes the role of a Data Provider that repeatedly sends data from a sensor to the EHS where it is then stored. A typical Data Consumer is the researcher that merely receives data for his research activities. Next, we need to elicit the accountability goals. Table 2 shows an excerpt of a Data Governance Lookup Table.

Table 2. An Excerpt of a Data Governance Lookup Table

Data governance responsibility	Responsible role
(D1) Data in the database	(R1) Data Owner (Doctor)
(D2) Update EHR data	(R2) Data Steward (Nurse)
(D3) Sensor maintenance	(R3) Data Custodian (Nurse)

Step 2. Identify Primary Compliance Requirements. The German Federal Data Protection Act (BDSG)[2] is relevant for an application of the proposed system, because it processes the personal information of the patient. The § 3 of the BDSG states that personal information can only be elicited, stored and processed for a

---

[2] Note: We will address the inclusion of further laws and resolving conflicts between them in the future and focus in this paper exclusively on the BDSG.

Table 3. Selected unwanted events and accountability goals

Unwanted events	Accountability goals
(E1) Patient is in an emergency and does not get help from the doctor.	(G1) Ability to reconstruct the root cause for the patient not receiving immediate help from the doctor during the medical emergency.
(E2) Patient received wrong treatment from the nurse.	(G2) Ability to analyze why the patient received a harmful treatment from the nurse.
(E3) Researcher access PII from Patient: name, disease, vital signs	(G3) Ability to identify the cause of the data leak of the Patient's PII

specific purpose and have to be anonymized if possible. In this case the doctor and nurse need to know the identity of the patient to be able to diagnose and administer treatment. Moreover, according to § 4 of the BDSG the patient has to provide an informed consent about the processing of her personal information and who will access it.

Step 3. Elicit Accountability Goals. We show three prototypical unwanted events and their respective accountability goals in Table 3. Note that for space reasons we omit the threat and hazard elicitation. We focus in our example on events that physically harm the patient or violate the patient's privacy. We derive accountability goals for each event that demand an accountability mechanism to trace the causes for this particular unwanted event within the scope of the STS.

Step 4. Create Accountability Graphs. We choose accountability goal (G1) *why the patient received a harmful treatment from the nurse* for the remainder of this example. We trace back all involved elements of the STS in the instance of the accountability model from the patient to the doctor and gather a sub graph of elements and their relationships. The resulting accountability graph is depicted in Fig. 4. All of these elements can cause the missing communication of the patient's emergency to the doctor. The elements were selected based on the information in the hazard and threat analysis. In our example for this particular accountability goal, we do not consider the researcher, because a data leak does not relate to this specific accountability goal.

Step 5. Design Accountability Architecture. We design an accountability architecture, which is comprised of several local accountability mechanisms. For each accountability goal all possible nodes and edges of the corresponding accountability graph need to be assessed for the demand of a separate accountability mechanism with respect to the accountability goal under consideration. The intention of this assessment is to find adequate parts of the accountability graphs so that individual causes can be localized with respect to the organizational needs. We are aware that there exist different types of accountability mechanisms, e.g. digital or analogue that are also potentially limited. We model each account-

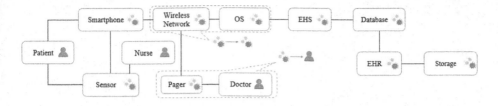

Fig. 4. Accountability graph for an eHealth scenario (2 interaction types shown)

ability mechanism as an STS, too. Following the policy-aware transaction log accountability mechanism, the patient needs to document every medicine ingestion. Both, analogue or digital logs, e.g. handwritten or via a tablet can be considered. A problem that arises in the design phase is the level of granularity that the information needs to have. The information should not be too detailed, because the identification of relevant details will take time and resources and the information should not be too abstract in order not to miss vital information for identifying the cause of the unwanted events.

Step 6. Identify Secondary Compliance Requirements. Humans have the right for transparency according to §§ 19,34 of the BDSG for any system that processes their personal information. In particular, transparency demands a detailed and complete report on the life cycle of the personal information. An accountability architecture as proposed in this work has the ability to provide this information with little effort. The accountability graph provides an abstract view of the flow of personal information in the system, which can be accompanied with detailed access logs of all persons reading or changing the personal information of the patient. These data provenance capabilities of the chosen accountability mechanism will improve the transparency of STS significantly, because all the foundations for providing detailed reports to affected persons will be available. However, these large amounts of personal data of the patient have to be protected from access of further actors in this scenario. For example, doctors or nurses that are employed by the hospital but are not involved in the treatment of the patient have to be prevented from gaining access to that data (BDSG §9). Moreover, the data has to be deleted after the purpose for its initial collection is not valid anymore, e.g., the patient is no longer treated (BDSG §§ 20, 35).

Step 7. Refine Accountability Architecture. We illustrate our resulting accountability architecture in Fig. 5. The architecture is comprised of individual accountability mechanisms that ensure the logging and monitoring of individual components and delivering these logs to the accountability evaluation part of our architecture. The evaluation takes care of analyzing the log files and answering the why and who questions of accountability. The resulting compliance requirements of Step 6 are incorporated into the architecture. For example, we have to incorporate access control mechanism for the data and a process that checks if the purpose for storing the data has not expired. This needs to be done for all accountability mechanisms to ensure a holistic solution for these problems.

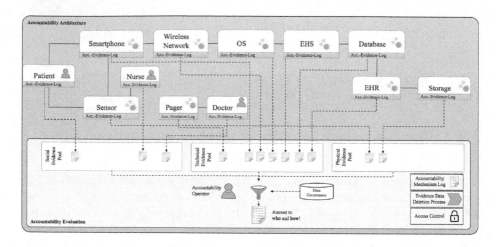

Fig. 5. Resulting accountability architecture for the eHealth scenario

We analyze all involved components of our accountability graph in detail for our exemplary unwanted event *Patient is in an emergency and does not get help from the doctor* and determine which data of the component has to be stored to be able to determine if this component was (part of) the root cause of the unwanted event. Additionally, for a policy-aware transaction log accountability mechanism for all components, it needs to be decided whether information that is forwarded to further components of the accountability graph needs to be stored in the component.

We consider in our example that the patient has a heart attack and the sensor monitoring his heart frequency should report this to the doctor. We choose the sensor as first component to consider in our accountability architecture. We need to log what information the sensor is capturing and at what time. The log can answer the first accountability question, did the sensor malfunction and did not record the correct heartbeat. We have to ensure that the log exists over time. Due to limitations of the sensor's memory capacity, the information has to be transported and stored on the smartphone. Each time a batch of information is transferred to the smartphone, the log file in the sensor stores a hash of the transported information and the date of transmission. This allows checking at the smartphone if all the data from the sensor has arrived at the smartphone. Moreover, the sensor transmits the heartbeat every 20 s to the smartphone. We have to implement a logging mechanism at the sensor that persists the information what was send to the smartphone at what time. This information allows us to decide if information was not send by the sensor or not received by the smartphone. Furthermore, we have to determine similar decisions for the smartphone, e.g., check that data was evaluated correctly and that the smartphone send an emergency message via the wireless network and the OS to the EHS and finally to the pager of the doctor.

So far all considered interaction types are of the machine-to-machine interaction type (see Sect. 5), which allowed us to specify automatic logging procedures. The doctor shares a human-machine interaction type with our system, which means that the she has to manually log her activities. For our example, a policy states that the doctor has to log all his reactions to received pager message. The purpose of this policy is that in a post mortem analysis it can be decided if the doctor reacted to all pager messages reasonably. A sensor's digital log file can be limited by its memory size. However, the sensor's log file can be sent and aggregated on the smartphone and still fulfil the accountability goal with respect to the organizational needs. Either you detect a node that is a machine, hence there needs to be a lookup in the Data Governance Lookup Table (see Table 2) who is responsible for the machine, or a role is detected and a lookup can be necessary, too.

In our use case, we find e.g. anomalies in the sensor log aggregated on the smartphone. This happened due to missing replacement of batteries from the data custodian of the sensor that is of the type machine. This is accounted by a missing manual log entry of the data custodian. Hence, we need a lookup in the Data Governance Lookup Table in order to find the responsible person for the machine. In this case, the nurse (R3) is responsible data custodian for the sensor maintenance. In addition to that, one could lookup who in the organization is the data owner of the sensor data, which in our case is the doctor. Hence, he also has a partial responsibility for the resulting problems of the patient. Further criminal investigations have to rule out any other causes such as the batteries have been robbed.

7   Conclusion and Future Work

This paper structures the accountability concept in socio-technical-systems (STS) by differentiating the four different types of interactions, namely human-human, human-machine, machine-human and machine-machine. Our approach is restricted to accountability on data. Consequently, we exclusively consider and analyze the flow of information, i.e. data, during the possible interactions. The interaction types allow a structuring of the various forms of accountability, offering an analytical way of defining accountability mechanisms considering relevant requirements arising from laws, SLAs, contracts, etc. Based on these insights we propose a structured method for deriving an accountability solution that incorporates functionalities for answering the questions of why an unwanted event did happen and who is responsible. We rely on previous work for data governance to answer the responsibility question and work on data accountability for answering the why question. We illustrate our approach by a case study in the eHealth domain. This proof of concept shows the applicability of our approach and is the baseline for our next steps, which are a more detailed conceptualization and implementation of these accountability mechanisms in an STS. Based on the proposed concept and differentiation, it is now possible to derive concrete accountability mechanisms based on data flow and information exchange. Trivially, these mechanisms need to be tailored to meet the requirements of a specific domain. We consider our approach as a step towards a unified understanding of data accountability, which can serve as a solid foundation for future research and applications.

348   K. Beckers et al.

Acknowledgments. This work is part of TUM Living Lab Connected Mobility (TUM LLCM) project and has been funded by the Bayerisches Staatsministerium für Wirtschaft und Medien, Energie und Technologie (StMWi).

References

1. Trist, E.L.: The evolution of socio-technical systems : a conceptual framework and an action research program. Ontario Quality of Working Life Centre On cover: Issues in the quality of working life : a series of occasional papers, no. 2 (1981)
2. Bonazzi, R., Hussami, L., Pigneur, Y.: Compliance management is becoming a major issue in is design. In: Information Systems: People, Organizations, Institutions, and Technologies: ItAIS: The Italian Association for Information Systems, pp. 391–398. Physica-Verlag HD, Heidelberg (2010)
3. Feigenbaum, J., Jaggard, A.D., Wright, R.N.: Towards a formal model of accountability. In: Proceedings of the 2011 Workshop on New Security Paradigms Workshop, NSPW 2011, pp. 45–56. ACM, New York (2011)
4. Weitzner, D.J., Abelson, H., Berners-Lee, T., Feigenbaum, J., Hendler, J., Sussman, G.J.: Information accountability. Commun. ACM **51**(6), 82–87 (2008)
5. Banks, D.L.: The health insurance portability and accountability act: does it live up to the promise? J. Med. Syst. **30**(1), 45–50 (2006)
6. Weill, P., Ross, J.: IT Governance: How Top Performers Manage IT Decision Rights for Superior Results. Harvard Business School Press, Boston (2004)
7. Khatri, V., Brown, C.V.: Designing data governance. Commun. ACM **53**(1), 148–152 (2010)
8. Buneman, P., Khanna, S., Tan, W.-C.: Why and where: a characterization of data provenance. In: Van den Bussche, J., Vianu, V. (eds.) ICDT 2001. LNCS, vol. 1973, pp. 316–330. Springer, Heidelberg (2000)
9. Bechtold, S., Perrig, A.: Accountability in future internet architectures. Commun. ACM **57**(9), 21–23 (2014)
10. Gössler, G., Le Métayer, D.: A general trace-based framework of logical causality. In: Fiadeiro, J.L., Liu, Z., Xue, J. (eds.) FACS 2013. LNCS, vol. 8348, pp. 157–173. Springer, Heidelberg (2014)
11. Halpern, J.Y., Pearl, J.: Causes and explanations: a structural-model approach – part 1: Causes. CoRR abs/1301.2275 (2013)
12. Eriksén, S.: Designing for accountability. In: NordiCHI 2002: Proceedings of the Second Nordic Conference on Human-Computer Interaction, pp. 177–186. ACM, New York (2002)
13. Gajanayake, R., Iannella, R., Sahama, T.: Sharing with care: an information accountability perspective. IEEE Internet Comput. **15**(4), 31–38 (2011)
14. Gajanayake, R., Sahama, T., Iannella, R.: Principles of information accountability: an ehealth perspective. Int. J. E-Health Med. Commun. **5**(3), 40–57 (2014)
15. Lankhorst, M.: Enterprise Architecture at Work: Modelling, Communication, and Analysis, 1st edn. Springer-Verlag New York Inc., New York (2005)
16. Jonkers, H., Lankhorst, M., Buuren, R.V., Bonsangue, M., Torre, L.V.D.: Concepts for modeling enterprise architectures. Int. J. Coop. Inf. Syst. **13**, 257–287 (2004)
17. Waltl, B., Reschenhofer, T., Matthes, F.: Data governance on ea information assets: logical reasoning for derived data. In: Persson, A., Stirna, J. (eds.) CAiSE 2015 Workshops. LNBIP, vol. 215, pp. 401–412. Springer, Heidelberg (2015)

Analysis of Imprecise Enterprise Models

Hector Florez$^{(\boxtimes)}$, Mario Sánchez, and Jorge Villalobos

Department of Systems and Computing Engineering,
Universidad de Los Andes, Bogotá, Colombia
{ha.florez39,mar-san1,jvillalo}@uniandes.edu.co

Abstract. Enterprises have a large amount of Information Technology (IT) elements for supporting their business. Enterprise models represent the state of IT and business elements and the relation between them in a certain moment. However, in some cases it is difficult to build models that accurately represent the enterprise because information may vary fast over time, or because the granularity of the model may be inadequate for its purpose. When models that are imprecise and do not represent accurately the enterprise are used to perform analysis, it is necessary to evaluate their suitability and determine whether they can be used or if better models have to be constructed. In this paper, we focus on this problem and propose an approach for evaluating the level of imprecision of enterprise models based on the *impact* and *sensitivity* of imprecise information regarding an analysis method.

Keywords: Enterprise modeling · Enterprise analysis · Models imprecision

1 Introduction

Enterprise models have become very important artifacts for organizations because they allow to coherently specify and describe business and IT components as well as their relations [1]. These enterprise models are simplified abstract representations of the reality [2,3]. On the one hand, they provide a holistic view of the enterprise for understanding its organizational and technological aspects [4,5]. On the other hand, they serve for documentation, communication, diagnosis, analysis, discussion, and design purposes [6]. Enterprise modeling is necessarily based on the use of modeling languages (e.g., BPMN, ArchiMate), for creating said enterprise models. The structure of such modeling languages is typically defined in metamodels, which provide the abstract syntax for the languages and identify the enterprise concepts and relations that are of relevance for the language. Nowadays, enterprise modeling languages cover domains such as IT infrastructure, IT architecture, as well as IT and business alignment [7,8].

The process for constructing an enterprise model is complex and requires demanding activities such as human observation, consulting sources of different

© Springer International Publishing Switzerland 2016
R. Schmidt et al. (Eds.): BPMDS/EMMSAD 2016, LNBIP 248, pp. 349–364, 2016.
DOI: 10.1007/978-3-319-39429-9_22

natures, and interpreting unstructured information [2]. On top of this, building a model that properly represents an enterprise has a high level of difficulty because of the increasing size and complexity of enterprises. Furthermore, there are uncontrolled factors such as the quality of the sources on information which results in lack of information [9]. Enterprise models are typically built by teams of modelers, who identify and classify the available sources of enterprise information. Supported by these sources, modelers obtain initial information that they use to begin the modeling process: using a modeling language they create an *incomplete* enterprise model which provides a partial representation of the enterprise. Later on, modelers validate and refine the model using new enterprise information gathered by consulting further enterprise sources. Nevertheless, building a model that represents the enterprise *completely* and *precisely* is unlikely for two main reasons. The former is that models represent the state of the enterprise in a given moment: enterprises change continuously, which implies that models are inherently imprecise (e.g., income of a product, availability of technology devices, application services processing and response time). The latter is that modelers have the responsibility to defined the level of detail (i.e., granularity). This means that depending on the purpose of the model, and the availability of resources, time and information, modelers have to decide what elements and details to leave in the model, and what to discard.

This makes enterprise models *imprecise*. This means that they correctly represents correctly, but include inexact approximations of the true state of certain elements of the enterprise [9]. Models' imprecision can include imprecise information which may be represented through imprecise attribute values. These include ranges of values, sets of values, sets of values where one or more elements are ranges of values, or tables of values.

Analysis of enterprise models can be used to support decision making processes, such as planning future states of the enterprise [10]; as a result, analysis processes have become a critical task because it contributes to the improvement of the business and IT elements. Typically, when one analyst wants to perform an analysis, he manipulates the enterprise model in order to extract information that is useful for evaluating the state of the enterprise. Enterprise analysis can be done by performing automated analysis methods, which are algorithms that extract information from the model and make calculations to provide results that can be introduced in the model or just presented through reports. Automating the procedures for extracting and calculating information makes it possible to work with all elements placed in the model.

However, if the enterprise model is imprecise, it might not be adequate because the analysis method might not be able to calculate accurate results or even the model might not be useful for performing analysis. For instance, if the analysis method *Application Component Availability* calculates the availability of the applications components based on the related infrastructure elements [11] and some infrastructure elements are imprecise, then it is likely that the analysis method cannot calculate the availability for every application component. Besides, if the calculated availability in application components is used for further analysis methods such as *Business Process Fault Susceptibility* [11], then it is possible that this second method cannot be performed because the first method does not calculate all required information. Then, before performing an analysis method, it might be necessary to evaluate the adequateness of the model regarding desired analysis methods in order to determine if it can be used to analyze the enterprise. Thus, if the model is not adequate, modelers must refine the model in order to improve its adequateness.

In this paper, we are focusing just on imprecise enterprise models which contain imprecise information placed in attributes as ranges of values. Based on these models, we evaluate imprecision through two mechanisms. The former is the *impact* that imprecise information in the model has on a desired analysis method. This impact is not a single value, but it is a visual representation. It means that given an analysis method, the impact of imprecision allows observing which imprecise information affects the elements involved in the analysis method. The latter is the *sensitivity* of elements involved in a desired analysis method based on the imprecise information in the model. The sensitivity allows identifying which elements are easily affected by imprecision. Based on these two mechanisms, which are presented in Sect. 4, analysts can identify whether or not a imprecise model is adequate to perform desired business analyses.

The rest of the paper is structured as follows. Section 2 argues the reasons for one model can be imprecise and presents our approach for creating imprecise enterprise models using ArchiMate as modeling language. Section 3 describes automated analysis based on business analysis methods. In Sect. 4, we present our proposal for analyzing imprecise enterprise models and evaluating the adequateness of models through measuring the impact and sensitivity of imprecision. Finally, Sect. 5 concludes the paper.

2   Modeling with Imprecise Information

Enterprise modeling requires consulting different sources that provide information about the enterprise. Based on the obtained information, modelers create an enterprise model. However, based on the purpose of the model, modelers decide its level of granularity. The term granularity deals with the construction, interpretation, and representation of granules [12], which is the result of dividing certain general elements of the model into more detailed and distinguishable elements. In addition, certain enterprise information might be imprecise by its own nature; then, modelers need to represent such information in a manipulable way.

For instance, consider a basic and small enterprise model of a commercial company that sells technology products such as laptops, tablets, and desktops. The model is created in order to obtain the annual profit, which is calculated based on the fixed and variable cost, price, and sales units per month. These attributes are numeric, but sales units is a range of values because this information has to be true based on the given timespan. Now, consider the following scenarios: (1) The modeler creates a coarse grained model that includes just one element for each product with the correspondent attributes' values. In this case, the model is imprecise and the results of calculating the annual profit might not be usable. (2) The model includes one element for each product with the correspondent attributes' values and countries as locations for selling the products, but it implies to include the attributes tax and transport cost; then, each product is related with every location. In this state of the model, it is not possible to calculate the profit because the sales units per month must be placed in every relation from products to countries. (3) The model includes products, countries, and the sales units per month in the correspondent relations. Then, the imprecision of calculating the annual profit is considerably reduced. (4) The model not only includes products, countries, and their relations, but also it includes cities per country, where each city needs to specify additional transport costs (usually cheaper that the country transport cost). In this case, this is a fine grained model. Thus, the results of calculating the annual profit might be even better than the previous scenario, but its improvement might not be significant. As a result, improving granularity might reduce imprecision in the results of one analysis; nevertheless, it is important to identify which level of granularity is adequate in order to obtain valuable results without exceeding effort in the modeling process. In addition, independent of the granularity of an enterprise model, usually some attributes in elements and relations might be imprecise because accurate information related with these attributes cannot be guaranteed by sources.

In order to illustrate imprecise models, we have built one enterprise scenario, which is one publisher of academic books. The enterprise model of this scenario has been built using ArchiMate as modeling language and has 184 business elements, 13 application elements, 13 infrastructure elements, 28 motivation elements, and 432 relations arranged in 12 views. In addition, it has some imprecise information in the attribute `availability` of some infrastructure elements. Figure 1 presents the layered view of the publisher scenario. This view of the model presents elements of the technology, application, and business layer. For illustration purpose, the following elements from the technology layer include the attribute `availability` with imprecise values: `Device` *Windows Server* and *Linux Server*; `SystemSoftware` *JBoss Application Server*, *Glassfish Application Server*, and *Apache Application Server*; and `InfrastructureService` *SQL Server Database Service*, *MySQL Database Service*, and *File Storage Service*.

For creating this model, we have used *iArchiMate*[1], which is our enterprise modeling and analysis tool [13]. The tool's core is a graphical editor based on Eclipse Modeling Framework Project (EMF) and Graphical Modeling Framework Project (GMF). *iArchiMate* allows creating imprecise models by assigning to numeric attributes a minimum and a maximum value that are placed in square braces, and separated by a dash (e.g., [1–2]). In addition, *iArchiMate* is capable of validating the model providing assistance to the user, in order to determine if the model fulfils the required information for running the desired automated analysis method. This validation is supported by Epsilon Validation Language (EVL); thus, when the desired analysis method is selected, *iArchiMate* generates an EVL script, for validating the model.

---

[1] http://iarchimate.virtual.uniandes.edu.co.

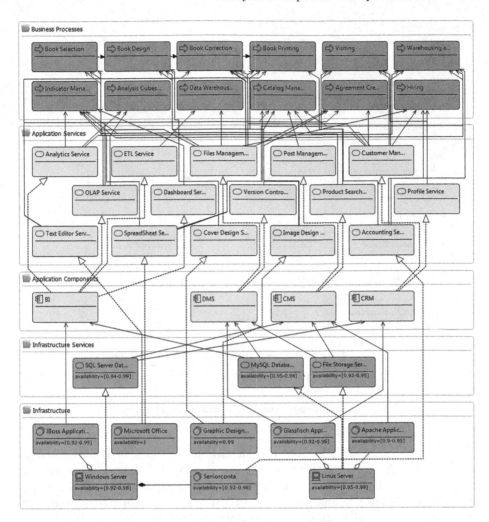

Fig. 1. Layered view of the Publisher Scenario.

3   Automated Analysis

Enterprise analysis is the application of property assessment criteria on enterprise models [14]. This means that, given a property and a criterion for assessing that property, doing model analysis requires evaluating said criteria using the information available in one model [15]. Analysis methods of all types have been studied and classified in the literature according to the concerns that they address and the kind of results that they provide [10].

Typically, model analyses are performed by humans with minimal technological support. Tools are used just to get access to the available data in an efficient way, and analysts have the entire responsibility of discovering information;

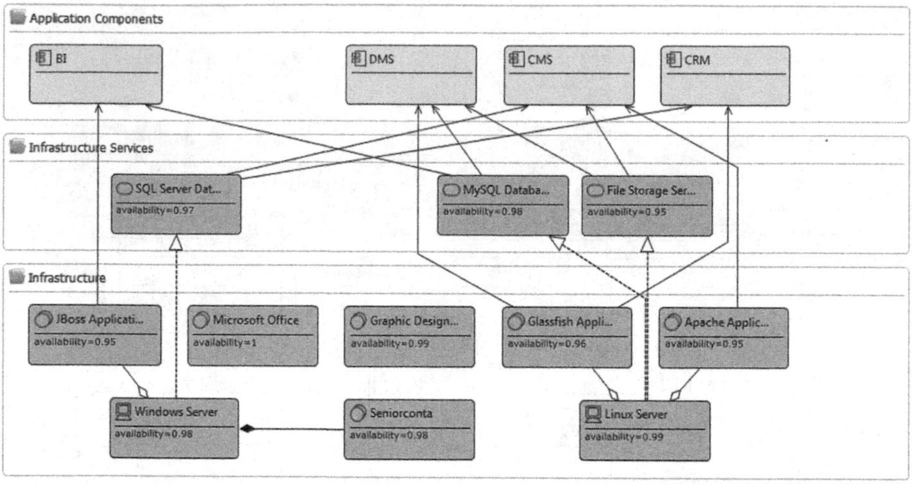

Fig. 2. Fragment of the Layered view of the Publisher Scenario.

Problems | Properties | iArchiMate Analysis Result

Concept: ApplicationComponent. Method: Availability

Element Type	Element Name	Attribute=Value	Associated Elements	Action
ApplicationComponent	DMS	availability=0.8846	(3) Glassfisch Application Server:availability=0.96. SQL Server Database Service:availability=0.97. ...	Created
ApplicationComponent	CMS	availability=0.931	(2) Apache Application Server:availability=0.95. MySQL Database Service:availability=0.98.	Created
ApplicationComponent	CRM	availability=0.931	(2) Apache Application Server:availability=0.95. MySQL Database Service:availability=0.98.	Created
ApplicationComponent	BI	availability=0.9215	(2) JBoss Application Server:availability=0.95. SQL Server Database Service:availability=0.97.	Created

Fig. 3. *Application component availability* results.

thus, the omission of elements and miscalculation of results, especially in large enterprise models, are very likely. As a result, automated analysis methods are becoming increasingly common and are being implemented on top of modeling tools. These methods follow predefined instructions to extract data from the models, to perform calculations, and to inform the results of the analysis.

In order to illustrate automated analysis, we use the enterprise model of the publisher scenario without imprecise information. Figure 2 presents a fragment of the layered view of the publisher scenario that contains infrastructure elements from the technology layer and application components from the application layer. Based on this model, we performed the quantitative automated analysis method *Application Component Availability* [11], which calculates the availability for every application component based on the related infrastructure elements i.e., the calculation of the availability for one application component is the product of the availabilities of all infrastructure services and system software related with the application component.

$$AC(a) = \prod_{i=1}^{n} IE(a)_i \tag{1}$$

Thus, the availability is calculated through the Eq. 1, where $AC =$ ApplicationComponent, $a =$ availability, $IE =$ InfrastructureElement (the collection of InfrastructureService plus SystemSoftware), and $n =$ amount of IE related with the AC. In addition, the calculated availability of each ApplicationComponent is propagated to every ApplicationService related by one Realization relation.

Figure 3 presents the results of performing the automated analysis method *Application Component Availability*. The results are deployed in a view of *iArchiMate* and present the calculated value of the attribute availability and the associated infrastructure elements for each application component in the model. The results of the automated analysis method are used by analysts because they have the skills to interpret these results and to make assessments aligned to the business goals of the enterprise. We consider that the way in which *iArchiMate* provides the results facilitates their interpretation not only individually (i.e., just taking into account the performed analysis method), but also in a general context (i.e., taking into account the analysis method into a general context or global analysis).

4   Analyzing Imprecise Enterprise Models

As described before, analysis methods require certain information in certain elements or relations in the model to be performed. This information is included through attributes. When one analysis method needs to make any mathematical operation, the attributes should accomplish a specific data type (e.g., double). However, when the model is imprecise, some attributes might include ranges of numeric values; then, it is not possible to perform the same calculations done for models without imprecise information. As a result, the algorithm of the analysis method must be upgraded to take into account the imprecise attributes required by the analysis method.

For instance, when the analyst desires to perform the automated analysis method *Application Component Availability*, the model should include the attribute availability with a double value in every InfrastructureService and SystemSoftware. However, if the model is imprecise, the attribute availability in some infrastructure elements might contain imprecise values such as range of values instead of a double value. Then, the analysis method performs the Algorithm 1, where AC=ApplicationComponent and IE=InfrastructureElement, in order to make the correspondent calculations using ranges of values. This algorithm makes all possible calculations applying the Eq. 1, for obtaining the minimum and maximum values.

Figure 4 presents the results of the analysis method *Application Component Availability* applied to the imprecise model of the publisher scenario presented in Fig. 1. The sub-figure (a) presents one report that deploys the results just for the application component *DMS*, showing all possible scenarios with the correspondent availability value, while the sub-figure (b) presents the application layer of the layered view with the correspondent calculated availability values.

Algorithm 1. Application Component Availability

---

for all ac in AC **do**
   for all ie in IE **do**
      for all r in $ie.relations$ **do**
         if $r.target = ac$ **then**
            $a \leftarrow ie.availability$
            $availability[] \leftarrow product(availability, a.min, a.max)$
         end if
      end for
   end for
   $range[0] \leftarrow getMinimum(availability)$
   $range[1] \leftarrow getMaximum(availability)$
   $ac.availability \leftarrow range$
end for

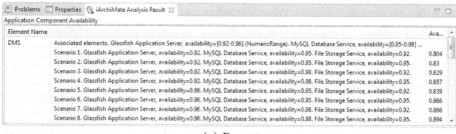

(a) Report

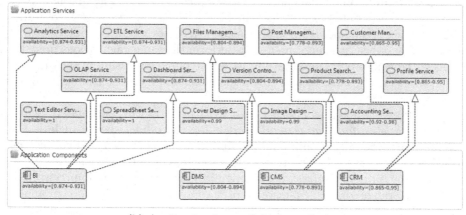

(b) Application layer of the layered view

Fig. 4. Results of the analysis method *Application Component Availability*.

The report presents multiple scenarios. One scenario corresponds to the calculation of the availability for one possible combination of the correspondent set of ranges of values. In addition, the range of values that is included in the attribute `availability` of the application component is composed by the minimum and the maximum value obtained in all scenarios.

For instance, as shown in Fig. 1, the `ApplicationComponent` *DMS* depends on the `ApplicationService` *MySQL Database Service*, *File Storage Service*, and the `SystemSoftware` *Glassfish Application Server*. However, the availability values in this infrastructure elements are respectively: *[0.95–0.98]*, *[0.92–0.95]*, and *[0.92–0.96]*. Then, there are eight scenarios for calculating the availability of the *DMS*. The minimum availability value is *0.804*, which is obtained by the scenario with the availability values: *Glassfish Application Server = 0.92*, *MySQL Database Service = 0.95*, and *File Storage Service = 0.92*. The maximum availability value is *0.894*, which is obtained by the scenario with the availability values: *Glassfish Application Server = 0.96*, *MySQL Database Service = 0.98*, and *File Storage Service = 0.95*. Thus, the range of values for the *DMS* is *[0.804–0.894]*. Finally, the *DMS* realizes the `InfrastructureService` *Files Management Service* and *Version Control Service*; consequently, the availability of these services is *[0.804–0.894]* as well.

The results of the analysis method enrich the model by including the calculated imprecise information in the correspondent element's attributes. Thus, now the model includes more imprecise information that can affect the execution of further analysis methods. The following sections discuss the *impact* and *sensitivity* that imprecise information has on one analysis method; thus, the impact and the sensitivity allow determining whether or not the model is adequate or the model needs to be refined.

4.1 Impact of Imprecision in Analysis Methods

The presence of imprecise information impacts on query evaluation, since the semantics are no longer obvious [16]. Then, it is of great value to any enterprise to incorporate processes for assessing, measuring, reporting, reacting to, and controlling the risks associated with unexpected data such as imprecise data [17]. Thus, the impact of imprecision to one analysis method is a measure that determines how much imprecise information placed in one element might provoke undesired results when an analysis method is performed. Then, if the impact is low to one analysis method, it can be performed and results can be used by analysts for making assessments. However, if the impact is high to one analysis method, it means that the model is not adequate for performing the analysis method.

Performing analysis methods using imprecise information generates imprecise results for every element that is target of the analysis. Thus, if the difference of the calculated maximum and minimum values is high, the result would not be useful. Then, the impact corresponds to the collection of the differences of the calculated maximum and minimum values for each target element involved in the analysis method.

Continuing with the publisher scenario and based on the results of the analysis method *Application Component Availability*, it is possible to perform further analysis methods such as *Business Process Fault Susceptibility* [11], which requires the attributes (a) `availability` in every `ApplicationService` related with at least one `BusinessProcess` and (b) `importanceLevel` in each `UsedBy`

Table 1. Importance level for `UsedBy` relations from application services to business processes

Source	Target:Value
Accounting Service	BP:0.1, H:0.9, WS:0.9
Analytics Service	IM:1, ACM:0.8, WS:0.1
Cover Design Service	BBP:1
Customer Management Service	IM:0.2, AC:0.1, BC:0.2, BS:0.1, H:0.2, V:0.1, WS:0.3
Dashboard Service	IM:0.8
ETL Service	DWM:1
Files Management Service	H:0.1, AC:1, BD:0.2, BC:1, BP:1, BS:1, DWM:0.1
Image Design Service	BC:0.1, BD:1
OLAP Service	ACM:1, CM:0.5, IM:0.8
Post Management Service	CM:1
Product Searching Service	ACM:1, V:02, BS:1, CM:1, IM:0.2, WS:0.9
Profile Service	ACM:1, CM:0.8, H:0.2, V:1
Spreadsheet Service	CM:0.1, IM:0.1, V:1, WS:0.1
Text Editor Service	AC:1, BS:1, BP:1
Version Control Service	ACM:1, AC:0.9, BC:1, BP:1, CM:1

relation from `ApplicationService` to `BusinessProcess`. Table 1 presents the values of the attribute `importanceLevel` for the *UsedBy* relations of the publisher scenario. The table uses the following acronyms: Book Selection BS, Book Design BD, Book Correction BC, Book Printing BP, Visiting V, Warehousing and Shipping WS, Indicator Management IM, Analysis Cubes Management ACM, Data Warehouse Management DWM, Catalog Management, Agreement Creation AC, Hiring H.

The analysis method *Business Process Fault Susceptibility* is intended to calculate one value for the attribute `faultSucceptibility` for each business process, and this value should be as small as possible. Equation 2 presents the formula for the calculation of the fault susceptibility for this analysis method, where BP = `BusinessProcess`, fs = `faultSusceptibility`, AS = `ApplicationService`, a = `availability`, $R_{AS \to BP}$ = `UsedBy` relation from `ApplicationService` to `BusinessProcess`, il = `importanceLevel`, and n = amount of AS related with BP.

$$BP(fs) = 1 - \frac{\sum\limits_{i=1}^{n} \left(AS(a)_i * R_{AS \to BP}(il)_i \right)}{\sum\limits_{i=1}^{n} R_{AS \to BP}(il)_i} \tag{2}$$

In this case, we are interested in observing, how the imprecise value of one infrastructure service affects the related business processes. Figure 5 presents

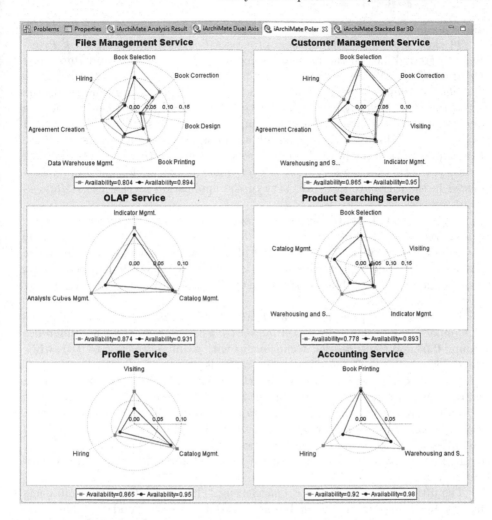

Fig. 5. Impact of imprecision in analysis method.

polar charts for analyzing the impact of imprecision in the analysis method *Business Process Fault Susceptibility*. There is one polar chart for each application service that has imprecise availability, where each axis represents one business process and the series represent the minimum possible result for the attribute faultSusceptibility based on the imprecise values (i.e., minimum and maximum values) of the attribute availability of the correspondent application service. In this figure, we illustrate just six polar charts, although *iArchiMate* allows selecting all desirable application services, for deploying the correspondent polar charts.

The first chart corresponds to the application service *Files Management Service*, which is used by seven business processes. This chart evidences that the difference of the minimum possible value for the attribute `faultSusceptibility` is: (1) close to 0.05 for the business processes *Book Selection* and *Book Printing*; (2) around 0.03 in the business processes *Book Correction* and *Agreement Creation*; and (3) close to zero in the business processes *Book Design, Data Warehouse Management,* and *Hiring.* As a result, imprecise information in the attribute `availability` of the application service *Files Management Service* has an important impact in just two business processes: *Book Selection* and *Book Printing.* However, there is an additional issue. The value of the attribute `faultSusceptibility` in the business process *Book Selection* is over 0.1 in both series; then, this business process is even more impacted for the imprecision placed in the attribute `availability` of the application service *Files Management Service.*

In the second chart, the application service *Customer Management Service* is analyzed. Just the business processes *Hiring* and *Warehousing and Shipping* have a difference close to 0.01. In the rest of the business processes, the difference tends to zero. Then, despite this service is used by seven business processes, its imperfection in the attribute `availability` does not affect in an important way any business process.

The third chart presents the application service *OLAP Service.* This service impacts the business process *Analysis Cubes Management* with a difference around 0.03 and the *Indicator Management* with a difference close to 0.01, while the difference in the *Catalog Management* tends to zero.

The application service *Product Searching Service* is illustrated in the fourth chart. This chart evidences an important impact for the business processes *Book Selection* and *Warehousing and Shipping* because the difference is around 0.05. For the business process *Catalog Management* the difference is close to 0.02 and the difference in the other business processes tends to zero. Furthermore, similar to the first chart, in the business process *Book Selection* the minimum value for both series is 0.1.

The fifth chart presents the *Profile Service.* This application service impacts just the business process *Visiting* because the difference is around 0.05. The other business processes that depend on this application service have a difference near zero.

In the sixth chart, the *Accounting Service* impacts the processes *Hiring* and *Warehousing and Shipping*; nevertheless, the difference of the impact over the first one is around 0.05 and over the second one is around 0.02.

Summarizing, these six charts present some interesting evidences regarding imprecision of the attribute `availability` placed in application services: (1) the *Files Management Service* impacts four out of seven business processes; (2) the *Customer Management Service* does not impact any of seven business processes; (3) the *OLAP Service* impacts just one of three business processes; (4) the *Product Searching Service* impacts three of five business processes, (5) The *Profile Service* impacts just one of three process, and (6) the *Accounting Service* impacts two of three processes.

4.2 Sensitivity of Analysis Methods for Imprecise Models

Another perspective for measuring the level of imprecision is sensitivity, which is also a measure that determines how much imprecision in certain elements in the model affects target elements of one desired analysis method. Thus, using this measure, analysts can assess which target elements of an analysis method are more sensitive to imprecision and which source elements of an analysis method affect more target elements. Sensitivity is calculated for each target element as the addition of the difference of the calculated minimum and maximum values of each source element involved in the analysis method.

Coming back to the analysis method *Business Process Fault Susceptibility*, before establishing a value for the attribute `faultSusceptibility` in every business process, it is useful to know how the imprecision in one required attribute (e.g., availability in application service) can affect the results of one analysis method.

Figure 6 presents a stacked bar chart, where each bar corresponds to one business process. This chart informs how sensitive is one business process regarding the imprecision included in the attribute `availability` of the related application services. In this case, we can observe two main facts. On the one hand, the process *Book Selection* is the most sensitive business process. On the other hand, the imprecision of the application service *Files Management Service* has the higher impact in the whole model because it impacts seven processes and, in five of those processes, the impact of this application service is the highest. However, there are another interesting facts in these results.

- Most of the business processes are sensitive to more than two application services, while just one business process is sensitive to one application service.
- The application service *Customer Management Service* impacts seven business processes; nevertheless, its impact in every process is very low.
- The application service *File Management Service* impacts seven business processes as well, but three of these processes have low sensitivity.

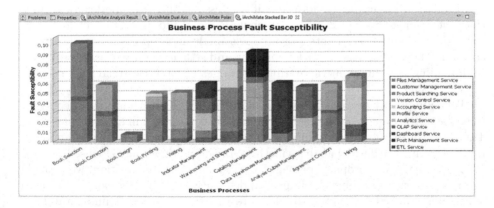

Fig. 6. Sensitivity of imprecision in business processes.

- The business process *Book Selection* is very sensitive to the imprecision of the application services *Files Management Service* and *Product Searching Service*; then, it is advisable to eliminate or at least to reduce the imprecision in the availability of these two application services.
- In most of the business processes, the sensitivity is lower than 0.06, which implies that the model can be used for performing the analysis method; nevertheless, the model can be refined in order to reduce imprecise results.
- The business process *Book Design* is sensitive just to the *File Management Service* and its sensitivity is very low.
- The *ETL Service* impacts just one process, which is *Data Warehouse Management*; however, its impact is high.
- The *Post Management Service* and the *Dashboard Service* impact just one process as well.

As conclusion of the sensitivity analysis results, imprecise information in the *Customer Management Service*, *Dashboard Service*, and *Post Management Service* would not affect the results of the analysis method; thus, the model could keep this imprecise information. In addition, imprecision in the *Files Management Service* and *Product Searching Service* should be reduced in order to avoid possible useless results of the analysis method. Finally, imprecision in the rest of the services i.e., *Accounting Service*, *ETL Service*, *Profile Service*, *OLAP Service*, *Version Control Service*, and *Analytics Service* have high impact in just one process; then, their imprecision might be acceptable.

Thus, due to the imprecision of application services comes by the analysis method *Application Component Availability*, the refinement of the imprecision in the technology layer decreases the impact in this analysis method.

5   Conclusions

Enterprise models are built using information provided by various and heterogeneous sources. It is very likely that these sources have inaccurate, incorrect, or obsolete information. In addition, it is possible that the modeler consider that the information obtained is not enough for including certain elements in the model. Consequently, enterprise models might not represent the enterprise correctly. Thus, imprecise models represent and structure imprecise information by including in some numeric attributes, ranges of values instead of a specific numeric value.

When imprecise models are used to perform analysis methods, it is necessary to measure the impact and sensitivity of imprecision to certain analysis methods. These measures allow determining whether the imprecise model is good enough for performing the desired analysis method. In this work, we have focused on providing mechanics to properly manage imprecise models. In addition, we provide techniques for determining the impact and sensitivity of imprecision to business analysis methods, by providing graphical results that allows easy understanding.

Based on the results of the impact and sensitivity analysis of imprecision, certain analysis methods might be performed. Thus, future work consists in providing analysis methods that can be performed based on imprecise information placed in the model, when the impact and sensitivity analysis results indicate that the model is adequate for analyzing the enterprise despite the existing imprecise information.

References

1. Jonkers, H., Lankhorst, M., Van Buuren, R., Hoppenbrouwers, S., Bonsangue, M., Van Der Torre, L.: Concepts for modeling enterprise architectures. Int. J. Coop. Inf. Syst. **13**(03), 257–287 (2004)
2. Bézivin, J.: On the unification power of models. Softw. Syst. Model. **4**(2), 171–188 (2005)
3. Ludewig, J.: Models in software engineering–an introduction. Softw. Syst. Model. **2**(1), 5–14 (2003)
4. Lankhorst, M.: Enterprise architecture at work: Modelling, communication and analysis. Springer, Heidelberg (2013)
5. Lagerström, R., Franke, U., Johnson, P., Ullberg, J.: A method for creating enterprise architecture metamodels-applied to systems modifiability analysis. Int. J. Comput. Sci. Appl. **6**(5), 89–120 (2009)
6. Kurpjuweit, S., Winter, R.: Viewpoint-based Meta Model Engineering. In: Proceedings of the 2nd International Workshop on Enterprise Modelling and Information Systems Architectures, pp. 143–161 (2007)
7. Frank, U.: Multi-perspective enterprise modeling: foundational concepts, prospects and future research challenges. Softw. Syst. Model. **13**(3), 941–962 (2014)
8. Avila, O., Goepp, V., Kiefer, F.: Understanding and classifying information system alignment approaches. J. Comput. Inf. Syst. **50**(1), 2–14 (2009)
9. Henricksen, K., Indulska, J.: Modelling and using imperfect context information. In: Proceedings of the Second IEEE Annual Conference on Pervasive Computing and Communications Workshops, 2004, pp. 33–37. IEEE (2004)
10. Buckl, S., Matthes, F., Schweda, C.M.: Classifying enterprise architecture analysis approaches. In: Poler, R., van Sinderen, M., Sanchis, R. (eds.) IWEI 2009. LNBIP, vol. 38, pp. 66–79. Springer, Heidelberg (2009)
11. Florez, H., Sanchez, M., Villalobos, J.: A catalog of automated analysis methods for enterprise models. SpringerPlus **5**, 1–24 (2016)
12. Holschke, O.: Impact of granularity on adjustment behavior in adaptive reuse of business process models. In: Hull, R., Mendling, J., Tai, S. (eds.) BPM 2010. LNCS, vol. 6336, pp. 112–127. Springer, Heidelberg (2010)
13. Florez, H., Sanchez, M., Villalobos, J.: iArchiMate: a tool for managing imperfection in enterprise models. In: 18th IEEE International Enterprise Distributed Object Computing Conference Workshops and Demonstrations (EDOCW), pp. 201–210. IEEE (2014)
14. Johnson, P., Johansson, E., Sommestad, T., Ullberg, J.: A tool for enterprise architecture analysis. In: 11th IEEE International Enterprise Distributed Object Computing Conference (EDOC 2007), pp. 142–142. IEEE, October 2007

15. Florez, H., Sanchez, M., Villalobos, J.: Extensible model-based approach for supporting automatic enterprise analysis. In: 18th IEEE International Enterprise Distributed Object Computing Conference (EDOC), pp. 32–41 (2014)
16. Morrissey, J.M.: Imprecise information and uncertainty in information systems. ACM Trans. Inf. Syst. (TOIS) 8(2), 159–180 (1990)
17. Loshin, D.: The practitioner's guide to data quality improvement. Elsevier, Amsterdam (2010)

Enterprise Process Modeling in Practice – Experiences from a Case Study in the Healthcare Sector

Snorre Fossland[1] and John Krogstie[2(✉)]

[1] eFaros Ltd, Oslo, Norway
snorre@efaros.com
[2] NTNU, Trondheim, Norway
John.krogstie@idi.ntnu.no

Abstract. In enterprise modeling it is customary to differentiate between the current, as-is situation and the future to-be situation and develop models of these to plan for how to fill the gap. In practice you are never able to implement the ideal to-be model, each to-be will be incremental steps on the way to a future best practice. So it will be useful to also maintain a separate ought-to-be model, to not forget the situation you strive for. A distinction between the ought-to-be, as-is, and the to-be model is necessary, and we have in this paper provided the basis for an approach for combining top-down ought-to-be and bottom-up as-is and to-be modelling to support the dynamic interplay between these models. The approach is illustrated through a practical application in the healthcare sector. The main results is that it is found beneficial to represent the to-be and ought-to-be models separately, to be able to discuss the long-term goals without being hampered by short-term technical and organizational limitations, but still have support for developing the next version of the organization.

Keywords: Enterprise process modelling · Case study · Ought-to-be model

1 Introduction

The clinical and administrative processes in today's healthcare environments are becoming increasingly complex and intertwined and the provision of clinical care involves a complex series of physical and cognitive activities. A multitude of stakeholders and healthcare providers with the need for rapid decision-making, communication and coordination, together with the steadily growing amount of medical information, all contribute to the view of healthcare as a complex cognitive work domain. The healthcare environment can also be characterized as very dynamic, in which clinicians rapidly switch between tasks. The process is partially planned, but at the same time driven by events and emergencies [4, 6].

To be able to cope with the dynamism and complexities in their environments, many organizations have been forced to restructure their operations and integrate complex business processes across functional units and across organizational boundaries [9]. This has in many cases led to the adoption of process-oriented approaches and enterprise process modeling to manage organizational activities. Process modeling is used within

© Springer International Publishing Switzerland 2016
R. Schmidt et al. (Eds.): BPMDS/EMMSAD 2016, LNBIP 248, pp. 365–380, 2016.
DOI: 10.1007/978-3-319-39429-9_23

organizations as a method to increase the focus on and knowledge of organizational processes, and function as a key instrument to organize activities and to improve the understanding of their interrelationships [24]. Today, there is a large number of modeling languages available. The first process modelling language was described as early as 1921 [11], and process modeling has been performed in earnest relative to IT and organizational development at least since the 70ties. Lately, with the proliferation of BPM (Business Process Management), interest and use of process modeling has increased even further.

A lot of research has been done in the field of enterprise process modelling, as well as on the subject of how to judge the appropriateness of the models and modeling languages [18, 20, 21, 23]. Much work is done on a theoretical level, but in order to better understand the mechanisms at work in the application of enterprise process models, real-life cases can provide interesting insights [12].

This paper presents experience from a case study on the use of process models in the healthcare sector. An overview of categories of process models are found in Sect. 2. How the interplay in particular between as-is, ought-to-be and to-be models can be exploited is presented in Sect. 3. In Sect. 4 we illustrate the approach in more through a case study in the healthcare sector. Discussion of lessons learned is found in Sect. 5 before conclusion and ideas on further work follow in Sect. 6.

2   Modeling of Business Processes in Enterprise Development

A model is not just a representation of something else; it is a conscious construction to achieve a certain goal. Based on the goals that the modeling is meant to support the achievement of and existing resources, persons gathers (physically or virtually, synchronously or asynchronously) to represent some area of interest using some modelling languages. The modeling activity is supported by tools resulting in models that are meant to help addressing the goals of modelling. In Table 1, we list relevant modeling situations, along the temporal and purpose axes:

Table 1. Categories of models according to temporal aspects and purpose [10]

	Past	Present	Future
Ideal model	Ideal model of the past	Reference model	*Ought-to-be model*
Simulated model (what-if)	Possible model of the past	Possible model	Possible model of the future
Model espoused	As-was model	As-is model	*To-be model*
Model in use	Actual as-was model	*Actual as-is model*	Workaround model [2]
Motivational model	Past burning-platform model	Burning platform model [5]	Burning platform model

Models can be of past situations, the present, or a potential future situation. We look here primarily at models of the present and future. Models can at all temporal stages be looked upon as being:

- Ideal: A model of a situation perceived to be ideal for an area, ignoring contextual restrictions such as current legacy systems and organizational practices.
- Simulated: A model that differs in some way to the actual state-of-things, e.g. to be able to play what-if analysis and other simulations.
- Model espoused: The official model of an area.
- Model in use: How the situation actually is (or was). This should ideally not be so different from the model espoused, but in practice these often differ.
- Motivational model: A simulation which depicts a defensive approach i.e. what happens if nothing is done. Also known as a burning platform scenario [5].

In total this gives 15 model types. It is important to realize that the to-be situation (both ideal and actual) is a moving target. When one implement a new solution (turning the to-be model into an (espoused) as-is model) both the ought-to-be and to-be have moved further. We will below look in particular on the interplay between the actual as-is model, the ought-to-be ideal future model, and the to-be model where contextual constraints are taken into account.

An organization is in a state including the existing processes, organization and IT-systems, and one often perceive future improved organizational states. The state of the organization is perceived (differently) by different persons through explicit or implicit models. This opens up for different goals for the use of models. This is an extension of the overview found in e.g. [18] as depicted in Fig. 1:

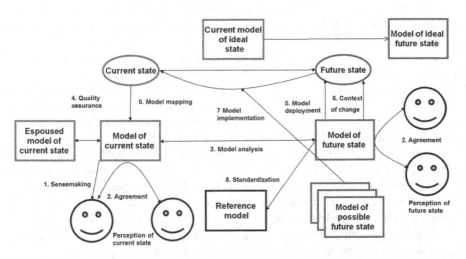

Fig. 1. Application of enterprise process modelling

0. Model mapping: Representation of the current situation in a model.
1. Human sense-making [29]: The development and use of a model of the current state can be useful for people to make sense of and learn about the current situation.
2. Communication to establishing agreement between people in the organization [3].
3. Model analysis: To gain knowledge about the organization through simulation or deduction.

4. Quality assurance, ensuring e.g. that the organization acts in compliance to a certified process typically represented as an espoused model.
5. Model deployment and activation: To integrate the model of the future state in an information system directly. Models can be activated in essentially three ways:
   a. Through people guided by manual process 'maps' [12].
   b. Automatically, e.g. through a workflow systems [30].
   c. Interactively, where the computer and the users co-operate [19].
6. To give the context for a traditional system development project.
7. Model implementation. In both usage area 5 and 6, it is the purpose to change the situation in the organization. In addition one often has to do other tasks (e.g. training) to have people work according to the new processes.
8. Standardization, influencing reference models external to the organization that others might need to relate to at a later stage.

3   Combining Top-Down and Bottom-up Modeling

Process modeling often starts with the company vision and business value to be achieved. It is important to develop both corporate future goals and target architecture in the form of a "Future Operating Model" (from hereon called the ought-to-be model), as well as detailed workflows with both as-is and to-be activities. To achieve this, one needs a combined top-down and bottom-up approaches. The ought-to-be model is best developed top-down describing how the organization ideally wants to operate, based on the Situated Best Practise.

The workflow model is often developed bottom-up model showing how the enterprise is operating with todays (as-is) and tomorrows (to-be) systems and organization. The main reason to have both the bottom-up and top-down approach, is because there is often a large gap between the long-term ambitions of the organization and what current installed base, technology and methods can deliver. To achieve value through process modeling, it is necessary to have a long-term perspective [12, 16].

The ought-to-be model describes the Situated Best Practice which are derived from previous experience, technological development and regulatory constraints etc., and shows the ambitions and plans - on a general level, how the enterprise is going to be operated in the future. The model is a generic model (cf the TOGAF continuum [27]). Together with an as-is model it can be used for basic analyses and help answer questions like: "What are our enterprise doing?", "Are we doing the right things?", "How are our main processes being performed?", "Could we redesign our basic processes?"

This is analysis that should be done before going into the details like: "Who / what does what?" (Human/machine), and "Which IT systems are used for what?"

Once these basic analyzes and decisions have been made, one can proceed with detailed workflow diagrams of the to-be situation.

We see in many cases that the business has little influence on the IT and often has to accept what the IT-systems have to offer, and adjust the processes to the IT-systems, even when it is not good practice. An Ought-to-be model will contribute to a common understanding of the business requirement to the systems and organization.

Process models are often structured in an hierarchical decomposition structure. The process hierarchy provides a full overview of the enterprise and what is agreed as situated best practice. Experience shows that it is in the transitions between activities in the value chain that it often slips, and this becomes explicitly in an overall end-to-end model. In this model it is also important to set the customer/client in focus and ensure that the customer interaction with the company is explicitly modeled.

As illustrated in Fig. 2, the top-down planning model shows the value-chains, but also value-shop and value-networks if relevant [26]. The to-be workflow model is a bottom-up implementation model, that contain the detailed workflow for defined parts of the value-chain. Further in Fig. 2 we see:

Top-down and Bottom-up modelling

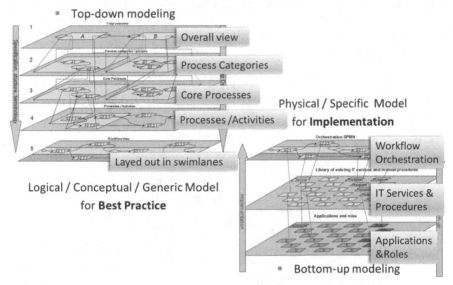

Fig. 2. Illustrating the interplay between top-down and bottom-up modeling

- On the left side a top-down process breakdown structure, from an "Overall view" detailed in several levels down to "Processes/Activities" layed out in swimlanes.
- The right side show a bottom-up workflow model which is built up from applications and roles, IT services and procedures for implementation (orchestration), up to a similar swimlane view.

Modeling a top-down generic architecture model [27], can be done with different notations, but we recommend using IDEF0 [15], which is regarded as best practice for generic process models with a process breakdown structure. IDEF0 also can be used to model all variants of value-chains, value-shops and value-networks. IDEF0 is a process modeling language and a method to model the decisions, actions, and activities of an organization or system. IDEF0 was derived from the Structured Analysis and Design

Technique (SADT) and standardized by NIST in 1993. IDEF0 makes it possible to represent the functions that are performed and what is needed to perform those functions.

The Process interactions in IDEF0 are usually known as ICOMs

- Input: Can be a trigger and is an input that is transformed to output in the process.
- Control: Guide or regulating activity. A main distinction between input and control is that inputs are transformed by the activity, whereas controls remain unchanged.
- Outputs: Results (products) of performing the process/activity.
- Mechanism: Resources needed to perform the activity. These can be people or roles, equipment, IT-systems or financial resources

As illustrated in Fig. 3, this top-down model shows not only the process-breakdown, but also the hierarchical breakdown of information (input/output), the logical applications and role/organizational and control structure.

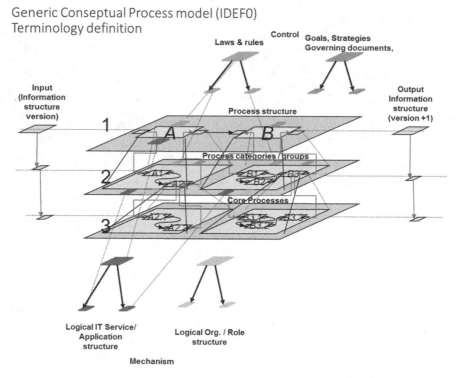

Fig. 3. Generic conceptual model of IDEF0

The bottom-up workflow-model in Fig. 2 is a specific architecture model using the TOGAF-vocabulary [27], describing detailed activities for each role and how the IT-systems are used for each activity. This gives a detailed overview of which roles, information objects and application functions that are used (as-is and to-be).

3.1   Combining Ought-to-be Models with as-is and to-be Models

Since process modelling combining IDEF0 and BPMN (Business Process Model and Notation [25]) can capture both an ought-to-be supply chain, as well as detailed work-flow diagrams, it makes the process of going from as-is to to-be, easier, more structured and efficient. By linking ought-to-be models with as-is and to-be models, it will be possible to analyse how close (or far) the current practice is from situated best practice (Fig. 4) (For detailed model-levels in the top-down and bottom-up models, see Fig. 2).

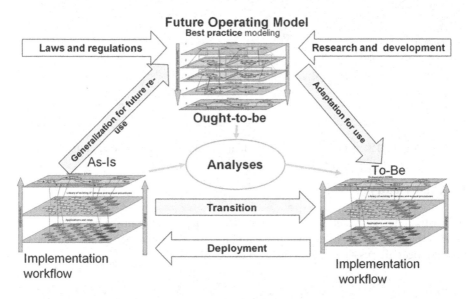

Fig. 4. The interplay between as-is, ought-to-be, and to-be models

The ought-to-be model is the result of the enterprise common understanding of the best practice of the ideal future state. It is a continuous updated "living" model, based on experience from current as-is, taking into account restrictions from laws and regulations beyond the control of the organization, but also including result from research and development. When to-be is fully implemented, it will be the new as-is.

As illustrated in Fig. 4, this imply that the analysis necessary to go from as-is to to-be, will be based on the common agreed best practice within the enterprise, unlike today, where the to-be often are controlled by the IT-professionals. A generic best practice model like this will give the business management and business architects more influence and control, and provide a better and more effective way of specifing how the IT systems should support the business processes.

When we get to a detailed level we often find common processes that are used in several value-chains. To avoid redundancy, we model these standard processes separate as stereotypes and make a link (relationship) from the value-chain process to the stereotype processes as illustrated in Fig. 5. The stereotypes can be aligned with a service catalog and might be seen as a specification for the services.

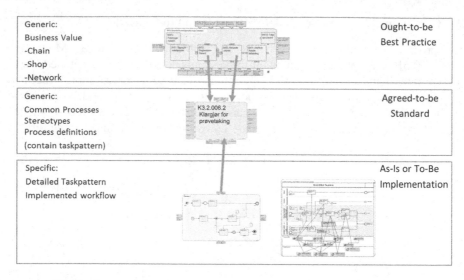

Fig. 5. Process states from generic future to specific implementation states

4   Case Study

To illustrate the approach, we use the experiences from a case in the healthcare sector. Parts of this case have earlier been presented in [10], but the experiences from the case have not been discussed in the same level of detail before.

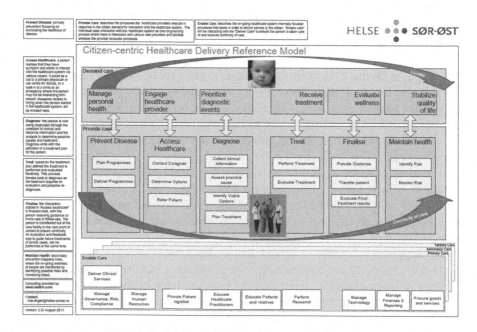

Fig. 6. Healthcare delivery reference model (from [8])

Health South East (HSØ) in Norway has been working with Clinical Pathway Processes (how patients are passed through the diagnose/treatment activities in the hospital) for many years, using different methods and notations and are developing a Citizen Centric Healthcare Delivery Reference Model [8], the top-level found in Fig. 6.

The modeling was done in three phases during the project of building a new hospital in Østfold Norway, replacing three existing old hospitals:

1. First an as-is model of current processes and systems was developed.
2. Second an ought-to-be model based on the existing reference model was built.
3. Third, a to-be model for the first installations necessary before opening the new hospital was made.

The research question for this paper is how the combination of top-down and bottom-up approach described in the previous section is experienced to be appropriate in the case setting. The experiences are primarily based on the work of one of the authors, working as a modeler and modeling facilitator through the project.

In the case we combined the use of IDEF0 and BPMN.

- The ought-to-be model is a generic planning model (in IDEF0) that represents value-chains, as well as value-shop and value-networks.
- The workflow models (as-is and to-be) are implementation models (in BPMN), that shows the detailed workflow for defined parts of the value-chain

Whereas the type of modelling that is described in this paper can be supported in a number of different tools, we used Troux Architect [28], based on a call for tender and procurement of tool and modelling services from HSØ. Troux Architect is a desktop visual modeling environment used to create models and analytical tools for communication and analysis of an enterprise. The content of Troux Architect visual models can be saved in the Troux repository being available for query, reporting, and alternative visualizations e.g. in web-based process navigators more familiar for users not being modeling experts. Likewise, repository-based content coming from other sources such as internal databases can be queried and incorporated into visual models.

The process modeling project for the new hospital was adjusted to the reference model and below we present some examples from this model. The models are in Norwegian, but we describe aspects specifically relevant for this paper in English below.

A top-level IDEF0 model from the case study is depicted in Fig. 7. This shows the sick patient as input and a cured patient as output. As controls on top the laws and regulations are modeled and as mechanisms at the bottom the main roles/skills and logical application systems are represented. Note that we do not need to represent a strict sequence-flow in IDEF0 (which you would mandate in e.g. a BPMN-model) since this would at this level be too restrictive to capture the variety of possible process patterns.

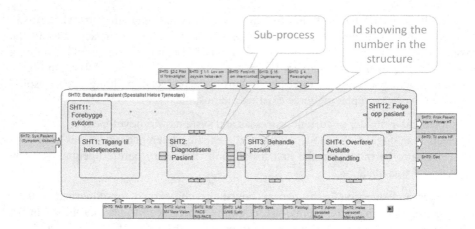

Fig. 7. Top-level IDEF0 model in case study

On lower levels one models the sub-processes in the pathway with more detailed inputs, controls, outputs and mechanisms. The processes and ICOM's are numbered according to the process breakdown structure. This top down generic model can be broken down in several levels. It is also important to include the patient's own processes in the model in order to put the patient in focus.

From this main process structure it is possible to make many different model views for various purposes and audiences. The processes can e.g. be presented in swimlanes representing main hospital units. A more detailed view is shown in Fig. 8. (In this figure the role-names are the most important, thus please ignore the activity names in Norwegian). This includes the patient processes with focus on the interactions between the healthcare organizations and the patient, highlighting the Line of Visibility (LoV) between the enterprise (hospital) and the customer (patient). In this view the ICOM's can be hidden.

These views can be made on several process levels, helping people from different professions with varying skills to get a common understanding of the enterprise processes.

We wanted to standardize the processes, and to make this more explicit, we made process definitions as stereotype-processes or standard reference processes (such as 'take blood sample') that can be used several places in the value-chain or in several value-chains (illustrated in Figs. 5 and 9). These process definitions represent the "layer" of common terms where the business meets IT, i.e. where it is agreed upon what names to use and define which stereotypes/workflows to be used.

Clinical Pathway – level 3

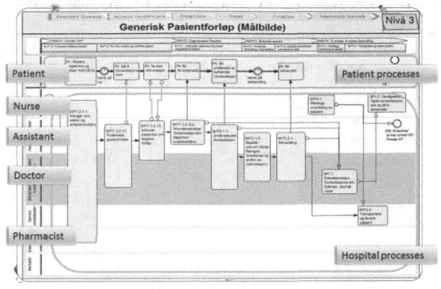

Fig. 8. Inclusion of both hospital and patient processes

Future Operating Model vs. Implementation models

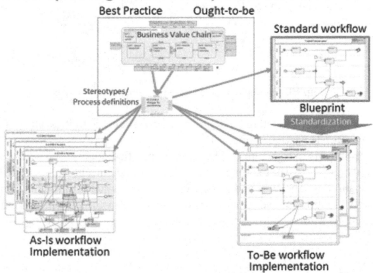

Fig. 9. Stereotypes as reusable process definitions

This generic, conceptual process can also be applied and be valid outside a hospital unit. There will be several similar clinical pathways in the municipal health service (local

doctor), emergency units (prehospital), and ambulance. It is important to see these similarities to be able to synchronize medical records information, supporting a situation of
BPM-in-the-large [13, 14].

When we come to the implementation models (as-is or to-be) we did go bottom up
from implemented systems (applications, application functions, information model) up
to activities in a workflow diagram using e.g. BPMN as illustrated in Fig. 10, linking
the implemented workflow activities to data model and application model.

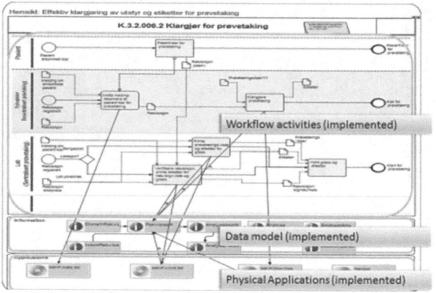

Fig. 10. Example of bottom-up implementation models

This is a specific architecture model referring to specific activities, applications and
information. (Ref TOGAF Continuum).

Going from as-is to to-be, guided by the best practice ought-to-be model will over
time close the gap between the long-term ambitions and current technical and organizational capabilities. The as-is and to-be will be different states of the implementation
model. A more long-term roadmap can be envisaged with a number of to-be models that
are expected to evolve into the ought-to-be model through reaching successive
plateaus [27].

4.1   Model Management and Use

The model(s) are created in Troux Architect and the diagrams can be stored in a shared
network drive or file solutions like Project place, SharePoint, or Dropbox.

The main architecture is illustrated in Fig. 11. The contents of the diagrams e.g.. the
BPMN objects like activities, data, events, gateways etc., are synchronized in a common

repository (TrouxSource) where they are maintained and reused and updated via a process portal (Process Navigator). The updates can then be synced back to the graphical diagrams.

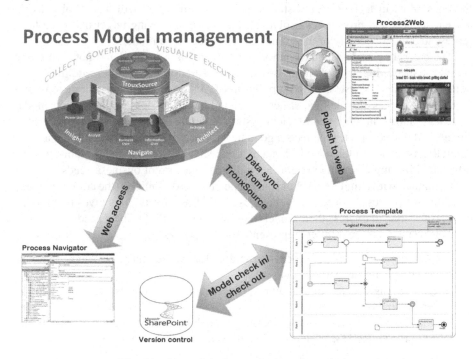

Fig. 11. Parts of the Troux Architecture solution

5   Discussion, Conclusion and Further Work

We have in this paper looked upon how to enhance the traditional practice with as-is and to-be models with an ought-to-be model representing the situated best practice – expressing the long-term ambitions for the enterprise. We describe the main lessons learned below:

BPMN has become the de facto standard for process modelling, but has also received much criticism for it's appropriateness as a general process modeling language [1]. BPMN is good for detailed bottom-up workflow-modeling, but less suited for top-down modeling of value-chain, -shop, -network processes. The purpose of this top-down modeling is often to get overview of the core business processes, and to make a process breakdown structure that is consistent through the different levels of details. Often it is desirable at this stage to leave open details about which roles can perform a task or process. i.e. a nurse can perform some tasks previously done by physicists or specialists. We see a similar division of notations in enterprise process models in other organizations [12].

Using BPMN will often result in cementing a specific way of implementing the process. In the hospital sector in Norway, it is clear that the wanted state and available technology is ten year or more ahead of current practice. In such a situation it is very important to be able to represent the ideal (ought-to-be) solution, although it will not be reached in a long time. The long-term ambitions have to give way in the short term for the constraints in technical/IT systems. Since the ambitions will also evolve as the current support and procedures evolve, this is clearly a moving target that need to be represented independently of the traditional as-is and to-be models.

The enterprise architects have not so far been using a method and notation that is suited for specifying an overall top-down process-model. For this purpose the IDEF0 notation, used in this case, is found superior to BPMN, in modelling a generic top-down model with process-breakdown structure, including a conceptual information model, logical organization model and logical system structure. This gives the expert professionals and the enterprise architects a good tool to specify the requirements at a level of detail that is in turn suitable for IT/System architects that is modeling bottom-up workflows used for implementing systems according to the current business needs.

Combined with a middle level of stereotype process definitions (the common understanding of a process between both enterprise and IT architects), this gives freedom for both professions to express their model in notations suitable for their needs.

This results in three different representations of processes.

1. Top-down (ought-to-be) process-structure for value-chain, - shop, -network modeled by enterprise architects.
2. Middle-out (agreed-to-be) stereotype process definitions representing the common understanding of a process.
3. Bottom-up (as-is, to-be) workflow activity models of implementation of the common agreed process for IT-, System- architects and operation managers.

Each of these types of models has their own lifecycles. A unifying overall process model like the ought-to-be model, makes it possible for people with various backgrounds, coming from different organizational units and disciplines, and who has worked in different ways in the past - to agree on common work processes and value chains. This contributes to developing a common terminology for processes, concepts and information objects. A generic overall model also contributes to the standardization of process modelling so that the work processes are described the same way in the different departments and disciplines, which is important for communication and reuse.

6   Conclusion and Further Work

Working with this approach hopefully will make it easier for the enterprise management and enterprise architects to express in more detail their ambitions, before the CIO and IT-architects come with their systems and limitations from current technology. A main learning from the case is that the ought-to-be models developed top-down, due to that they are not to be immediately implemented makes it possible to describe ideas and ambitions on a generic level, avoiding both 'accidental' organizational and technical limitations, but also terminological constraints making it easier to be innovative and learn from others without this being experienced as threatening to current practice.

As a case study this work is limited to a certain phase of the specification and building of a new hospital in HSØ, threating external validity (i.e. generalization of results). It is also primarily based on the experiences from one of the main modelers, threating the internal validity of the results since other stakeholders might have different perceptions. In further work we will be able to follow the use of the models over time through several iterations of updates. We also hope to be able to use this approach on other cases in other domains, potentially also supported by other tools, investigating what is particular for this case and what is to a larger degree generalizable.

In the investigation of the approach so far, we have used traditional process modelling languages such as IDEF0 and BPMN for the top-down and bottom-up modeling. We note that it is often needed to combine languages also from other perspectives [17]. In future work we will experiment with combining this with the use of approaches such as AKM [13], DEMO [7] and ArchiMate [22]. AKM for instance is believed to be better for supporting the agile use and evolution of the enterprise process knowledge captured in the model, in particularly when capturing knowledge bottom-up directly from work practice.

References

1. Aagesen, G., Krogstie, J.: Analysis and design of business processes using BPMN. In: vom Brocke, J., Rosemann, M. (eds.) Handbook on Business Process Management. International Handbook on Information Systems, pp. 213–235. Springer, Berlin (2010)
2. Alter, S.: A workaround design system for anticipating, designing, and/or preventing workarounds. In: Gaaloul, K., Schmidt, R., Nurcan, S., Guerreiro, S., Ma, Q. (eds.) BPMDS 2015 and EMMSAD 2015. LNBIP, vol. 214, pp. 489–498. Springer, Heidelberg (2015)
3. Bråten, S.: Model Monopoly and communications: systems theoretical notes on democratization. Acta Sociologica J. Scand. Sociological Assoc. 16(2), 98–107 (1973)
4. Clancy, T.R., Effken, J.A., Pesut, D.: Applications of complex systems theory in nursing education, research, and practice. Nurs. Outlook 56(5), 248–256 (2008)
5. Conner, D.: Managing at the Speed of Change. Random House, New York (1992)
6. Dahl, Y., Sørby, I.D., Nytrø, Ø.: Context in care–requirements for mobile context-aware patient charts. Stud. Health Technol. Inform. 107(Pt 1), 597–601 (2004)
7. Dietz, J.: Enterprise Ontology. Springer, Berlin (2006)
8. Engen, R., Viljoen, S.: Citizen-centric Heathcare Delivery Reference Model (2015)
9. Fawcett, S.E., Cooper, M.B.: Process integration for competitive success: benchmarking barriers and bridges. Benchmarking: Int. J. 8(5), 396–412 (2001)
10. Fossland, S., Krogstie, J.: Modeling as-is, ought-to-be and to-be – experiences from a case study in the health sector. In: PoEM 2015, Valencia, Spain (2015)
11. Gilbreth, F.B., Gilbreth, L.M.: Process Charts. American Society of Mechanical Engineers, New York (1921)
12. Heggset, M., Krogstie, J., Wesenberg, H.: Understanding model quality concerns when using process models in an industrial company. In: Gaaloul, K., Schmidt, R., Nurcan, S., Guerreiro, S., Ma, Q. (eds.) BPMDS 2015 and EMMSAD 2015. LNBIP, vol. 214, pp. 395–409. Springer, Heidelberg (2015)

13. Houy, C., Fettke, P., Loos, P., van der Aalst, W.M.P., Krogstie, J.: BPM-in-the-large – towards a higher level of abstraction in business process management. In: Janssen, M., Lamersdorf, W., Pries-Heje, J., Rosemann, M. (eds.) EGES/GISP 2010. IFIP AICT, vol. 334, pp. 233–244. Springer, Heidelberg (2010)

14. Houy, C., Fettke, P., Loos, P., van der Aalst, W.M.P., Krogstie, J.: Business process management in the large. Bus. Inf. Syst. Eng. **3**(6), 385–388 (2011)

15. IDEF0 (2016). http://www.idef.com/IDEF0.htm. Accessed 1 Mar 2016

16. Krogstie, J., Dalberg, V., Moe Jensen, S.: Process modeling value framework. In: Manolopoulos, Y., Fillipe, J., Constantopoulos, P., Cordeiro, J. (eds.) Enterprise Information Systems. LNBIP, vol. 3, pp. 309–321. Springer, Heidelberg (2008)

17. Krogstie, J.: Integrated goal, data and process modeling: from TEMPORA to model-generated work-places. In: Johannesson, P., Söderström, E. (eds.) Information Systems Engineering From Data Analysis to Process Networks, pp. 43–65. IGI, Hershey (2008)

18. Krogstie, J.: Model-Based Development and Evolution of Information Systems: A Quality Approach. Springer, London (2012)

19. Lillehagen, F., Krogstie, J.: Active Knowledge Modeling of Enterprises. Springer, Heidelberg (2008)

20. Moody, D.L.: Theorethical and practical issues in evaluating the quality of conceptual models: current state and future directions. Data Knowl. Eng. **55**, 243–276 (2005)

21. Nelson, H.J., Poels, G., Genero, M., Piattini, M.: A conceptual modeling quality framework. Softw. Qual. J. **20**, 201–228 (2011)

22. Open Group Archimate 2.1 Standard. http://pubs.opengroup.org/architecture/archimate2-doc/toc.html. Accessed 30 Mar. 2016

23. Price, R., Shanks, G.: A semiotic information quality framework: development and comparative analysis. J. Inf. Technol. **20**(2), 88–102 (2005)

24. Recker, J.C., et al.: Business process modeling : a comparative analysis. J. Assoc. Inf. Syst. **10**(4), 333–363 (2009)

25. Silver, B.: BPMN Method and Style. Cody-Cassidy Press, Aptos (2012)

26. Stabell, C.B., Fjeldstad, Ø.D.: Configuring value for competitive advantage: on chains. Shops Netw. Strateg. Manag. J. **19**, 413–437 (1998)

27. TOGAF (2016). https://www.opengroup.org/togaf/. Accessed 1 Mar 2016

28. Troux Architect (2016). http://www.troux.com/. Accessed 1 Mar 2016

29. Weick, K.: Sensemaking in Organisations. Sage, London (1995)

30. Weske, M.: Business Process Management: Concepts, Languages, Architectures. Springer Verlag Inc., New York (2007)

Information and Process
Model Quality

How Modeling Language Shapes Decisions: Problem-Theoretical Arguments and Illustration of an Example Case

Alexander Bock[(✉)]

Research Group Information Systems and Enterprise Modeling,
University of Duisburg-Essen, Essen, Germany
alexander.bock@uni-due.de

Abstract. To facilitate decision making and problem solving in organizations, numerous modeling approaches have been advanced in various research fields. Many of them are grounded on the idea that problem situations can be structured by means of designated sets of modeling concepts. A critical, yet often implicit, assumption in parts of the literature concerns the view that a given set of modeling concepts can capture the problem situation "as it is". Considering arguments about the constructive nature of problems, the paper illustrates a practical example case in which different modeling approaches are used to describe a single decision situation, to the effect that the formative role of decision modeling languages becomes apparent. Theoretical and practical implications for the field of conceptual modeling are outlined, and directions for future research are drawn.

Keywords: Conceptual modeling · Decision making · Problem solving · Problem construction · Decision models · Modeling concepts

1 Introduction

In recent decades, various research fields have felt called upon to offer help for dealing with "the more intractable organizational problems" of a "connected and turbulent world" [1, p. 1]. Many of them operate on the basis of *models*. *Decision analysis* and decision theory aim to construct formal models on which various kinds of optimizing procedures can be performed (e.g., [2–4]). Stemming from a background in operations research, sociology, and political studies, *problem structuring* methods employ conceptual models as part of procedures to guide multiple actors in discussing problem situations (e.g., [1,5,6]). More recently, decision making has started to receive increasing attention in the field of *conceptual modeling*. A variety of conceptual decision modeling languages have been set forth, all with slightly different ideas of how to aid decision making (e.g., [7–9]). Furthermore, the area of conceptual goal modeling has found particular use in the analysis of decision situations (see [10, pp. 676–678]). In sum, the organizational decision maker can pick from a rich assortment of modeling languages helping her or him deal with problems.

© Springer International Publishing Switzerland 2016
R. Schmidt et al. (Eds.): BPMDS/EMMSAD 2016, LNBIP 248, pp. 383–398, 2016.
DOI: 10.1007/978-3-319-39429-9_24

But there are intricacies in the application of decision modeling languages which are rarely discussed (though see [11,12]). They relate to the idea that what is provided in a decision modeling approach is a meaningful linguistic structure to decompose and analyze 'problems' as they appear. For example, authors in the field of decision analysis have likened their task to "the provision of a framework in which to think and communicate" [3, p. 1]. Likewise, approaches in the field of conceptual modeling such as the industry-driven OMG Decision Model and Notation (DMN) include on their part a "basic structure, from which all decision models are built" [7, p. 23]. With regard to the use of these frameworks, it is often admitted that particular constructed models are not at once complete or authoritative, but need gradual refinement and careful interpretation—*within* the given frame (e.g., [3, p. 1] [13, pp. 29–32]).

However, what is usually *not* at issue, or recognized as a bias at all, is the possibility that "the (modeling) framework" or "the (modeling) language" *itself* imposes upon the problem or decision situation a particular way of seeing. And, it is the argument of the paper, through this way of seeing decision modeling approaches do not merely bring to the fore "the" problem, but they determine what is seen and recognized *as part of a problem* in the first place. The practical importance of this point seems clear: The use of different decision modeling approaches might not only advocate different courses of action as (re-)solutions for problems, but they might lead to different interpretations of what are the problems, and, *ipso facto*, possible (re-)solutions altogether.

The purpose of this paper is (1) to clarify and direct attention at this crucial role of decision modeling languages in shaping problem constructions and decisions, (2) to illustrate its tangible and practical effects by means of an example case, and (3) to outline possible directions for the field of conceptual modeling to take note of these conditions. The observations of the paper have come to the fore during a conceptual analysis of existing model-based decision aids [14, p. 318], and the presented arguments are intended to contribute, and provide fuel for thought, to decision-related efforts currently underway in the field of conceptual (enterprise) modeling (e.g., [7–9,15]).

The argument of the paper proceeds as follows. The next section summarizes theoretical basics of decision and problem solving research. In Sect. 3, the main point of the paper is practically illustrated by an example case, using modeling approaches from various fields. Implications and possible routes for future research are discussed in Sect. 4. The paper closes in Sect. 5.

2   On Decisions and Problems

Decision making and problem solving are studied in many disciplines. Selected theoretical positions as to key concepts of these themes are outlined below.

The Basic Choice View. The most common idea of a 'decision', as found in many fields, is that of a *choice* (see, e.g., [16, pp. 656–657] [17, pp. 568–572]). A common framework, found both in descriptive and prescriptive research, assumes that a decision situation is characterized by a set of *alternative courses of action* (from which only one can be chosen), a set of possible *future states* (for which it is uncertain which one will emerge), a set of *outcomes* linked to all alternatives and future states, and a set of *goals* or values in the light of which the outcomes are evaluated (see, e.g., [4, pp. 1–3] [16, pp. 656–657]). Thus, in this view, a decision maker is confronted with an uncertain choice and wishes to obtain the most desirable outcome, much like in a "gambling decision" [17, p. 569].

From Decisions to Problems. But already a long time ago it has objected that such views "falsify decision by focusing on its final moment," and "ignore the whole lengthy, complex process of alerting, exploring, and analyzing that precede that final moment" [18, p. 40]. The gradual exploration and analysis of the situation, then, is assumed to happen only in the course of *decision processes* ([19, pp. 322–323] [18, pp. 40–44]). A great deal of prototypical process schemes has been proposed in the literature (for an overview, see, e.g., [20]). While details vary, it is commonly recognized that especially early stages include activities such as 'problem exploration' or 'problem definition' (cf. [20, pp. 855–856]). It is in this process view that activities of decision making come to be seen as closely related, if not identical, to activities of *problem-solving* (cf. [21, pp. 1489–1490] [20, pp. 855–856] [19, pp. 322–324]). Thus, generally speaking, a "pre-decisional phase of defining or formulating a problem" [19, p. 322] can be assumed.

Basic Aspects of Problem Solving. Basic psychological definitions of 'problem' state that, e.g., "a problem is an unknown in some context [...] (the difference between a goal state and a current state)", and "someone believes that it is worth finding the unknown" ([22, p. 2680]; for further definitions, see [23, pp. 5–13]). Solving a problem, then, according to elementary psychological insights first "requires the construction of a mental representation of the problem situation" [22, p. 2680]. More precisely, research indicates that representations are typically modified many times, and perhaps radically, in the process [24, pp. 414–420]. Second, problem solving includes various modes of reasoning, operating on the representations (e.g., [22, pp. 2682–2683] [24, pp. 414–420]). Both aspects are believed to be moderated by many factors, including (domain-specific) 'knowledge', 'epistemological beliefs', 'socio-cultural roles & expectations', and environmental 'information' and 'resources' (see [23, pp. 20–23] [22, pp. 2681–2683]).

The Constructive Nature of Problems. Considering the above basic positions, it can be asked whether being a 'problem' is an inherent and definite property of a situation. Although such an "objectivist view" can be detected in parts of the decision research literature [25, pp. 319–323], several authors have contended that "problems are not objective entities in their own right" ([19, p. 322]; see also [26, p. 119] [21, p. 1491]). This is related to various lines of argument. First, as noted above, a 'problem' is ordinarily defined in relation to what an individual takes as (un)desirable (as indicated by concepts such as 'goal'). In this sense, a situation cannot be said to be a problem independently of an individual process of valuation (cf. [21, p. 1491]). Second, what might become part of a mental representation of a perceived situation is ordinarily assumed to depend on factors such as (domain-specific) knowledge, obtained information, and basic beliefs (see above). If this position is accepted, then it cannot be said that a ("real-world") situation objectively hosts a specific problem, because there is no single way to interpret "the" situation in the first place (cf. [26, pp. 120-122]). Against the backdrop of these and further arguments (e.g., arguments related to epistemological debates, or arguments related to the idea of the social construction of reality; see [25, 26]), various authors have come to regard problems as 'constructions' or 'constructs' (e.g., [21, p. 1491] [5, p. 63] [26, p. 121]).

In sum, the short synthesis indicates that decision processes, rather than being limited to reasoning in the confines of given choice frames, are concerned with analyzing and *constructing* representations of perceived situations.

3 How Language Shapes Decisions: An Example Case

In order to illustrate how modeling languages exert influence on the construction of problems, in this section, an example decision case is considered through the lens of different modeling approaches. The scenario is broadly based on a real-world IT outsourcing case which has been reported in the literature [27, pp. 228–230]. Parts of the case are modified and shortened for purposes of illustration. The overall intention is to discuss a typical real-world decision process; the concern is not with specifics of the given institution or professional area.

Four decision modeling approaches have been selected for the following illustration. These have been deliberately chosen so as to cover a spectrum of research fields. However, the aim of this paper is not to conduct an exhaustive review of existing approaches. Overviews of decision aids from various research fields are available elsewhere (e.g., [4,5,10]). From among the selected approaches, *decision matrices* and *influence diagrams* stem from the field of decision analysis [4,28], *cognitive maps* have been developed in the realm of "soft" operational research [5,6], and *goal modeling* languages are a central subject in the field of *conceptual modeling* [10,29]. The disciplinary variety is intended to help illustrate markedly different perspectives on a single situation. The point of this paper, however, is not limited to specifics of these example approaches.

In line with the purpose of this paper, the following discussion places emphasis on the modeling *concepts* (i.e., static language constructs) of the approaches. The discussion abstracts from procedural components that guide model construction and (formal) analysis, just as from other influences such as group behavior or organizational regulations. Finally, note that the presented models are *possible* models; they are not supposed to convey a single way of interpretation.

3.1 Scenario Setting

The decision scenario concerns a central Regional Health Authority (RHA) and its local sub-units (cf., here and below, [27, pp. 228–230] for a description of the original case). The RHA is responsible for the overall healthcare coordination in a national region, while the sub-units (e.g., hospitals or other medical offices) offer local healthcare services. Recently, the government has advanced an initiative to reorganize the national public healthcare system, which advises to run the health administration in a more competitive manner, to privatize where possible, and to decrease overall costs. To enforce a more flexible administration, RHAs are further prompted to decentralize overall managerial responsibilities and to establish "purchaser-provider contractual relationships" [27, p. 228] among their various (sub-)units. These directives, in turn, have implications for the IT organization. They encourage IT cost reductions, externalization of IT services, establishing IT purchaser-provider contracts in the (sub-)units, and redistributing IT management responsibilities. The RHA, which itself feels a shortage of money, needs to consider options for outsourcing the IT unit and its local subsidiaries.

At first sight, four options suggested themselves. *"One was to contract the whole [IT] unit out to an outsourcing vendor. The others were to convert the unit into a trading agency to sell its services in the region [...], to contract out parts of the unit, or go for a management buyout."* [27, pp. 228–229]. But assessing these options turns out to be complicated. First, IT services are provided and used in a distributed and heterogeneous environment, making it hard to determine what could be outsourced at all. Some services are provided centrally by the RHA (e.g., an expiring mainframe system offering basic applications for the sub-units), others are run by the sub-units themselves. Second, each sub-unit has a different level of expertise both with respect to specific technologies and with respect to IT management in general. As a result, some sub-units would continue to depend on legacy systems (e.g., the central mainframe) because they cannot run systems locally; other sub-units, in turn, do have these capabilities. Moreover, due to the directive to redistribute (IT) management responsibilities, any outsourcing option would possibly have to be managed by the local (sub-)units alone. But because experience and external vendor maturity vary greatly, it is uncertain whether the (sub-)units would have the skills necessary to negotiate IT outsourcing contracts. Finally, strategic issues need to be considered. It is felt that some IT services should not be externalized because they involve critical healthcare knowledge. Also, of course, outsourcing would be regarded as useless if the resulting IT service quality would not exceed in-house service quality.

In sum, the RHA would appear to face a typical 'problematic' situation: There is a clear need to act, but a confusing range of issues, constraints, aims, options, and developments present themselves. Let us consider what aspects of these affairs can be untangled through the lens of different modeling languages.

3.2   Perspective 1: The Situation as a Choice

A first possible approach to consider decision situations are *decision matrices* (e.g., [4, pp. 2–3] [3, pp. 21–24]). Different variants exist. One common variant enables to view situations through the lens of the classical decision analysis concepts *alternatives*, *environmental states*, *goals (criteria)*, and *outcomes*

	Goal criteria	Ratio of externalized IT services		Degree of IT management responsibility distribution		Number of IT services obtained via provider-purchaser contracts		Legacy IT service availability maintained (e.g., central mainframe)		Critical IT knowledge kept in-house		Fit between IT management responsibilities and capabilities	
	Environmental states*	s_1	s_2	s_1	s_2	s_1	s_2	s_1	s_2	s_1	s_2	s_1	s_2
Alternatives	Fully outsource to one vendor	1	1	...	...	...	...	...	...	0	0	.4	.8
	Outsource parts of the IT services	.5	.5	:		:		:		1	1	.5	.825
	Transform into an independent firm	.575	.575	:		:		:		.8	.8	.6	.88
	Management buyout	1	1	:		:		:		.9	.9	.7	.91

* where s_1 = *Local sub-units overwhelmed by IT management responsibilities*, s_2 = *Local sub-units successfully deal with new responsibilities*.

Fig. 1. A possible decision matrix for the decision situation

(see, e.g., [30, p. 807]; for a conceptual reconstruction, see [8, pp. 311–312]). In essence, the situation is apprehended as a *choice* in view of multiple criteria.[1]

What can be seen through this lens? Fig. 1 presents an example. First of all, a clear-cut set of alternatives is listed, covering the four options indicated in the case description. However, as becomes apparent, the formulated alternatives need to be placed at a high level of abstraction because the circumstances permit a great deal of adjustments and strategies for all alternatives (e.g., determining *which* parts of the IT should be outsourced). Next, six example goals (or criteria) to evaluate alternatives are found in the decision matrix. These criteria can be seen as some of the more operational goals that can be interpreted in the case. Other, more general or distant goals in the case (e.g., overall cost estimates or overall IT service efficacy) may be seen as functions of the listed criteria (see [30, p. 810] for notes on this procedure, and see Sect. 3.4 for a detailed discussion of goals). Finally, decision analysis advises to assess alternatives with respect to concrete *outcomes* for each criterion. To account for uncertainty, different future *environmental states* may be defined. One uncertainty in the case concerns the question whether or not the sub-units will have the IT management skills to handle outsourcing contracts. This is condensed into two states (s_1 and s_2), which affect outcomes for the sixth criterion. Example outcomes, normalized on a scale from 0 to 1, are indicated for other criteria as well.

In sum, decision matrices enable to structure particular options and their possible future outcomes in detail, serving as a basis for a great deal of analytic procedures (see [4, 30]). However, when considering the given case, it would

---

[1] A different variant of decision matrices is provided by the OMG DMN (see [7, pp. 70–87]). In this variant, outcomes (which then also represent alternatives) are mapped to combinations of input values [7, pp. 70–72]. These decision tables are helpful mainly for "operational decisions" [7, p. 27] where it is known ex ante how to act under what circumstances. This is not the case in the more complex example case.

seem that the concern is not so much with assessing outcomes for given options. Rather, there is a need to assemble possible externalization options in the first place, while considering a host of technical constraints and governmental directives. The choice-centric lens of decision matrices does not directly permit to see these background conditions, motivations, and conflicts.

3.3 Perspective 2: The Situation as an Entanglement of Causes

Some 'problem structuring methods' utilize so-called 'cognitive maps' (e.g., [13, pp. 26–37]). Many variants of 'causal maps' are found in the literature (see [5, pp. 139–140]). Here the focus rests on a variant supposed to convey "a person's assertions about his beliefs with respect to some limited domain" ([31, p. 72], see also [6, p. 6]). Cognitive maps essentially consist of two kinds of constructs: '*Concepts*' and '*causal links*' (see [8, pp. 314–315]). A 'concept', in cognitive maps, is a broad construct. Authors have suggested to use it to express, e.g., 'assertions', 'assumptions', 'ideas', 'actions', 'issues', 'concerns', 'strategies', 'goals', and 'missions' (see [5, pp. 142–143] [6, pp. 4, 36–38]). The links, in turn, are supposed to convey "a causal assertion" [31, p. 72], meaning that more of one concepts is assumed to lead to more of another (or vice versa, dependent on the sign) (cf. [5, pp. 143–145]). Figure 2 presents an example cognitive map for the case.

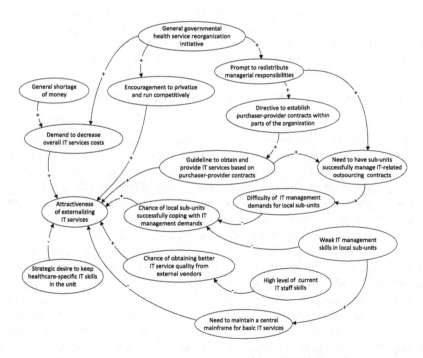

Fig. 2. A possible cognitive map for the decision situation

It is clear at a glance that the scope of what is covered by the cognitive map is broader than what is captured by the decision matrix. For example, the cognitive map includes various elements that describe the background developments and stimuli of the decision situation (e.g., the governmental initiative; top of the figure). More generally, the map describes different kinds of factors, demands, and interrelations centering around the idea of 'externalizing IT services'. As a result, the indicated causal relations go beyond the single relation of "alternative A_i may cause outcomes O_{ij}", as it can be expressed in decision matrices. The cognitive map is also more abstract than the decision matrix in that it may describe general ideas or factors rather than concrete states or outcomes. In fact, the example map does not even include hints concerning possible courses of action (although, admittedly, it could); it rather summarizes, from an individual perspective, factors affecting the attractiveness of externalizing IT services.

Despite this broadness, there are aspects which cannot be seen through the eyes of a cognitive map. Consider, for instance, the concepts *'Need to have sub-units successfully manage IT-related outsourcing contracts'* and *'Weak IT management skills in local sub-units'* (bottom of Fig. 2). While the outgoing causal links of the concepts indicate that these factors together constitute an issue, it is not possible to express that directly because it is not a 'causal' relationship. The same goes for other conceivable relationships types, such as teleological ones (e.g., expressing that one goal is valued more than another). Finally, many facets of the situation remain concealed due to the low level of semantics of the construct of a 'concept'. Although conventions exist (see, e.g., [5, pp. 145–146] [6, pp. 36–38]), it is not possible to express through linguistic constructs whether a 'concept' in a cognitive map is supposed to describe, say, a 'goal' or an 'idea'.

3.4 Perspective 3: The Situation as a Goal Conflict

While several goals of the RHA, broader and more specific ones, have been indicated in the preceding perspectives already, the modeling approaches used there do not provide the conceptual means to consider goals and their relationships at a more subtle level. For that purpose, various goal modeling languages have been advanced in recent years (for overviews, see [10, pp. 675–676] [29, pp. 13–14]). In order to analyze the given case, the modeling language MEMO GoalML [29, 32] is used. This language is selected because it offers a particularly nuanced set of inter-goal relationships. Figure 3 presents a possible goal system for the case.

When interpreting the diagram, it appears that the whole situation can in fact be understood as an array of goal conflicts. At the top of the diagram, several more final goals are found, whereas the lower goals correspond to the more operational goals already listed in Sect. 3.2. In a related sense, some goals of the RHA constitute rather broad orientations (e.g., to maximize IT long-term competitiveness), whereas others demand for more specific conditions to be fulfilled (e.g., to maximize the number of purchaser-provider contracts). To express this distinction, the GoalML offers two goal concepts. *Symbolic goals* (represented by a lighthouse symbol) describe goals mostly inspirational in nature, while *engagement goals* (represented by a target symbol) are linked to concrete

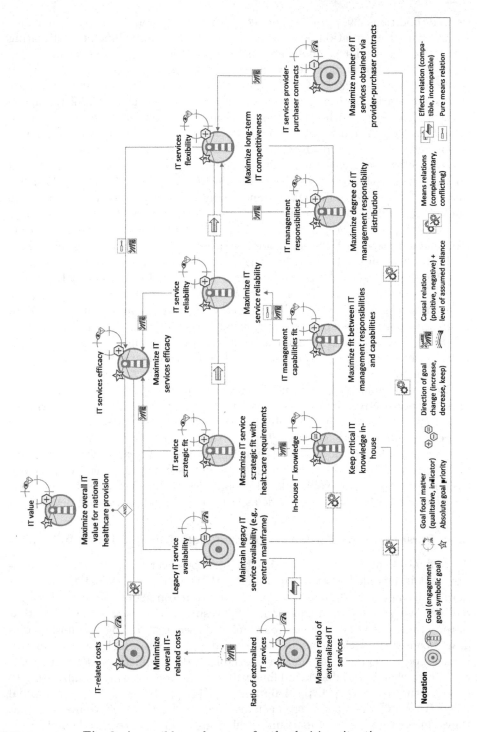

Fig. 3. A possible goal system for the decision situation

desired conditions [29, pp. 4–7]. The recognition of symbolic goals partly explains why viewing the situation as a choice (Sect. 3.2) has been difficult—estimating concrete outcomes for symbolic goals seems precarious, if possible at all.

Most importantly, however, viewing the situation through the lens of goal relations makes explicit some of the value dilemmas the RHA is facing. First, several conflicts arise because possible means to achieve one goal will likely impede actions to achieve another (expressed through the *conflicting means* relation; cf. [32, p. 233]). For example, actions to externalize IT services will probably oppose attempts to keep knowledge in-house. In a related, but subtly different way, effects of goals might also be incompatible with respect to a third goal (expressed through the *effects incompatible* relation; cf. [32, p. 233]). For instance, externalizing IT services and maintaining legacy system availability might not conflict in terms of means, but together they can be expected to affect incompatibly the goal of decreasing IT costs (because having an external vendor host legacy systems could be a cost driver). Another aspect which could not be expressed explicitly using the previous approaches are goal priorities (shown in the stars attached to each goal). Finally, the GoalML permits to make visible selected further assumptions of the modeler. For example, the diagram shows many *causal* relations between goals (meaning that actions to reach one goal are assumed to contribute to reaching another; cf. [32, p. 231]). For the relation between IT externalization and cost savings (left-hand side), it is explicitly expressed that the assumed reliability is in fact weak (see the dotted arrow symbol).

However, again, there are many aspects which remain unseen in a goal-oriented perspective. In essence, goal models focus on *value* and intermediate *means-ends* statements. Goals express *"oughts"*, and these can neither be reduced to, nor be derived from, *"factual"* (in the sense of "fact-related") statements [33, pp. 46–47]. In consequence, it is not possible to express through goals, e.g., developments in the decision environment (see the previous perspective), assumptions about non-value-related probabilistic influences (see the next perspective), and other domain-specific aspects (e.g., IT infrastructure details).

3.5   Perspective 4: The Situation as a Net of Probabilistic Influences

As a final example, recall that in the first perspective (Sect. 3.2) the decision matrix included a set of environmental states (s_1, s_2). What the decision matrix did not permit to see, however, were environmental *factors* and *probabilistic influences* that might eventually result in the listed states. For that purpose, influence diagrams can be employed (see, e.g., [12, 34]). Influence diagrams consist of decision nodes (representing variables which can be varied), chance nodes (representing aleatory variables), deterministic nodes (representing non-stochastic variables), and final value nodes (representing variables sought to be optimized) (see [34, pp. 4–6]; and see [14, pp. 312–314] for a conceptual reconstruction). Influence diagrams may seem similar to cognitive maps in that they can be used to describe assumed relations among non-controllable factors. But they are different in that they, apart from providing more kinds of nodes,

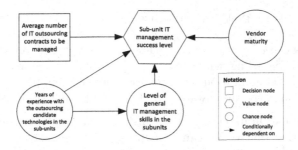

Fig. 4. A possible influence diagram for the decision situation

describe probabilistic relations: A link between variables expresses conditional dependence. Figure 4 shows an example influence diagram for the case.

Note that the example influence diagram shows only those factors which might, according to the case description (Sect. 3.1), result in the different environmental states for sub-unit IT management success. It would also be possible to create a more encompassing influence diagram. But it becomes apparent from the brief consideration already that influence diagrams, again, yield another way of seeing the situation. In a way, they are a more abstract counterpart to decision matrices. Rather than enabling to see particular alternatives or environmental states, they permit to describe related variables or factors (for instance, the states s_1 and s_2 from the decision matrix would be concrete possible values for the middle node in the diagram). Then again, influence diagrams have also blind spots. For example, it is not possible to express details of value dispositions (e.g., preferences among values as in the goal perspective), or any other factual aspects other than those concerned with variables and probabilistic dependencies (e.g., broad notions such as "encouragement to privatize" found in the cognitive map).

4 Discussion and Implications

Having considered a single decision situation through the lens of different modeling languages, several insights, concerns, and implications suggest themselves. These are considered below, together with suggestions for future research.

Ways of Seeing. First of all, the example case has illustrated how each modeling language, by means of its concepts, provides the ability to see and describe selected aspects of a decision situation, while necessarily omitting others. This observation corresponds to Birrer's position that all decision aids offer "fixed structures (fragmentation schemes) [...] for decomposition of complexity [which] cannot be considered merely as 'neutral' instruments" [11, p. 49]. Modeling languages appear to possess *enabling* and *restrictive* qualities, whose specific implications for the modeling of 'problems' are considered below. On a more general level, raising awareness of language-specific lenses alone seems to be an important responsibility for decision modeling research. Sometimes, the customary rhetoric can underemphasize, or downright conceal, this influence. For example, when its authors state that "the influence diagram [...] is [...] a formal description of the problem" [28, p. 127], it is implicitly suggested that the approach can capture the essence of a 'problem'—although it is but one perspective. In the field of enterprise modeling, where authors advocate a "reflective [...] perspective that accounts for the limitations of enterprise models" [35, p. 47], it would seem appropriate to similarly raise awareness of the limitations of decision models.

Enabling Qualities. On its face, it is a truism to note that a modeling language enables to describe that which can be expressed by means of its concepts. But there are important implications. First, as has been indicated in Sect. 2, 'problems' can essentially seen as mental representations constructed from perceived 'situations'. Aside from the fact that the possible richness of these constructions are likely captured only incompletely (restrictive facets are considered below), a complicated interplay between (a) what is perceived, (b) what is constructed mentally, and (c) what is modeled can be assumed. There is no need to speculate on specifics here (cognitive aspects of modeling are discussed elsewhere; e.g., [24, pp. 424–425]). But in this interplay, the 'enabling' qualities of modeling approaches (cf. c) may also be assumed to have 'direction-giving' qualities, affecting how attention is distributed (cf. a) and thus what emerges as a mental representation (cf. b). Similarly, Birrer argues that "tools often'invite' ('seduce') the user to certain kinds of use, which almost invariably leads to some problem shift" [11, p. 49]. In an even more pessimistic, but practically grounded view, decision analysts have remarked that "the too-frequently-encountered methods expert [...] looks for problems that can be made to fit methods on hand" [12, p. 101]. As a conclusion for the field of conceptual modeling, it would seem important to further study the desirability and controllability of the formative role of modeling languages, and to test ways of making it explicit or to mitigate it.

Related to that thought, and as a second implication, it follows that the enabling qualities of languages may also affect the selection of strategies to *resolve* the problem. For example, the decision matrix essentially suggests to concentrate on picking an *optimal means* to defined ends; the goal system suggests to devise a course of action that *mediates between* conflicting goals; and the cognitive map might suggest to devise a course of action that considers *underlying causes* of relevant developments. In another way, the resultant models could be distinguished into supporting what has been described in the psychological literature as either "an attitude of *problem-mindedness* [or] the usual attitude of *solution-mindedness*" [36, p. 151]. The presented cognitive map and the goal model focus on the former[2] (because they concentrate on "exploring the character of the problem situation" [36, p. 151]), while the other tools focus on the latter (because they include possible solutions ex ante). Whatever the specifics, it can be assumed that some courses of actions are unlikely to be discovered through the modes of reasoning stimulated by a certain model. As a consequence, it would

---

[2] Note, however, that cognitive maps and goal models could be created such that they include solutions as well. In fact, this is what is done in parts of the literature on goal model analysis (see [10, pp. 680–681]). But, as has been pointed out, these "procedures can only find alternatives already in the model" [10, p. 681].

appear helpful to foster a variety of modes of reasoning in decision modeling. This is not a new point. In fact, it is a classical "recommendation [...] to keep an open mind in the early steps of structuring a decision problem and consider multiple alternative structures" [12, p. 86]. But in the field of conceptual modeling, the study of ways and methods to specifically promote such 'pluralistic' decision support has not yet received the attention it would seem to deserve.

Restrictive Qualities. Besides enabling facets, it has also been noticed that through each language a good deal of situational aspects remain unseen. This would seem to occur in at least two ways. For one thing, it may happen that an approach simply does not provide the concepts to describe certain aspects. For example, influence diagrams do not possess a concept to describe, say, 'conflicting means-ends' goal relations, as can be found in the GoalML [29]. Consequently, it might occur, for each modeling language, that aspects of an individual mental representation of a problem cannot be expressed, forcing the individual to focus on other aspects and thus, as indicated above, possibly contributing to "problem shift" [11]. But there is another way in which expressiveness can be limited. It might be that a modeling language offers generic concepts which can be used to describe a broad range of phenomena. This is essentially what general-purpose modeling languages do (see [35, pp. 26–29]). Here, such a genericity can be noticed, e.g., for 'cognitive maps'. Almost any idea (e.g., an event or a goal) can be modeled as a 'concept' in a cognitive map. But then there is nothing which distinguishes the use of a 'concept' in the sense of a 'goal' or an 'event'. Hence, the way of seeing conveyed by the language can be said to be restrictive in the sense that it does not permit to "see a difference"—the level of semantics of the modeling constructs is low; they allow for a too wide range of possible interpretations (cf. [35, pp. 7,13]). From a practical perspective, this seems critical not only for interpersonal model interpretability, but also when an individual would prefer to structure a situation at a higher level of detail than is enabled. In sum, it appears warranted to study more thoroughly how possible undesired effects of restrictive qualities of modeling languages could be made explicit and, perhaps, mitigated—if only by increasing conceptual richness (see below).

Enhancing Both Generic and Domain-Specific Conceptual Richness. Finally, it has been indicated above that fostering conceptual richness in decision modeling may be helpful. This can happen along a spectrum of specificity. On one end of the spectrum, there are *generic* concepts. A selection of such generic concepts is essentially provided by the considered approaches (e.g., 'alternative', 'concept', 'goal', or 'chance'; cf. [14, p. 317]). All these concepts have a basic plausibility and intuitiveness. But there is nothing which implicates that this selection is already sufficiently comprehensive. Thus, it seems useful to investigate whether further generic concepts to describe situations can be found, or perhaps combined, in decision aids. On the other end of the spectrum, decision modeling might profit greatly from *domain-specific* concepts. For example, the considered case is essentially concerned with the domain of IT infrastructures. A modeler might therefore wish to describe domain-specific concepts such as 'mainframe' and 'software'. But, naturally, none of the considered general-purpose decision aids include concepts from these domains. Considering that psychological research over time has increasingly emphasized the role of domain-specific knowledge in decision making [17], it appears promising to examine possible ways of extending or combining general-purpose decision aids with domain-specific modeling constructs, so that a richer overall modeling experience could be achieved.

5   Conclusions

When contemplating puzzling real-world decision situations, a variety of modeling approaches can be used to dissect the perceived problem. The task of this paper was to show that each of these linguistic structures imposes upon the situation a particular way of seeing, and that this way of seeing affects what is considered as a part of the problem, and, *ipso facto*, a desideratum of a possible solution. Based on a practical illustration of an example case and basic problem-theoretical views, several implications and routes for the field of conceptual modeling have been outlined. In particular, future research could beneficially (1) raise critical awareness about the limits and formative role of decision modeling languages, (2) investigate possibilities for aiding a greater variety of modes of reasoning, perhaps in the form of 'pluralistic' decision modeling approaches, and (3) examine ways of enhancing the conceptual richness of decision modeling languages, both in terms of generic and domain-specific modeling constructs.

References

1. Rosenhead, J., Mingers, J.: A new paradigm of analysis. In: Rosenhead, J., Mingers, J. (eds.) Rational Analysis for a Problematic World Revisited, pp. 1–19. Wiley, Chichester (2001)
2. Raiffa, H.: Decision Analysis: Introductory Lectures on Choices under Uncertainty. Addison-Wesley, Reading (1970)
3. French, S. (ed.): Readings in Decision Analysis. Chapman and Hall, London (1989)
4. Triantaphyllou, E.: Multi-Criteria Decision Making Methods: A Comparative Study. Kluwer, Dordrecht (2000)
5. Pidd, M.: Tools for Thinking: Modelling in Management Science, 2nd edn. Wiley, Chichester (2003)
6. Bryson, J.M., Ackermann, F., Eden, C., Finn, C.B.: Visible Thinking: Unlocking Causal Mapping for Practical Business Results. Wiley, Chichester (2004)
7. Object Management Group: Decision Model and Notation: Beta 1. OMG Document dtc/2014-02-01
8. Bock, A.: Beyond narrow decision models: toward integrative models of organizational decision processes. In: Proceedings of the 17th IEEE Conference on Business Informatics (CBI 2015), pp. 181–190. IEEE Computer Society (2015)
9. Plataniotis, G., de Kinderen, S., Proper, H.A.: EA anamnesis: an approach for decision making analysis in enterprise architecture. Int. J. Inf. Syst. Model. Des. 5(3), 75–95 (2014)
10. Horkoff, J., Yu, E.: Analyzing Goal Models - Different Approaches and How to Choose Among Them. In: Chu, W. (ed.) Proceedings of the 2011 ACM Symposium on Applied Computing (SAC 2011), pp. 675–682. ACM, New York (2011)

11. Birrer, F.A.J.: Problem drift. eliciting the hidden role of models and other scientific tools in the construction of societal reality. In: DeTombe, D.J. (ed.) Analyzing Complex Societal Problems, pp. 43–55. Hampp, München (1996)
12. von Winterfeldt, D., Edwards, W.: Defining a decision analytic structure. In: Edwards, W., Miles Jr, R.F., von Winterfeldt, D. (eds.) Advances in Decision Analysis, pp. 81–103. Cambridge University Press, Cambridge (2007)
13. Eden, C., Ackermann, F.: SODA-the principles. In: Rosenhead, J., Mingers, J. (eds.) Rational Analysis for a Problematic World Revisited, pp. 21–41. Wiley, New York (2001)
14. Bock, A.: The concepts of decision making: an analysis of classical approaches and avenues for the field of enterprise modeling. In: Ralyté, J., España, S., Pastor, Ó. (eds.) PoEM 2015. LNBIP, vol. 235, pp. 306–321. Springer, Heidelberg (2015). doi:10.1007/978-3-319-25897-3_20
15. van der Linden, D., van Zee, M.: Insights from a study on decision making in enterprise architecture. In: Short Paper Proceedings from the 8th IFIP WG 8.1 Working Conference on the Practice of Enterprise Modelling. pp. 21–30 (2015)
16. Hastie, R.: Problems for judgment and decision making. Ann. Rev. Psychol. **52**, 653–683 (2001)
17. Goldstein, W.M., Weber, E.U.: Content and discontent: indications and implications of domain specificity in preferential decision making. In: Goldstein, W.M., Hogarth, R.M. (eds.) Research on Judgment and Decision Making, pp. 566–617. Cambridge University Press, Cambridge (1997)
18. Simon, H.A.: The New Science of Management Decision. Harper & Row, New York (1960)
19. Dery, D.: Decision-making, problem-solving and organizational learning. Omega **11**(4), 321–328 (1983)
20. Lang, J.R., Dittrich, J.E., White, S.E.: Managerial problem solving models: a review and proposal. Acad. Manag. Rev. **3**(4), 854–866 (1978)
21. Smith, G.F.: Towards a heuristic theory of problem structuring. Manag. Sci. **34**(12), 1489–1506 (1988)
22. Jonassen, D.H., Hung, W.: Problem Solving. In: Seel, N.M. (ed.) Encyclopedia of the Sciences of Learning, pp. 2680–2683. Springer, New York (2012)
23. Frensch, P.A., Funke, J.: Definitions, traditions, and a general framework for understanding complex problem solving. In: Frensch, P.A., Funke, J. (eds.) Complex Problem Solving, pp. 3–25. Lawrence Erlbaum, Hillsdale (1995)
24. Bassok, M., Novick, L.R.: Problem solving. In: Holyoak, K.J., Morrison, R.G. (eds.) The Oxford Handbook of Thinking and Reasoning. Oxford Library of Psychology, pp. 413–432. Oxford University Press, Oxford (2012)
25. Landry, M.: A note on the concept of 'problem'. Org. Stud. **16**(2), 315–343 (1995)
26. Eden, C., Sims, D.: On the nature of problems in consulting practice. Omega **7**(2), 119–127 (1979)
27. Willcocks, L., Fitzgerald, G.: Market as opportunity? case studies in outsourcing information technology and services. J. Strateg. Inf. Syst. **2**(3), 223–242 (1993)
28. Howard, R.A., Matheson, J.E.: Influence diagrams. Decis. Anal. **2**(3), 127–143 (2005)
29. Overbeek, S.J., Frank, U., Köhling, C.: A language for multi-perspective goal modelling: challenges, requirements and solutions. CS&I **38**, 1–16 (2015)
30. Keeney, R.L.: Decision analysis: an overview. Oper. Res. **30**(5), 803–838 (1982)
31. Axelrod, R.: The analysis of cognitive maps. In: Axelrod, R. (ed.) Structure of Decision, pp. 55–73. Princeton University Press, Princeton (1976)

32. Köhling, C.A.: Entwurf einer konzeptuellen Modellierungsmethode zurUnterstü-tzung rationaler Zielplanungsprozesse in Unternehmen. Cuvillier, Göttingen (2013)
33. Simon, H.A.: Administrative Behavior: A Study of Decision-Making Processes in Administrative Organization, 3rd edn. Free Press, New York (1976)
34. Howard, R.A.: From influence to relevance to knowledge. In: Oliver, R. (ed.) Influence Diagrams, Belief Nets and Decision Analysis, pp. 3–23. Wiley, New York (1990)
35. Frank, U.: Multi-Perspective Enterprise Modelling: Background and Terminological Foundation. ICB Research Report 46, University of Duisburg-Essen, Essen
36. Maier, N.R.F., Solem, A.R.: Improving solutions by turning choice situations into problems. Pers. Psychol. **15**(2), 151–157 (1962)

On Suitability of Standard UML Notation for Relational Database Schema Representation

Drazen Brdjanin[1]([✉]), Slavko Maric[1], and Zvjezdan Spasic Pavkovic[2]

[1] Faculty of Electrical Engineering, University of Banja Luka,
Patre 5, 78000 Banja Luka, Bosnia and Herzegovina
`bdrazen@etfbl.net`, `ms@etfbl.net`
[2] RT-RK Institute, Patre 5, 78000 Banja Luka, Bosnia and Herzegovina
`zvjezdanspasic@gmail.com`

Abstract. The suitability of the standard UML notation for representation of relational database schema has been considered in this paper. Unlike the existing approaches using specialized notation (UML profiles), in this paper we propose an alternative approach for representation of relational database schema by standard UML class diagram. Apart from the analysis of the suitability of the isID (meta)attribute, we propose an alternative representation of composite keys by using class operations. The main idea of the proposed approach is based on the fact that the standardized order of operation parameters can be used to represent the order of key segments. The proposed approach is illustrated by a simple model in forward engineering of relational database.

Keywords: Relational database schema · UML · Class diagram · Profile · Eclipse-topcased · Forward engineering · Primary key · Foreign key

1 Introduction

The relational database model [1] has come a long way since the 1970s and it has become the dominant model of database organization. The description of the overall structure, relationships between data and corresponding constraints in a relational database (RDB) are specified by the relational database schema (RDBS). In an RDB design process, the RDBS constitutes a transitional model between the conceptual database model/schema (CDM) and target physical RDB. In the context of model-driven RDB development, the CDM represents a platform independent model (PIM), while the RDBS represents a platform specific model (PSM), since the RDBS represents platform specific implementation details such as platform specific data types, etc. In the existing findings, there is no single or standardized approach to RDBS representation. The existing tools for database design (commercial and open source, as well) use different notations for RDBS representation, where the traditional notations (IE [2], IDEF1X [3], etc.) are dominant. Since such models have a very limited portability, the Unified Modeling Language (UML) has been widely used in database design process.

© Springer International Publishing Switzerland 2016
R. Schmidt et al. (Eds.): BPMDS/EMMSAD 2016, LNBIP 248, pp. 399–413, 2016.
DOI: 10.1007/978-3-319-39429-9_25

UML [4] is a standard(ized) graphical language for the design, specification, visualization and documentation of software systems in different phases of their life cycle. The first reason for its widespread use is the rich notation that is independent of the modeling process. The second reason lies in the concept of openness, since the standard UML notation can be extended and specialized for a specific domain. A set of extensions of the standard notation for use in a particular domain is called a profile.

Given that the RDBS constitutes a PSM, the majority of the proposed UML-based approaches and implemented tools use a specialized UML notation (i.e. profiles) for RDBS representation. In this paper we propose an approach for RDBS representation by standard UML notation, eliminating the need for defining and applying domain-specific profile, which directly results in a more simple and more effective database design process.

The paper is organized as follows. The second section follows the introduction and presents the related work. The suitability of the isID (meta) attribute for representation of the primary key is analyzed in the third section. In the fourth section we propose an alternative representation of (composite) keys by using class operations. The forward RDB engineering, based on the proposed approach, is illustrated in the fifth section. The final section concludes the paper and gives the directions for future research.

2   Related Work

The UML-based RDBS representation has been the subject of research since the beginning of UML development. Although OMG (Object Management Group) issued a request for proposal for a UML profile dedicated to database modeling [5] ten years ago (2005), there is still no standardized approach to UML-based RDBS representation.

The first important industrial implementation (Rational Software Corp.) of a UML profile for database design was presented by Naiburg and Maksimchuk [6]. The majority of all subsequent proposals for UML-based RDBS representation follows that initial UML profile specification, which assumes that: (a) tables are represented by stereotyped classes (≪table≫), (b) columns are represented by class attributes, (c) keys are represented by stereotyped (≪PK≫, ≪AK≫, ≪FK≫) class attributes, but the column ordering in composite keys is not visible, (d) relationships between tables are represented by stereotyped (composite) associations (≪identifying≫/≪non-identifying≫), (e) RDBS constraints are represented by stereotyped operations (≪PK≫, ≪FK≫, etc.), but all other necessary information about constraints is represented by tagged values, (f) table indices are represented by stereotyped operations (≪index≫), etc.

Li and Zhao [7] proposed the usage of specific tagged values to designate the keys. They proposed the modeling of relationships between tables by stereotyped dependencies, but they proposed specific tagged values to designate referential integrity constraints and column ordering in composite keys. Further, the platform specific details were not considered.

Ambler [8] proposed the specific stereotypes for modeling of columns that represent or belong to the keys, but the column ordering in the composite key is specified by tagged values. The relationships between tables are represented by stereotyped associations (like in [6]), while the indices are modeled by stereotyped (≪index≫) classes and connected to the corresponding tables by dependencies.

Lo and Hung [9] proposed a UML profile for modeling database retrieval. Although it allows RDBS modeling, it is not focused on efficient RDBS visualization, but on modeling of data retrieval (queries).

Marcos et al. [10] proposed a methodological approach for object-relational database design using UML, covering the whole process from conceptual to physical level. In the context of UML-based RDBS representation, they follow the previous proposals, except for representation of indices (they also propose the usage of stereotyped classes, like in [8]).

By following Naiburg and Maksimchuk [6], Tomic et al. [11] proposed a UML profile where the typical RDBS constraints are represented by stereotyped operations (≪PK≫, ≪FK≫, etc.), too. However, they proposed the usage of the standardized order of operation parameters to specify and visualize the order of segments in complex keys. The proposed approach enables: (a) reduction of additional properties in stereotypes representing the keys, and (b) better RDBS visualization. That idea for the representation of complex keys also constitutes a basis for this paper, since it eliminates the need for UML profile and enables the representation of fundamental RDBS concepts by standard UML notation.

Apart from the previous papers focusing on specification of the UML profile for RDBS modeling, in the existing literature there are also many papers (e.g. [12–14]) using a limited set of UML extensions of the standard UML class diagram for modeling some RDBS concepts (typically tables and primary keys). However, such a class diagram is closer to a conceptual database model than to an implementation model (RDBS).

3   Suitability of isID Meta-attribute

The UML specifications prior the UML 2.4 [15] did not contain explicit support for representation of identifiers – there was no explicit way to indicate that some class attribute uniquely identifies each particular instance of the given class.

From the UML 2.4 specification, (meta) class Property has an additional (meta) attribute called isID. The default value of this logical (meta) attribute is false, which implies that class attributes are not identifiers by default. The fact that the value of this (meta) attribute for some class attribute is set to true, implies that the given attribute can be used to uniquely identify an instance of the containing class. In the official specification [15], it is stated that some class attribute may be marked as being (part of) the identifier (if any) for a class of which it is a member. The interpretation of this possibility is left open, but it is suggested that this could be mapped to implementations such as primary keys for RDB tables. The fact that some attribute is marked as an identifier (i.e. `isID=true`) is represented by the `{id}` modifier for the given attribute. If multiple class attributes are marked as `{id}`, then the combination of corresponding $(property, value)$ tuples logically provides the uniqueness for any instance. In [15] it is also stated that there is no need for additional specification of order and it is possible for some (but not all) property values to be empty, as well. The instance identity may be specified as $ID = \{(p_1, v_1), \cdots, (p_k, v_k)\}$, where $p_1, \cdots, p_k$ are the `{id}` attributes and $v_1, \cdots, v_k$ are corresponding attribute values, respectively. The subsequent UML specifications (2.4.1 [16] and 2.5 [4]) did not introduce additional semantics for the `isID` (meta) attribute. The UML 2.4 metamodel excerpt, which is relevant for representation of classes is shown in Fig. 1.

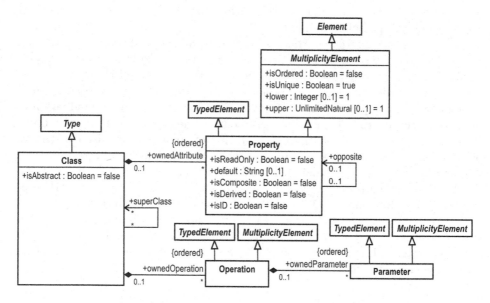

Fig. 1. UML 2.4 metamodel [15] excerpt relevant for representation of classes

Let us now consider the possibility of using the `isID` (meta) attribute to represent the primary key in RDB tables (i.e. relation schemas). A RDBS S is a set of relation schemas, i.e. $S = \{R_1, \cdots, R_m\}$, and a set of integrity constraints IC. A relation schema R, denoted by $R(A_1, \cdots, A_n)$, is a set of attributes $R = \{A_1, \cdots, A_n\}$. A relation (or relation instance) r of the relation schema $R(A_1, \cdots, A_n)$, also denoted by $r(R)$, is a set of n-tuples $r = \{t_1, \cdots, t_n\}$ representing instances of the given entity type (i.e. class). A relational database instance DB of S is a set of relation instances, i.e. $DB = \{r_1, \cdots, r_m\}$ such that each $r_i \in DB$ satisfies the integrity constraints specified in IC.

The subset $SK \subseteq R$ is called a superkey of the relation schema R if for any two distinct tuples t_1 and t_2 in a relation $r(R)$, we have $t_1[SK] <> t_2[SK]$, where $t_x[SK]$ refers to subtuple of values from t_x corresponding to the attributes contained in SK. Every relation has at least one superkey (the set of all its attributes). A minimal superkey, i.e. a superkey from which we cannot remove any attributes in order to keep a superkey, is called a candidate key. Although every candidate key can be chosen as the primary key of the relation, it is usually better to choose a primary key with single attribute or a small number of attributes. A primary key consisting of a single attribute is called a simple primary key (single-column key), while a primary key consisting of several attributes is called a composite primary key (multi-column or multisegment key). The order of the segments constitutes a very important property of the primary key. This order is to be specified in the corresponding RDB constraint and it directly affects the corresponding index structure in a physical RDB. Consequently, the primary key PK is a sequence of segments, i.e. $PK = (p_1, \cdots, p_k)$, where $p_1, \cdots, p_k \in R$. Furthermore, for any two distinct tuples $t_1, t_2 \in r(R)$, it must be $t_1[PK] <> t_2[PK]$.

Apart from the constraint that candidate/primary key values must be unique for every tuple in any relation instance of the given relation schema, an entity integrity constraint is to be satisfied, as well. The entity integrity constraint states that values of all primary key attributes are not to be null. This constraint comes from the fact that the value(s) of the primary key attribute(s) must uniquely identify each n-tuple in the given relation, since the unique identification of n-tuples would not be possible if the primary key has a null value, or if its segments have null values. According to the SQL standard [17], each column participating in the primary key is implicitly defined as NOT NULL, and most of DBMSs require explicit specification of primary key attributes as NOT NULLs. Since in [15] it is stated that it is possible that some (but not all) {id} attributes have no values, it can be concluded that the UML specification of the {id} attribute is in contradiction to the entity integrity constraint. To conclude, all {id} attributes in the relation schema must have values, i.e. the lower multiplicity of each {id} attribute must be 1^1.

In the case of a simple primary key defined on the relation schema R, the $PK = (p_1, \cdots, p_k)$ expression becomes $PK = (p)$, i.e. $PK = p$, where $p \in R$. The built-in mechanism (isID meta-attribute), excluding optionality, enables appropriate representation of the primary key. Namely, the {id} modifier enables appropriate visualisation of the primary key – all attributes representing primary keys in relation schemas are explicitly designated in the class diagram representing RDBS. The automatic generation of the corresponding primary key constraint in the target RDB is very simple, as well. In order to illustrate, Fig. 2 (top) shows the class with one {id} attribute, which represents a relation schema with simple primary key (in accordance with UML 2.4 specification) and appropriate DDL (Data Definition Language) script for creation of the corresponding table in the target RDB. The corresponding excerpt from the Acceleo [18] transformation program for automatic generation of the given DDL script is shown in Fig. 2 (bottom).

---

[1] This fact is not explicitly represented in class diagrams, since the 1..1 multiplicity is default. Allowed null values would require that the lower multiplicity of attributes is 0, which would be shown in the class diagram in the form of [0..1].

```
      Class
─────────────────────       ⇒
+ a1 : <type> {id}
+ a2 : <type> [0..1]
─────────────────────
```

```
CREATE TABLE Class
(
    a1 <type> NOT NULL PRIMARY KEY,
    a2 <type>
);
```

```
...
[for (aClass:Class | aPack.ownedElement) after('\n')]
CREATE TABLE [aClass.name/]
(
   [for (aProp:Property | aClass.ownedAttribute)
   separator(',\n') after('\n')]
     [aProp.name/] [aProp.type.name/][if (aProp.lower>0)]
  NOT NULL[/if][if (aProp.isID)] PRIMARY KEY[/if][/for]
);
[/for]
...
```

Fig. 2. Class representing relation schema with simple primary key and DDL script for creation of the corresponding table (top); corresponding excerpt from the Acceleo [18] transformation program for automatic generation of DDL script (bottom)

Although it is possible that each relation in RDB has a simple primary key (a relation schema may possess an attribute that uniquely identifies each n-tuple, or a designer can introduce a surrogate key – a synthetic attribute that uniquely identifies each n-tuple), composite primary keys are common in practice, as well. Therefore, it is necessary to consider the possibility of representation of composite primary keys by using the isID (meta) attribute.

Given the fact, according to [15], that the unique identification of class instances is specified by (unordered) set $ID = \{(p_1, v_1), \cdots, (p_k, v_k)\}$, and the primary key is specified by (ordered) sequence $PK = (p_1, \cdots, p_k)$, the question arises whether the given set of attributes $\{p_1, \cdots, p_k\}$ may represent the primary key of the relation schema modeled by given class. The answer to this question lies in the serialization of UML models. Namely, if the physical order of class attributes (serialized model) is equal to the logical order (visualized on the diagram), then the answer to the previous question is affirmative and the set of {id} attributes can be used as a primary key.

Serialization of UML models (i.e. MOF-based models) is standardized by the XMI (XML Metadata Interchange) specification. XMI specifications [19,20] preceding the UML 2.4 [15], define serialization of composite elements in the form of XML elements (association of a class and its attributes is a composite, as it can be seen in Fig. 1). Additionally, the order of corresponding XML elements in the model corresponds to the order of attributes in the diagram, because the OwnedAttribute property is qualified as {ordered} (see Fig. 1). The previously described serialization is retained in subsequent XMI specifications [21,22], as well. To conclude, XMI-based serialization of UML models makes it possible that set of {id} class attributes can be used as a primary key of the relation schema represented by given class, since the set of {id} attributes is ordered, i.e. $\{p_1, \cdots, p_k\} = (p_1, \cdots, p_k)$. In order to illustrate, Fig. 3 (top) shows a class representing relation schema with composite primary key consisting of two attributes and appropriate DDL script for creation of the corresponding

```
                                    CREATE TABLE Class
┌──────────────────────┐           (
│        Class         │               p1 <type> NOT NULL,
├──────────────────────┤               a <type>,
│ + p1 : <type> {id}   │    ⇒          p2 <type> NOT NULL,
│ + a : <type> [0..1]  │               PRIMARY KEY (p1, p2)
│ + p2 : <type> {id}   │           );
└──────────────────────┘
```

```
...
[for (aClass:Class | aPackage.ownedElement) separator('\n')]
CREATE TABLE [aClass.name/]
(
   [for (aProp:Property | aClass.ownedAttribute)]
       [aProp.name/] [aProp.type.name/][if (aProp.lower>0)]
  NOT NULL[/if],
  [/for]
  [if (aClass.ownedAttribute->select(o | o.isID)->notEmpty())]
       PRIMARY KEY [for
  (aProp:Property | aClass.ownedAttribute->select(o | o.isID))
  before('(') separator(', ') after(')\n')][aProp.name/][/for][/if]
  );
  [/for]
  ...
```

Fig. 3. Class representing relation schema with composite primary key and DDL script for creation of the corresponding table (top); corresponding excerpt from the Acceleo [18] transformation program for automatic generation of DDL script (bottom)

table. The corresponding excerpt from the Acceleo [18] transformation program for automatic generation of the given DDL script is shown in Fig. 3 (bottom).

4 Alternative Representation of Keys

As shown in the previous section, UML specifications (starting from the version 2.4) have a built-in mechanism to represent the identifier, or to indicate that some class attribute uniquely identifies or belongs to the unique identifier of each instance of the given class, which provides an adequate representation of the primary key in the class diagram representing RDBS. However, some modeling tools, including some open source development platforms, still do not support the recent UML specifications and do not allow representation of the primary key by applying the isID (meta) attribute. This raises the question of whether there are alternative options for the primary key representation by applying standard UML notation, without extending the UML metamodel, i.e. specialization of the standard notation. An alternative mechanism should ensure adequate representation of the primary key, regardless of whether it is simple or composite (taking into account the order of its segments), with desirable visualization of the primary key and as simple as possible automatic generation of corresponding constraints in the target RDB.

The primary key can be specified by using OCL [23] (Object Constraint Language), i.e. by OCL invariants. However, such approach has several limitations. Although OCL invariants can be specified and automatically generated in a unified way, automatic interpretation of invariants and generation of corresponding DDL script is not simple and requires appropriate OCL interpreter, e.g. [24]. Another very important disadvantage is inappropriate visualization of the primary key. OCL invariants are not typically shown in the diagram, but they are only available during the modeling as additional class properties. However, the primary key constitutes a very important and unavoidable concept in database design process and its visualisation represents mandatory functionality of modeling tools and notations for RDB design. Consequently, the UML/OCL combination allows specification of the primary key, but without possibility for its adequate (especially desired) visualization in relational schema. To conclude, a different mechanism for simple and efficient specification as well as visualization of the primary key is required.

Our recent research [11] suggests that standard UML possesses another inherent (but not explicit) mechanism that also provides the possibility for very simple and efficient representation of the primary key. Namely, by analyzing the UML metamodel (Fig. 1), it can be concluded that parameters of the class operation represent the sequence, i.e. class operation parameters are ordered ({ordered} constraint on the OwnedProperty end of the composite association Operation←Property)[2]. Since the primary key is sequence $PK = (p_1, \cdots, p_k)$, the previous conclusion allows us to represent the primary key by appropriate class operation $PK(op_1, \cdots, op_k)$, where the primary key segments $p_1, \cdots, p_k$ are represented by corresponding operation parameters $op_1, \cdots, op_k$, respectively.

The idea to represent the primary key by class operation is not new. Class operations have been used for the representation of primary keys from the initial proposals for UML database profiles, but these operations are used only to indicate primary keys – they do not have any parameters. In this approach, not only does PK operation denote the primary key, but it also serves to specify the primary key segments and their order within the primary key. For instance, if a relation schema Exam has a composite primary key consisting of the course_id and date attributes, then this key can be represented by the following operation:

PK(course_id:text,date:date).

The proposed approach allows simple and complete representation of the primary key, because all segments of the primary key and their order within the primary key (which had previously been a special challenge) can be represented by applying standard UML elements – class operation and its parameter(s). In this way, there is no need for specialized notation that additionally indicates the attributes belonging to the primary key.

In a similar way, we can also represent alternate keys (taking into account the operations naming). For instance, the fact that a relation schema has two alternate keys can be represented by the following operations:

AK_key1(ak_11,...,ak_1p),
AK_key2(ak_21,...,ak_2q),

where ak_11,...,ak_1p and ak_21,...,ak_2q are specifications of operation parameters representing the components of alternate keys key1 and key2, respectively.

---

[2] This conclusion is valid for all UML 2.x specifications.

Apart from the primary key and corresponding constraints, a foreign key and corresponding referential integrity constraint are concepts of great importance for RDBS representation. The referential integrity constraint is a constraint that is specified between two relations and it is used to maintain consistency among tuples of these two relations. It states that a tuple in one relation that refers to another relation must refer to an existing tuple in that relation. It is said that a foreign key $FK(R_2) = \{f_1, \cdots, f_k\} \subseteq R_2$ from relation schema R_2 references or refers to the relation schema R_1 if it satisfies the following rules: (i) the $FK(R_2)$ attributes have the same domain as the $PK(R_1)$ attributes of relation schema R_1, and (ii) a value of FK in a tuple $t_2 \in r_2(R_2)$ either occurs as a value of PK for some tuple $t_1 \in r_1(R_1)$ or is null. In the former case, we have $t2[FK] = t1[PK]$, and it is said that the tuple t_2 references or refers to the tuple t_1. Foreign key may also be simple or composite, referencing the corresponding single/composite primary key in the referenced table. Unlike a primary key, which is unique for the relation schema, some relation schema may have several foreign keys referencing to the same or different relation schemas. The referential integrity constraint is additionally characterized by appropriate actions for ensuring referential integrity in the target RDB. All of the above constitute an even greater challenge in UML representation of RDBS.

The proposed approach for modeling the order of the (primary) key segments, based on the standardized order of the operation parameters, allows us to represent foreign keys by standard UML notation, as well. For the sake of illustration, suppose that the **Assessment** relation schema has two foreign keys. The first foreign key (column **student_id**) references to the primary key (column id) in the **Student** table. Other foreign key (consisting of the **exam_course_id** and **exam_date** columns) references to the primary key in the **Exam** table (earlier used to illustrate the composite primary key). These foreign keys can be represented by following operations:

```
FK_Student(student_id:text),
FK_Exam(exam_course_id:text,exam_date:date).
```

From the given examples it is evident that the proposed approach allows easy modeling of the order of segments in composite foreign keys, and the existence of multiple foreign keys in one relation schema can be resolved using operations with different names (like modeling alternate keys). The introduction of the appropriate prefix in the operation name facilitates the forward engineering of DDL constraints in the target RDB – corresponding foreign key constraints are to be generated based on the FK-prefixed operations, etc. To be able to generate a particular DDL constraint based on the operation representing the foreign key, it is still necessary to adequately specify the referenced table. As it can be seen from the previous examples, the name of the referenced table can be specified as a part of the operation name. Obviously, the operation FK_Student represents a foreign key referencing to the **Student** table, and FK_Exam to the **Exam** table. In case that some table has several foreign keys referencing to the same table, the names of the FK operations (and the corresponding constraints in the target RDB) should be different. For instance, the fact that the **Student** table has two foreign keys (the first one represents the place of residence ID, and the second one represents the place of permanent residence ID) that reference to the primary key `id` in the `City` table, can be represented by the following operations:

```
FK_City_residence(city_residence_id:int),
FK_City_permResidence(city_permResidence_id:int).
```

Additionally, the standard mechanisms for specification of referential actions for foreign keys are to be considered. The standard UML notation allows specification of operation constraints. A set of constraints (`ownedRule`) can be specified for each operation. In our case, the appropriate constraint, which specifies the necessary referential integrity actions, is to be defined for each operation representing the foreign key. Although constraints can be specified by using the OCL, they can be specified in another way, as well. The easiest way is to directly specify the DDL statement part, e.g.

`ON DELETE RESTRICT ON UPDATE CASCADE.`

The proposed method for the representation of a foreign key is sufficient to represent all important aspects related to the foreign key. Additionally, it is desirable to visualize connections between related tables in an appropriate manner, in order to achieve more intuitive visualized RDBS (class diagram). In the existing literature, there are several different proposals (which sometimes violate the standard UML semantics) for visualization of relationships between tables. The simplest (and semantically correct) way to visualize relationships between tables is the use of dependency directed from referencing to the referenced relation schema. It is also desirable that the dependency name matches the name of the operation representing the corresponding foreign key.

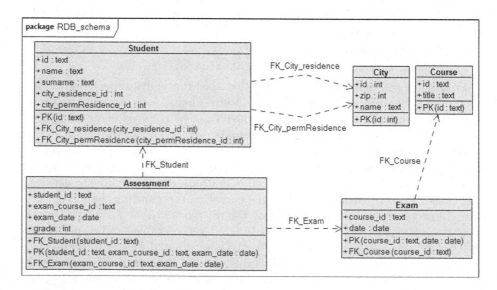

Fig. 4. Illustrative example of RDBS in accordance with proposed approach

Figure 4 shows a simple RDBS in accordance with the proposed approach – shown RDBS includes examples that were used to illustrate the approach.

5 Forward Engineering of Relational Database

Forward engineering of relational database (Fig. 5) goes through three phases: (i) conceptual modeling, (ii) mapping conceptual model to relational model, (iii) creation of target physical database schema, i.e. DDL script generation.

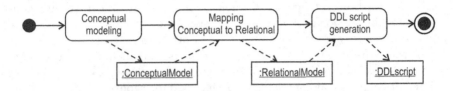

Fig. 5. Forward engineering of relational database

The initial phase in database design is conceptual modeling. The main goal of conceptual modeling is to provide an overall description of data on a high level of abstraction in the entire system. The corresponding model is usually called the conceptual model/schema and it represents a semantic data model (data structure, relations between data, semantics, constraints). Figure 6 shows a UML class diagram representing simple conceptual model, which will be used as the starting model to illustrate forward engineering of relational database according to the proposed approach. Since the given conceptual schema is very simple and intuitive, its detailed description is omitted.

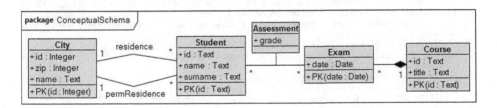

Fig. 6. Sample conceptual UML database model

After conceptual modeling, given conceptual schema should be mapped into corresponding RDBS. A relatively simple set of mapping rules [25] can be automated. In our case, the tool is implemented as an Eclipse-Topcased [26] plug-in. After selection of the input conceptual schema, specification of the path and name of the output RDBS, selection of the target DBMS and setting default platform specific data types, the implemented tool processes the input (XMI-serialized) conceptual UML schema, applies transformation rules and generates the output (XMI-serialized) UML RDBS. The visualization of the automatically generated UML RDBS, i.e. generation of the corresponding .umldi file, is performed using the built-in Topcased functionality for automatic visualization. Due to the paper length limitation, further implementation details are not provided.

By applying the implemented tool on the input conceptual schema (Fig. 6) we obtain the RDBS as shown in Fig. 4 (the obtained RDBS was shown in the previous section with the intent to better illustrate the proposed approach).

Automatically generated RDBS can be further manually improved and adapted. For instance, migrated attributes may be renamed (their names are generated in the form of <source_schema>_<source_attribute_name> or <source_schema>_<rolename>_<source_attribute_name>). In that case, segments of the corresponding foreign keys are to be renamed. Since the implemented tool applies the default DBMS data types for integers, floats, strings and dates, it might be also necessary to change data type for some attributes (apart from the package representing the RDBS, the tool also generates the package containing all data types supported by the selected DBMS).

Finally, the corresponding DDL script is to be generated, i.e. a physical database schema is to be created in the target DBMS. The existing CASE tools for database design have built-in functionality for automatic generation of DDL

```
[comment encoding = UTF-8 /]
[module generate('http://www.eclipse.org/uml2/3.0.0/UML')]
[template public generateElement(aPackage : Package)]
[comment @main/]
[file (aPackage.name.concat('.ddl'), false, 'UTF-8')]
CREATE SCHEMA  [aPackage.name/];
  [for (aClass:Class | aPackage.ownedElement) after('\n')]
CREATE TABLE [aPackage.name.concat('.').concat(aClass.name)/]
(
  [for (aProperty:Property | aClass.ownedAttribute) separator('\n') after('\n')]
  [aProperty.name/] [aProperty.type.name/][if (aProperty.lower>0)] NOT NULL[/if],[/for]
  [for (aOperation:Operation | aClass.ownedOperation->
          select(o | o.name.startsWith('FK') or o.name.equalsIgnoreCase('PK')) )
    separator(',\n') after('\n')]
    [if (aOperation.name.equalsIgnoreCase('PK'))]
  PRIMARY KEY ([for (op:Parameter | aOperation.ownedParameter)
    separator(', ')][op.name/][/for])[/if]
    [if (aOperation.name.startsWith('FK'))]
  CONSTRAINT [aOperation.name/]
    FOREIGN KEY ([for (op:Parameter | aOperation.ownedParameter)
    separator(', ')][op.name/][/for])
    REFERENCES [for (aDependency:Dependency | aPackage.ownedElement)]
      [if (aDependency.name.equalsIgnoreCase(aOperation.name))]
[aPackage.name.concat('.')/][aDependency.supplier.name/][/if][/for]
[for (r:Constraint | aOperation.ownedRule)] [r.name/][/for][/if]
    [/for]
);
[/for]
[/file]
[/template]
```

Fig. 7. Acceleo transformation program for automatic generation of DDL script based on UML RDBS

```
CREATE SCHEMA  RDB_schema;
CREATE TABLE RDB_schema.Course
(
  id text NOT NULL,
  title text NOT NULL,
  PRIMARY KEY (id)
);
CREATE TABLE RDB_schema.Student
(
  id text NOT NULL,
  name text NOT NULL,
  surname text NOT NULL,
  city_residence_id int NOT NULL,
  city_permResidence_id int NOT NULL,
  PRIMARY KEY (id),
  CONSTRAINT FK_City_residence
    FOREIGN KEY (city_residence_id)
    REFERENCES RDB_schema.City ON UPDATE CASCADE ON DELETE RESTRICT,
  CONSTRAINT FK_City_permResidence
    FOREIGN KEY (city_permResidence_id)
    REFERENCES RDB_schema.City ON DELETE RESTRICT ON UPDATE CASCADE
);
CREATE TABLE RDB_schema.Assessment
(
  student_id text NOT NULL,
  exam_course_id text NOT NULL,
  exam_date date NOT NULL,
  grade int NOT NULL,
  CONSTRAINT FK_Student
    FOREIGN KEY (student_id)
    REFERENCES RDB_schema.Student ON DELETE RESTRICT ON UPDATE CASCADE,
  PRIMARY KEY (student_id, exam_course_id, exam_date),
  CONSTRAINT FK_Exam
    FOREIGN KEY (exam_course_id, exam_date)
    REFERENCES RDB_schema.Exam ON UPDATE CASCADE ON DELETE RESTRICT
);
CREATE TABLE RDB_schema.City
(
  id int NOT NULL,
  zip int NOT NULL,
  name text NOT NULL,
  PRIMARY KEY (id)
);
CREATE TABLE RDB_schema.Exam
(
  course_id text NOT NULL,
  date date NOT NULL,
  PRIMARY KEY (course_id, date),
  CONSTRAINT FK_Course
    FOREIGN KEY (course_id)
    REFERENCES RDB_schema.Course ON DELETE RESTRICT ON UPDATE CASCADE
);
```

Fig. 8. Generated DDL script

code. Topcased development environment makes it easy to automatically generate DDL code using specialized M2T (Model To Text) transformation languages, such as Acceleo [18]. Figure 7 shows an Acceleo transformation program for automatic generation of DDL script based on UML RDBS. By executing the transformation program (Fig. 7), based on the given RDBS (Fig. 4), we get the DDL script as shown in Fig. 8.

6 Conclusion

The suitability of the standard UML notation for RDBS representation has been considered in this paper. Firstly, the suitability of the built-in mechanism (isID meta-attribute) for representation of primary keys has been analyzed. It has been shown that the isID meta-attribute can be used to represent the primary key, regardless of whether it is simple or composite.

Regardless of the ability to represent the primary key by the isID meta-attribute, UML still has no built-in mechanism dedicated to representation of other important RDBS concepts, such as foreign keys and corresponding constraints. Furthermore, some modeling tools still do not support the recent UML specifications and do not allow representation of the primary key by applying the isID meta-attribute. Motivated by the previous facts, we tried to identify other inherent UML concepts that would enable simple and efficient representation of keys. As a result, an alternative representation of composite keys by using class operations has been proposed. The main idea of the proposed approach is based on the fact that the standardized order of operation parameters can be used to represent the order of key segments. It has been shown that the proposed approach can be used not only to represent the primary key, but also the alternate and foreign key, as well. The representation of the order of attributes in composite keys has been previously recognized as a special challenge, which has been mainly addressed by specialization of the standard UML notation, i.e. by UML profiles.

The proposed approach has several direct advantages compared to existing approaches: (i) RDBS is represented by standard UML notation and there is no need to define and apply the specific profile; (ii) modeling of RDBS by applying standard UML notation is easier and faster than using the specialized notation; (iii) visualization of RDBS is better, since the standard notation is used without specific stereotypes; (iv) forward database engineering is easier and more efficient, since it is easier to generate the corresponding DDL script for creation of the target physical database schema based on standard UML models.

The preliminary results of applying the proposed approach show that the fundamental RDBS concepts can be represented by standard UML notation in an easy and intuitive way. Furthermore, the implemented tool provides good visualization of the automatically generated and/or manually modeled RDBS, as well as simple generation of DDL script and forward database engineering for several contemporary DBMSs. Someone may find that the strict naming of class operations constitutes the main limitation for adopting this proposal for manual RDBS modeling. However, automatic RDBS generation eliminates these difficulties.

The future work will be focused on: (i) further evaluation and validation of the proposed approach by using empirical experiments based on more complex and extensive case studies, (ii) further analysis of the suitability of standard UML notation to represent other important RDBS-related concepts (indices, views, queries, triggers, etc.), and (iii) integrating the implemented tool in the ADBdesign tool [27, 28] for automated business model-driven design of relational databases.

References

1. Codd, E.: A relational model of data for large shared data banks. Commun. ACM **13**(5), 377–387 (1970)
2. Martin, J.: Information Engineering. Prentice Hall, Englewood Cliffs (1990)
3. NIST: FIPSP 184 - Integration Definition for Information Modeling (IDEF1X). NIST, Gaithersburg (1993)
4. OMG: Unified Modeling Language (OMG UML), v2.5. OMG (2015)
5. OMG: Request for Proposal Information Management Metamodel (IMM). OMG (2005)
6. Naiburg, E., Maksimchuk, R.: UML for Database Design. Addison-Wesley, Reading (2001)
7. Li, L., Zhao, X.: UML specification of relational database. J. Object Technol. **2**(5), 87–100 (2003)
8. Ambler, S.W.: Agile Database Techniques. John Wiley and Sons, Indianapolis (2003)
9. Lo, C.M., Hung, H.Y.: Towards a UML profile to relational database modeling. Appl. Math. Inf. Sci. **8**(2), 733–743 (2014)
10. Marcos, E., Vela, B., Cavero, J.M.: A methodological approach for object-relational database design using UML. Softw. Syst. Model. **2**, 59–72 (2003)
11. Tomic, I., Brdjanin, D., Maric, S.: A novel UML profile for representation of a relational database schema. In: Proceedings of EUROCON 2015, pp. 1–6. IEEE (2015)
12. Muller, R.J.: Database Design for Smarties: Using UML for Data Modeling. Morgan Kaufmann Publishers, San Francisco (1999)
13. Armonas, A., Nemuraite, L.: Pattern based generation of full-fledged relational schemas from UML/OCL models. Inf. Technol. Control **35**(1), 27–33 (2006)
14. Lo, C.M., Huang, S.J.: MDA-based rapid application framework. Int. J. Advancements Comp. Tech. **4**(8), 307–314 (2012)
15. OMG: Unified Modeling Language: Infrastructure, v2.4. OMG (2010)
16. OMG: Unified Modeling Language: Infrastructure, v2.4.1. OMG (2011)
17. ISO, IEC: ISO/IEC FDIS 9075-1 Information technology - Database languages - SQL - Part 1: Framework (SQL/Framework). ISO/IEC (2011)
18. ISO, IEC: Acceleo. http://www.eclipse.org/acceleo/
19. OMG: MOF 2.0/XMI Mapping Specification, v2.1. OMG (2007)
20. OMG: MOF 2 XMI Mapping Specification, v2.1.1. OMG (2007)
21. OMG: OMG MOF 2 XMI Mapping Specification, v2.4.1. OMG (2013)
22. OMG: XML Metadata Interchange (XMI) Specification, v2.5.1. OMG (2015)
23. ISO: Information technology - Object Management Group Object Constraint Language (OCL). ISO/IEC 19507: 2012 (2012)
24. Heidenreich, F., Wende, C., Demuth, B.: A framework for generating query language code from OCL invariants. ECEASST **9**, 1–10 (2008)
25. Embley, D.W., Mok, W.Y.: Mapping conceptual models to database schemas. In: Embley, D.W., Thalheim. B., (eds.) Handbook of Conceptual Modeling, pp. 123–163. Springer, Heidelberg (2011)
26. TOPCASED Project: Toolkit in OPen-source for Critical Application & SystEms Development, v5.3.1. http://www.topcased.org
27. Brdjanin, D., Maric, S., Gunjic, D.: ADBdesign: an approach to automated initial conceptual database design based on business activity diagrams. In: Catania, B., Ivanović, M., Thalheim, B. (eds.) ADBIS 2010. LNCS, vol. 6295, pp. 117–131. Springer, Heidelberg (2010)
28. Brdjanin, D., Maric, S.: An approach to automated conceptual database design based on the UML activity diagram. Comput. Sci. Inf. Syst. **9**(1), 249–283 (2012)

Meta-modeling and Domain Specific Modeling and Model Composition

Towards a Meta-Model for Networked Enterprise

Gabriel Leal[1,2,3(✉)], Wided Guédria[3(✉)], Hervé Panetto[1,2], and Erik Proper[3]

[1] CNRS, CRAN UMR 7039, Vandœuvre-lès-Nancy, France
[2] Université de Lorraine, CRAN UMR 7039, Boulevard des Aiguillettes, B.P. 70239, 54506 Vandœuvre-lès-Nancy, France
{gabriel.da-silva-serapiao-leal,herve.panetto}@univ-lorraine.fr
[3] ITIS, TSS, Luxembourg Institute of Science and Technology (LIST), 5, Avenue des Hauts-Fourneaux, 4362 Esch-sur-Alzette, Luxembourg
{gabriel.leal,wided.guedria,erik.proper}@list.lu

Abstract. To deal with challenges as globalization and fast-changing environments, enterprises are progressively collaborating with others and becoming a Networked Enterprise (NE). In this context, Enterprise Interoperability (EI) is a crucial requirement that needs to be verified by enterprises when starting a relationship to avoid interoperability problems. The concepts of NE and EI are not easy to understand due the variety of interpretations that exist in the literature. Having a clear and shared understanding of the NE and the different interoperations between partners is a necessity to manage the interoperability development. In order to reach such an objective, this research work defines a meta-model for NE based on a systemic approach. Concepts related to EI are taken into account to highlight the importance of this ability (i.e. Interoperability), seen as a requirement, within a system to attain its targeted goals. Finally, a real case study is proposed to validate the defined meta-model.

Keywords: Networked enterprise · Enterprise interoperability · Meta-model · Systemic approach

1 Introduction

Contemporary enterprises face a variety of challenges in the increasingly dynamic socio-economic environment where they evolve. Challenges such as globalization, novel technologies, financial crisis, the need for cost reduction and new markets are change-drivers that require transformation within companies and their environments. These challenges can be illustrated by the growing number of start-ups around the world; the rapid evolution of information and communication technologies (ICT) that offers, paradoxically, opportunities (e.g. ease the long-distance communications) and threats (e.g. incompatibilities between communication protocols); the boost of customized products demand, etc. In order to deal with these challenges, enterprises are progressively collaborating with each other and participating to a so-called Networked Enterprise (NE) [1–5]. The concept of NE is commonly confused with Collaborative Network [6], Enterprise Networks [7, 8] and Value Network [9, 10]. In the NE context, interoperability [11–13], is a crucial requirement having to be

© Springer International Publishing Switzerland 2016
R. Schmidt et al. (Eds.): BPMDS/EMMSAD 2016, LNBIP 248, pp. 417–431, 2016.
DOI: 10.1007/978-3-319-39429-9_26

verified by enterprises when starting a relationship with others to attain shared goals [14, 15]. As soon as this requirement is not achieved when systems or system's elements need to operate together, interoperability becomes a problem that must be solved [16]. Many research works were proposed in the literature to study Enterprise Interoperability (EI) and propose related frameworks such as: the Athena Interoperability Framework (AIF) [17], the IDEAS Interoperability Framework [12], the Framework for Enterprise Interoperability (FEI) [18, 19], the Classification Framework for Interoperability of Enterprise applications [20], the Ontology of Enterprise Interoperability (OoEI) [16, 21] etc. Among these, the particularity of the OoEI is its basis on the other cited researches and its unicity in defining the EI concepts in a systemic approach [22]. Having a systemic view is very important and widely used in Enterprise Modelling (EM) [23] because it provides a component-oriented view, which reflects closely the reality of enterprise functioning. According to Giachetti [24], an enterprise is *a complex, socio-technical system that comprises interdependent resources of people, information, and technology that must interact with each other and their environment in support of a common mission.* As part of a network, an enterprise can also be seen as part (i.e. System element or component) of a more complex system: the network. Having a clear and shared understanding of the NE and the different interoperations between partners is a necessity to manage the interoperability development, including the detection and prediction of problems at the early stage. Thus, the following research question is raised: *How can we establish a common and clear understanding of the NE and its interoperations?* To answer this question, an analysis of the different perspectives of both concepts (i.e. NE and EI), as well as, the representation of the relations between them are required. This raises a new research question: *How can we design the interoperability in the context of Networked Enterprise?*

The main objective of this work is to develop a common understanding of the Networked Enterprise domain and the interoperability issues involved in the design of such network. This is tackled through the proposition of a meta-model for Networked Enterprise (NE), that we call the "Networked Enterprise Meta-MOdel" (NEMO). This meta-model is defined based on the Design-Science Research (DSR) methodology [25, 26] and uses a systemic approach to describe the NE elements. The identification of the NE elements and characteristics are based on the definitions and interpretations proposed in the literature [1–10]. Concepts related to the interoperability domain are mainly taken into account based on the OoEI [16, 21].

The reminder of this paper is as follow – Sect. 2 gives an overview of the research methodology applied for this research. Section 3 presents the relevant related work. This is followed by Sect. 4 where the NEMO is proposed. Section 5 illustrates a real case study based on an active NE in the field of marketing and communication in Luxembourg. The conclusion and future work are brought forward in Sect. 6.

2   Research Methodology

In order to answer the research question and to achieve the research objective, this work is based on a simplification of the design-science research (DSR) as proposed by [25, 26]. The methodology applied is divided according to the two processes (*Build* and *Evaluate*) and the research outcome [27]. The *Build* process is composed by two stages: The *conceptual definition* where we proceed with the literature study on Networked Enterprise interpretations together with Enterprise Interoperability concepts. Also, at this stage, the identification and definition of the concepts that are presented in Sect. 3 are performed. The second stage is the *construction* of the meta-model presented in Sect. 4. An analysis of the relation between NE and EI concepts is required in this stage to understand the proposed meta-model. The *Evaluate* process is done based on the observational case study. This is illustrated through a real case study in Sect. 5.

3   Conceptual Definition - Related Work and Positioning

This section presents some of the different definitions and interpretations that have been found in the literature about Networked Enterprise. This will allow the identification of the main properties that need to be considered in this domain and propose a general definition that can serve as a consensus and be used in different contexts. The ability to interoperate, as a key factor within the NE, is also studied through the OoEI and the interoperability requirements that should be satisfied to reach the objectives of the network. The concepts identified in the following subsections are then used to describe interoperability and related properties in the proposed meta-model.

3.1   Networked Enterprise

The notion of "Networked Enterprise" is ubiquitous, but hard to understand due the variety of definitions and interpretations. In [1], NE is defined as *"any coordinated undertaking that involves at least two autonomous parties that interact using informa-tion and communication technology (ICT)"*. NE is also considered as *"loosely coupled, self-organizing network of enterprises that combine their output to provide products and services offerings to the market. Partners in the networked enterprise may operate independently through market mechanisms or cooperatively through agreements and contracts"* [2]. In [5], the authors define NE as *"linked companies that collaboratively aim at enabling or implementing the collective Business Model by means of offering service and product and/or sharing resources and competencies"*. In [6], the expression "collaborative network" is used to define *"a network consisting of a variety of entities (e.g. organizations and people) that are largely autonomous, geographically distrib-uted, and heterogeneous in terms of their operating environment, culture, social capital and goals, but that collaborate to better achieve common or compatible goals, thus jointly generating value, and whose interactions are supported by computer network"*. In [7], the authors use the term "enterprise network" to define *"two or more participating enterprises are engaged in the supply and receipt of goods or services on a regular and on-going basis. Within enterprise networks, partners rely on each other and the supply of goods (or services) will be constrained by the associated logistics, manufacturing commitments and the operating dynamics of the participating enterprises"*. In [10], the author use the term "Value Network" to define *"a dynamic network of actors working together to generate customer value and network value by means of a specific service offering, in which tangible and intangible value is exchanged between the actors involved"*.

Although, these definitions are based on different context and have different point of views (e.g. technological, manufacturing, industrial, etc.), we can notice that some similar characteristics are considered among these work, such as: the necessity of a NE to be composed by **at least two autonomous** enterprises and the **ability to collaborate** to achieve a **shared objective**.

When adopting a systemic view and being inspired by these common character-istics, we define a **Networked Enterprise** as: *"a system composed of at least two autonomous systems (enterprises) that collaborate during a period of time to reach a shared objective"*.

3.2   The Ontology of Enterprise Interoperability

In the past years, researchers and practitioners have proposed numerous definitions for interoperability [11–13, 17–19, 28]. In this research work, we consider a general systemic approach of interoperability, where interoperability is first viewed as a problem to solve: *An interoperability problem appears when two or more incompatible systems are put in relation* [29]. Then, when taking the view of interoperability as a goal to reach, we can also write: *Interoperable systems operate together in a coherent manner, removing or avoiding the apparition of related problems* [30]. To have a clear understanding about the Enterprise Interoperability, we need to study the core concepts and elements of the EI and the operational entities where interoperations take place within an enterprise. These are mainly defined by the OoEI, where interoperability is seen as a problem caused when incompatible systems are put in relation. Its main purposes are to have a common under-standing about interoperability and to diagnose *a priori* and *a posteriori* [31] interopera-bility problems and propose solutions. The EI problems and solutions concepts are related to the three Interoperability dimensions, as defined in the FEI [18, 19]. These are: *Inter-operability aspects* (conceptual, organizational and technical), *Interoperability concerns* (business, process, service, and data) and *Interoperability approaches* (integrated, unified and federated). The OoEI includes a systemic model, having a systemic core centered on the notion of the system and its properties, and a decisional model that constitutes the basis to build a decision-support system for EI.

Aligned with the systemic approach used by the OoEI, an enterprise can be decom-posed into three main sub-systems [32]: *an operating or physical system*; *a decisional or pilot system*; and *an information system*. In [33], the authors used the GRAI Integrated Methodology [34] to represent the enterprise sub-systems as depicted in Fig. 1.

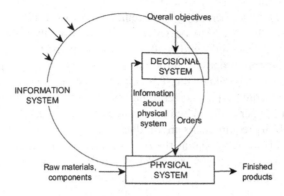

Fig. 1. The three subsystems from an enterprise [34]

In the Fig. 1, the decisional system ensures the overall objectives of the enterprise taking them as inputs to send orders to the physical system. Furthermore, to determine how to control the operating system in order to successfully achieve the system goals and objectives, the pilot system communicates with the environment relating to the system's goals, accepting orders, making commitments and exchanging any other information with the environment that is necessary. The decisional system relies on models of the physical system to make its decisions. However, for these models to reflect reality to a sufficient degree, the decisional system must receive information, or feedback, from the physical system.

As the main objective of this research is to define a meta-model for NE while taking into account the different interoperations between stakeholders, the OoEI and the Enterprise-as-Systems concepts seems to be perfect candidates to be considered in the development of the proposed meta-model since they are grounded in systemics and have a problem-solving perspective.

3.3 Interoperability Requirements

Interoperability is a crucial requirement having to be verified by systems when being in relationship with other systems in order to assume a common mission [15]; where systems are considered as enterprises or parts of enterprises that need to interact in a collaborative and common process with other enterprises or part of enterprises to achieve a common goal [15]. Considering this perspective, the authors in [14] proposed an approach based on the requirement engineering [35, 36] that can be used to describe and structure interoperability requirements that are related to any interoperability problem that may obstruct a collaborative process. The definition proposed is the following: "*an Interoperability Requirement is a statement that specifies a function, ability or characteristic, related to the capacity of a partner to ensure its partnership regarding compatibility, interoperation, autonomy, and reversibility, which it must satisfy*" [14]. In [21], a list of 48 best practices, which can be understood as requirements, were proposed. These best practices describe the "what to do" in broad terms so that enterprises are left great leeway in creatively implementing the "how to do it".

As soon as these interoperability requirements are not fulfilled, interoperability becomes a problem that needs to be solved. To deal with that, evaluations can be performed to assess the strengths and weaknesses of the considered system. Numerous assessment methods were proposed in the literature such as: the Compatibility Matrix [37], the formal metrics to evaluate the semantic interoperability between systems [38], the Interoperability Score [39] and several maturity models [31, 40–43]. This stays out of the scope for this paper and will be investigated in future work.

The interoperability requirements are fundamental assets to support the management of the interoperability development as they can be used as indications to identify interoperability problems. Hence, the interoperability requirements and related concepts will be also considered in the design of the proposed meta-model.

4   Construction Stage - The Networked Enterprise Meta-Model

In this section we define relevant concepts and definitions used to build the "Networked Enterprise Meta-Model" (NEMO).

Based on related work, we have defined a networked enterprise as: "a system composed of at least two autonomous systems (enterprises) that collaborate during a period of time to reach a shared objective". (C.f. Sect. 3.1).

In this context, the **Objective** represents the system's goal (NE goal) at a given time [16]. This Objective should be compatible with the objectives of the **Enterprise members** that compose the NE and their businesses. This Objective can be described as a *short-term objective*, where there is a temporary alliance to seize a particular business opportunity or *long-term objective*, where enterprises have a stable collaboration that is not limited by only one business opportunity. The objective of the NE should also be aligned with its **Function** (i.e. Business), which represents the set of actions that the system can execute in its environment, to achieve its objectives [16]. Based on that, the NE can have different organizations, called also **Classification** [6, 7, 44–47].

A Networked Enterprise has its **Lifecycle** representing the different phases that a given networked enterprise may pass through. We define five stages based on [6, 48]: (a) *Creation* is the stage when the networked enterprise is started. It includes the strategic planning, the recruiting, the organizational structure constitution and the setting up; (b) *Operation* is the operating stage of the networked enterprise; (c) *Evolution* is the stage when small changes in membership, roles and work methods happen; (d) *Transformation* is the stage when significant changes in objectives, principles and membership happen, leading to a new form of organization; (e) *Decomposition* is the stage when the networked enterprise ceases to exist.

To be part of the NE there are defined **Requirements** specifying the ability or characteristic that must be satisfied in a given context [35, 36] to avoid problems, mainly the ones related to interoperability. The **Interoperability Requirements** concept adopted here refers to the ability of partners to ensure the compatibility, interoperation, autonomy and reversibility requirements of a NE [14]. Where a compatibility requirement specifies a function considered to be invariable throughout the collaboration and related to interoperability barriers for each interoperability concern. An interoperation requirement specifies a function considered to be variable during the collaboration, related to the performance of the interaction. An autonomy requirement specifies a function related to the capacity of partners to perform their governance and maintain their operational capacity during collaboration. A reversibility requirement specifies a function related to the capacity of a partner to go back to its original state after collaboration. These requirements are also related to the life cycle stages i.e. each stage has its requirements that need to be fulfilled. The compatibility requirements are mainly related to the creation stage of a NE. The autonomy and interoperation requirements are related to the operation stage. The reversibility requirements are essentially related to the decomposition stage. Figure 2 illustrates an overview of the NEMO model taking into account the concepts defined above.

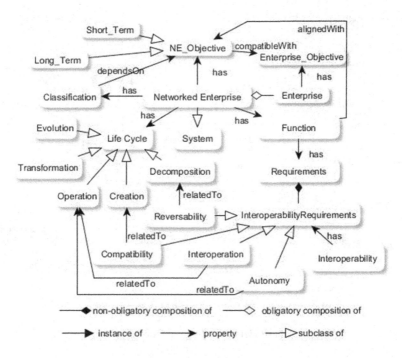

Fig. 2. The NEMO meta-model.

The meta-model gives an extensive view of a Networked Enterprise and its constituents. However it is not enough to realize an accurate characterization of the EI domain because it represents interoperability only as a requirement of a system's function but, as mentioned before, as soon as this requirement is not achieved, interoperability becomes a problem that must be solved. Hence, we combine the OoEI elements because it also considers interoperability from a problem-solving perspective. Therefore, we adopt the following concepts: *EnterpriseInteroperability, EnterpriseInteropDimensions, InteroperabilityAspect, InteroperabilityConcern, InteroperabilityApproach, InteroperabilityBarrier, Problem, ExistenceCondition, Incompatibility,* and *Solution.*

Solution uses interoperability approaches to remove interoperability barriers and solve problems. Figure 3 shows the OoEI concepts (identified by the prefix *"OoEI:"*, and the grey color) integrated into the NEMO (elements in white color). Based on the proposed meta-model, we can clearly see both views of the interoperability concept: the interoperability as a requirement between systems willing to collaborate and as a problem when the requirement is not fulfilled.

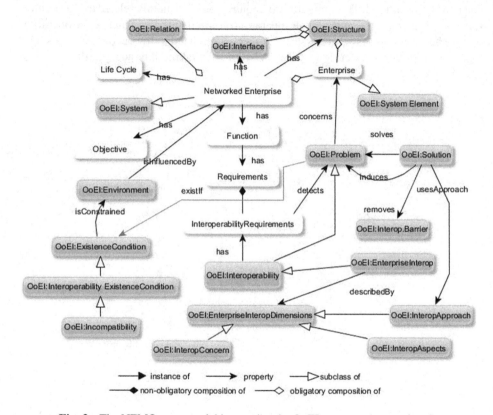

Fig. 3. The NEMO meta-model integrating the OoEI concepts (grey colored).

Considering the Enterprise as System concepts [33] (c.f. Sect. 3.2), Fig. 4 shows the integration of these systemic concepts (identified by the prefix *"OoEI:"*, and colored in grey) in the NEMO meta-model (elements in white color).

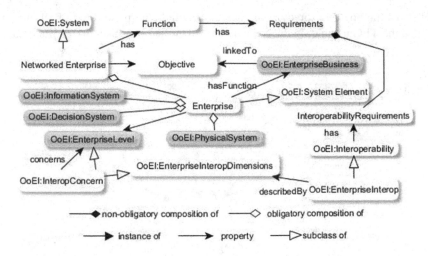

Fig. 4. NEMO meta-model with the enterprise as system concepts (grey colored).

The *PhysicalSystem* is concerned with the interoperation of physical facilities. The *DecisionSystem* is mainly concerned with operational, administrative and strategic decisions; and the *InformationSystem*'s interoperability concerns the exchange of information between two systems [33]. The *EnterpriseBusiness* denotes the enterprise function such as delivery of products and services to customers. *EnterpriseLevel* represents the layers of enterprise in general. Thus, the four interoperability concerns are also subclasses of this concept. These enterprise-as-systems concepts facilitate analyses on specific systems without influencing the network as a whole.

5   Evaluation Using a Case Study

As part of the research approach, this section illustrates the evaluation of the proposed meta- model using a real case study based on The Factory Group (TFG) [49], an active NE in the field of marketing and communication in Luxembourg. TFG brings together independent companies linked by their capital structure or by joint venture agreement. This NE is composed of five distinct enterprises:

1. Concept Factory [50]: Full-service communications consulting agency.
2. Interact [51]: Provider for multimedia information technology services.
3. Exxus [52]: Innovation and strategy consulting agency.
4. Sustain [53]: Service provider for sustainable development projects and corporate social responsibility.
5. Quest [54]: Market Research Company.

It is worth noting that, for some reasons (that stays confidential), Quest has the intent to leave the NE; consequently, we do not consider this company in this analysis. The information used to define the scenario were gathered through interviews and analysis of provided documents by the different enterprises. The selected interviewees are members of the board of directors of each considered enterprise. First of all, we have modelled the TFG using only the NE concepts identified (c.f. Sect. 4), as illustrated in Fig. 5.

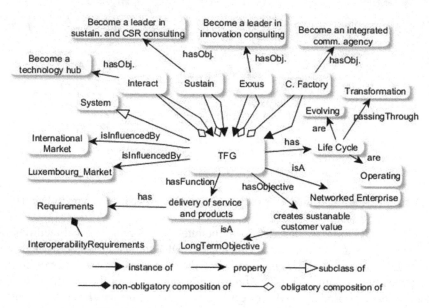

Fig. 5. TFG representation using NE concepts.

Considering the gathered information, the TFG *is composed of* Exxus, Sustain, Concept Factory and Interact. Where the four enterprises collaborate to achieve the TFG goals but remain autonomous to operate and pursue their individual goals. The individual objectives of each enterprise are the following: Exxus *has the objective* to become a leader in innovation consulting, Sustain *has the objective* to become a leader in sustainability and corporate social responsibility (CSR) consulting, Interact *has the objective* to become a technological hub and the Concept Factory *has the objective* to become an integrated communication agency, offering both digital and printed products. The NE as a whole *has the objective* of "creates sustainable customer value". To achieve this goal, the NE *has* functions related to their domain of activity (marketing and communication), for example TFG *has the function* of delivery services and products to its customers. The TFG is located in Luxembourg, and the majority of its clients are from Luxembourg, however, the number of international clients, in the past few years, is increasing. Hence, the TFG *is influenced by* the Luxembourgish and International markets. The TFG is *passing through* three stages in its life cycle. While the group is operating, small changes in the work methods are happening constantly (i.e. they are evolving). TFG are also going through a transformation changing some fundamental principles and roles. For example, Interact are becoming an IT specialized agency rather than a digital marketing agency. In order to provide sustainable products and services, the group has the interest to stay together for a long period of time. Thus, the objective identified hereinabove can be classified as a long-term objective. In order to execute functions to achieve its objectives, a given number of requirements need to be achieved (i.e. each function *has* its requirements). These requirements are *composed of* interoperability requirements.

Even though the NE elements are well described and consider some concepts related to interoperability, using only the NE concepts to model TFG does not allow to represent the importance of the interoperability concept and its properties. For instance, it is not possible to represent an interoperability problem, its existence condition (i.e. why this problems is happening) and which enterprise level (i.e. business, process, service and data) it is affecting. Without these concepts, it may become difficult to identify the cause and location of the problem, which makes the selection of an appropriate solution rapidly harder. Further, it is important to represent the enterprise interoperability dimension (i.e. Interoperability aspects, concerns and approaches) and the interoperability barrier concept. These four concepts (c.f. Sect. 3.2) describe the main interoperability elements related to an enterprise. As mentioned before (c.f. Sect. 4), to fill this gap related to the interoperability representation, we use OoEI elements. Considering the different concepts that need to be taken into account in the OoEI and in the NE context, we have designed the TFG using NEMO, as depicted by Fig. 6. The specific OoEI elements are colored in grey.

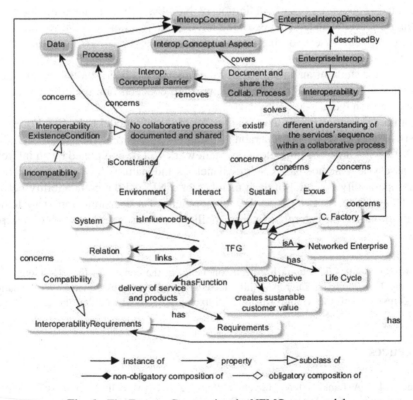

Fig. 6. The Factory Group using the NEMO meta-model.

In Fig. 6, we illustrate the following interoperability problem: *"the different understanding of the services' sequence within collaborative processes"*. This problem *concerns* all partners within the NE. A potential cause (*Existence Condition*) of this incompatibility is the fact that there is no collaborative processes documented or shared within the TFG. Consequently, information is not clear to all employees. This incompatibility *is concerned* with the data and process concerns and the conceptual aspect of an enterprise. This problem is considered as a *conceptual barrier*, because it is concerned with semantics and syntactic problems in the process and data levels of the NE. A potential solution to *solve* this problem is to document and share the TFG collaborative processes within the NE.

Applying the NEMO has allowed us to identify and relate the main elements of The Factory Group. Having this real use case was useful to validate the NEMO meta-model.

6   Conclusion and Future Work

In this paper, we have defined the Networked Enterprise Meta-Model (NEMO). Prior to that, an investigation about the different definitions and interpretations about Networked Enterprise (NE) has been done to identify the core concepts related to this domain and propose a systemic definition of NE. The proposed meta-model aims at providing a common understanding of the NE domain. Within this context, interoperability is a key factor to seize business opportunities. Thus, concepts from EI related work was considered.

A real case study of an active NE in Luxembourg has been studied to validate the proposed meta-model, by illustrating the main NE concepts and the different interoperations between them.

As future work, we intend to extend the NEMO meta-model to build a Framework for Networked Enterprise Interoperability using enterprise modelling approaches such as UEML [55], CIMOSA [56], etc. This framework will be completed by an interoperability assessment method based on formal metrics and maturity levels which will tackle the interoperability potential of each member of a NE and the compatibility between them. This will serve as basis to the development of a decision-support system for preventing and solving enterprise Interoperability problems in the Networked Enterprise context.

Acknowledgements. This work has been conducted in the context of the PLATINE project (PLAnning Transformation Interoperability in Networked Enterprises), financed by the national fund of research of the Grand Duchy of Luxembourg (FNR), under the grant C14/IS/8329172/R2.

References

1. Steen, M.W.A, Lankhorst, M.M., van de Wetering, R.G.: Modelling networked enterprises. In: Proceeding of the 6th International Enterprise Distributed Object Computing Conference (EDOC), pp. 109–119. IEEE Computer Society (2002)

2. Li, Q., Zhou, J., Peng, Q.-R., Li, C.-Q., Wang, C., Wu, J., Shao, B.-E.: Business processes oriented heterogeneous systems integration platform for networked enterprises, integration and information in networked enterprises. Comput. Ind. **61**(2), 127–144 (2010)
3. Chung, A.A.C., Yam, A.Y.K., Chan, M.F.S.: Networked enterprise: a new business model for global sourcing. Int. J. Prod. Econ. **87**, 267–280 (2004)
4. Bellini, E.: Which collaboration strategy for the networked enterprise in wine industry? technological and organizational challenges. In: Anastasi, G., Bellini, E., Di Nitto, E., Ghezzi, C., Tanca, L., Zimeo, E. (eds.) Networked Enterprises. LNCS, vol. 7200, pp. 17–30. Springer, Heidelberg (2012)
5. Solaimani, S.: The alignment of business model & business operations within networked-enterprise environments. Ph.D. thesis (2014)
6. Camarinha-Matos, L.M., Afsarmanesh, H., Galeano, N., Molina, A.: Collaborative networked organizations – concepts and practice in manufacturing enterprises. Comput. Ind. Eng.: CAIE: Int. J. **57**(1), 46–60 (2009)
7. Jagdev, H.S., Thoben, K.-D.: Anatomy of enterprise collaborations. Prod. Plan. Control: Manage. Oper. **12**(5), 437–451 (2001)
8. Basole, R.C., Rouse, W.B., McGinnis, L.F., Bodner, D.A., Kessler, W.C.: Models of complex enterprise networks. J. Enterp. Transform. **1**(3), 208–230 (2011)
9. De Reuver, M.: Governing mobile service innovation in co-evolving value networks. Ph.D. dissertation, Delft University of Technology, The Netherlands (2009)
10. Allee, V.: Reconfiguring the value network. J. Bus. Strategy **21**(4), 36–39 (2000)
11. Institute of Electrical and Electronics Engineers. IEEE standard computer dictionary: a compilation of IEEE standard computer glossaries (1990)
12. Interoperability Development for Enterprise Application and Software (IDEAS), IDEAS Project Deliverables (WP1-WP7), Public Reports (2003)
13. Boudjlida, N., Panetto, H.: The basics of interoperability: a curricula. In: 5th China - Europe International Symposium on Software Industry Oriented Education, CEISIE 2009, pp. 185–189. Wiley-ISTE (2010)
14. Mallek, S., Daclin, N., Chapurlat, V.: The application of interoperability requirement specification and verification to collaborative processes. Comput. Ind. **63**(7), 643–658 (2012)
15. Roque, M., Chapurlat, V.: Interoperability in collaborative processes: requirements characterisation and proof approach. In: Camarinha-Matos, L.M., Paraskakis, I., Afsarmanesh, H. (eds.) PRO-VE 2009. IFIP AICT, vol. 307, pp. 555–562. Springer, Heidelberg (2009)
16. Naudet, Y., Latour, T., Guedria, W., Chen, D.: Towards a systemic formalisation of interoperability. Comput. Ind. **61**(2), 176–185 (2010). Integration and Information in Networked Enterprises
17. Advanced Technologies for Interoperability of Heterogeneous Enterprise Networks and their Application (ATHENA): Deliverable Number: D.A4.2: Specification of Interoperability Framework and Profiles, Guidelines and Best Practices (2007)
18. ISO: Advanced automation technologies and their applications — Part 1: Framework for enterprise interoperability, International Organization for Standardization, ISO 11354, ISO/TC 184/SC 5 (2011)
19. Chen, D., Daclin, N.: Framework for enterprise interoperability. In: Interoperability for Enterprise Software and Applications: Proceedings of the Workshops and the Doctorial Symposium of the Second IFAC/IFIP I-ESA International Conference: EI2N, WSI, IS-TSPQ, pp. 77–88 (2006)

20. Panetto, H.: Towards a classification framework for interoperability of enterprise applications. Int. J. Comput. Integr. Manuf. **20**(8), 727–740 (2007). Taylor & Francis: STM, Behavioural Science and Public Health Titles
21. Guedria, W.: A contribution to enterprise interoperability maturity assessment. Ph.D. thesis (2012)
22. Bertalanffy, L.V.: General System Theory: Foundations, Development. Applications. Georges Braziller Inc., New York (1968)
23. Vernadat, F.: Enterprise modeling in the context of enterprise engineering: state of the art and outlook. Int. J. Prod. Manage. Eng. [S.l.] **2**(2), 57–73 (2014)
24. Giachetti, R.E.: Design of Enterprise Systems: Theory, Architecture, and Methods, 1st edn. CRC Press Inc., Boca Raton (2010)
25. Winter, R.: Design science research in Europe. Eur. J. Inf. Syst. **17**(5), 470–475 (2008)
26. Hevner, A., March, S., Park, J., Ram, S.: Design science in information systems research. MIS Q. **28**(1), 75–105 (2004)
27. Guédria, W., Gaaloul, K., Proper, H.A., Naudet, Y.: Research methodology for enterprise interoperability architecture approach. In: Franch, X., Soffer, P. (eds.) CAiSE 2013 Workshops. LNBIP, vol. 148, pp. 16–29. Springer, Heidelberg (2013)
28. Interoperability Research for Networked Enterprises Applications and Software (INTEROP): Deliverable DI.3 Enterprise Interoperability Framework and knowledge corpus (2007)
29. Naudet, Y., Latour, T., Chen, D.: A systemic approach to interoperability formalization. In: IFAC WC 2008, Invited Session on Semantic-Based Solutions for Enterprise Integration and Networking, Seoul, Korea (2008)
30. Guédria, W., Naudet, Y., Chen, D.: Interoperability maturity models – survey and comparison –. In: Meersman, R., Tari, Z., Herrero, P. (eds.) OTM-WS 2008. LNCS, vol. 5333, pp. 273–282. Springer, Heidelberg (2008)
31. Guédria, W., Naudet, Y., Chen, D.: Maturity model as decision support for enterprise interoperability. In: Meersman, R., Dillon, T., Herrero, P. (eds.) OTM-WS 2011. LNCS, vol. 7046, pp. 604–608. Springer, Heidelberg (2011)
32. Le Moigne, J.-L.: La theorie du systeme general, Theorie de la modelisation. Les Classiques du Reseau Intelligence de la Complexite (1994)
33. Naudet, Y., Guedria, W.: Extending the Ontology of Enterprise Interoperability (OoEI) using enterprise-as-system concepts. In: Mertins, K., Bénaben, F., Poler, R., Bourrières, J.-P. (eds.) I-ESA 2014, Proceedings of the International Conference on Interoperability for Enterprise Software and Applications. Springer, Heidelberg (2014)
34. Chen, D., Vallespir, B., Doumeingts, G.: GRAI integrated methodology and its mapping onto generic enterprise reference architecture and methodology. Comput. Ind. **33**, 387–394 (1997)
35. Wiesner, S., Peruzzini, M., Hauge, J.B., Thoben, K.D.: Requirements engineering. In: Concurrent Engineering in the 21st Century, chap. 5, pp. 103–132 (2015)
36. Hull, E., Jackson, K., Dick, J.: Requirement Engineering, pp. 1–20. Springer, London (2011)
37. Chen, D., Vallespir, B., Daclin, N.: An approach for enterprise interoperability measurement. In: Model Driven Information Systems Engineering: Enterprise, User and System Models, Montpellier, France, vol. 341, pp. 1–12, June 2008
38. Yahia, E., Aubry, A., Panetto, H.: Formal measures for semantic interoperability assessment in cooperative enterprise information systems. Comput. Ind. **63**(5), 443–457 (2012)
39. Ford, T., Colombi, J., Graham, J., Jacques, D.: The interoperability score. In: Proceedings of the 5th Annual Conference on Systems Engineering Research, Hoboken, N.J (2007)
40. Department of Defense: C4ISR Architecture Working Group Final Report - Levels of Information System Interoperability. LISI), Washington, DC (1998)

41. Clark, T., Jones, R.: Organizational interoperability maturity model for c2. In: Proceedings of the Command and Control Research and Technology Symposium, Washington (1999)
42. ATHENA Integrated Project. Framework for the Establishment and Management Methodology, ATHENA Deliverable DA1.4 (2005)
43. Tolk, A., Muguira, J.A.: The levels of conceptual interoperability model. In: Fall Simulation Interoperability Workshop, USA (2003)
44. Spekman, R., Davis, E.W.: The extended enterprise: a decade later. Int. J. Phys. Distrib. Logistics Manage. **46**(1), 43–61 (2016)
45. Lau, W., Li, Y., Sinanceur, S.: Ecosystem for virtual enterprise. In: Camarinha-Matos, L.M., Afsarmanesh, H. (eds.) PRO-VE 2003. IFIP, vol. 134, pp. 111–120. Springer, Boston (2004)
46. Baum, H, Schütze, J.: A model of collaborative enterprise networks. In: Procedia CIRP, vol. 3, pp. 549–554. Elsevier Amsterdam (2012)
47. Livieri, B., Kaczmarek, M.: Modeling of collaborative enterprises - CSFs-driven high-level requirements. In: IEEE 17th Conference Business Informatics, vol. 1, pp. 199–208 (2015)
48. Jagdev, H.S., Thoben, K.-D.: Typological issues in enterprise networks. Prod. Plann. Control: Manage. Oper. **12**(5), 421–436 (2001)
49. The Factory Group. http://www.thefactorygroup.com/
50. Concept Factory. http://conceptfactory.lu/
51. Interact. http://interact.lu/
52. Exxus. http://exxus.lu/
53. Sustain. http://sustain.lu/
54. Quest. http://quest.lu/
55. Anaya, V., Berio, G., Harzallah, M., Heymans, P., Matulevičius, R., Opdahl, A.L., Panetto, H., Verdecho, M.J.: The unified enterprise modelling language—overview and further work. Comput. Ind. **61**, 99–111 (2010)
56. Vernadat, F.B.: The CIMOSA languages. In: Bernus, P., Mertins, K., Schmidt, G. (eds.) Handbook of Information Systems, pp. 243–263. Springer, Berlin (1998)

PoN-S: A Systematic Approach for Applying the Physics of Notation (PoN)

Maria das Graças da Silva Teixeira[1(✉)], Glaice Kelly Quirino[1],
Frederik Gailly[2], Ricardo de Almeida Falbo[1], Giancarlo Guizzardi[1],
and Monalessa Perini Barcellos[1]

[1] Ontology and Conceptual Modeling Research Group (NEMO),
Federal University of Espírito Santo, Vitoria, ES, Brazil
`maria.teixeira@ufes.br`,
`{gksquirino,falbo,gguizzardi,monalessa}@inf.ufes.br`
[2] Faculty of Economics and Business Administration,
Ghent University, Ghent, Belgium
`frederik.gailly@ugent.be`

Abstract. Visual Modeling Languages (VMLs) are important instruments of communication between modelers and stakeholders. Thus, it is important to provide guidelines for designing VMLs. The most widespread approach for analyzing and designing concrete syntaxes for VMLs is the so-called Physics of Notation (PoN). PoN has been successfully applied in the analysis of several VMLs. However, despite its popularity, the application of PoN principles for designing VMLs has been limited. This paper presents a systematic approach for applying PoN in the design of the concrete syntax of VMLs. We propose here a *design process* establishing activities to be performed, their connection to PoN principles, as well as criteria for grouping PoN principles that guide this process. Moreover, we present a case study in which a visual notation for representing Ontology Pattern Languages is designed.

Keywords: Concrete syntax · Design process · Visual Modeling Language · Physics of Notation · Ontology-Pattern Languages

1 Introduction

Visual Modeling Languages (VMLs) are important instruments of communication between modelers and stakeholders. The quality of a VML influences the results of a modeling task [1]. Thus, it is relevant to provide guidelines for designing VMLs. Basically, a VML comprises an abstract syntax, which defines the modeling elements (constructs) of the language, and a concrete syntax, which defines the representational elements (symbols) of the language [2]. The concrete syntax can be constituted of one or more dialects, which are different symbol sets to represent the same abstract syntax. These different dialects reflect variations in the language users' profile and modeling task application [3]. Complementary to these alternative syntaxes, there are representation strategies for managing model complexity, which identify mechanisms to visualize large (or complex) models. Our focus is on the design of concrete syntaxes for VMLs.

© Springer International Publishing Switzerland 2016
R. Schmidt et al. (Eds.): BPMDS/EMMSAD 2016, LNBIP 248, pp. 432–447, 2016.
DOI: 10.1007/978-3-319-39429-9_27

Thus, we are interested in, given an abstract syntax, how to design the correspondent concrete syntax, its complementary dialects and representation strategies to manage model complexity.

The most widespread work in the area of designing visual aspects of modeling languages is the *Physics of Notation (PoN)* [3]. PoN defines an approach that is supposed to be used for designing cognitively effective visual notations, i.e. notations that are optimized for being processed by the human mind. PoN consists of nine principles that are based on theories and empirical evidence from a wide range of fields [4]. However, PoN does not prescribe any method or process for systematically applying its principles [5, 6].

In this paper, we present a systematic approach for bridging the theory and practice of PoN in the design of concrete syntaxes for VMLs. We term this approach *PoN-Systematized (PoN-S)*.[1] The process establishes an ordered set of tasks and suggests when to apply the PoN principles. It takes into account a way of grouping these principles. Also, we describe a case study applying the approach in the design of the concrete syntax of a visual language for modeling *Ontology Pattern Languages* [7].

This paper is structured as follow. Section 2 presents the foundations of PoN. Section 3 describes the proposed process for systematizing the application of PoN. Section 4 presents the case study. Section 5 discusses some related works. Finally, Sect. 6 presents our final considerations.

2 Fundamentals of the Physics of Notation (PoN)

PoN defines a set of principles for designing cognitively effective visual notations. The approach considers information visualization and pragmatic theories in order to improve the cognitive effectiveness of VMLs, which is defined as "the speed, ease, and accuracy with which a representation can be processed by the human mind" [3].

Following the tradition in the literature, PoN [3] considers the following elements as the ingredients of a visual notation: a set of graphical symbols (visual vocabulary), a set of compositional rules for forming valid expression (visual grammar), and semantic definitions for each symbol (visual semantics). The set of symbols and compositional rules form the visual (concrete) syntax. Graphical symbols are used to signify or symbolize (perceptually represent) semantic constructs, typically defined by a meta-model. An expression in a visual notation is called a visual sentence or diagram. Diagrams are composed of instances of graphical symbols arranged according to the rules of the visual grammar [8].

PoN identifies nine principles for designing cognitively effective visual notations, namely [3]: (i) *Semiotic Clarity*: "There should be a 1:1 correspondence between a meta-model construct and a graphical symbol"; (ii) *Semantic Transparency*: "Use symbols whose appearance suggests their meaning"; (iii) *Perceptual Discriminability*: "Symbols should be clearly distinguishable from one another"; (iv) *Complexity*

---

[1] PONS is a region of the brainstem with neural pathways that carry sensory signals including those related to *eye movement*. Etymologically, the term from Latin also means *bridge*.

Management: "Include explicit mechanisms for dealing with complexity"; (v) *Cognitive Integration*: "Include explicit mechanisms to support integration of information from different diagrams"; (vi) *Visual Expressiveness*: "Use the full range and capacities of visual variables"; (vii) *Graphic Economy*: "The number of different graphical symbols should be cognitively manageable"; (viii) *Dual Coding*: "Use text to complement graphics"; (ix) *Cognitive Fit*: "Use different visual dialects for different tasks and audiences".

There are in the literature a number of concrete cases of the application of PoN, for instance: (i) in [8], the authors describe the evaluation and redesign of a visual notation for the i* language. The publication describes a notation redesign effort and a description of the redesign process employed, which adds operational characteristics to PoN. This work is used as a study in [4], which analyzes the influence of model readers on the language concrete syntax; (ii) in [6], the authors evaluated the UCM visual notation and present not only the evaluation and redesign proposal, but their impressions concerning the application of the approach.

An important consideration is that the principles in PoN influence one another. Knowledge of these influence relations can be used to spot tradeoffs (where principles conflict with one another), as well as synergies (where principles complement or reinforce one another). In [3], Moody presents detailed descriptions of each principle and the influence relations between them. However, when a designer is applying these principles in a design process, s/he needs further design guidance. For instance, when should s/he apply a principle? In which sequence should principles be applied? Which principles should be applied in tandem? In order to overcome this limitation, we propose a systematic process for applying PoN.

3   PoN Systematized (PoN-S)

In this section, we present an approach for systematizing the application of PoN in the design of concrete syntax of VMLs. The methodology we followed is based on the Design Science approach discussed in [9]. The steps carried out here were: (i) we identified the design questions that a developer needs to answer; (ii) we established how the PoN principles are related to these design questions; (iii) in order to systematize the process, we described groups of PoN principles; (iv) we added ordering relations between the design tasks constituting the design process. Our approach is based not only on the foundations of PoN [3], but also on works that have applied PoN for analysis or (re)design of VMLs, as [4–6, 8].

Design questions: When designing the VML's concrete syntax, we should deal with concerns at different levels. First, we need to decide whether different dialects for the same abstract syntax are needed. The motivation for creating more than one dialect should be clearly identified (e.g., the fact that the language must be suitable for more than one stakeholder profile, modeling task or problem domain characteristics). Second, at the language level, we need to determine the symbols to be used in the concrete syntax. Finally, at the instance level, we are concerned with the development of diagrams using the proposed concrete syntax. Table 1 presents the concerns for these different levels as design questions and identifies the PoN principles that can be applied to answer them. Answering these design questions helps the designer to understand the rationale behind the application of each principle, acting as initial guidance. However, this is not enough for guiding the design effort. To do so, we propose a way of grouping the principles and a process for applying them when designing a concrete syntax.

Table 1. Answering to some basic design questions with PoN principles

Design question	Related PoN principles
Dialect set	
Do we need different dialects for the abstract syntax? If so, which dialects should we consider?	Cognitive fit
For each dialect	
Language level	
Which symbol(s) do we need to create?	Semiotic clarity
How to create each symbol?	Semantic transparency
How to relate differents symbols? To what extent two or more symbols should be similar/different?	Perceptual discriminability
How visual variables (such as shape, color and texture) and text should be applied in order to aid the identification of each representational element?	Visual expressiveness
	Graphic economy
	Dual coding
Instance level	
Which procedures should we create to support the development of a (some) diagram(s)? (Depending on the answer to this question, it may be necessary to create new symbols, affecting decisions at the language level.)	Complexity management
	Cognitive integration

Grouping the Principles: Moody describes a number of influence relations between pairs of principles [3]. However, most of the time these principles act in group. This perception is fundamental to guide the VML design process. Thus, we suggest grouping PoN principles into the following groups:

- *Group 1 – Basic principles.* This group comprises three principles: Semiotic Clarity, Semantic Transparency and Perceptual Discriminability. These principles are considered basic principles, because they should be applied at some extent in the design of any concrete syntax. They are complementary in the sense that we need to create a symbol to each construct (Semiotic Clarity), and each of these symbols should be clearly identifiable (Semantic Transparency), and yet clearly distinguishable from other symbols in the language (Perceptual Discriminability). So, these principles should be applied together in the design of each dialect of the concrete syntax, and the level they should be in compliance with can vary in each dialect. Semiotic Clarity acts as a guarantee that the mapping between abstract and concrete syntaxes is complete, avoiding possible anomalies, i.e., that all necessary symbols are defined. Perceptual Discriminability is concerned with whether such symbols are adequately different from (or similar to, depending on the case) the others. Finally, Semantic Transparency is concerned with whether each symbol has its meaning easily inferred.

- *Group 2 – Information complexity management principles.* This group comprises two principles: Complexity Management and Cognitive Integration. These principles are commonly applied when dealing with large or complex diagrams. They are complementary, since the former deals with how to organize the information in a model (probably separating them in several diagrams), and the second refers to how to keep connection and traceability of the information spread in different diagrams. Thus, they should be applied together. Basically, this group of principles will be applied at the level of individual diagrams, giving rise to representation strategies for managing model complexity. Ideally, the way of addressing information complexity management should be the same (or very similar) in any dialect of the concrete syntax. Finally, it is worth pointing out that the application of these two principles can demand the creation of new symbols, hence, affecting the language level (Group 1).
- *Group 3 – Supporting principles.* The principles in this group can somehow affect principles of groups 1 and 2. The support principles are: Visual Expressiveness, Graphic Economy and Dual Coding. Visual Expressiveness is connected to the other principles (except Dual Coding), in the sense that it provides the mechanisms (as visual variables) for implementing the other principles. Graphic Economy is also connected to the other principles, since it establishes a way to control them, trying to keep them as simple as possible. We consider here Dual Coding to refer only to redundant textual representational support.
- *Group 4 – Dialect set principle.* This group, in fact, contains only one principle: Cognitive Fit. This principle has an indirect connection to the other principles, because the other principles are applied to each dialect of the concrete syntax at a time, while Cognitive Fit is about defining the set of dialects.

The principles of a group can interact with principles of another group, as in Group 3 (which influences groups 1 and 2). Furthermore, the principles inside a group can interact with each other. Typically, this intra-group relationship is stronger than the inter-group relationships. This is a reason for grouping the principles in such way.

Design process: The design questions and groups of principles give us some guidance for designing the concrete syntax. However, to truly systematize the application of PoN, we need a *design process* for guiding this. Figures 1, 2, 3 and 4 present the proposed design process. This process is structured according to the concerns shown in Table 1, starting by the concern related to the dialect set, and in the sequel addressing concerns related to the language level and then the instance level, for each dialect. Each figure presents a part of the process, including inputs, outputs, tasks and decisions to be made. The process is represented by means of an extension the UML activity diagrams notation, introducing a new modeling element: *PoN principle*. A PoN principle can be seen as a guideline to perform an activity and it is represented by means of an ellipse, which is connected by a line to the activity that applies the principle at hand.

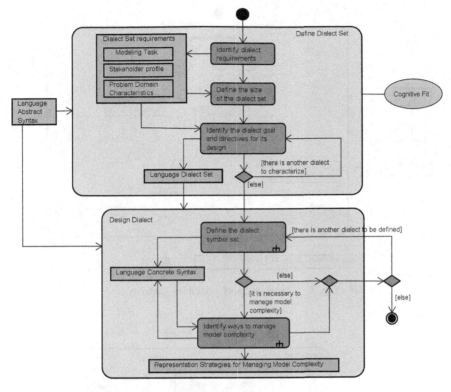

Fig. 1. PoN-S design process overview

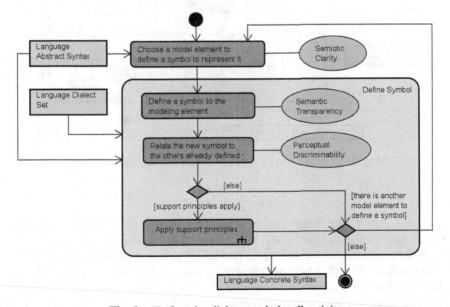

Fig. 2. "Define the dialect symbol set" activity

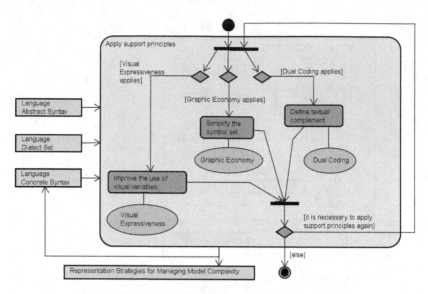

Fig. 3. "Apply support principles" activity

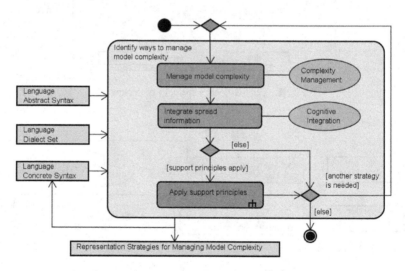

Fig. 4. "Identify ways to manage model complexity" activity

Figure 1 depicts the whole design process, which comprises two main stages: *Define dialect set* and *Design dialect*. In the stage *Define dialect set*, the designer shall identify the requirements for the VML (modeling task, stakeholder profile, problem domain characteristics) that help to *Define the size of the language dialect set*. Next, in the activity *Identify the dialect goal and directives for its design*, each dialect should be characterized, establishing its goal and directives for its design. In this task, the designer should take into account the influence relations (conflicts or synergies) that exist among PoN principles (see [3]). It is not possible to establish the same level of compliance to all principles. So, the designer should choose the principles to highlight in each dialect. The stage *Define dialect set* should be performed considering language abstract syntax (as input) and the Cognitive Fit principle (as method).

In the second stage (*Design dialect*), each dialect identified should have its set of symbols (representational elements) defined in accordance to the goal and directives previously identified. This stage has two complex tasks: (i) *Define the dialect symbol set*, which is responsible for defining representational elements for the model elements identified by the abstract syntax; and (ii) *Identify ways to manage model complexity*, an optional task performed when the amount of elements requires managing model complexity. The input for these tasks are the language dialect set and the abstract syntax. The output is the concrete syntax, and optionally some representation strategies to deal with size and complexity of the models. These two complex tasks are further detailed in Figs. 2 and 4, respectively.

Figure 2 depicts the steps for defining the dialect symbol set for each dialect previously identified. This activity starts by choosing a model element to be represented. This task is guided by Semiotic Clarity principle to ensure that each model element will be represented by exactly one symbol, unless this situation is required due to the directives established for the dialect. Once the model element to be represented is chosen, we need to define a symbol for it (task *Define a symbol to the modeling element*). This activity is guided by the Semantic Transparency principle in order to establish a clear meaning to the symbol. Also, we should relate the chosen symbol to the other symbols already defined in the concrete syntax, following the Perceptual Discriminability principle. This task aims at evaluating the visual distance between the new symbol and the other symbols already defined. These two tasks can be supported by the application of supporting principles (see Fig. 3). They are performed in a loop until all the representational elements of that dialect have been defined.

The *Apply supporting principles* activity depicted in Fig. 3 deals with the possible application of three supporting principles: Visual Expressiveness, Graphic Economy and Dual Coding. The designer can apply each principle as much as s/he deems necessary. There is no pre-defined order to be followed. The inputs are the language abstract and concrete syntaxes and the characteristics of the dialect. The output can be an update of the language concrete syntax or an update of some representation strategy for managing model complexity.

The *Improve the use of visual variables* activity is guided by the Visual Expressiveness principle. In this task, the designer shall review the symbol(s) (or strategies), possibly updating the visual variables values to maximize their expressiveness. The designer can do this individually (per symbol) or considering the whole symbol set. The *Simplify the symbol set* activity is guided by the Graphic Economy principle. In this task, the designer may also review the symbol(s) (or strategies), now with the goal of simplifying the dialect. Finally, in the *Define textual complement* activity, by applying the Dual Coding principle, the designer should evaluate when it is useful to introduce redundancy through the use of text. This can be necessary when the designer deems that the text will increase symbol expressiveness.

After defining the symbols of a dialect, the designer must decide if it is necessary to manage the model complexity in diagrams developed using this dialect. Therefore, the *Identify ways to manage model complexity* activity is an optional activity whose importance increases as the language grows in size and complexity.

Figure 4 details the complex activity *Identify ways to manage model complexity*. The inputs for this activity are the language abstract and concrete syntaxes as well as the characteristics of the dialect set. The outputs are the language concrete syntax (in case it suffers some update) and representation strategies for managing model complexity (as many as the designer deems necessary).

The first task is *Manage model complexity*, which is guided by the Complexity Management principle. In this task, representation strategies for managing the complexity of diagrams written in that dialect shall be established. An example is the use of modularization. As a complement to this task, there is the task *Integrate spread information*, which is guided by the Cognitive Integration principle. This task is responsible for establishing ways to trace information spread in several diagrams and strategies for connecting them. It is important to say that these two tasks can be applied in parallel, resulting in a single representation strategy that is in accordance with both aspects of complexity management (organization and integration of information). In fact, both tasks are applied in independent loops until deemed sufficient by the designer. Usually, each cycle results in a representation strategy for managing model complexity, which is complemented by new concrete syntax elements, when necessary.

4   Applying PoN-S: A Case Study

In a preliminary evaluation of PoN-S, a case study was performed aiming at designing a visual notation for representing Ontology Pattern Languages (OPLs). An OPL is a network of interrelated domain-related ontology patterns that provides holistic support for solving ontology development problems for a specific domain. It contains a set of interrelated domain-related ontology patterns, plus a process providing explicit guidance on what problems can arise in that domain, informing the order in which these problems should be addressed, and suggesting one or more patterns to solve each specific problem [7, 10]. For adequately representing OPLs, two types of models are necessary: a structural model, showing the patterns and the dependency relationships between them, and a process model, showing, among other things, the activities of applying the patterns, decision points, and entry and end points in the OPL process.

Regarding the process model, in a nutshell, its meta-model is an extension for representing OPLs of the meta-model of the UML activity diagram [9]. For this reason, its concrete syntax is based on the UML notation for activity diagrams. This has the advantage of benefiting users who are familiar with this notation. Due to space limitations, however, in this paper we do not discuss the design of the visual notation for the process model. Our focus here is on discussing the application of PoN-S, and thus we concentrate in the design of the visual notation for the structural model.

Figure 5 shows the meta-model of the language concerning the OPL structural model. This model is composed of *OPL Structural Elements*. There are two types of *OPL Structural Elements*: *Pattern* and *Pattern Group*. A *Pattern* represents a domain-related ontology pattern, i.e., a small and reusable fragment of an ontology conceptual model, extracted from a reference ontology [11]. A *Pattern Group* is a way of grouping related patterns and other pattern groups. Thus, a *Pattern Group* is composed by *OPL Structural Elements*. A special type of pattern group is the *Variant Pattern Group*, which is a set of (variant) patterns that solve the same problem, but each in a different way. Only one pattern from a *Variant Pattern Group* can be used at a time. *Patterns* that compose a *Variant Pattern Group* are variants of each other, giving rise to the derived relationship *variantOf* between patterns.

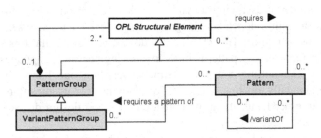

Fig. 5. OPL structural meta-model

Patterns may depend on other patterns, i.e., for applying a pattern *p2* another pattern *p1* has to be applied first. An OPL should be able to represent dependencies between patterns or between a *Pattern Group* and a *Pattern*. The *requires* relationship captures this dependency. In the case of a dependency between a *Pattern Group* and a *Pattern*, the following rule applies: If a pattern *p1* is part of a pattern group *pg* and *pg* requires a pattern *p2*, then *p1* requires *p2*. Finally, a *Pattern* may require the application of a pattern from a *Variant Pattern Group*.

As Fig. 1 shows, the design process started by identifying the dialect requirements, which includes: (i) *Domain characteristics:* the visual notation for representing OPLs. Each OPL can refer to a different domain. Thus, this is the case of a domain-independent language; (ii) *Stakeholder profile*: OPLs are typically used by ontology engineers (both beginners and experienced); (iii) *Modeling task*: developing domain ontologies by reusing domain-related ontology patterns.

Although there are stakeholders with different levels of experience, the OPL visual notation should be simple and intuitive for all kinds of stakeholders. Thus, the designer established that only one dialect is enough. The goal of this dialect is to provide a simple and intuitive visual notation for ontology engineers to develop domain ontologies by reusing ontology patterns [7, 10]. The notation should contain symbols to represent all OPL constructs without ambiguity. Moreover, in case of the use of colors, it should be possible to print the diagrams in gray scale without denting their comprehensibility.

The next step is to define the dialect symbol set. A loop was performed, in which each model element was characterized and had a symbol defined for it. This loop was guided by the principles of Semiotic Clarity, Semantic Transparency, Perceptual Discriminability as well as the supporting principles. Initially, considering the abstract

442   Maria das Graças da Silva Teixeira et al.

syntax defined by the meta-model shown in Fig. 5, and taking into account the Semiotic Clarity principle, a 1:1 correspondence between the meta-model constructs and graphical symbols was defined. This otherwise isomorphic mapping has two exceptions: the designer decided that it was not necessary to assign a symbol to the *OPL Structural Element* construct (an abstract modeling element), but only to its (concrete) subtypes (*Pattern* and *Pattern Group*). Moreover, symbols should be assigned to the relationships between these constructs, except for the *variantOf* relationship, since it is a derived association. Thus, symbols should only be assigned to the constructs shown in gray in Fig. 5 and to the regular associations between them.

The designer started assigning a symbol to the *Pattern* construct. Since s/he was dealing with a domain-independent language, s/he decided to represent patterns by rectangles (an abstract sign). This choice was done considering that this is a common symbol used for representing patterns in Software Engineering Languages (e.g., UML class diagrams). Concerning Semantic Transparency, on one hand, this symbol is considered semantically opaque, since it does not inform its meaning directly [3]. However, on the other hand, it can be considered a good design decision, given that this symbol is easily recalled [3].

Pattern Groups are represented by figures closed by straight solid lines (solid polygons). For representing *Variant Pattern Groups*, the same notion was applied, but now using dashed lines. This decision was taken considering the Perceptual Discriminability principle, aiming at guaranteeing that symbols representing groups have a small visual distance. Furthermore, the visual variables *texture* and *color* were used to differentiate them. The lines of *Variant Pattern Groups* are dashed and red, while the lines of *Pattern Groups* are solid and blue.

For representing the relation between *Patterns* and *Pattern Groups*, the designer chose the notion of spatial containment: *Patterns* that are part of a *Pattern Group* represented as spatially enclosed by the symbol representing the latter. This choice affords the so-called *inferential free-rides* to the language, i.e., visual querying and reasoning operations of minimal cognitive costs [12]. Moreover, it is noteworthy that there is a visual variable that qualifies *Patterns* and *Pattern Groups*: size. The region that represents the group encompasses several patterns. Thus, the size of this region is greater than the rectangle representing the pattern.

Regarding the dependency relations *requires* and *requires a pattern of*, both are represented by an arrow from the dependant to the dependee. For differentiating between them, arrows representing the *requires* association are symbolized with solid lines, in contrast to the dashed lines for the *requires a pattern of* association. This decision is in line with the one of representing *Pattern Groups* using solid lines, and *Variant Pattern Groups* using dashed lines. Thus, it takes the Perceptual Discriminability principle into account. So, these symbols have small visual distances.

It is worthwhile to point out that supporting principles were also applied for making the aforementioned choices. Regarding the Visual Expressiveness principle, the proposed visual notation uses the following visual variables: shape, texture and size. Color values are used as a redundant encoding, because variation in color disappears when a diagram is printed in grayscale. The designer decided not to apply other visual variables, keeping the notation as simple as possible.

The Graphic Economy principle did not play a strong role in this case study. This is because PoN advocates the use of up to six elements in a dialect and the structural meta-model considered here has only four classes and three regular associations. Nevertheless, some decisions were taken aiming at making the language as simple as possible. In summary, no symbol was assigned to the following meta-model elements: *OPL Structural Element* construct, since it is an abstract class in the meta-model (i.e., it cannot be directly instantiated); whole-part relationship between *Pattern Group* and *Pattern*, since the notion of containment used to represent *Pattern Groups* also addresses this relation; and the derived association *variant of*, since it is also derived from the representation for *Pattern* and *Pattern Group*.

Finally, the Dual Coding principle, which deals with the use of text as an information supplement, was not applied. This is because, according to the designer: there´s a small amount of constructs to represent, their semantic are clear enough without textual redundancy and use of textual values can be better applied to distinguish between instances (as instance labels).

After defining an initial version of the concrete syntax, it is time to evaluate if the language demands representation strategies for managing model complexity. If this is the case, we should apply the principles of Complexity Management and Cognitive Integration. The Complexity Management principle emphasizes the importance of managing the diagrammatic complexity, which is measured by the number of elements in a diagram, among others. In the case of this case study, the designer recognized the need for managing complexity. Although the proposed language for representing OPLs is simple, the models that may be built using it tend to be large. Thus, to increase the speed and accuracy of understanding the diagrams, the designer decided to introduce a symbol for representing *Pattern Groups* (including *Variant Pattern Groups*) that encapsulates the *Patterns* that comprise it. Following the Perceptual Discriminability principle, the designer chose to represent these alternative forms by means of rectangles decorated by the following icon (⬚), indicating that this element is detailed in another diagram[2].

Table 2 shows the final concrete syntax for representing OPL structural models.

Figure 6 shows an example of a structural model of an OPL: Service OPL (S-OPL). This OPL, which provides ontology patterns for service modeling, is discussed in details in [10]. As shown in this figure, S-OPL is organized in three groups: *Service Offering, Service Negotiation and Agreement* and *Service Delivery*. The *Service Offering Group* is composed by three patterns (*SOffering, SODescription* and *SOCommitments*) and two groups of variant patterns (*Provider Variant Group* and *Target Customer Variant Group*). The patterns *SODescription* and *SOCommitments* as well as the *Provider* and *Target Customer Variant Groups* require the pattern *SOffering. SOffering*, in turn, requires patterns of both *Provider* and *Target Customer Variant Groups. Provider* and *Target Customer Variant Groups* are both composed of seven variant patterns each. The *Service Negotiation and Agreement Group* is composed by four patterns (*SNegotiation, SADescription, HPCommitments* and

---

[2] This icon is commonly used by UML to represent that an element represented by the decorated construct encapsulates further elements. A similar symbol is used by ARIS.

Table 2. Symbols of the visual notation for OPL structural models

Structural Model	
Element	**Symbol**
Pattern	
Pattern Group (expanded format)	
Pattern Group (black box format)	
Variant Pattern Group (expanded format)	
Variant Pattern Group (black box format)	
Relation "requires"	
Relation "requires a pattern of"	

SCCommitments) and three groups of variant patterns (*Agreement Variant Group*, *Hired Provider Variant Group* and *Service Customer Variant Group*). The *Agreement Variant Group* is composed by three patterns: *SNegAgree*, *SOfferAgree*, and *SAgreement*. The first two of these patterns as well as the SNegotiation pattern require *SOffering*. The patterns *SADescription*, *HPCommitments* and *SCCommitments* require a pattern of the *Agreement Variant Group*. The *SAgreement* pattern requires patterns of both *Hired Provider* and *Service Customer Variant Groups* (shown as black boxes in Fig. 6). These two variant groups, in turn, require the *SAgreement* pattern. Finally, the *Service Delivery Group* (shown as a black box in Fig. 6) requires the *Service Negotiation and Agreement Group*.

5   Related Work

In a brief literature review, executed to identify how concrete syntax of conceptual modeling languages have been evaluated and designed, we identified PoN as the most widespread approach for analysis and design of VML concrete syntax [5, 13]. Also, we noticed that studies discussing efforts in analyzing modeling languages (with associated redesign suggestions) (e.g., [6, 14]) are more common than those describing efforts in language design (e.g., [15]).

The need for improving the design process involving PoN has been identified by many researchers, including Moody himself. In [8], Moody *et al.* discuss operational issues of PoN when presenting the analysis and redesign of i* (a language in the Requirements Engineering field). However, these issues are discussed individually for each principle, i.e., the authors do not define a process involving all principles. In [4], a work complementing the i* evaluation described in [8], the authors added the idea of

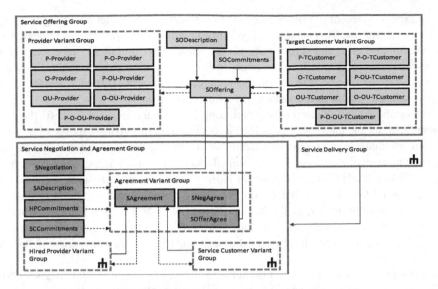

Fig. 6. S-OPL structural model

PoN operationalization, highlighting the importance of considering stakeholder profiles during language design. It is a clear contribution towards considering pragmatic issues for notation analysis and design. However, once more, they did not define a design process.

In [5], Storrle and Fish criticize PoN judging that it still needs improvements towards operationalization. In that article, the authors propose ways for operationalizing PoN focusing on the analysis task of modeling languages. Moreover, they established a series of measures that complement the PoN original proposal. However, they also do not propose a design process.

The work proposed in this paper contributes to this collective effort of proposing *operationalizable* techniques for the design of visual languages. In particular, PoN-S is a methodological contribution that supports language designers in the application of PoN through the definition of a design process, a gap that has been identified in the literature.

6   Final Considerations

This paper focused on the elaboration of PoN-S, a design process for applying the PoN principles in practice. The elements involved in this process model are: inputs, outputs, tasks, task ordering and procedures (the PoN principles). This process was applied in a case study aiming at developing a visual notation for Ontology Pattern Languages. The case study was the first validation of PoN-S. As a result of its execution, the design process was refined, for example, by identifying the need for a *Identify dialect requirements* task. This study also indicated that PoN-S is easy to follow. We are currently conducting an on going survey whose preliminary findings indicate that the OPLs' users approve the resulting concrete syntax.

An expected benefit of PoN-S is the establishment of a path that modeling language designers can follow. This is particularly helpful mainly for novice language designers. When defining a clear and simple path we are reducing the possibility of errors during the process. The need to reduce the effort of PoN application is a recognized problem [6]. Also, a systematic process aids in the standardization of the language design, which facilitates future maintenance tasks and facilitates teamwork.

The establishment of the tasks constituting PoN-S take into account: (i) the PoN principles, assuring that every principle is considered; (ii) visual aspects of a VML (symbol set, dialects, representation strategies for manage model complexity). A current limitation of the process is the level of details in which some tasks have been defined. For example, in *Identify ways to manage model complexity* task, we state that strategies should be defined, but we do not identify how to create these strategies.

We are planning to extend the design process of PoN-S to provide more directed and complete guidelines for the language designer. In particular, ontological theories such as the ones discussed in [12] are the basis for such future extensions of PoN-S.

Acknowledgments. This research is funded by the Brazilian Research Funding Agency CNPq (National Council for Scientific and Technological Development) (Processes 461777/2014-2 and 206255/2014-4).

References

1. Krogstie, J., Solvberg, A.: Information Systems Engineering: Conceptual Modeling in a Quality Perspective. Draft of Book, Information Systems Groups, NTNU, Trondheim, Norway (2000)
2. Ruiz, M., Costal, D., España, S., Franch, X., Pastor, Ó.: Integrating the goal and business process perspectives in information system analysis. In: Jarke, M., Mylopoulos, J., Quix, C., Rolland, C., Manolopoulos, Y., Mouratidis, H., Horkoff, J. (eds.) CAiSE 2014. LNCS, vol. 8484, pp. 332–346. Springer, Heidelberg (2014)
3. Moody, D.L.: The "physics" of notations: toward a scientific basis for constructing visual notations in software engineering. IEEE TSE **35**(6), 1–22 (2009)
4. Caire, P., Genon, N., Heymans, P., Moody, D.L.: Visual notation design 2.0: towards user comprehensible requirements engineering notations. In: Requirements Engineering Conference (RE), pp. 115–124. IEEE Computer Society (2013)
5. Störrle, H., Fish, A.: Towards an operationalization of the "Physics of Notations" for the analysis of visual languages. In: Moreira, A., Schätz, B., Gray, J., Vallecillo, A., Clarke, P. (eds.) MODELS 2013. LNCS, vol. 8107, pp. 104–120. Springer, Heidelberg (2013)
6. Genon, N., Amyot, D., Heymans, P.: Analysing the cognitive effectiveness of the UCM visual notation. In: Kraemer, F.A., Herrmann, P. (eds.) SAM 2010. LNCS, vol. 6598, pp. 221–240. Springer, Heidelberg (2011)
7. Falbo, R.A., Barcellos, M.P., Nardi, J.C., Guizzardi, G.: Organizing ontology design patterns as ontology pattern languages. In: Cimiano, P., Corcho, O., Presutti, V., Hollink, L., Rudolph, S. (eds.) ESWC 2013. LNCS, vol. 7882, pp. 61–75. Springer, Heidelberg (2013)
8. Moody, D.L., Heymans, P., Matulevicius, R.: Visual syntax does matter: improving the cognitive effectiveness of the i* visual notation. Requir. Eng. **15**, 141–175 (2010)

9. Wieringa, R.J.: Design Science Methodology for Information Systems and Software Engineering. Springer, London (2014)
10. Falbo, R.A., Quirino, G.K., Nardi, J.C., Barcellos, M.P., Guizzardi, G., Guarino, N., Longo, A., Livieri, B.: An ontology pattern language for service modeling. In: Proceedings of the 31th Annual ACM Symposium on Applied Computing - ACM-SAC 2016 (2016)
11. Falbo, R.A., Guizzardi, G., Gangemi, A., Presutti, V.: Ontology patterns: clarifying concepts and terminology. In: Proceedings of the 4th Workshop on Ontology and Semantic Web Patterns (2013)
12. Guizzardi, G.: Ontology-based evaluation and design of visual conceptual modeling languages. In: Reinhartz-Berger, I., Sturm, A., Clark, T., Cohen, S., Bettin, J. (eds.) Domain Engineering. Product Lines, Languages and Conceptual Models, p. 345. Springer, New York (2013)
13. Genon, N., Heymans, P., Amyot, D.: Analysing the cognitive effectiveness of the BPMN 2.0 visual notation. In: Malloy, B., Staab, S., van den Brand, M. (eds.) SLE 2010. LNCS, vol. 6563, pp. 377–396. Springer, Heidelberg (2011)
14. Figl, K., Derntl, M.: The impact of perceived cognitive effectiveness on perceived usefulness of visual conceptual modeling languages. In: Jeusfeld, M., Delcambre, L., Ling, T.-W. (eds.) ER 2011. LNCS, vol. 6998, pp. 78–91. Springer, Heidelberg (2011)
15. Miske, C., Rothenberger, M.A., Peffers, K.: Towards a more cognitively effective business process notation for requirements engineering. In: Tremblay, M.C., VanderMeer, D., Rothenberger, M., Gupta, A., Yoon, V. (eds.) DESRIST 2014. LNCS, vol. 8463, pp. 360–367. Springer, Heidelberg (2014)

How Cognitively Effective is a Visual Notation? On the Inherent Difficulty of Operationalizing the Physics of Notations

Dirk van der Linden[(✉)], Anna Zamansky, and Irit Hadar

Department of Information Systems, University of Haifa, Haifa, Israel
{djtlinden,annazam,hadari}@is.haifa.ac.il

Abstract. The Physics of Notations [9] (PoN) is a design theory presenting nine principles that can be used to evaluate and improve the cognitive effectiveness of a visual notation. The PoN has been used to analyze existing standard visual notations (such as BPMN, UML, etc.), and is commonly used for evaluating newly introduced visual notations and their extensions. However, due to the rather vague and abstract formulation of the PoN's principles, they have received different interpretations in their operationalization. To address this problem, there have been attempts to formalize the principles, however only a very limited number of principles was covered. This research-in-progress paper aims to better understand the difficulties inherent in operationalizing the PoN, and better separate aspects of PoN, which can potentially be formulated in mathematical terms from those grounded in user-specific considerations.

Keywords: Visual notations · Cognitive effectiveness · Physics of Notations · Operationalization

1 Introduction

Conceptual modeling is a widely used technique in software engineering and information systems development to capture and reason about a particular domain of interest. Visual notations used in such modeling tasks have often been designed without eliciting and considering empirical evidence for what fits best the potential users and the task at hand. Some of the most widespread visual notations used in practice, such as UML, are affected by this limitation (cf. [8]). Some work has attempted to alleviate this by more explicitly tracing design to its rationale (cf. [14,20]), but such work remains on the level of the domain, not the notation itself.

The main issue with visual notations developed in this way is a lack of focused attention on ensuring their cognitive effectiveness, namely the ease with which people can read and understand diagrams written in the notation. Given that visual languages are often used for their convenience over textual languages, they should be designed and analyzed "from the perspective of languages that are cognitively usable and useful." [12]

© Springer International Publishing Switzerland 2016
R. Schmidt et al. (Eds.): BPMDS/EMMSAD 2016, LNBIP 248, pp. 448–462, 2016.
DOI: 10.1007/978-3-319-39429-9_28

Over the years, several frameworks have been proposed (e.g., Cognitive Dimensions [3], SEQUEL [7], GoM [15]) that, at least partially, paid attention to this aspect and provided notation designers with guidelines on how to improve the quality of visual notations. Recently, one such framework focusing exclusively on the cognitive effectiveness of visual notations, the Physics of Notations (PoN) [9], has become relatively widespread. Its adoption by researchers is evident by the ever growing number of analyses using it [18], including having been applied to e.g., BPMN, UML, i*, WebML, as well as the increase in the number of works citing it over other frameworks [2].

Moody positions the PoN's nine principles as constituting a type V prescriptive theory in terms of Gregor's [4] taxonomy of theory in IS [9, p. 775]. He states that these principles "can be used to evaluate, compare, and improve existing visual notations as well as to construct new ones". This effectively means that instead of considering endless possibilities when coming up with a new visual notation, one may opt for those possibilities which best comply with PoN. We refer to the activity of checking the compliance of a visual notation with a PoN principle as an *operationalization* of that principle.

Unfortunately, to the best of our knowledge no concrete guidelines on practical operationalization of PoN principles have been proposed thus far. Moreover, there has been criticism aimed towards their formulation as informal, though well-described and thorough, guidelines. In particular, the feasibility of verifying whether they can be verified in a replicable and systematic way has been criticized (cf. [1,5,16,17,19]). The latter authors have further argued that the PoN's principles in their current state are "neither precise nor comprehensive enough to be applied in an objective way to analyze practical visual software engineering notations".

One natural direction toward operationalization of PoN principles, proposed by [16], is their formalization (or formulation in mathematical terms). However, they encountered a number of challenges while attempting to formalize the first two (out of nine) principles of the PoN. Information needed to formalize the principles was posited, while acknowledging that "[we] do not yet have empirical evidence to support our assumption" [16, p. 116]. The authors similarly acknowledged that the application or formalization of a number of principles requires a base in other existing theories [16, p. 118].

In this paper we aim to better understand the inherent difficulties behind operationalization, and in particular of formalization of PoN principles. Clearly, we cannot expect to have an algorithm for computing compliance to PoN of every newly introduced visual notation. It is not only due to the fact that visual notations usually do not have fully formalized representations, but also that some PoN principles rely on information that can only be obtained from cognitive theories and/or empirical data from users of the particular new notation. This leads to the question to what extent aspects of the PoN can be formalized. As a starting point, we define the notion of visual notation in set-theoretical terms, which provide a formal ground for our analysis. We then use these terms to answer the following research questions:

RQ1. What elements are involved in operationalizing each PoN principle with respect to a given visual notation?

RQ2. What effect do these elements have on the feasibility of operationalizing each principle into a well-defined mathematical question?

RQ1 will be addressed by analyzing the different PoN principles, analyzing them for the basic elements required for their employment. These findings will be used for investigating RQ2, where we will discuss the way in which the identified elements can be used to address the operationalization of the principles in a structured mathematical way. Finally, we further reflect on what the identified challenges mean in terms of needed research efforts.

By addressing the above questions, this paper takes a first step towards grounding the PoN in more formal and operational foundations.

2   PoN Principles Overview

This section provides a brief overview of the principles of the PoN. Table 1 presents the nine principles of the PoN together with their high-level descriptions.

Table 1. Overview of the PoN's nine principles.

Principle	Explanation
Semiotic clarity	There should be a one-to-one correspondence between elements of the language and graphical symbols
Perceptual discriminability	Different symbols should be clearly distinguishable from each other
Semantic transparency	The use of the visual representations whose appearances suggest their meaning
Complexity management	Notation includes explicit mechanisms for dealing with complexity
Cognitive integration	Notation include explicit mechanisms to support the integration of information from different diagrams
Visual expressiveness	The use of the full range and capacities of visual variables
Dual coding	Use of text to complement graphics
Graphic economy	The number of different graphical symbols should be cognitively manageable
Cognitive fit	Use of different visual dialects for different tasks and audiences

From the above descriptions it becomes clear that the principles involve different types of elements, which have a direct impact on their operationalization.

The first principle of semiotic clarity, e.g., mentions *language elements* and *graphical symbols*. Given the language elements, graphical symbols and a mapping between them, it is an easy mathematical question to determine whether the mapping is 1:1.

The second principle, perceptual discriminability, again speaks of graphical symbols, but this time requires their *distinguishability*. Note, however, that given two graphical symbols, establishing their distinguishability is not a mathematical question. Symbols distinguishable for a typical human, may not be distinguishable for a color-blind one. And even after determining the target user, we need to know the values of the parameters of the representation medium in which the notation is used (such as number of pixels of the presented UI, texture, and color difference, in computer aided environment) which may affect distinguishability. But if, for instance, we know that a difference of more than 10 pixels is distinguishable, then given two shapes establishing their distinguishability becomes a mathematical question.

The semantic transparency principle, however, seems not to fall even in the latter category, as it speaks in terms of appearances (symbols) suggesting their meaning. How do we know, given a symbol, that it suggests its meaning? Suggests to whom? In what sense? How sensitive is it to, for example, cultural differences? And can this be verified?

In what follows we propose some notions which provide a formal ground for making the above distinctions in a more systematic way.

3 An Analysis of PoN Operationalization

3.1 A Set-Theoretical Framework for PoN

The basic element of our framework is a *graphical symbol*. Each graphical symbol g has an *appearance*, which can be represented using appearance *variables* (such as size, shape, texture, etc.) which may assume different values from associated ranges. We shall identify appearance of a given graphical symbol $\mathsf{Ap}(g)$ with some assignment of values to appearance variables.

Note that $\mathsf{Ap}(g)$ is an abstraction of the actual symbol g; thus, for example the following three symbols have the same appearance in terms of variables shape, color and size, although they can be distinguished by texture and line style (Fig. 1):

In addition to appearance, each graphical symbol g also has an associated meaning $\mathbf{I}(g)$ which takes the form of a semantic construct.

Fig. 1. Symbols equivalent in terms of shape, color and size

The above can be formalized as follows:

Definition 1. *A visual notation is a triple* $\mathbf{V} = (\mathcal{G}(\mathbf{V}), \mathcal{L}(\mathbf{V}), \mathcal{R}(\mathbf{V}))$ *where* $\mathcal{G}(\mathbf{V})$ *is a set of graphical symbols,* $\mathcal{L}(\mathbf{V})$ *is a set of textual symbols (letters) and* $\mathcal{R}(\mathbf{V})$ *is a set of rules for the composition of elements from* $\mathcal{G} \cup \mathcal{L}$ *into models in* $\mathbf{V}$*. The closure of* $\mathcal{G} \cup \mathcal{L}$ *under* R *provides the set of possible models that can be constructed over* $\mathbf{V}$*, denoted by* $\mathsf{Models}(\mathbf{V})$*.*

The set of appearance variables of $\mathcal{G}$*, together with their associated ranges, is denoted by* $\mathsf{ApVars}(\mathbf{V}) = \{\mathsf{Ap}(g) \mid g \in \mathcal{G}(\mathbf{V})\}$*.*

An interpretation for $\mathbf{V}$ *is a mapping* $\mathbf{I} : \mathsf{Models}(\mathbf{V}) \to \mathbf{C}$*, where* $\mathbf{C}$ *is a set of semantic constructs.*

For example, consider the following excerpt from the BPMN 2.0 OMG standard [13] introducing the concept of web task. The graphical symbols discussed here are a rectangle with rounded corners and a rectangle with rounded corners that has a marker in its left corner. These two symbols are mapped to the semantic constructs Task and Web Task respectively (Fig. 2):

Service Task

A **Service Task** is a **Task** that uses some sort of service, which could be a Web service or an automated application.

A **Service Task** object shares the same shape as the **Task**, which is a rectangle that has rounded corners. However, there is a graphical marker in the upper left corner of the shape that indicates that the **Task** is a **Service Task** (see Figure 10.11).

A **Service Task** is a rounded corner rectangle that MUST be drawn with a single thin line and includes a marker that distinguishes the shape from other **Task** types (as shown in Figure 10.11).

Figure 10.11 - A Service Task Object

Fig. 2. Excerpt from the BPMN 2.0 OMG standard [13, p. 158]

The above set-theoretical terms will be useful in the sequel to make the meaning of PoN more precise and make a clearer distinction between the principles. In particular, we will distinguish between the following levels of notation:

- Level 1: principles considering only symbols from $\mathcal{G}(\mathbf{V})$.
- Level 2: principles considering symbols from $\mathcal{G}(\mathbf{V})$ together with the mapping $\mathbf{I}$ to semantic constructs.
- Level 3: principles considering elements from $\mathsf{Models}(\mathbf{V})$ as a whole (which consist of symbols from $\mathcal{G}(\mathbf{V})$, as well as from $\mathcal{L}(\mathbf{V})$).

3.2   Operationalization Analysis

In what follows we analyze each of the PoN principles in terms of their operationalization. In other words, given a visual notation, we ask what it takes to check whether a certain principle applies to it. In addition to the levels of notation specified above, we consider also an additional dimension: the extra information (e.g., particular thresholds, measures, definitions or evaluation) that is needed for operationalization.

Semiotic clarity: requires a visual notation to have a 1:1 correspondence between semantic constructs and graphical symbols. This principle implies that when there is a graphical symbol in the notation (e.g., a stickman), it is used for representing solely one meaningful semantic construct or thing from the universe of discourse (e.g., a person). The PoN provides a number of exact instructions to ensure this, based on ontological literature. Concretely, the following situations should be avoided: òne construct represented by multiple graphical symbols, multiple constructs represented by the same graphical symbol, graphical symbols that do not correspond to any construct, and constructs that do not have any graphical symbols. While ontological theory has been used to ground the instructions given for this principle, the given simple rules require no acquaintance with other theoretical frameworks. An example of a notation that does not satisfy the criteria is i*, which has 27 semantically distinct relationships, but only five graphically distinct graphical symbols for relationships [10].

Set-theoretical Formulation: Let V be a visual notation and I an interpretation for V. We say that V enjoys semiotic clarity if the restriction of I to $\mathcal{G}(V)$ is 1:1.

Classification: The operationalization of this principle requires both $\mathcal{G}(V)$ and the semantic mapping I (level 2 of notation). Once the sets of graphical symbols, semantic constructs and the mapping between them are established, checking whether the mapping is 1:1 does not require any extra information. The main challenge here remains the required explicit specification of all needed constructs.

Dual coding: requires a visual notation to use text to complement graphics. For example, using commonly understood and agreed upon words to complement graphical symbols to further ensure they are interpreted unambiguously. The PoN suggests using both annotations (i.e., including textual explanations in analog to comments in source code) and hybrid symbols (i.e., textual reinforcement of visual symbol meaning). Further requirements placed upon such text are not fully clarified. For example, it is not clear whether the use of free form natural language is preferred over, e.g., a controlled or structured natural language (e.g., SBVR[1]), or whether there should be limits to the length of text (i.e., concrete string limits). Many modeling languages satisfy the core criteria of dual coding by letting users place textual annotations. ORM 2.0 [6] could be a good example of potential further operationalization, providing its textual

---

[1] http://www.omg.org/spec/SBVR/.

annotation of ternary fact types being written in a way that follows the structure and layout of the related visual elements.

Set-theoretical Formulation: Let $\mathbf{V}$ be a visual notation. We say that $\mathbf{V}$ enjoys dual coding if there are models in $\mathsf{Models}(\mathbf{V})$ which include elements from both $\mathcal{G}(\mathbf{V})$ and $\mathcal{L}(\mathbf{V})$.

Classification: This principle involves $\mathsf{Models}(\mathbf{V})$ (level 3 of notation). Interpreting the question of dual coding in the Boolean sense, it requires no extra information. However, it seems that the intended meaning here is more than just Boolean (a yes/no question); additional external information could give more valuable insights into further constraints placed on the text like e.g., cognitive limits on the amount of text that is efficiently parsed. The vague formulation of this principle leaves room for a variety of interpretations, and the extent to which text should be combined with symbols should be clarified before operationalization can be made possible.

Graphic economy: requires the visual notation to make economical use of graphical symbols. The size of the notation's visual vocabulary should not exceed the cognitive limit of how many distinct visual symbols can be effectively recognized. The PoN references existing and widely known work, re-iterating that people can discriminate between around six different visual graphical symbols, and therefore proposes to not exceed this number. Regardless of *how* this is achieved, for which the PoN gives a number of different strategies and instructions, operationalizing this criteria and verifying whether it holds is simple, requiring only the visual notation itself to check how many distinct graphical symbols it has. An example of a visual notation that likely satisfies most operationalizations would be petri nets and ER-diagrams, both consisting of very few visually distinct elements. Petri net models indeed appear out of only three elements (four, if one includes tokens): places, transitions and arcs. Of course, the more specialized a visual notation becomes, the harder it typically is to keep the total number of graphical symbols down; for example, the total number of graphical symbols in BPMN has grown to be over 50 [11].

Set-theoretical Formulation: Let $\mathbf{V}$ be a visual notation. We say that $\mathbf{V}$ is graphically economic with respect to a threshold n if $|\mathcal{G}(\mathbf{V})| < n$.

Classification: This principle involves only $\mathcal{G}(\mathbf{V})$ (level 1 of notation), and given the threshold n requires no extra information for operationalization.

Complexity Management: This is similar to graphic economy, except that the formulation here is on a diagram (or model) level. Visual complexity of entire diagrams often becomes high due to a large number of elements in a diagram. The PoN grounds itself in literature showing that the number of diagram elements that a person can comprehend at a time is limited by working-memory capacity, and should this limit be crossed, the degree of comprehension decreases significantly. To be cognitively effective, a visual notation should thus avoid such situations from occurring. While the PoN clearly states that complexity management is about preventing a particular threshold of comprehension being crossed, it does not offer values for such a threshold.

Set-theoretical Formulation: Given such threshold n and a way to establish the size of a model in $\mathbf{V}$, this principle can be taken to mean that for every $m \in$ Models($\mathbf{V}$), $|m| < n$.

Classification: This principle involves Models($\mathbf{V}$) (level 3 of notation). If the question of complexity management is understood in a Boolean way, no extra information is required. However, it seems that checking a Boolean assertion that this threshold is never crossed is not useful, and one needs to check that the notation offers good enough mechanisms to ensure it can be dealt with, such as having semantic constructs for subsystems, decomposable constructs, and relevant syntactical diagrammatic conventions for decomposing diagrams. Thus also in this case the abstract formulation of the principle leaves room for many interpretations and should be further clarified. Therefore, the extra information required here is what is what exactly is understood by "complexity management mechanisms".

Cognitive integration: requires a visual notation to incorporate explicit mechanisms to support the integration of information from different diagrams. For example, in ArchiMate where an enterprise is described by the three layers of business, application, and technology, models can exist for each separate layer, but the information therein has to be able to be directly related to each other. Short of its extensive description of potential implementations, the concrete features that the PoN argues a visual notation needs to have are: "Mechanisms to help the reader assemble information from separate diagrams into a coherent mental representation of the system", and "Perceptual cues to simplify navigation and transitions between diagrams." However, the problem is that while ostensibly only the visual notation is needed in order to check whether such mechanisms exist, the PoN describes what can be done to implement these requirements in a visual notation only as suggestions, not as hard requirements. For example, to implement contextualization, the PoN reasons that one can "include all directly related elements from other diagrams (its "immediate neighborhood") as foreign elements."

Set-theoretical Formulation: this principle can be taken to mean that R has integration mechanisms.

Classification: This principle is formulated in terms of Models($\mathbf{V}$) (level 3 of notation). As in the previous principle, although a Boolean condition could be formulated here, it seems to be not useful enough, and the vague formulation of the principle should be further elaborated, providing as extra information a working definition of "integration mechanisms".

Perceptual discriminability: requires a visual notation to have clearly distinguishable symbols. This means that the main visual elements used are not strongly similar, or difficult of being discriminated. The PoN operationalizes this as having to investigate the visual distance between symbols, basing it on existing discriminability thresholds. The primary suggestions given are to use the shape of symbols as their primary discriminant, to introduce redundant coding in the sense of employing multiple visual variables to distinguish between graphical symbols (e.g., shape and color), ensuring a perceptual pop out by having each visual element have at least one unique visual variable (e.g., a particular concept is always, and uniquely visualized as a square), as well as using textual differentiation. In order to verify this principle, the visual notation and its specification are needed, complemented with suitable additional information grounding the choice for discriminability thresholds.

Set-theoretical Reformulation: Let $Disc$ be a discriminability relation on $\mathcal{G}(\mathbf{V})$. We say that a visual notation $\mathbf{V}$ enjoys perceptual discriminability if for every $g_1, g_2 \in \mathcal{G}$, $Disc(g_1, g_2)$ holds.

Classification: This principle uses only $\mathcal{G}(\mathbf{V})$ (level 1 of notation). The extra information required here is the measure $Disc$. As discriminability thresholds are published and referenced explicitly by the PoN, defining such measures in a natural way seems feasible. Complications here might stem from a need to validate that the used additional information accounts for potentially expected complications in discriminability thresholds, such as for instance colorblind users of a modeling language who cannot distinguish between some used colors, thereby potentially reducing the overall discriminability (e.g., if red and green are used to distinguish elements, for a colorblind user the discriminability would not be achieved).

Visual expressiveness: concerns the number of visual variables used in the notation, such as color, shape and texture. The PoN recommends that notation designers: use color (though only for redundant coding); ensure that form follows content, meaning that the choice of visual variables should not be arbitrary but rather match the properties of the visual variables to the properties of the information to be represented. This is operationalized in more detail by explaining that (1) the power of the visual variable (nominal, ordinal, interval) should be greater than or equal to the measurement level of the information; and, (2) the capacity defined as the number of perceptible steps ranging from two to infinity should be greater than or equal to the number of values required.

Set-theoretical Reformulation: Let $WellUsed$ be an expressiveness predicate defined on the set $\mathsf{ApVar}(\mathbf{V})$. We say that a visual notation $\mathbf{V}$ enjoys visual expressiveness if for every $v \in \mathsf{ApVar}(\mathbf{V})$, $WellUsed(v)$ holds.

Classification: This principle uses only $\mathcal{G}(\mathbf{V})$ and their visual variables (level 1 of notation). The extra information required for operationalization of this principle is the availability of the expressiveness $WellUsed$ predicate. This is not trivial, as the PoN provides many examples for the range of visual expressiveness, including what elements contribute and detract (e.g., use of color, positioning, size, brightness), but does not detail hard values for minimum or maximum thresholds. The PoN provides data on the total capacity of different visual variables in terms of how distinctive they are for human observers (e.g., orientation yielding four distinct variables), but does not explicitly say to what degree to use it. Thus, determining the parametric values for the expressiveness predicate, which itself is to be built on measuring the different visual variables requires interpretation of relevant literature to determine suitable values.

Semantic transparency: deals with ensuring that visual representations suggest their meaning via their appearance. The PoN describes it as a continuum of meaning, arguing that it "formalizes informal notions of "naturalness" or "intuitiveness" that are often used when discussing visual notations" [9].

Set-theoretical Reformulation: We say that a visual notation $\mathbf{V}$ enjoys semantic transparency if for every $g \in \mathcal{G}(\mathbf{V})$, $\mathbf{I}(g)$ is "suggested".

Classification: This principle uses both $\mathcal{G}(\mathbf{V})$ and $\mathbf{I}$ (level 2 of notation). The crucial extra information we need here is a more precise characterization of what it means for a semantic meaning to be "suggested" by a graphical symbol. This of course cannot be determined *a priori* and needs empirical evaluation. The PoN describes a range of how suggestive visual symbols can be characterized, from fully transparent (i.e., conveying its intended meaning) to perverse (i.e., conveying a different, incorrect meaning). Empirical work directly involving the user is needed to determine how well a particular symbol suggests its intended meaning.

However, instead of providing a formal notion, the PoN suggests avoiding situations where novice readers would likely infer a different meaning from appearance, and further advocates the use of icons as symbols that perceptually resemble the concepts they represent. This principle seemingly can only be performed by directly involving users. Furthermore, cultural and temporal ("zeitgeist") dependency of such suggested meaning would make it more challenging to generalize findings from users. While some icons and symbols might have meaning for a group of people, few of them are universal. Furthermore, the meaning of icons or symbolism changes over time, making operationalizations also temporally bound. A practical example of how suggested meaning is clearly culture bound can be found in an application of the PoN to i* [10], a goal modeling notation. In this notation, it is proposed to distinguish different kinds of acting entities, where agents are proposed to be depicted with "black sunglasses and a pistol", arguing that users would make "an association of the 007 kind." This presupposes a shared cultural knowledge between the designer and user of the notation that needs empirical grounding.

Cognitive fit: concerns personalizing the visual notation to the target audience and ensuring that it "fits" with the cognitive background and skills of different users and tasks it is used for. For example, when people with different backgrounds and skill sets use the notation, it is important they can all use it at a minimum level of proficiency. The PoN recommends focusing on taking into account at least (1) expert-novice differences, and (2) the representational medium. While particular instructions are given for how to optimize a notation for either expert or novice, the principle itself centers on ensuring that the visual notation does not exhibit visual monolinguism. In a way, only the visual notation is needed to verify whether this principle holds: one can check whether different dialects for particular users or tasks exist. However, the core difficulty of the principle is that for a given notation these differences need to be identified first. Thus, users have to be directly involved, leading to the same challenges described for other principles, such as semantic transparency, requiring direct user involvement. For example, say that the visual notation of some process modeling language uses realistic pictograms in order to clearly visualize what things are needed for a particular task. Specifically, a realistic pictogram of a wrench is used for a task of 'screwing down bolts'. If this notation has the requirement that it can be drawn on paper, how do we actually verify whether needing to draw a wrench is difficult or not? Without knowing the users, one cannot postulate their artistic skill, or their inclination to spend time drawing realistic depictions. Regardless of whether it was intended, BPMN is an example of a language, which, in practical use seems to satisfy what cognitive fit aims to achieve. It has been viewed as consisting of a number of 'sets' of functionality, a common core, extended core, specialist set, overhead in use by people of varying levels of expertise and focus. [11]

Set-theoretical Reformulation: this principle seems to us to be the most vague of all, and no set-theoretical reformulation in the terms defined in this paper can be suggested.

Classification: this principle uses Models(**V**) (level 3 of notation). The starting point for extra information required here is providing more concise characterization of the elements involved in the formulation of this principle.

4   Summary and Identified Concerns

The above discussion provides a number of new insights into the inherent difficulty of operationalizing PoN principles. First of all, two dimensions emerge from our analysis, which may provide indications on the feasibility of operationalization of the principles. The first is the distinction between the different layers of visual notation addressed by each principle. Some principles are targeted at the level of an individual symbol and its structure, others at the interplay of the symbols with their semantic constructs, and some target the interplay of many symbols (i.e., a model). These different levels as referenced in Sect. 3.1 increase the challenge of clearly operationalizing, as the increase in elements that have to be considered make clear and precise verification more challenging.

The second is the distinction between the different types of extra information needed for operationalization of the principle. Sometimes additional information is needed that is both simple to gather and interpret, such as widely published accounts of how many distinct graphical symbols the human mind can perceive at a time. However, when more information has to be distilled from more complicated literature (e.g., scientific theory), an additional challenge arises of ensuring the correct selection and interpretation of that information. Finally, when information specific to users is needed (e.g., to determine what meaning is 'suggested' by a symbol), a whole new challenge appears with the need to design empirical work, argue for the validity of elicited information, and reason how it either generalizes or applies to the intended users of the visual notation.

Table 2 provides an overview of our findings. For each principle it presents the notation level, a set-theoretical formulation of the principle, and the extra information that is needed to achieve operationalization.

Table 2. Summary of PoN operationalization analysis

Principle	Set-theoretical Desc.	Elements used	Extra info required
SemCl	$\mathbf{I}|\mathcal{G}$ is 1-1	$\mathcal{G}(\mathbf{V}) + \mathbf{I}$ (level 2)	-
PerDisc	$\forall g_1, g_2 \in \mathcal{G}(\mathbf{V}) : Disc(g_1, g_2)$	$\mathcal{G}(\mathbf{V})$ (level 1)	**measure** $Disc$ **on** $\mathcal{G}(\mathbf{V})$
SemTr	$\forall g \in \mathcal{G}(\mathbf{V})$: g **"suggests"** $M(g)$	$\mathcal{G}(\mathbf{V}) + \mathbf{I}$ (level 2)	**evaluation of "suggestiveness"**
CmpMng	R has **"compl. management"**	Models(**V**)(level 3)	**defn. of "compl. management"**
CogInt	R allows **"integration"**	Models(**V**)(level 3)	**defn. of "integration"**
VisExp	$\forall v \in$ ApVar(**V**), WellUsed(v)	$\mathcal{G}(\mathbf{V})$ (level 1)	**measure** WellUsed **on** ApVar(**V**)
DualC	Some $m \in$ Models(**V**) combine symbols & text	Models(**V**)(level 3)	-
GrE	$\mathcal{G}(\mathbf{V}) < n$	$\mathcal{G}(\mathbf{V})$	**threshold** n
CogFit	?	Models(**V**)(level 3)	**evaluation** of "cog. fit"

To the extent of our knowledge, dedicated operationalization efforts so far address only two principles out of nine, focusing on semiotic clarity and perceptual discriminability [16]. These two principles are arguably among the best candidates for operationalization as they provide clear, quantitative judgement criteria, and involve the lowest degree of subjective interpretation[2]. Indeed, our classification of the principles supports this view. Another good candidate for formalization, according to our classification, seems to be visual expressiveness. The most challenging principle, according to Table 2, seems to be cognitive fit. The most vague principles, requiring a reformulation in precise terms, are complexity management and cognitive integration.

---

[2] Nonetheless, existing work [16] seems to take debatable choices, such as seemingly arbitrary weights for distinguishing visual distance variables, whose objective nature can also be discussed.

460 D. van der Linden et al.

Below we summarize a number of further concerns that should be addressed in the context of PoN operationalization:

Vague Satisfaction Criteria. A significant problem in operationalization of the PoN is the vague satisfaction criteria of many principles. While it is clearly stated what a principle should do, or achieve, the exact details on how to achieve that are left up to the theory's wielder. For example, for cognitive integration we can check a Boolean assertion that structures exist to support e.g., modularization or clustering. However, this says little about how successfully such structures will be used, as their design in itself is also subject to cognitive factors. Thus a degree-based approach is more appropriate here.

Relative Impact of Satisfying a Principle Is Unclear. Given that some principles are defined in such a way that their satisfaction is almost trivial (e.g., dual coding not saying anything about the *kind* or *structure* of complementary text), how much each individual principle contributes to the overall cognitive effectiveness of a visual notation is unclear. This also makes it harder to know what principles to focus, or spend most time on should they prove challenging for a particular notation.

Operationalization Interrelations. An additional complication arises from the relationships that exist between the different principles. Given that multiple principles have been documented to have positive or negative influence on each other (for example, increasing graphic economy can decrease semiotic clarity), operationalization of one principle may involve having to operationalize multiple principles concurrently. For example, when considering semiotic clarity, one should also take into account graphic economy, which requires taking visual expressiveness into account, which in its turn requires additional external information. Gaining a better understanding of the interrelations between the principles is thus crucial for their operationalization.

5 Concluding Outlook

This paper presented a preliminary analysis of PoN principles with respect to difficulties of their operationalization. The main contribution of this work is establishing a formal ground for distinction of different aspects that pose difficulties for operationalization of PoN principles. Using this distinction, different types of efforts can be directed at different principles, e.g., reducing vagueness of formulations, providing concrete mathematical metrics and/or methods for empirical evaluation.

Our most immediate direction for future research is using empirical methods to establish the relative importance of each principle for users of particular modeling domains (e.g., software architecture, business processes). Such empirically grounded data can be used to more clearly operationalize domain-specific 'instantiations' of the PoN, and also show where principles that are mathematical in their nature, but afford for more complex evaluation given the involvement of additional elements, can and should be raised to a higher level of evaluation.

References

1. Giraldo, F.D., España, S., Pineda, M.A., Giraldo, W.J., Pastor, O.: Conciliating model-driven engineering with technical debt using a quality framework. In: Nurcan, S., Pimenidis, E. (eds.) CAiSE Forum 2014. LNBIP, vol. 204, pp. 199–214. Springer, Switzerland (2015)
2. Granada, D., Vara, J.M., Brambilla, M., Bollati, V., Marcos, E.: Analysing the cognitive effectiveness of the webml visual notation. Softw. Syst. Model., 1–33 (2013)
3. Green, T.R.G., Petre, M.: Usability analysis of visual programming environments: a cognitive dimensions framework. J. Vis. Lang. Comput. **7**(2), 131–174 (1996)
4. Gregor, S.: The nature of theory in information systems. MIS Q. **30**, 611–642 (2006)
5. Gulden, J., Reijers, H.A.: Toward advanced visualization techniques for conceptual modeling. In: Proceedings of the CAiSE Forum 2015 Stockholm, Sweden, 8–12 June, 2015
6. Halpin, T.: ORM 2. In: Meersman, R., Tari, Z. (eds.) OTM-WS 2005. LNCS, vol. 3762, pp. 676–687. Springer, Heidelberg (2005)
7. Krogstie, J., Sindre, G., Jørgensen, H.: Process models representing knowledge for action: a revised quality framework. Eur. J. Inf. Syst. **15**(1), 91–102 (2006)
8. Moody, D., van Hillegersberg, J.: Evaluating the visual syntax of UML: an analysis of the cognitive effectiveness of the UML family of diagrams. In: Gašević, D., Lämmel, R., Wyk, E. (eds.) SLE 2008. LNCS, vol. 5452, pp. 16–34. Springer, Heidelberg (2009)
9. Moody, D.L.: The physics of notations: toward a scientific basis for constructing visual notations in software engineering. IEEE Trans. Softw. Eng. **35**(6), 756–779 (2009)
10. Moody, D.L., Heymans, P., Matulevičius, R.: Visual syntax does matter: improving the cognitive effectiveness of the i* visual notation. Requirements Eng. **15**(2), 141–175 (2010)
11. zur Muehlen, M., Recker, J.: How much bpmn do you need (2008). http://www.bpm-research.com/2008/03/03/how-much-bpmn-do-you-need
12. Narayanan, N.H., Hübscher, R.: Visual language theory: towards a human-computer interaction perspective. In: Marriott, K., Meyer, B. (eds.) Visual Language Theory, pp. 87–128. Springer, New York (1998)
13. (OMG), O.M.G.: Business process model and notation (BPMN) version 2.0. Technical report, January 2011. http://taval.de/publications/BPMN20
14. Plataniotis, G., de Kinderen, S., Proper, H.A.: Ea anamnesis: an approach for decision making analysis in enterprise architecture. Int. J. Inf. Syst. Model. Des. (IJISMD) **5**(3), 75–95 (2014). http://dx.doi.org/10.4018/ijismd.2014070104
15. Schuette, R., Rotthowe, T.: The guidelines of modeling - an approach to enhance the quality in information models. In: Ling, T.-W., Ram, S., Lee, M. (eds.) ER 1998. LNCS, vol. 1507, pp. 240–254. Springer, Heidelberg (1998)
16. Störrle, H., Fish, A.: Towards an operationalization of the "Physics of Notations" for the analysis of visual languages. In: Moreira, A., Schätz, B., Gray, J., Vallecillo, A., Clarke, P. (eds.) MODELS 2013. LNCS, vol. 8107, pp. 104–120. Springer, Heidelberg (2013)
17. van der Linden, D., Hadar, I.: Cognitive effectiveness of conceptual modeling languages: examining professional modelers. In: Proceedings of the 5th IEEE International Workshop on Empirical Requirements Engineering (EmpiRE). IEEE (2015)

18. van der Linden, D., Hadar, I.: Evaluating the evaluators - an analysis of cognitive effectiveness improvement efforts for visual notations. In: Proceedings of the 11th International Conference on Evaluation of Novel Approaches to Software Engineering. INSTICC (2016)

19. van der Linden, D., Hadar, I.: User involvement in applications of the PoN. In: Proceedings of the 4th International Workshop on Cognitive Aspects of Information Systems Engineering (COGNISE). Springer (2016)

20. Van Zee, M., Plataniotis, G., van der Linden, D., Marosin, D.: Formalizing enterprise architecture decision models using integrity constraints. In: 2014 IEEE 16th Conference on Business Informatics (CBI), vol. 1, pp. 143–150. IEEE (2014)

Modeling of Architecture and Design

Modeling and Analysis of Vehicles Platoon Safety in a Dynamic Environment Based on GSPN

Mohamed Garoui[✉]

IRTES-SET, UTBM, 90010 Belfort Cedex, France
garouimohamed2010@gmail.com

Abstract. Throughout the transport system mission, the system and its environment can be in an unsafe state that are can be caused by a set of disturbing events which can be internal or external. To avoid this unsafe state, it is necessary to studying the vehicles platoon safe in order to identify the factors that have an important impact on platoon safe. Many researches have been conducted to study the safety of the smart vehicles platoon in their environment. In this paper, our aim is to propose a formal model based on Generalized Stochastic Petri Nets models for modeling and analyzing the impact of same factors on the system safety. Within this context, same results are obtained by evaluate the vehicles platoon safety according to the factors: the number of faulty vehicles per platoon, the failure rate in the system and appearance frequency of external perturbation.

1 Introduction

The automotive field was subject for much research in recent years to develop new solutions and to deal with the increasing traffic, improve the road safety and also to address new environmental and societal needs. Special interest has focus on the use the technologies of embedded computing and communication to meet those needs. In this context, much work has been focused on the modeling the vehicle platoon system and analyze their dependability and performances without considering its environment.

Platoon concept is grouping vehicles for fulfill a transportation mission. It is a method of increasing the capacity of roads. A vehicle platoon might be one of the technological benefits of self-driving (autonomous vehicles), but it does not come without its problems.

In this article, we address the problem of the safety analysis for vehicle platoon. therefore, our aim is to help the specialist to develop and install comprehensible and valid safe system (platoon) for a given application in a given environment. Our work focuses on developing formal models to assess the platoon safety implemented in a context of transportation system in a urban environment. Our aims is to validate the platoon behavior in order to ensure the system safety and to estimate the impact some related factors such as the impact of

© Springer International Publishing Switzerland 2016
R. Schmidt et al. (Eds.): BPMDS/EMMSAD 2016, LNBIP 248, pp. 465–478, 2016.
DOI: 10.1007/978-3-319-39429-9_29

faulty vehicle par platoon on platoon safety. We can define the system safety by *the system complete its mission without any disturbance which causes dangerous state to the system and its environment*. The safety characterizes the confidence that can support the absence of system failures that can cause catastrophic consequences for the environment, for example the loss of human lives. Several events such as the occurrence of accidental faults, and frequent loss of communications between the system entities are taken into account. We evaluate the measures quantifying the risk of accidents in a driving context, to estimate the impact of new technologies on system and environment safety. We regard as a case study the vehicles platoon architectures developed in the framework of a project of intelligent transport systems, ANR VTT SafePlatoon[1].

The proposed architectures are based on the implementation of automatic maneuvers to ensure the safety of vehicles platoon in the presence of perturbation events from the environment. The maneuvers are also planned to ensure the smooth functioning of the system following the occurrence of failures affecting the vehicles, their environment or inter-vehicle communication.

In this article, our contribution is to propose and illustrate a safety modeling and analysis approach based on Generalized Stochastic Petri Nets (GSPN) to evaluate the probability of failure maneuvers implemented to ensure the safety of a vehicle platoon. To build our GSPN model, we start by identify different failure modes, the associated maneuvers and then identify the catastrophic situation that can affect one vehicle or multiple vehicle of the same platoon.

Our paper is structured as follow: Sect. 2 gives same related works to our research and Sect. 3 gives an overview about the formal method, Generalized Stochastic Petri Nets, for modeling and quantitative analysis. Section 4 define the vehicle platoon in a dynamic environment. In detail, we have shown the failure modes that can disturb the mission of urban vehicle platoon and affect the system safe. In Sect. 5, we show our GSPN model for evaluating the vehicle platoon safety, based on the proposed failure modes and their recovery maneuvers. Section 6 summarizes the results obtained and discusses their impact on the safety of platooning system. Finally, Sect. 7 concludes the paper and presents the envisaged extensions.

2   Related Works

Among current works in safety assessment, in [1], *de Albuquerque et al.* suggest an approach for modeling and evaluating supply chains based on Generalized Stochastic Petri Net (GSPN) [2] components. The suggested modeling process assures desirable model properties that start from a set of predefined modules for typical supply chain entities. In [3], *Ossama Hamouda et al.* report to safety modeling and evaluation of automated highway systems, based on the use of platoons of vehicles driven by automated agents. He analyzes the effect on safety of the strategy used to manage the operations of the vehicles, inside each platoon and between platoons, when trucks go in or leaving the platoon, or when

---

[1] http://web.utbm.fr/safeplatoon/.

maneuvers are carried out to recover from failures affecting the vehicles or their communication.

In [4], Nawel Gharbi et al. suggest an method for modeling and analyzing finite-source retrial systems with several customers classes and servers classes with the Colored Generalized Stochastic Petri Nets (CGSPNs) [5]. This high-level mathematical model is appropriate for describing and analyzing the performance of systems presenting concurrency and synchronization, maybe with diverse components. In [6], the writers intend a method to assess the Emergency Response System (ETRS) performances based on SPN and Markov Chain (MC) to model and analyze the main procedure of ETRS. Moreover, through isomorphic MC of the SPN model, the performance of ETRS was studied. Other works in [7], presents a Stochastic Petri Net-based approach for performance assessment of ad hoc networks. The method abuses the symmetry in ad hoc networks, and precisely models the semantics of activities of a node. The arithmetical results of the Stochastic Petri Nets (SPN) model show a good match to simulation results, especially under full traffic. Since of its easiness, the suggested approach has great advantage to the design and implementation of ad hoc networks.

In [8], the authors propose an extension for Architecture Analysis and Design Language (AADL) formalism and build so-called component-based modeling approach to system-software co-engineering of real-time embedded systems as aerospace system. The aims are then subject to different kinds of formal analysis such as model checking, safety and dependability analysis and performance evaluation.

In [9], El Zaher Madeleine propose a modeling and verification approach. It presents a compositional verification method adapted to a wide range of Reactive Multi-Agent Systems (RMAS) applications. This method is appropriate for the verification of safety properties. The application considered in this paper is a platoon of vehicles with linear configuration. The safety property to be verified is the non collision between platoon vehicles. In her safety verification process, he use SAL tool-kit [10] by applying Symbolic Analysis Laboratory (SAL) model checkers. The verification method bases on a compositional verification rule.

3   Generalized Stochastic Petri Nets for Modeling and Quantitative Analysis

The Petri Nets formalism [11], in their various forms, have been used for studying the qualitative and quantitative properties of systems displaying concurrency and synchronization characteristics. The use of PN-based model for the quantitative study of systems necessitates the outline of temporal specifications in the basic, un-timed models. Generalized Stochastic Petri Net is an extension of PN. Formally, a GSPN model is define as $GSPN = (P, T, Pre, Post, R, M_0)$, with

- $P = \{p_1, p_2, \ldots, p_n\}$ a set of places,
- $T_t = \{t_1, t_2, \ldots, t_m\}$ a set of timed transition,
- $T_i = \{t_{i1}, t_{i2}, \ldots, t_{im}\}$ a set of immediate transition,

- $T = T_t \cup T_i$ a set of transition,
- $Pre = P \times T \to \mathbb{N}$, a incidence before function with $Pre(p,t)$ contains the integer value n associated to the weight of arc from p to t.
- $Post = P \times T \to \mathbb{N}$, a back incidence function with $Post(p,t)$ contains the integer value n associated to the weight of arc from t to p.
- $R: T \to \mathbb{R}^+$, {firing rate expressions}, associated to timed transitions
- M_0: set of initial marking ($\{m_{01}, m_{02}, \ldots, m_{0n}\}$)

GSPN, as showing in Fig. 1, is formed by a set of places represented by circles, a set of timed transitions represented by white rectangles, a set of immediate transitions represented by black rectangles and a set of direct arcs. The directed arcs connecting places to transitions and transitions to places. Places may contain tokens are represented by small black circles.

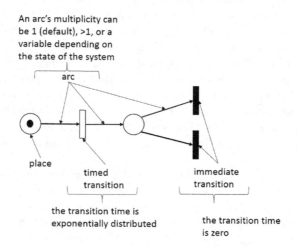

Fig. 1. Generalized Stochastic Petri Net models

The necessary step in the dependability evaluating of a tolerant system is the construction of one (or several) model(s) for a system. A model is a representation based on a mathematical construct for the study. The behavior of a system is characterized by the execution state of delivers service (operational states of the model) and the interruption state (temporary or permanent) of this service (the failed states of the model), as well as the transitions between these states.

Moreover, the GSPN models is also involved to Dependability analysis. Since dependability studies are a essential prerequisite for the design of a secure desired system and provides help in the decision on identifying risks and verifying proper operation achieving safety goals. Among the dependability attributes [12] which we are interested, the *Safety S(t)* witch defined by *"the ability of a system to avoid the appearance of critical events in given conditions"*. A system is in safe state if it does not entail no adverse consequences in the presence of faults. Formally, the safety is *S(t) = Prob(E has no catastrophic failure between 0 and t)*, Where E is an entity (component, system, function).

4   Vehicles Platoon System in a Dynamic Environment

Our case study is a vehicle platoon which as a set of vehicle that have shared its one mission. Our vehicle platoon adopts a column configuration during its movement with the lateral inter-distance (i.e. between vehicle) is null and the longitudinal inter-distance is varied in the interval $[0.5m, 3m]$. The vehicle platoon will have a fixed number of vehicles. During the case study, we are interested to verify and analysis the safety system in her dynamic environment as the urban environment. This environment can also be seen as a dynamic. A set of external and/or internal events may intervene to disrupts the platoon movement. It is made up of several mobile entities and road infrastructures. Mobile entities may be a walker, vehicle, animal, etc. Road infrastructures are several in traffic domain, we cite such STOP sign, traffic lights, etc.

The internal events affect the modules of the embedded system in one vehicle. Those modules are *Perception module* which is responsible for the perception of traffic lane. the *Communication module* able to control the continuity of communication and exchange of information between vehicle platoon. The *Command module* control the vehicle command quality and the *module Functional* (or *Dynamic*) which monitors the dynamics of the vehicle components. *Decision module* is used to help the embedded system to decide and select the action to be launched.

Failure Modes and Repair Maneuvers. During the vehicle platoon movement, there are many failures modes that can affect the vehicles in the same vehicle platoon. Depending on the failures modes, different maneuvers can be considered to ensure the platoon safety into its environment. Some maneuvers need to stop the faulty vehicle or to help him to leave the platoon, as soon as possible. In the cases where the failures have a minor effect on the platoon safety, the exit of faulty vehicle from the platoon can be considered without any assistance of other vehicles.

By relying on the studies of the platoon scenarios presented in [13], Table 1 provides the failure modes and these associated maneuvers to ensure the service continuity in safe conditions despite the presence of these failure modes. The maneuvers considered for fault tolerance are respectively five potential failure modes have been identified, presented in Table 1. This table shows for each failure mode, some examples of causes witch lead to the failure mode, priority class, and the maneuver that ensures the safe continuity of service despite the presence of failures. When simultaneous failures affect one vehicle in the platoon, the maneuver with the highest priority was applied. The success of an operation depends on many factors, for example, the state of faulty vehicles in the platoon. The successive failure modes can eventually lead to a state where no maneuvers are available to recover the faulty situation.

For example, assume that the vehicle v1 is faulty and must make immediate Disjoint maneuver. If another vehicle is already done this maneuver with a higher priority, the operation requested by v1 will be denied. Thus, v1 ask another maneuver a higher priority until the requested operation is accepted. Similarly, when a maneuver failed, the system evolves towards a degraded failure mode and its related maneuvers should be attempted to put the system in a safe state. The failure of successive maneuvers can ultimately lead to a catastrophic state. The catastrophic situations leading the system to a dangerous situation of the occurrence of simultaneous failures affecting several vehicles of the same platoon.

We consider a platoon composed of autonomous vehicles. Each has a moved capabilities, perceives its environment and interact with his surrounding. Disturbances may occur during the movement of a vehicle platoon. These can be classified into two categories: temporary or permanent. In our case, we consider a number of disturbance that has a direct impact on the security of the vehicle platoon. We model our system taking into account the failure modes, related maneuvers and catastrophic situations presented in both Table 1.

Table 1. The failure modes and associated maneuvers

Failure mode	Priority	Cause	Maneuver
FM_Perception	1	Wrong information: Wrong measure of distance obstacles	Immediate Stop (IS)
FM_Com	2	Loss of communication flow between the vehicles	Restart Network Card (RNIC)
FM_Cmd	3	Dysfunction of electronic components	Immediate Disjoint (DI_Cmd)
FM_F	4	Low battery	Immediate Disjoint (DI_F)
FM_Decision	4	Slow decision	Repeat Decision (RD)

5   The GSPN Model for Vehicles Platoon System

Here, we regard a GSPN model for modeling and analyzing the faulty behavior for vehicle platoon in its dynamic environment. During the vehicle platoon movement, there are events that disrupt the platoon mission. Those events are classified into two categories: external events (external perturbation) and internal events (internal perturbation) such as detailed in the previous section. For this, the generic GSPN model (Fig. 2) for vehicle platoon system consists of three main blocks: the first named *Normal Mvt* which models the normal movement of the system who finished its mission safety. The second, *External perturbation*, specify the faulty movement and its treatment with the associated maneuvers a of the system in the presence of an external events which disturbs the platoon mission and led to unsafe state (accident). The third block, *Internal perturbation*, models the faulty movement and its treatment with the associated maneuvers

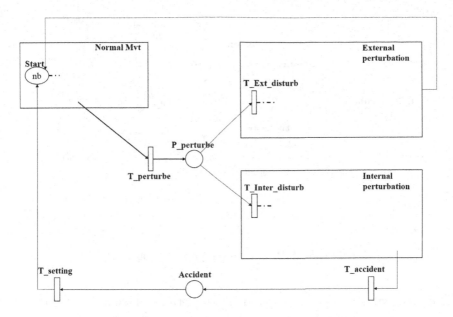

Fig. 2. Global architecture of GSPN model

in the presence of an internal perturbation related to the embedded system of the vehicle and the platoon capabilities.

In the GSPN model for the platoon behavior, Fig. 2, the place *Start* initializes the regular movement and the mission of the vehicle platoon. The number of vehicles in the platoon is set by a global variable that indicate the number of marking in the place *Start*. In our model, the vehicles number is declared as a variable (nb) which is used later in the evaluation experiments. The GSPN model concerns the normal movement (see Fig. 3) depict the normal platoon movement without any perturbations. Each vehicle starts its mission by triggering the immediate transition *T_launch* and completes mission safely (the place

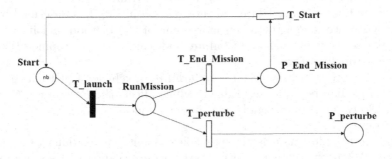

Fig. 3. The GSPN of normal movement block

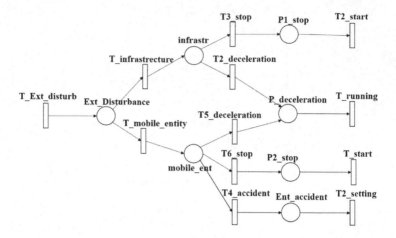

Fig. 4. The GSPN of external perturbation block

P_End_Mission). As soon as the platoon complete its mission, it return to the sate ready (place *Start*).

The regular behavior of the platoon is disturbed as soon as the transition *T_perturbe* is triggered. The perturbation can be either by an external (the transition *T_Ext_disturb*) either an internal events (the transition *T_Inter_disturb*). When the transition *T_Ext_disturb* (see Fig. 4) fire that means there is an external event, such as infrastructure or an entity in the road presented respectively by the transitions *T_infrastrecture*, *T_mobile_entity* and the places *infrastr* and *mobile_ent*. The details of the block *External perturbation* is illustrated by the Fig. 4. Soon as the presence of an infrastructure, each vehicle into the vehicle platoon must slow down or stop completely which are modeled by the two transitions *T2_deceleration* and *T3_stop*. In the case of presence of an entity (such as an obstacle or a pedestrian) in the road, the vehicles should either stop the wheels or slow down the speed of the platoon to avoid an accident with the obstacle. The third case is an accident with the road entity which requires to make the setting (transition *T2_setting*) of all the train.

The second block, named *Internal perturbation* in Fig. 2, is described by the Fig. 5 in which we model the failed movement of the platoon and its vehicles along the presence of internal faults, failure modes and/or catastrophic situations which are outlined in Table 1.

As soon as the transition *T_Inter_disturb* fire, a vehicle of platoon can have one or multiple failure mode (see Table 1) which affects the *Perception*, *Communication*, *Decision*, *Dynamic* and/or *Command* modules in each vehicle of the same platoon.

Each failure mode is modeled by a timed exponential transitions, as described in Table 2. We choose the exponential distribution because it arises in practice as the distribution of the waiting time until some event occurs.

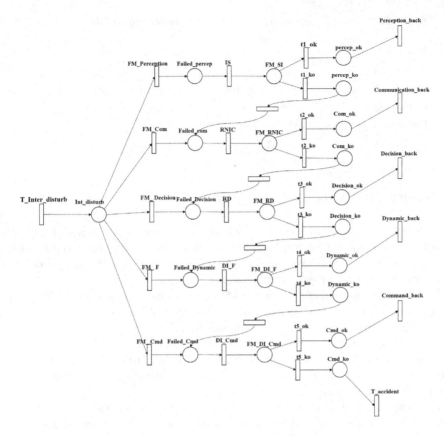

Fig. 5. The GSPN of internal perturbation block

The failure mode *FM_Perception* disrupt the Perception module of the vehicle and can cause a unsafe situation to the vehicle and subsequently a dangerous situation to the platoon. In order to avoid the unsafe state, the system triggered a repair maneuver *Immediate Stop* (IS), transition *IS* in the Fig. 5. After repairing, the Perception module can be in two states either the perception module is successfully repaired (the place *percep_ok*) or not. In the case of success, the concerned module returns to operational status by triggering the transition *Perception_back*.

In the case of failure maneuver, the system triggers the next maneuver associated to the failure mode which has the low degree of priority when multiple failure affect the vehicle. In Fig. 5, when the perception maneuver is failed (transition *t1_ko*), the maneuver transition *RNIC* is fired in order to avoid the communication failure.

Table 2. Rates of all exponential transitions associated to failure modes

Module	Failure mode	Transition	Rate
Perception	FM_Perception	FM_Perception	λ_{FM_Percep}
Communication	FM_Comm	FM_Comm	λ_{FM_Comm}
Command	FM_Cmd	FM_Cmd	λ_{FM_Cmd}
Functional (or Dynamic)	FM_F	FM_F	λ_{FM_F}
Decision	FM_Decision	FM_Decision	$\lambda_{FM\_Decision}$

6   Safety Analysis and Results

We consider a platoon with NB_v vehicle in traffic lane. Vehicle in each platoon can control itself and the surrounding environment. We model this system, taking into account the internal events as the five failure modes and their associated maneuvers which are presented in Table 1.

The measure assessed corresponds to the probability that the modeled system is in one of the catastrophic situations as a function of time (t). This measure is referred to as system unsafe, and is denoted by $\overline{S(t)}$.

As we said, several factors need to be considered when studying the impact of failures on the safety of a platooning system. In particular, the success or failure of a recovery maneuver depends on the state of the adjacent vehicles contributing to the maneuver.

The results obtained from the treatment of the models presented in the previous sections to observe the influence of various system parameters on the safety of the platoon. The analysis focuses on the influence of the failure modes rates (λ) associated to vehicle failure modes, the number of vehicles nb on the system safety. The aim is to illustrate the type of results that can be obtained from the proposed GSPN model and analyze trends rather than to obtain realistic estimates of safety system.

In this section, we illustrate the results obtained from the evaluation of the GSPN model presented in previous section, and display the analysis with respect to several parameters affecting the vehicle platoon safety.

The unsafety estimation $\overline{S(t)}$ corresponds to the probability to have a token in the place $Unsafe\_v$ of Fig. 2. The studies concentrate on the impact on $\overline{S(t)}$ of the failure rates associated with the failure modes of each functional module in each vehicle, the maximum number of faulty vehicles per platoon and the appearance frequency of external perturbation.

We assume that all the processes represented by timed activities in the SAN models have exponential distributions (i.e., have constant occurrence rates).

We set the failure rate λ as global variable in our model. The others rate relating to the failure modes are defined in terms of $\lambda = 10^{-1}/hr$. Such as $\lambda_{FM\_Per} = \lambda$, $\lambda_{FM\_Com} = 2 * \lambda$, $\lambda_{FM\_Decision} = 3 * \lambda$, $\lambda_{FM\_F} = \lambda$ and $\lambda_{FM\_Cmd} = \lambda$.

The values of rates associated with the maneuvers (μ_{IS}, μ_{RNIC}, μ_{DI_Cmd}, μ_{DI_F}, μ_{RD}) are set by the developer and the builder of the system.

We suppose that the platoon start with NB_v vehicles at any time each vehicle can be affected by one or several failure modes.

The numerical values used are inspired from real life similar situations. However, these values can be easily modified. The results illustrated in the next part have been attained, using the simulator provided by the TimeNet tool.

6.1   The Impact of the Number of Faulty Vehicles (n) per platoon Unsafe $\overline{S(t)}$

We first show in Fig. 6 the impact of number of faulty vehicles (n) per platoon on $\overline{S(t)}$, for travel durations varying from 0 to 18 h.

This figure displays that:

1. For a given value n, the probability of reaching the unsafe state increases by one order of magnitude when the travel duration increases from 0 to 18 h.
2. For a given instant of time in travel duration, increasing n leads to increase of $\overline{S(t)}$.

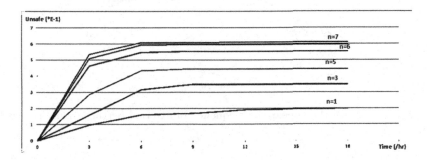

Fig. 6. $\overline{S(t)}$ versus time and depending number of faulty vehicle (n) per platoon

6.2   Failure Rate Impact on Platoon Unsafe $\overline{S(t)}$

The impact of the failure rate is illustrated in Fig. 7 considering three values for λ. We notice that the probability of reaching an unsafe state is very related to the value of the failure rate. For example, increasing the failure rate from $10^{-8}/hr$ to $10^{-6}/hr$ leads to an increase of system unsafe in an travel duration.

6.3   Appearance Frequency of External Perturbation Impact on Platoon Unsafe $\overline{S(t)}$

The impact of appearance frequency ($Freq$) of external perturbation on platoon Unsafe $\overline{S(t)}$ is illustrated in Fig. 8 by considering three values for $Freq$. The figure

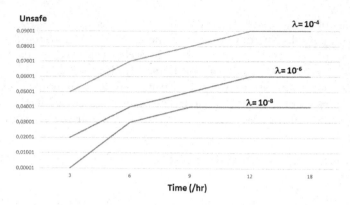

Fig. 7. $\overline{S(t)}$ depending λ

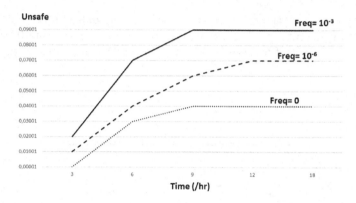

Fig. 8. $\overline{S(t)}$ depending the appearance frequency of external perturbation $Freq$ (with $\lambda = 10^{-6}$)

shows the appearance frequency impact of an external perturbation, due to the environment. Such disturbances may be either a pedestrian or a fixed/mobile object. The figure helps us to illustrate that the frequency of occurrence has an important impact on the platoon unsafe.

For a null occurrence frequency ($Freq = 0$), $\overline{S(t)}$ takes value that due to the other disturbances impact. Moreover, we note that increasing the frequency of occurrence from 0 to 10^{-3} leads to an increase of system unsafe.

7   Conclusion

This work present a formal approach for modeling and evaluate the safety for vehicular systems in a dynamic environment. In this approach, we propose a formal model for analysis the quantitative estimation related to vehicle platoon safety in its environment. During the construction of our model, we have based on some scenario collected in our research laboratory and we have used the formalism GSPN witch is well used in quantitative analysis.

The proposed GSPN model is relative to any vehicle platoon system in its environment. The model is composed of three main blocks: the first named *Normal Mvt* which models the normal movement of the vehicle platoon. The second, *External perturbation*, specify the faulty platoon movement and the associated maneuvers of the system in the presence of external events which disturbs the platoon mission and led to unsafe state. The third block, *Internal perturbation*, models the faulty movement and its associated maneuvers in the presence of an internal perturbation related to the vehicle of the platoon. The list of maneuvers are used to keep the safe of the system in presence of failure modes.

To illustrate the approach and the type of results that can be achieved, we consider simple examples. We have conducted studies of the sensitivity to analyze the impact of some parameters on the platoon safety: the rates associated with a failure modes affecting vehicles, the number of faulty vehicles per platoon and the appearance frequency of external perturbation *Freq*. In particular, the analyzes that we performed allowed us to quantify and analyze the gain in terms of safety that can be expected with the studied system.

Future works will be devoted to several key points aimed at verifying and analysis the safety properties for vehicular systems in a dynamic Military environment. The second point is to develop our models with an other formal method, such as Stochastic Activity Network (SAN), in order to improve the models for vehicle platoon system.

Acknowledgment. Works exposed in this paper are done in collaboration with Systems and Transportation Laboratory (IRTES-SET) with the support of the French ANR (National research agency) through the ANR-VTT *Safeplatoon* project (ANR-10-VPTT-011).

References

1. de Albuquerque, G.A., Maciel, P.R.M., Lima, R.M.F., Zimmermann, A.: Automatic modeling for performance evaluation of inventory and outbound distribution. IEEE Trans. Syst. Man Cybern. Part A **40**(5), 1025–1044 (2010)
2. Marsan, M.A., Balbo, G., Conte, G., Donatelli, S., Franceschinis, G.: Modelling with Generalized Stochastic Petri Nets, 1st edn. John Wiley and Sons Inc., New York (1994)
3. Hamouda, O., Kaaniche, M., Kanoun, K.: Safety modeling and evaluation of automated highway systems. In: DSN, pp. 73–82. IEEE (2009)
4. Gharbi, N., Dutheillet, C., Ioualalen, M.: Colored stochastic petri nets for modelling and analysis of multiclass retrial systems. Math. Comput. Model. **49**(78), 1436–1448 (2009)
5. Chiola, G., Dutheillet, C., Franceschinis, G., Haddad, S.: Stochastic well-formed colored nets and symmetric modeling applications. IEEE Trans. Comput. **42**(11), 1343–1360 (1993)

6. Yuanchun, Y., Haoxue, L., Yong, Z.: Evaluation emergency transport performances in response for hazardous materials road transportation accident based on spn: a case of jiangsu. In: 2010 International Conference on Optoelectronics and Image Processing (ICOIP), vol. 1, pp. 509–513, November 2010
7. Chen, L., jun Jiang, C., Fang, Y., Liu, F.: Performance evaluation of ad hoc networks based on spn. In: Proceedings of the 2005 International Conference on Wireless Communications, Networking and Mobile Computing, vol. 2, pp. 816–819, September 2005
8. Bozzano, M., Cimatti, A., Katoen, J.P., Nguyen, V.Y., Noll, T., Roveri, M.: Safety, dependability and performance analysis of extended aadl models. Comput. J. **54**(5), 754–775 (2011)
9. El Zaher, M.: Approche reactive pour la conduite en convoi des vehicules autonomes: Modelisation et verification. Ph.D. thesis, Université de Technologie de Belfort-Montbeliard (2013)
10. Bensalem, S., Ganesh, V., Lakhnech, Y., Munoz, C., Owre, S., Rue, H., Rushby, J., Rusu, V., Saidi, H., Shankar, N., Singerman, E., Tiwari, A.: An overview of sal (2000)
11. Haddad, S., Mairesse, J., Nguyen, H.-T.: Synthesis and analysis of product-form petri nets. In: Kristensen, L.M., Petrucci, L. (eds.) PETRI NETS 2011. LNCS, vol. 6709, pp. 288–307. Springer, Heidelberg (2011)
12. Laprie, J.C., Avizienis, A., Kopetz, H. (eds.): Dependability: Basic Concepts and Terminology. Springer-Verlag New York Inc., Secaucus (1992)
13. Contet, J., Derutin, J., Koukam, A., Gruer, P.: Projet anr- 10-vptt-011, safeplatoon, prsentation des scnarii, livrable d41. Technical report, Rapport de recherche, IRTES-SeT, UTBM (2012)

Use Case Diagrams for Mobile and Multi-channel Information Systems: Experimental Comparison of Colour and Icon Annotations

Sundar Gopalakrishnan and Guttorm Sindre[✉]

Department of Computer and Information Science,
Norwegian University of Science and Technology (NTNU),
7491 Trondheim, Norway
{sundar, guttors}@idi.ntnu.no

Abstract. In mobile information systems, it may be more important to capture where a user is supposed to perform an activity, as well as what type of device is going to be used, than what is the case for traditional, stationary information systems. Yet, mainstream diagram notations like use case diagrams seldom capture such information. In previous papers we have proposed some adaptations to use case diagrams to be able to include location and equipment requirements, but these adaptations have not been evaluated experimentally. This paper reports on a student experiment comparing two different notations, one using colour and the other using symbolic icons. The experiment also includes a task where the models contained both location and equipment information at the same time. In that case, one alternative used colour for locations and icons for equipment, while the other used icons both for colour and equipment. The results showed no significant difference between the two treatment groups, neither in the quality of answers to the experimental tasks, the time needed to perform the tasks, nor in their opinions given in post-task questionnaires about the notations they were exposed to.

Keywords: Mobile · Multi-channel · Information system · Process model · Diagram notation · Visual communication

1 Introduction

Use cases [1] have proven helpful for the elicitation of, communication about and documentation of requirements. While use cases can be written in pure textual form [2], the addition of diagrams has been found to enhance modelers' and stakeholders' understanding by providing a visual overview [3]. Use case diagrams have also been found to complement other notations in the UML family, such as class diagrams, enhancing the discovery of requirements [4], and seem to be understandable to a wider range of stakeholders than the other sub-languages of UML. Much of this can be attributed to the simplicity of use case diagrams, as well as their closeness to the user (depicted as intuitive stick figure "actors"), dealing with the system's external features rather than internal technical details.

© Springer International Publishing Switzerland 2016
R. Schmidt et al. (Eds.): BPMDS/EMMSAD 2016, LNBIP 248, pp. 479–493, 2016.
DOI: 10.1007/978-3-319-39429-9_30

A core goal for use case diagrams has been simplicity, so that diagrams shall be easy to understand. Normally, the location of the user while performing the various actions of a use case has not been considered important enough for inclusion in use case diagrams, and neither has the type of equipment to be used for interacting with the information system. These are understandable omissions in use cases for a traditional information system. The location of the user (e.g., whether in city A or B, office D or E) is not of much importance for the functional requirements and subsequent design of the computerized information systems. Nor is it of much importance to document the type of equipment used – often it would be some standard desktop PC, and if not, it might also be considered a premature design decision to specify the type of equipment in a use or misuse case, which should rather focus on the requirements level. However, for mobile and multi-channel information systems, the capture of location and equipment might be relevant already at an early stage [5]. Also when coming up with new versions of mobile information systems, the key change may sometimes be to make the functionality available in new places, using new equipment, rather than really introducing new features.

This makes it interesting to investigate specialized use case diagram notations that do show location and equipment-related information in diagrams. Of course it is not necessarily a good idea to cram all kinds of information into a diagram, which could make the use case diagrams lose their current strong point of being simple. It will always be the case that the underlying textual use cases will contain a lot of details that will not be in the diagrams. However, since users often do prefer working with diagrams over other forms of representation [6], and location/equipment may be more important to visualize in some projects than in others [7], it appears worthwhile to investigate if it is feasible to adapt use case diagrams to depict such information.

In previous work [8] we looked at templates to capture location and equipment information in textual use cases, and in [9] we outlined an extensive range of possibilities for including location and equipment information in use case diagrams, and evaluated these alternatives analytically using a framework based on Moody's 9 principles for visual notations [10]. However, none of the possible notations have been evaluated experimentally. In the current paper, we will compare two of the notations that seemed most promising from the analytical evaluation. The research questions are as follows:

RQ1: Which adapted notation will the participants understand best, one that applies colour or one that applies icons to show location or equipment in use case diagrams?
RQ2: Which adapted notation will the participants understand best, one that applies colour & icons or one that applies icons and icons to show location & equipment in use case diagrams?
RQ3: Will there be any difference in user preference for the alternative notations?

For RQ2, it might have appeared cleaner to try colour for both location and equipment, vs. icons for both. However, as will be explained later in the paper, using colour for several aspects at the same time was simply not found feasible – whereas using icons for several aspects was found feasible if using intuitive icons. Also, it might have been possible to try the opposite mix (colour for equipment, icons for locations) but we had limited time for the experiment and these two alternatives were thought not to be very different – as will be explained in the next section.

The rest of the paper is structured as follows: Sect. 2 motivates the choice of notation alternatives to be tried experimentally, and details the hypotheses for the experiment. Section 3 describes the experimental design, and Sect. 4 reports on the results. Section 5 gives a discussion and threats to validity, followed by related work in Sect. 6, whereupon Sect. 7 concludes the paper.

2 Motivation and Hypotheses

Bertin [11] presents 8 different visual variables that might convey meaning in two-dimensional diagrams, namely the *two planar variables* horizontal position and vertical position (an object could be placed above, below, left, right of another, or also inside or partly overlapping another object, to convey various meanings), and the *six retinal variables*: size, brightness (="value"), texture, colour, orientation, and shape. In our previous work [8] these were used to outline a number of possible ways to represent location or equipment in use case diagrams. If wanting to represent *both* location and equipment in the same diagram, combinatorial explosion generates a vast number of alternatives for *possible* visual syntaxes. However, most of these would be very poor, and since experiments are time-consuming to design, conduct and analyze, one cannot try all strange alternatives empirically. Hence it makes sense to use common sense and analytical evaluations to prune the search space, discarding notations that would be likely failures, and concentrate on those that seem promising. Figure 1 shows the notation alternatives that were considered analytically in [8], and Table 1 shows the result of the analytical evaluation. Most of the criteria (columns in Table 1) were based on Moody's proposal from [10]. This figure just shows possibilities for showing three different locations for a use case, in the example a a home care assistant logging a visit to a patient. If the use case were to be performed at the patient's home, the leftmost column would apply, if in the parked car just afterwards, the middle column would apply, and if later in the office, the rightmost column would apply. Similar notation variables could be used if it was instead equipment we wanted to capture, e.g., changing "at patient's home" in the first three rows to "using smart phone", changing pool names in the bottom row accordingly, and changing the icons in rows 4 and 5 (e.g. house icon into a smart phone icon), other rows could remain the same just with a change of legend.

Table 1 shows the results of the analytical evaluation, the order of rows corresponding to that of Fig. 1. The column Sum (second to the right) gives the total score if all criteria are considered equal. However, the first three criteria (Semiotic Clarity, Perceptive Discriminability, Semantic Transparency) may appear more essential than the other ones. Hence, the rightmost column WS shows a weighted sum, where these particular columns have been given double score. According to this column, some of the best scoring notations that are Pools (+6), Icon in node (+5), Iconic note (+4) and Fill colour (+4). However, a limitation of the display of alternatives in Fig. 1 is that it only shows one location for each use case node. In many mobile information systems, a key success criterion might be that some activities should be possible to perform in several different locations. The same would be the case for equipment − for many use cases it would be necessary to leave this to user preference. If needing to show several

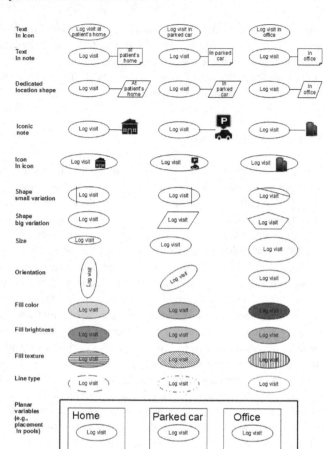

Fig. 1. Possible ways of showing location, reproduced from [8] (Color figure online)

different options for one node, pools might have a weakness that either they must be made to overlap, which soon breaks down with multiple overlaps, or use cases must be replicated in all relevant pools. Also, pools have another problem that they may be confused with system boundaries. Hence it was decided not to try this alternative in the experiment, but rather go for one using icons (and then preferring the one with icons directly at nodes rather than linked to them, as this avoids the extra edge), and one using colour fills. Of course, these could also have challenges with several locations/equipment types for the same node, but this was one of the things we wanted to explore in the experiment.

Two obvious comparisons, then, are (i) showing locations with colour, or with icons, and (ii) showing equipment with colour, or with icons. But what if we want to show both locations and equipment in the same diagram? Here it was considered that icons could still fly. A user would intuitively think about a house icon as indicating some location, and a smart phone icon as equipment. Besides, confusion could also be

Table 1. Results of analytical evaluation [8]

Location by	SC	PD	ST	CM	VE	DC	GE	MC	Sum	WS
Text in icon	−	−			−	+	+	+	0	−2
Text in note	−	−		−	−	+	+	+	−1	−3
Location shape				−	−	+		+	0	0
Iconic note	+	+	+	−		−	−	+	+1	+4
Icon in node	+	+	+	+		−	−		+2	+5
Shape, small v.	+	−	+	−		−		+	0	0
Shape, big var.	+	+	−	+	−		−	+	+1	+2
Size	+	−	−			−		+	−1	−3
Orientation	+	−	−			−		+	−1	−3
Fill colour	+	+		+	+			−	+3	+4
Fill brightness	+	+		+	+			−	+3	+4
Fill texture	+	+		+	+	−		−	+2	+3
Line type	+	−	−	+				−	−1	−2
Pools		+	+		−	+	+	+	+4	+6

mitigated by placing e.g. location icons in the left side of nodes and equipment icons in the right side. Something similar could be imagined with colour, e.g., giving the left half of a use case node the colour for some location, and the right half the colour for some equipment. However, this was quickly dismissed as infeasible. First of all the nodes would look quite messy, especially if needing to show several types of location and several types of equipment. Moreover, there is a limit to how many different colours you may use meaningfully in diagrams before they become difficult to distinguish. E.g., red, blue, yellow, green would likely be fine, but if needing a larger number of colours, like orange, purple, turquoise, you quickly get into situations where some get difficult to distinguish. Using colour both for location and equipment in the same diagram would immediately double the needed number and easily break down rather quickly. On the other hand, icons are more robust to an increase in the number, as can be seen from the hundreds of different traffic signs (e.g., triangular danger signs that are distinguished only by the icons inside, and which are still quite understandable to most drivers). Hence, it was decided that location icons + equipment icons in the same diagram was an alternative worth exploring while colours + other colours not. To have something to compare the double icon alternative against, we instead opted for using colour for locations but icons for equipment. The opposite choice (icons for locations, colour for equipment) could also have been made, but if we had to choose between the two, it seemed most intuitive to use icons for equipment, since the devices are physical objects with an intuitive look, whereas some locations may be more abstract, not having an obvious visual representation.

This left us with three different things to compare experimentally. Since the analytical evaluations differed just by 1 point and are quite uncertain, we did not know exactly what to expect, so hypotheses were not given any up-front assumption about direction. Our main null hypotheses were as follows:

- $H1_0$: There will be no difference in task score between participants getting use case diagrams with colour for locations and participants getting use case diagrams with icons for locations.
- $H2_0$: There will be no difference in task score between participants getting use case diagrams with colour for equipment types and participants getting use case diagrams with icons for equipment types.
- $H3_0$: There will be no difference in task score between participants getting use case diagrams with colour for location but icons for equipment types, and participants getting use case diagrams with icons both for locations and equipment types.

In addition to the task score, we also wanted to look at the time spent by the participants to solve the tasks, and their opinion about the notations through responses to a TAM (Technology Acceptance Model) inspired questionnaire survey. We had similar null hypothesis for these also, i.e., assuming no difference in time spent on any of the three tasks, and no difference in opinion for any of the three questionnaires. For space reasons we do not list these additional six null hypotheses here.

3   Experiment Design

The experiment was designed to have the following tasks:

- Task 1: Looking at a Figure with two use case diagrams, one for the current and one for the proposed future situation of a system, and answering 10 True/False questions about the models. Treatment Group A would get diagrams showing locations by colour fill, while Treatment Group B would get diagrams showing locations by icons. Measures to be obtained: Score on the test (1 point per correct answer), plus time spent.
- Q1: Answering a questionnaire giving their opinion about the notation they had just been exposed to. The questionnaire contained 8 questions, whereof 3 related to Perceived Ease of Use (PEOU), 3 to Perceived Usefulness (PU), and 2 to Intention to Use (ITU).
- Task 2: Similar to Task 1, but the diagrams would now include equipment types, not locations. Again, Group A would get colour filled diagrams, Group B diagrams with icons.
- Q2: Similar to Q1, asking the opinion about the notation just seen.
- Task 3: Looking at a Figure with one use case diagram (proposed solution), now containing both locations and equipment types. This time, the task was to find defects in the diagram by comparing it to a prose text (approx. one A4 page) giving requirements for the system in plain natural language text. A number of errors were deliberately seeded into the figure, so a perfect score would entail that the participant found all the errors while not reporting any false positives. Here, Group A would get a diagram with colour for locations and icons for equipment, while B would have icons for both.
- Q3: Similar to previous questionnaires, giving the opinion about the notation they were just exposed to in Task 3.

Participants were recruited from a class for approximately 50 informatics students close to finishing their third year of study. Each participant was randomly assigned to either Group A or Group B, which in the end gave 22 participants in Group A and 25 in Group B. The experiment was performed online, so the tasks were developed as anonymous tests in the LMS (Learning Management System) *It's learning* currently used at our university, and links for Variant A and B of the test were emailed to the students in the respective groups. As it was impossible to find one time or place fitting for everybody, each participant was allowed to perform the experiment at a time of his / her own liking within a 72 h interval − but with strict instructions to do it individually and not discuss it with any classmates before or after, until the end of the 72 h slot (to ensure independent observations and avoid diffusion of treatments). Of course, this is not an ideal solution, it would have been better if the time slot was more limited and students performed the experiment in a place where they were observed by the researchers. However, we have no indication that the students broke the stated rules.

Figure 2 shows diagrams for Task 1 and Task 2. As can be seen, some of the use cases had several locations and equipment types allowed. For the colour version this was done by putting several coloured ovals inside each other, which was found go better with the textual labels than, e.g., splitting colours vertically or horizontally inside the oval. The True/False questions were typically about possible locations and equipment for specific use cases, differences between the current and proposed solution in this respect, and some about a diagram as a whole (e.g., how many use cases could be done by smart phone, or whether some equipment was more or less used in the proposed situation than in the current one). Figure 3 shows the diagrams for Task 3. Here, for space reasons, we only show the diagram for Group B, which had icons, both for locations (on the left side of each use case oval) and for equipment (on the right side of use case ovals). The corresponding diagram for Group A had icons for equipment placed in the exact same way as in Fig. 3, but colours for locations (similar to Fig. 2, top). For this task there was a textual description of stakeholder requirements, including wanted equipment and locations for the use cases, formulated as a plain prose paragraph (i.e., somewhat unstructured rather than neatly itemized). The diagrams were seeded with some errors (exactly the same for each group) and the students' task was to identify these errors. This was done via 2 questions for each use case ("When it comes to locations/equipment, use case X..."), each with four alternatives ("is correctly shown", "lacks something required", "adds something not required", "both adds and lacks something"). This was preferred over having students write down found errors in a list, where mistyping etc. could give data that was harder to analyze.

Students performed the experiment online via the LMS, so the experiment questions were responded to in an online questionnaire. For each participant, the following variables could be measured for each task:

Task Score: Number of correct answers for the task
Time: How many minutes was used for the task? (self-reported)

Perceived Ease of Use (PEOU), Perceived Usefulness (PU) and **Intention to Use (ITU),** measured by a post-task questionnaire inspired by the Technology Acceptance Model (TAM) [12] and the related Method Evaluation Model (MEM) [13]. For each

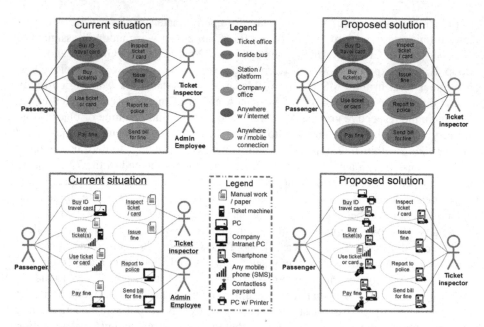

Fig. 2. Diagrams for group A, task 1 (top), group B, task 2 (bottom) (Color figure online)

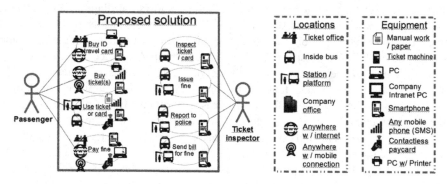

Fig. 3. Diagrams for Group B, Task 3

task there were 8 questions, 3 for PEOU, 3 for PU, and 2 for ITU. All questions were scored on a 5 point Likert scale from Strongly agree (5), via agree (4), neutral (3), disagree (2), and strongly disagree (1). Questions 4, 5, 6, 7, and 10 were formulated in the opposite direction of the others to mitigate risk of "auto-pilot" responding by participants. These 5 questions were of course reversed in the scoring for the analysis.

4   Results

The student responses were automatically exported from the LMS to Excel. Since the data were fairly simple to analyze, the statistical analysis was performed directly in Excel (2007, SP3 MSO) rather than moving to a dedicated statistics tool. Significance was tested by t-tests after first running F-tests to check equality of variances.

Results for the performance of the two groups are shown in Table 2. Times are given with decimals, not minutes and seconds. On average Group A (using coloured diagrams) scored slightly better than Group B (icons) on all the three tasks, and was also faster on all three tasks (although on Task 1, the difference was microscopic, just 6 s for an average duration of more than 7 min). However, *none of the differences were significant*. The one that came closest was the score on Task 1 ($p = 0.16$) but this is quite far away from our required $p < 0.05$.

Table 2. Performance of the two groups.

Compared variable	Gr A (colour, N = 22)		Gr B (icons, N = 25)		Difference	Effect size	Sign.? Y/N (p = ...)
	Mean	SD	Mean	SD			
Score, Task 1	9.59	0.67	9.28	0.79	0.31	0.42	No (0.16)
Time, Task 1	7.27	2.96	7.38	4.36	−0.11	0.02	No
Score, Task 2	9.64	0.58	9.36	0.95	0.28	0.34	No (0.23)
Time, Task 2	5.55	1.44	6.28	3.05	−0.73	0.30	No (0.29)
Score, Task 3	12.59	3.17	12.20	2.77	0.39	0.13	No
Time, Task 3	18.64	6.28	20.32	10.56	−1.68	0.19	No

Although Task 1 and Task 2 had a very similar structure and level of difficulty (except colours/icons were used for locations in Task 1, equipment in Task 2), there was a clear reduction in average time on Task 2 vs. Task 1. This is not surprising, as there may be several types of learning effects (e.g., understanding the concept of adapted use case diagrams, getting familiar with the question format and how to respond in the LMS). For Group A, the speed up was significant ($p < 0.02$) while for Group B it was not ($p = 0.3$).

Table 3 shows the results from the post-task questionnaire. To save space we show only the overall opinions from the questionnaires (i.e., aggregates of PEOU, PU, and ITU − all 8 questions), plus PEOU from T3, which was the variable with the greatest difference. Even this one was far from being significant, thus it is not so interesting to look at the detailed statistics for all the other variables. As can be seen, overall scores were very close to equal for Task 1 and Task 2 (using either colour or icons for location and equipment, respectively). For Task 3, there was a small to medium effect in favour of the colour alternative (i.e., colour for locations, icons for equipment) versus the alternative using icons for both locations and equipment. The difference was most evident for PEOU, but even there the difference was not significant.

Table 3. Overall post-task questionnaire opinions

Compared variable	Gr A (color, N = 22)		Gr B (icons, N = 25)		Difference	Effect size	Significance? Y/N (p =)
	Mean	SD	Mean	SD			
Opinion, T1	4.13	0.46	4.11	0.33	0.02	0.04	No
Opinion, T2	3.94	0.59	3.96	0.57	−0.02	0.03	No
Opinion, T3	3.33	0.80	3.06	0.78	0.27	0.34	No (0.4)
PEOU, T3	3.39	0.94	3.05	0.90	0.34	0.38	No (0.3)

Since the questionnaire scores went from 1 (strongly negative) to 5 (strongly positive), both treatment groups can be seen as quite positive towards the notations they were exposed to. For Task 1 and 2 with average scores around 4, the opinions are significantly better than the neutral 3 ($p < 10^{-6}$). For Task 3 the opinions about the notation were somewhat less enthusiastic, and though still above the average 3 they were no longer significantly different from 3 though Group A came fairly close (Group A, p = 0.07, Group B, p = 0.4). The drop in opinion from Task 2 to Task 3 was significant, for Group A with p < 0.01, for Group B with p < 0.0001. This might be caused by the more complex diagram notations of Task 3 (showing both location and equipment in the same diagram, while the two former tasks showed only one or the other), but could also have other reasons, such as increased complexity of the task (finding defects rather than answering True/False questions, cf. time spent on Task 3 being more than both the previous tasks taken together) or effects of boredom or exhaustion of the participants towards the end of the experiment.

Summing up on the results, there were no significant findings related directly to our a priori hypotheses. Thus, none of the hypotheses could be rejected, so it appears that the two notations are about equally good, at least for the tasks that were tried in this experiment. The only significant findings were as follows:

1. Participant opinions about the notations (both groups) were significantly better than the neutral 3 for Tasks 1 and 2. This could be interpreted in the way that they liked the notational adaptations as long as only one extra aspect was added to the normal use case diagram notation.
2. Group A participants spent significantly shorter time on Task 2 than Task 1, although the amount of work was presumably the same. If seeking a positive interpretation, this might indicate that they learnt the notational concept (using color to add extra information about a use case) quite rapidly.
3. Participant opinion about the notations in Task 3 (both groups) was significantly lower than for the two previous tasks, so maybe attempts to add two different types of information to use case nodes at the same time, is less feasible than adding just one.

However, these are not the only possible interpretations, as will be discussed in the next section about Threats to Validity.

5 Discussion and Threats to Validity

Although we did not have any significant findings related to our hypotheses, and thus do not make any strong claims, it could be worth discussing Threats to Validity, which is done according to the much used categories construct validity, conclusion validity, internal validity, and external validity.

Construct validity: Do our measures correctly represent what we wanted to measure? A typically considered threat in this category is *mono-method bias*, which we would have had if we had only one type of task (e.g., only True/False questions). Adding another type of task, namely finding defects, makes it slightly better, but of course is still far from a complete coverage of the concept of understandability, as there would be many other ways also to test various aspects of this (e.g., have them look at a diagram and explain what it means). Another relevant threat is so-called *hypothesis guessing*. Although the students would then have guessed wrongly (as our hypotheses were about possible differences between two notation alternatives, not about the absolute goodness of any of them), students may have believed that the researchers were hoping for a positive score for the notational adaptation they were exposed to, thus answering the questionnaire more positively than their honest opinion. To mitigate this, it was clearly stated at the start of each questionnaire that there were no right or wrong answers for these questions (unlike the T/F questions) and that we were just seeking their honest opinion. This, however, does not guarantee that we got honest opinions.

Conclusion validity: Without any significant claims for the hypotheses, there is no point asking if we have enough statistical power for our conclusions. However, it could be asked whether we failed to observe effects that were really there, by having too few participants in our experiment. The difference closest to being significant was score on Task 1. If effect size and standard deviation were to remain unchanged, power calculations indicated that this would have been significant if we had 50 students in each treatment group, and for the Task 2 scores and Task 3 opinions, about 100 participants would have been needed in each group. However, unless new experiments are made with such larger groups, it remains a mere speculation whether these effects would really materialize or were due to chance.

Internal validity: Were observed effects due to the difference in treatment, or something else? Since we did not observe any effect, there is little to say here, except to the bullet items 1–3 towards the end of the previous section, those were not related to our hypotheses. Here, (1) positive responses about the notations could be partly from other reasons than really liking them, e.g., fear of appearing stupid if responding that a notation was hard to learn. Our mitigation towards such effects was to have the test entirely anonymous and since it was performed online, the researchers were not present while participants performed the test. (2) Reduced time on Task 2 vs. Task 1 could be due to learning the notation quickly, but could also be simply due to learning the question format or the technicalities of responding through the LMS. Our choice of using the LMS that was well known to them beforehand can still be seen as a mitigation here compared to using some other test or survey tool, as an unfamiliar tool might have caused more variance between participants. (3) The drop in opinion from Task 2 to Task 3 could be due to more negative feelings about the notations given in Task 3 (showing both locations and equipment at the same time) than the notations used in Task 1 and 2, and this appears as an intuitive explanation. But there could also be other factors, such as boredom or exhaustion effects towards the end of the experiment session, plus the fact that the task given was more complicated (finding errors rather than just answering True/False questions), cf. that the average time spent for Task 3 was more than the time for Task 1 + Task 2 together. Hence, if really wanting to investigate whether this notation is less preferred, more experiments would be needed, for instance adding True/False questions for the Task 3 notation and error detection questions for the Task 1–2 notations, plus mixing the order of appearance of notations to control for effects related to this. In the current experiment this was not done because our research questions did not concern the relative merits of the Task 3 notation vs. the Task 1–2 notations, rather the relative merits of the Group A vs. Group B notations.

External validity: The experiment population was a class of Norwegian IT students with a fairly homogeneous age and background in terms of previous education, so findings may not apply to students with different backgrounds. Nor need they apply to practitioners, not even these students themselves when they have become practitioners in a year or two, since they may then have developed different skills and preferences in modeling, although several sources indicate that the distinction between students and practitioners need not be the most serious problem for an experiment [14–16]. Even more limiting to its generalizability is maybe the fact that it only looked at some fairly small and simple tasks, while modeling tasks in industry may be much more complex, among other things relying on communication with stakeholders rather than just reading material and answering questions.

6 Related Work

There have been several proposals for modification of use case diagram notation for various purposes, such as misuse cases for security [17] and safety [18], found to work well with stakeholders due to their intuitive visual appeal [19]. In [20] Saleh and El-Attar find that the readability of misuse case diagrams can be further enhanced by adding colour and icons, though not for indicating locations and equipment as proposed in this paper. Berenbach and Borotto [21] propose to use both colour and icons in use cases, but for a quite different purpose than ours. Colour indicates the status of the use case (e.g., draft, preliminary, accepted) and special icons are used to indicate terminal (leaf node) use cases, namely gear icons for functional requirements and oil cans for non-functional requirements. Hence both the underlying conceptual meaning and the notational adaptation is quite different from ours.

There have been a number of other proposed adaptations to the visual notation of use case diagrams, using other means than colour or icons in the ovals, and for other purposes than adding location or equipment types. For instance, notational adaptations have been used to capture inconsistency [22] or refinement relationships [23], distinguish business use cases from normal use cases [24], model variability for product family development [25, 26], and to propose several more advanced role concepts [27].

To our knowledge, however, there has been little previous work by others to add location or equipment information to use case diagrams.

In previous work we have looked at notational adaptations for adding location information to UML activity diagrams, evaluating various alternatives both analytically [28, 29] and experimentally [30, 31]. Apart from looking at a different diagram type, our current experiment with use case diagrams also goes beyond these previous works in two other respects: (i) it tries out not only the addition of location information, but also of equipment types – both separately and in the same diagram. (ii) it uses examples where some use cases had several alternative locations (or equipment types) attached, whereas the experiments with activity diagrams only tried one location per activity node. Our analytical evaluations have been strongly inspired by Moody's presented "physics" of notations [10], which has also been used for a more comprehensive evaluation of the entire UML notation [32], whereas our work has been more focussed on the specific evaluation of our proposed notation adaptations.

7 Conclusions and Further Work

Although it might be said that it was reasonably promising for the proposed notation adaptations (both options) that the students did quite well on the tasks and leant towards the positive end of the scale for the questionnaires, it can be seen as somewhat disappointing that the experiment did not come up with any significant findings related to the hypotheses. One explanation could be that the two notation alternatives are indeed about equally good. However, it might also be that the experimental tasks were too simple, and thus did not challenge the participants enough to elicit any difference between the notations. Hence, more experiments are needed, with a broader range of participants and tasks, some of them more complex and thus more realistic with respect to industrial modelling practice. Even then, there are limits to the realism of controlled experiments; so larger case studies would also be needed.

References

1. Jacobson, I., et al.: Object-Oriented Software Engineering: A Use Case Driven Approach. Addison-Wesley, Boston (1992)
2. Cockburn, A.: Writing Effective Use Cases. Addison-Wesley, Boston (2001)
3. Gemino, A., Parker, D.: Use case diagrams in support of use case modeling: deriving understanding from the picture. J. Database Manag. (JDM) 20(1), 1–24 (2009)
4. Siau, K., Lee, L.: Are use case and class diagrams complementary in requirements analysis? An experimental study on use case and class diagrams in UML. Requirements Eng. 9(4), 229–237 (2004)
5. Walderhaug, S., Stav, E., Mikalsen, M.: Experiences from model-driven development of homecare services: UML profiles and domain models. In: Chaudron, M.R. (ed.) MODELS 2008. LNCS, vol. 5421, pp. 199–212. Springer, Heidelberg (2009)
6. Figl, K., Recker, J.: Exploring cognitive style and task-specific preferences or process representations. Requirements Eng. 21, 1–23 (2014)

7. Veijalainen, J.: Developing mobile ontologies; who, why, where, and how? In: 2007 International Conference on Mobile Data Management, pp. 398–401. IEEE, Manheim, Germany (2007)
8. Gopalakrishnan, S., Krogstie, J., Sindre, G.: Extending use and misuse cases to capture mobile information systems. In: Fallmyr, T. (ed.) Norsk Konferanse for Organisasjoners Bruk av Informasjonsteknologi (NOKOBIT). Tapir, Trondheim (2011)
9. Gopalakrishnan, S., Krogstie, J., Sindre, G.: Extending use and misuse case diagrams to capture multi-channel information systems. In: Zeki, A., Zamani, M., Chuprat, S., El-Qawasmeh, E., Abd Manaf, A. (eds.) ICIEIS 2011, Part I. CCIS, vol. 251, pp. 355–369. Springer, Heidelberg (2011)
10. Moody, D.L.: The "physics" of notations: toward a scientific basis for constructing visual notations in software engineering. IEEE Trans. Softw. Eng. **35**, 756–779 (2009)
11. Bertin, J.: Semiology of Graphics: Diagrams, Networks, Maps. University of Wisconsin Press, Madison (1983)
12. Davis, F.D., Bagozzi, R.P., Warshaw, P.R.: User acceptance of computer technology: a comparison of two theoretical models. Manag. Sci. **35**(8), 982–1003 (1989)
13. Moody, D.L., et al.: An instrument for empirical testing of frameworks for conceptual model quality frameworks. In: Seventh CAiSE/IFIP8.1 International Workshop on Evaluation of Modeling Methods in Systems Analysis and Design (EMMSAD-02), Toronto, Canada (2002)
14. Berander, P.: Using students as subjects in requirements prioritization. In: Proceedings of the 2004 International Symposium on Empirical Software Engineering, ISESE 2004 (2004)
15. Runeson, P.: Using students as experiment subjects – an analysis on graduate and freshmen student data. In: Linkman, S. (ed.) Proceedings of the 7th International Conference on Empirical Assessment and Evaluation in Software Engineering (EASE 2003), pp. 95–102. Keele University, Staffordshire, UK (2003)
16. Carver, J., et al.: Issues in using students in empirical studies in software engineering education. In: Proceedings of the Ninth International Software Metrics Symposium (2003)
17. Sindre, G., Opdahl, A.L.: Eliciting security requirements by misuse cases. In: TOOLS Pacific 2000. IEEE CS Press, Sydney (2000)
18. Sindre, G.: A look at misuse cases for safety concerns. In: Ralyté, J., Brinkkemper, S., Henderson-Sellers, B. (eds.) Situational Method Engineering: Fundamentals and Experiences. IFIP, vol. 244, pp. 252–266. Springer, Heidelberg (2007)
19. Alexander, I.F.: Initial industrial experience of misuse cases in trade-off analysis. In: 10th Anniversary IEEE Joint International Requirements Engineering Conference (RE 2002). IEEE, Essen, Germany (2002)
20. Saleh, F., El-Attar, M.: A scientific evaluation of the misuse case diagrams visual syntax. Inf. Softw. Technol. **66**, 73–96 (2015)
21. Berenbach, B., Borotto, G.: Metrics for model driven requirements development. In: Proceedings of the 28th International Conference on Software Engineering. ACM (2006)
22. Hausmann, J.H., Heckel, R., Taentzer, G.: Detection of conflicting functional requirements in a use case-driven approach: a static analysis technique based on graph transformation. In: Proceedings of the 24th International Conference on Software Engineering. ACM (2002)
23. Whittle, J.: Specifying precise use cases with use case charts. In: Bruel, J.-M. (ed.) MoDELS 2005. LNCS, vol. 3844, pp. 290–301. Springer, Heidelberg (2006)
24. Johnston, S.: Rational UML Profile for business modeling. IBM Developer Works (2004). http://www.ibm.com/developerworks/rational/library/5167.html
25. von der Maßen, T., Lichter, H.: Modeling variability by UML use case diagrams. In: Proceedings of the International Workshop on Requirements Engineering for Product Lines. Citeseer (2002)

26. Bühne, S., Halmans, G., Pohl, K.: Modelling dependencies between variation points in use case diagrams. In: Proceeding of 9th International Workshop on Requirements Engineering. Citeseer (2003)
27. Wegmann, A., Genilloud, G.: The role of "Roles" in use case diagrams. In: Evans, A., Caskurlu, B., Selic, B. (eds.) UML 2000. LNCS, vol. 1939, pp. 210–224. Springer, Heidelberg (2000)
28. Gopalakrishnan, S., Sindre, G.: Diagram notations for mobile work processes. In: Johannesson, P., Krogstie, J., Opdahl, A.L. (eds.) PoEM 2011. LNBIP, vol. 92, pp. 52–66. Springer, Heidelberg (2011)
29. Gopalakrishnan, S., Krogstie, J., Sindre, G.: Capturing location in process models: comparing small adaptations of mainstream notation. Int. J. Inf. Syst. Model. Design 3(3), 24–25 (2012)
30. Gopalakrishnan, S., Krogstie, J., Sindre, G.: Adapting UML activity diagrams for mobile work process modelling: experimental comparison of two notation alternatives. In: van Bommel, P., Hoppenbrouwers, S., Overbeek, S., Proper, E., Barjis, J. (eds.) PoEM 2010. LNBIP, vol. 68, pp. 145–161. Springer, Heidelberg (2010)
31. Gopalakrishnan, S., Krogstie, J., Sindre, G.: Adapted UML activity diagrams for mobile work processes: experimental comparison of colour and pattern fills. In: Halpin, T., Nurcan, S., Krogstie, J., Soffer, P., Proper, E., Schmidt, R., Bider, I. (eds.) BPMDS 2011 and EMMSAD 2011. LNBIP, vol. 81, pp. 314–331. Springer, Heidelberg (2011)
32. Moody, D., van Hillegersberg, J.: Evaluating the visual syntax of UML: an analysis of the cognitive effectiveness of the UML family of diagrams. In: Gašević, D., Lämmel, R., Van Wyk, E. (eds.) SLE 2008. LNCS, vol. 5452, pp. 16–34. Springer, Heidelberg (2009)

Author Index

Printed in the United States
By Bookmasters